Fundamentals of Electrical Engineering and Technology

William D. Stanley
John R. Hackworth
Richard L. Jones

THOMSON
DELMAR LEARNING

Australia Brazil Canada Mexico Singapore Spain United Kingdom United States

Fundamentals of Electrical Engineering and Technology

William D. Stanley, John R. Hackworth, Richard L. Jones

Vice President, Technology and Trades ABU:
David Garza

Director of Learning Solutions:
Sandy Clark

Executive Editor:
Stephen Helba

Senior Product Manager:
Michelle Ruelos Cannistraci

Marketing Director:
Deborah S. Yarnell

Senior Channel Manager:
Dennis Williams

Marketing Coordinator:
Mark Pierro

Director of Production:
Patty Stephan

Senior Production Manager:
Larry Main

Content Project Manager:
Jennifer Hanley

Art & Design Coordinator:
Francis Hogan

Technology Project Manager:
Linda Verde

Senior Editorial Assistant:
Dawn Daugherty

COPYRIGHT © 2007 Thomson Delmar Learning, a division of Thomson Learning Inc. All rights reserved. The Thomson Learning Inc. logo is a registered trademark used herein under license.

Printed in the United States of America
1 2 3 4 5 XX 08 07 06

For more information contact Thomson Delmar Learning
Executive Woods
5 Maxwell Drive, PO Box 8007,
Clifton Park, NY 12065-8007
Or find us on the World Wide Web at
www.delmarlearning.com

ALL RIGHTS RESERVED. No part of this work covered by the copyright hereon may be reproduced in any form or by any means—graphic, electronic, or mechanical, including photocopying, recording, taping, Web distribution, or information storage and retrieval systems—without the written permission of the publisher.

For permission to use material from the text or product, contact us by
Tel.   (800) 730-2214
Fax   (800) 730-2215
www.thomsonrights.com

Library of Congress Cataloging-in-Publication Data:

Stanley, William D.
  Fundamentals of electrical engineering and technology / William D. Stanley, John Hackworth, Richard L. Jones.
   p. cm.
  Includes bibliographical references and index.
  ISBN 1-4180-0020-5 (alk. paper)
  1. Electric engineering—Textbooks.
  2. Electronic circuits—Textbooks.
  3. Electronics—Textbooks. I. Hackworth, John R. II. Jones, Richard L. III. Title.
  TK145.S73 2006
  621.3—dc22
                              2006017432

NOTICE TO THE READER

Publisher does not warrant or guarantee any of the products described herein or perform any independent analysis in connection with any of the product information contained herein. Publisher does not assume, and expressly disclaims, any obligation to obtain and include information other than that provided to it by the manufacturer.

The reader is expressly warned to consider and adopt all safety precautions that might be indicated by the activities herein and to avoid all potential hazards. By following the instructions contained herein, the reader willingly assumes all risks in connection with such instructions.

The publisher makes no representation or warranties of any kind, including but not limited to, the warranties of fitness for particular purpose or merchantability, nor are any such representations implied with respect to the material set forth herein, and the publisher takes no responsibility with respect to such material. The publisher shall not be liable for any special, consequential, or exemplary damages resulting, in whole or part, from the readers' use of, or reliance upon, this material.

Table of Contents

PREFACE vii

PART I BASIC CIRCUIT ANALYSIS

1
Basic DC Circuits 3
Overview and Objectives 3
1-1 DC Circuit Variables 3
1-2 A Very Simple DC Circuit 8
1-3 Measuring DC Circuit Variables 12
1-4 Power and Energy 14
1-5 Kirchhoff's Laws 15
1-6 Equivalent Resistance 17
1-7 Single-Loop or Series Circuit 19
1-8 Single Node-Pair or Parallel Circuit 23
1-9 Voltage and Current Divider Rules 28
Problems 30

2
General DC Circuit Analysis 33
Overview and Objectives 33
2-1 Mesh Current Analysis 33
2-2 Node Voltage Analysis 36
2-3 Source Transformations 39
2-4 Thevenin's and Norton's Theorems 41
Problems 47

3
Transient Circuits 51
Overview and Objectives 51
3-1 Basic Circuit Parameters 52
3-2 Boundary Conditions for Energy Storage Elements 58
3-3 First-Order Circuits with DC Excitations 64
3-4 Inductance and Capacitance Combinations 69
Problems 71

PART II ALTERNATING CURRENT CIRCUITS

4
AC Circuits I 77
Overview and Objectives 77
4-1 Sinusoidal Functions 78
4-2 Phasors 80
4-3 AC Voltage-Current Relationships 85
4-4 AC Impedance 87
Problems 92

5
AC Circuits II 95
Overview and Objectives 95
5-1 DC and RMS Values 95
5-2 Complex Power in AC Circuits 99
5-3 Power Factor Correction 103
5-4 Admittance 106
5-5 Resonance 110
Problems 115

PART III ELECTRONIC DEVICES AND LINEAR ELECTRONICS

6
Diodes and Their Applications 119
Overview and Objectives 119
6-1 Semiconductor Concepts 120
6-2 Junction Diode 120

- 6-3 Diode Circuit Models 122
- 6-4 Diode Rectifier Circuits 126
- 6-5 Rectifier Circuits with Filtering 132
- 6-6 Survey of Other Diode Types 136
- 6-7 Zener Regulator Circuits 137
- Problems 140

7
Transistors 143
- Overview and Objectives 143
- 7-1 Introduction 143
- 7-2 Bipolar Junction Transistor 144
- 7-3 BJT Operating Regions 146
- 7-4 BJT Characteristic Curves 148
- 7-5 Brief Look at a Classical BJT Amplifier Circuit 153
- 7-6 BJT Switches 154
- 7-7 Field Effect Transistor Family 157
- 7-8 Junction Field Effect Transistors 158
- 7-9 Brief Look at Classical JFET Amplifier Circuit 163
- 7-10 MOSFETs 164
- 7-11 FET Switches 166
- 7-12 FET Schematic Summary 168
- Problems 169

8
Operational Amplifiers 173
- Overview and Objectives 173
- 8-1 Amplifier Properties 173
- 8-2 Operational Amplifiers 179
- 8-3 Operational Amplifier Circuit Analysis 183
- 8-4 Inverting Amplifier Circuit 184
- 8-5 Noninverting Amplifier Circuit 188
- 8-6 Operational Amplifier Controlled Sources 191
- 8-7 Circuits That Combine Signals 194
- 8-8 Integration and Differentiation Circuits 196
- 8-9 Active Filters 199
- 8-10 Voltage Regulation with an Op-Amp 204
- Problems 205

PART IV DIGITAL ELECTRONICS

9
Digital Circuits: Basic Combinational Forms 209
- Overview and Objectives 209
- 9-1 Binary Number System 209
- 9-2 Negative Binary Numbers 212
- 9-3 Subtraction Using Twos-Complement Numbers 213
- 9-4 Hexadecimal Number System 214
- 9-5 Octal Number System 216
- 9-6 Binary Coded Decimal 217
- 9-7 Fundamental Logic Operations 218
- 9-8 NAND and NOR Operations 219
- 9-9 Boolean Algebra and Switching Functions 220
- 9-10 Boolean Reduction and Karnaugh Maps 222
- 9-11 K-Maps with Don't Care Conditions 228
- 9-12 DeMorgan's Theorem 228
- 9-13 Boolean Expansion 229
- 9-14 Exclusive OR and Exclusive NOR 230
- 9-15 Logic Diagrams and Combinational Logic 230
- Problems 231

10
Digital Circuits: Advanced Combinational Forms 235
- Overview and Objectives 235
- 10-1 Adders 235
- 10-2 Decoders 237
- 10-3 Encoders 239
- 10-4 Multiplexers 241
- 10-5 Integrated-Circuit Digital Logic Families 242
- Problems 243

11
Digital Circuits: Sequential Forms 245
- Overview and Objectives 245
- 11-1 Introduction 245

- 11-2 Latch and Flip Flop Initialization 246
- 11-3 Latches 246
- 11-4 Flip Flops 247
- 11-5 Parallel Registers 251
- 11-6 Serial (Shift) Registers 251
- 11-7 Counters 252
- 11-8 Memories 254
- Problems 257

PART V POWER SYSTEM FUNDAMENTALS

12
Magnetic Circuits 261
- Overview and Objectives 261
- 12-1 Magnetism and Electromagnetism 261
- 12-2 Flux 262
- 12-3 Flux Density 263
- 12-4 Magnetomotive Force 264
- 12-5 Reluctance 264
- 12-6 Determining Magnetic Field Direction of an Electromagnet 265
- 12-7 Closed (Circular) Cores 265
- 12-8 Magnetic Field Intensity 266
- 12-9 Permeability 266
- Problems 270

13
Three-Phase Circuits 271
- Overview and Objectives 271
- 13-1 Introduction to Three-Phase Theory 271
- 13-2 Phase Voltages and Line Voltages 274
- 13-3 Three-Phase Sequencing 277
- 13-4 Three-Phase Wye and Delta Connections 277
- 13-5 Three-Phase Wye and Delta Load Connections 278
- 13-6 Three-Phase Power Calculations 279
- 13-7 Comparison of Power Flow in Single- and Three-Phase Systems 281
- Problems 283

14
Transformers 285
- Overview and Objectives 285
- 14-1 Introduction 285
- 14-2 The Ideal Transformer 286
- 14-3 Transformer Power Losses 289
- 14-4 Transformer Efficiency 290
- 14-5 Effect of Power Factor on Transformer Performance 291
- 14-6 Impedance Reflection 292
- 14-7 Impedance Matching Transformers 293
- 14-8 Transformer Construction 294
- 14-9 Three-Phase Transformers 295
- 14-10 Autotransformers 297
- 14-11 Instrument Transformers 300
- 14-12 Useful Transformer Tips 303
- Problems 303

PART VI ELECTRICAL MACHINES

15
DC Machines 307
- Overview and Objectives 307
- 15-1 Introduction 307
- 15-2 Magnetic Induction and the DC Generator 308
- 15-3 Shunt and Compound DC Generator 313
- 15-4 Motor Action and the DC Motor 319
- 15-5 Shunt, Series, and Compound DC Motor 322
- 15-6 Dynamic Braking of DC Motors 326
- Problems 326

16
AC Machines 329
- Overview and Objectives 329
- 16-1 Introduction 329
- 16-2 AC Generator (Alternator) 330
- 16-3 Three-Phase Induction Motor 337
- 16-4 Single-Phase Induction Motor 341
- 16-5 Reluctance Motor 346
- 16-6 Universal Motor 348
- Problems 348

PART VII CURRENT ADDITIONAL TOPICS

17
Programmable Logic Controllers 351
Overview and Objectives 351
17-1 Introduction 351
17-2 Introduction to Machine Controls 351
17-3 Machine Control Components 352
17-4 Ladder Diagrams 359
17-5 Latch Circuit 361
17-6 PLC Types 362
17-7 PLC Configuration 363
17-8 PLC Operation 364
17-9 Fundamental PLC Programming 365
17-10 PLC Programs with Timers 369
17-11 Advanced PLC Features 371
17-12 PLC Wiring 371
Problems 375

18
Digital Communications 377
Overview and Objectives 377
18-1 Introduction and Terminology 378
18-2 Encoding and Transmission of Textual Data 381
18-3 Sampling an Analog Signal 384
18-4 Pulse Code Modulation 386
18-5 Analog-to-Digital and Digital-to-Analog Conversion 388
18-6 Modulation Methods 395
18-7 *M*-ary Encoding 396
Problems 398

PART VIII MULTISIM LABORATORY
Notational Differences 401
Introduction 401
Primer 403
MultiSIM Examples for Chapter 1 410
MultiSIM Examples for Chapter 2 415
MultiSIM Examples for Chapter 3 422
MultiSIM Examples for Chapter 4 427
MultiSIM Examples for Chapter 5 431
MultiSIM Examples for Chapter 6 435
MultiSIM Examples for Chapter 7 439
MultiSIM Examples for Chapter 8 443
MultiSIM Examples for Chapter 9 449
MultiSIM Examples for Chapter 10 453
MultiSIM Examples for Chapter 11 461
MultiSIM Examples for Chapter 13 469
MultiSIM Examples for Chapter 14 470
MultiSIM Examples for Chapter 15 473
MultiSIM Examples for Chapter 16 474
MultiSIM Examples for Chapter 17 476
MultiSIM Examples for Chapter 18 478

Appendix A Solving Simultaneous Equations 485

Appendix B Complex Numbers 489

B-1 Addition and Subtraction of Complex Numbers 492
B-2 Multiplication and Division 494

Answers to Selected Odd-Numbered Problems 497

Index 503

Preface

This book has been written to achieve two separate but related goals: (1) to provide a treatment of the basic fundamentals of electrical engineering and electrical engineering technology in a clear, student-friendly manner, and (2) to provide some reasonable samples of the numerous modern applications of the technology.

To achieve goal 1, a broad spectrum of the basic concepts has been included. These concepts are those that have proved to be as relevant today as they were a half-century ago, and to the extent that can be predicted, will still be considered relevant many years into the future. The achievement of goal 2, however, requires more discretion in selecting topics, so the choice has been to select some of the common ones that are significant in the modern world.

The book is suitable for non-electrical and electronic majors, such as mechanical and civil engineering or technology majors, or it is suitable for electrical and electronics engineering technology majors as a broad overview prior to the more in-depth treatment within the specialized courses. It can also be used for science majors (for example, physics, computer science, and chemistry majors) or for any technical or engineering personnel in industry who desire a broad overall coverage of the fundamentals of the electrical/electronics field and common areas of application.

Sometimes the term "survey" carries certain negative connotations, because the assumption is often made that there is no depth in such coverage. However, we believe that there can be considerable depth in such a book if developed properly. Yet it must be classified as a survey book based on the breadth of coverage and the scope of topics considered.

Most of the survey books currently on the market tend to fall into one of the following two categories: (1) engineering-level texts assuming a strong background in mathematics including several terms of calculus and differential equations and (2) one- and two-year level texts assuming no calculus at all. This text will be aimed toward a "midpoint" between these two divergent objectives. Therefore, calculus will be used as appropriate but will be carefully explained and integrated into any relevant treatments from a practical point of view. The only chapter that could be considered "calculus-intensive" is Chapter 3, which deals with transient analysis, and much of the material there is covered from an intuitive viewpoint rather than from a formal mathematical viewpoint.

An underlying philosophy is that at the presumed level of students who use the book, the treatment of basic circuit analysis should be sufficient to ensure that the student can deal with all of the pertinent application areas. Therefore, the first five chapters are devoted to the fundamentals of dc circuits, basic transient circuits, and steady-state ac circuits. Once this background is established, the text will move into the areas of linear and digital electronics, with emphasis on the "modular" approach rather than many of the traditional transistor-level treatments, which are believed by the authors to be too detailed and irrelevant for the intended purpose of the text. Emphasis will then be directed toward some coverage of the electro-mechanical areas of the field, including magnetic circuits, three-phase systems, power transformers, and dc and ac machines. These topics have particular significance in the mechanical and manufacturing areas. Programmable controllers, which have assumed a major role in many industrial areas, will be covered and their applications will be studied. Finally, an introductory treatment of digital communication and data systems will be provided.

The book provides enough material for a broad two-semester or three-quarter course in the general field of electrical engineering or technology. However, because many parts of the book are uncoupled from others, a one-term treatment may be easily achieved, with various emphasis areas ranging from linear electronics, digital electronics, electrical power, and electrical machines. Some guidelines on choices will be discussed later.

A very unique feature of the book is that one complete section (Part VIII) is devoted solely to the application of the circuit analysis program MultiSIM® to form a virtual laboratory to support the remainder of the book.

Some Topics Treated Differently than in Other Texts

At the risk of creating a negative reaction, we wish here to state a few topics covered in other books of this type and why our coverage has been altered from the norm. Before making a judgment, be sure that you understand the reasoning and what has been provided to compensate in some cases.

Topic: Solving Simultaneous Circuit Equations

In dealing with circuit analysis in Chapter 2, only a few circuits requiring three simultaneous equations are considered. We don't believe that the intended audience for this book will encounter many situations requiring the analysis of complex arrays of circuit equations. Moreover, if such a situation arises, the availability of powerful mathematical software packages on modern computer systems would be the choice for the analysis. However, Appendix A does provide a brief treatment of simultaneous equations and their solutions.

Topic: Transistor Amplifier Circuits

In the treatment of transistors in Chapter 7, only two short discussions of transistor amplifiers are made. Instead, the emphasis is on the classifications, terminal characteristics, and the applications of transistors as switches for interfacing with various devices. Unless one is an electronics specialist or engaged in designing integrated circuits, there is little need to discuss how a discrete transistor amplifier works in today's world. Instead, most applications of this type, at least at low and moderate frequencies, can be much more easily achieved with operational amplifiers, which are covered in some detail in Chapter 8.

Topic: Communications

In the treatment of communications in Chapter 18, none of the traditional analog modulation methods such as AM and FM are covered. Although AM and FM are both alive and well, most situations that will be encountered by the intended book audience will be based on digital communication concepts, and they are emphasized in the chapter along with data conversion modules such as analog-to-digital converters and digital-to-analog converters.

Order of Topics and Choices

For a two-semester or three-quarter course sequence with a typical format of three hours of lecture per week, virtually all of the book can be covered. However, where only one term is available, topics must be selected according to the course objectives.

Initially, the authors planned to list several one-term suggested groupings as guides to instructors. Upon considering that task in detail, however, it became clear that there are so many possibilities that the result could be more confusing than helpful. Therefore, the strategy was changed to discuss the essential prerequisites for each portion of the book, in which case instructors could decide for themselves according to the course objectives, the backgrounds of the students, and other factors.

It is recommended that all course structures begin with Chapters 1 and 2, in which the basic circuit laws and analysis methods are covered. This material is fundamental to just about everything else in the book and should not be shortchanged.

Chapter 3, dealing with transient circuits, depends on the eventual course objectives. For an electronics emphasis, it should be covered in detail. For a power emphasis, the basic properties of inductance and capacitance should be covered, but some of the details of transient analysis and first-order circuits could be treated lightly.

Chapters 4 and 5, dealing with ac circuits, are essential as a background for power and electrical machines. For an electronics emphasis, some parts of these two chapters could be treated lightly.

Chapters 6, 7, and 8 constitute an overall treatment of electronic devices and linear electronics, and if this is a major objective of the text, they should be covered in detail. Likewise, Chapters 9, 10, and 11 constitute an overall treatment of digital electronics. In a one-term treatment, it would be difficult to cover everything in both of these broad areas plus the earlier basic material, so some selectivity may be necessary.

If power system fundamentals are to serve as a major objective, Chapters 12, 13, and 14 should be covered. Further power system topics are those of dc and ac machines, which are covered in Chapters 15 and 16.

Last, but not least, are the topics of programmable logic controllers (PLCs) in Chapter 17 and digital data systems and communications in Chapter 18. A treatment of PLCs is enhanced by a knowledge of basic combinational logic as covered in Chapter 9, along with a knowledge of transistor switches as covered in Chapter 7. The treatment of digital data systems and communications in Chapter 18 presupposes an understanding of the fundamentals of the binary number system as covered in Chapter 9.

Organization of Book

Part I: Basic Circuit Analysis

Chapter 1: Fundamentals of DC Circuits

The definitions and forms of circuit variables and parameters are introduced. Basic circuit laws and theorems are developed using dc circuit forms. Topics include Ohm's law, Kirchhoff's voltage and current laws, single-loop (series) circuits, single node-pair (parallel) circuits, and voltage and current divider rules. Solutions of simple circuit forms to obtain voltages, currents, and power are emphasized.

Chapter 2: General DC Circuit Analysis

Circuits having more complex configurations than considered in Chapter 1 and the means for solving them are covered. The general methods provided are mesh current analysis and node voltage analysis. Thevenin's and Norton's theorems and their importance in simplifying circuits are covered.

Chapter 3: Transient Circuits

Inductance and capacitance are introduced and the voltage-current relationships for these elements are developed. Special emphasis is made on piece-wise linear functions, because differentiation and integration can be performed with simple arithmetic calculations. Initial transient and final steady-state models are provided. First-order circuits are introduced, and the means for determining and sketching the complete response forms with dc excitations are developed.

Part II: Alternating Current Circuits

Chapter 4: AC Circuits I

Sinusoidal voltage and currents are studied and the phasor representations of these variables are developed. Impedance forms and phasor diagrams for all the basic circuit elements are provided using complex numbers. Complete solutions for the ac steady-state responses of circuits with sinusoidal excitations are covered.

Chapter 5: AC Circuits II

The principles of steady-state ac analysis covered in Chapter 4 are extended to include rms values, ac power, power factor, and complex power analysis. Admittance is introduced and conversion between impedance and admittance forms is covered. Some of the applications considered are power factor correction in single-phase systems, series resonance, and parallel resonance.

Part III: Electronic Devices and Linear Electronics

Chapter 6: Diodes and Their Applications

The semiconductor diode is introduced and various models to describe the external operating characteristics are presented. Simple circuits utilizing diodes are analyzed. The major applications emphasized for diodes are those of rectifiers and power-supply circuits. Various forms of diodes are surveyed and discussed. The design of zener voltage reference circuits is covered.

Chapter 7: Transistors

The bipolar junction transistor (BJT) is introduced and the terminal characteristics are studied in some detail. Models for cutoff, saturation, and active operation are covered. The design of BJT switches as used in interfacing circuits is considered. The field effect transistor (FET) is introduced and the various families of junction field-effect transistors (JFETs) and metal-oxide semiconductor field-effect transistors (MOSFETs) are surveyed. The behavior of FETs as switches is also considered.

Chapter 8: Operational Amplifiers

General properties of all linear amplifiers are first discussed. The integrated circuit operational amplifier (op-amp) is then introduced and its various specifications and circuit models are discussed. Numerous linear applications of op-amps are developed throughout the chapter. This includes various amplifier forms, controlled sources, linear combination circuits, active filters, and voltage regulator circuits. The emphasis throughout is on simple design procedures. Some design data for low-pass and band-pass active filter design using op-amps are provided.

Part IV: Digital Electronics

Chapter 9: Digital Circuits: Basic Combinational Circuits

Number systems and methods to convert between systems are introduced, followed by binary arithmetic operations including two's complement notation. The basic logic gates, AND, OR, NAND, NOR, and INVERT, are explained. Boolean algebra is introduced, including postulates and theorems, and the Boolean operations are related to the logic gate functions. Reduction techniques are demonstrated, including Karnaugh maps, followed by methods of converting between Boolean switching functions and logic circuits.

Chapter 10: Digital Circuits: Advanced Combinational Circuits

Chapter 10 introduces more complex (MSI) combinational logic functions including adders, decoders, encoders, and multiplexers. The chapter concludes with a brief investigation of the operating characteristics of the two most popular logic families, TTL and CMOS.

Chapter 11: Digital Circuits: Sequential Forms

Sequential logic functions are introduced, including RS latches, D latches, and the D, T, and JK flip flops. Excitation tables for each of the flip flops are shown and explained. This is followed by investigation of more complex sequential logic functions, including asynchronous and synchronous up, down, and up/down counters, shift registers, and memory devices.

Part V: Power System Fundamentals

Chapter 12: Magnetic Circuits

Chapter 12 begins with an introduction of magnetics and electromagnetics, including the concepts of flux, flux density, and retentivity. It then continues with the subject areas of magnetomotive force, reluctance, permeability, and magnetic field intensity, including the introduction of B-H curves. Throughout the chapter, mathematical concepts are shown that demonstrate the relationship between magnetic quantities, and the parallel relationship between magnetic circuits and electrical circuits.

Chapter 13: Three-Phase Circuits

The reader is introduced to the concept of three-phase power systems. Relationships between line and phase voltages and currents in three-wire delta and three- and four-wire wye systems are shown. Methods of calculating power in both delta and wye systems are demonstrated.

Chapter 14: Transformers

Chapter 14 begins with the introduction of the ideal transformer and the relationship between turns ratio, voltage, current, and power in a transformer. The internal transformer power losses are covered and their effect on transformer efficiency is stressed. In addition to coverage of the single-phase and three-phase transformer, specialty transformers are discussed, including autotransformers, current transformers, and potential transformers.

Part VI: Electrical Machines

Chapter 15: DC Machines

After a brief introduction to Faraday's Law and Lenz's Law, the relationship between a magnetic field and the velocity of a conductor in the field to produce generator action is introduced. The fundamental parts of a dc generator are shown, and the related mathematical analysis is demonstrated. The chapter then moves on to motor action, the dc motor, and the various way of connecting a dc motor, including the operating characteristics of each.

Chapter 16: AC Machines

An introduction of the three-phase alternator is provided, including the calculation of internal power losses, output frequency, and efficiency. The methods to connect an alternator to a power grid and remove an alternator from a power grid are discussed. The three-phase synchronous motor is presented. The operating theory of the simple three-phase induction motor is presented, followed by its operating characteristics and mathematical relationships between speed, slip, and efficiency. The operating principles of the single-phase motor are discussed, including the various types of single-phase motors and the operating characteristics of each. Specialty motors, including the reluctance motor and universal motor, are introduced.

Part VII: Current Additional Topics
Chapter 17: Programmable Logic Controllers

Chapter 17 first introduces the reader to the electrical controls ladder diagram, including many of the component schematic symbols. PLC hardware is introduced with coverage of the various types of PLC configurations, and the components that make up a PLC. PLC operating principles are explained showing how a PLC executes a program and performs I/O updates, followed by ladder logic programming techniques. The chapter concludes with an introduction of methods of wiring a PLC into a control system.

Chapter 18: Digital Communications

The fundamentals of digital data and communication systems are introduced, including terminology, definitions, and data rate measurements. Analog-to-digital (A-D) and digital-to-analog (D-A) conversion are covered, including the specifications required for selecting a converter. The sampling theorem is introduced and the techniques for representing analog signals in pulse-code modulation (PCM) form are developed. Digital signal transmission in various formats is discussed and the means for achieving higher data rates by using so-called *M*-ary encoding are introduced.

Part VIII: MultiSIM Laboratory

Unlike the other seven parts, Part VIII is not divided into chapters. Rather it is related to all but one chapter elsewhere in the book and has sections that refer to those chapters. These sections provide numerous examples of circuit simulations using MultiSIM®.

MultiSIM, a product of Interactive Technologies Ltd.*, is comprehensive circuit analysis software that permits the modeling and simulation of a wide variety of electrical and electronic circuits. It offers a very large component database, schematic entry, analog/digital circuit simulation, and many other features, including seamless transfer to printed circuit board (PCB) layout packages. The program has evolved from the company's earlier Electronics Workbench (EWB) program. The circuit simulation portion of the program is based on the popular SPICE program (*S*imulation *P*rogram with *I*ntegrated *C*ircuit *E*mphasis). SPICE was developed at the University of California at Berkeley and was a batch-oriented program. However, MultiSIM is interactive and offers many user-friendly features. National Instruments, Inc. now owns the company.

---

*Interactive Image Technologies, Ltd.
111 Peter Street, Suite 801
Toronto, Ontario, Canada M5V 2H1

The authors believe that, in the broadest sense of modern technology, MultiSIM is a type of laboratory that can be used to serve virtually all of the same educational functions as a traditional teaching laboratory. In fact, we will take that statement one step further and say that there are important operations that can be studied with MultiSIM that are very difficult to implement in a traditional laboratory. One example is the study of the effect of worst-case parameter variations in a circuit design, a task readily implemented with the program, but one that might be difficult in a traditional laboratory.

The circuit components parallel those of actual laboratory units and must be "wired" into the circuit in essentially the same fashion as in an actual laboratory. Thus, many laboratory skills can be taught with a computer and software without damaging any components or instruments.

Appendices

Two appendices are provided to assist students with certain mathematical procedures:

- Appendix A: Solving Simultaneous Equations
- Appendix B: Complex Numbers

CD

The CD enclosed with the text contains two folders, each of which has MultiSIM files for all circuits considered in Part VIII of the text. One folder contains circuits in Version 8 and the other contains the same circuits in Version 9.

Supplements

The e.resource available to educators upon adoption of the text includes the following teaching tools:

- A solutions manual that contains the solutions to all problems.
- All images from the textbook that can be used to create or customize slides and tests.
- Several hundred questions in the computerized test bank.

About the Authors

William D. Stanley is an Eminent Professor Emeritus at Old Dominion University and was a recipient of the State Council of Higher Education in Virginia Outstanding Faculty Award. He received his Bachelor of Science degree from the University of South Carolina and the Master of Science and Ph.D. degrees from North Carolina State University. All degrees were in Electrical Engineering. Dr. Stanley is a Registered Professional Engineer in the Commonwealth of Virginia. He is also a Senior Member of the IEEE and Life Member of ASEE.

John R. Hackworth is an Associate Professor and Program Director for the Electrical Engineering Technology program at Old Dominion University. He also holds the special designation of University Professor, which is awarded to faculty for special teaching recognition. He received a Bachelor of Science Degree in Electrical Engineering Technology and a Master of Science Degree in Electrical Engineering, both from Old Dominion University. Before joining the Old Dominion University faculty, he had approximately 20 years of industrial experience in test engineering and plant automation. He is a member of ASEE and a senior member of IEEE.

Richard L. Jones is an Instructor for the Electrical Engineering Technology program at Old Dominion University. He received a Bachelor of Science Degree in Electrical

Engineering Technology at Oklahoma State University and a Master of Science Degree in Electronics Engineering at the Naval Postgraduate School in Monterey, California. He is a retired United States Navy Submarine Service Lieutenant Commander with subspecialties in ballistic missile, torpedo, and sonar systems. Mr. Jones has previously taught Mechanical Engineering Design at the United States Military Academy and Electrical Engineering at the United States Naval Academy.

Acknowledgments

The authors would like to express their deep appreciation to the following people for their important contributions in the development and production of this book:

From Thomson Delmar Learning:

    Dave Garza, *Vice President Technology & Trades*

    Steve Helba, *Executive Editor*

    Michelle Ruelos Cannistraci, *Senior Product Manager*

    Dennis Williams, *Senior Channel Manager*

    Jennifer Hanley, *Content Project Manager*

    Francis Hogan, *Art & Design Coordinator*

    Dawn Daugherty, *Senior Editorial Assistant*

From Interactive Composition Corporation:

    Paromita Das, *Production Project Manager*

Freelance supplement development provided by Allyson Powell.

Finally, both the authors and Thomson Delmar Learning would like to thank the following reviewers for their valuable suggestions:

    Charles Bunting, *Oklahoma State University*

    Joseph A. Coppola, *SUNY Morrisville*

    Thomas Hartley, *University of Akron*

    Joseph C. McGowan, *United States Naval Academy*

    Andrew Mayers, *Pennsylvania State University*

    Lt. Aaron Still, *United States Naval Academy*

    Thomas Young, *Rochester Institute of Technology*

    Ron Zammit, *California Polytechnic State University*

BASIC CIRCUIT ANALYSIS

1 Basic DC Circuits

2 General DC Circuit Analysis

3 Transient Circuits

Basic DC Circuits

OVERVIEW AND OBJECTIVES

In this chapter, we introduce the basic circuit variables and their physical and mathematical relationships. We will consider simple circuits primarily using **direct current (dc)** variables in the development. We will present ideal models used in circuit analysis and will provide numerous examples to illustrate the computations involved.

Objectives

After completing this chapter, the reader should be able to

1. Distinguish between **dc** and **ac** and discuss some of the basic properties of each.
2. State and describe the basic circuit variables including **charge**, **voltage**, **current**, **power**, and **energy**.
3. Define **resistance** and **conductance** and show the schematic symbol.
4. State the properties of the **ideal voltage source** and **ideal current source** models and show their schematic symbols.
5. State and apply **Ohm's law**.
6. Discuss voltage and current measurements and how they are made.
7. State and apply **Kirchhoff's voltage law (KVL)**.
8. State and apply **Kirchhoff's current law (KCL)**.
9. Determine equivalent resistance using **series** and **parallel** combinations.
10. Analyze a **single-loop (series)** circuit to determine all the variables.
11. Analyze a **single node-pair (parallel)** circuit to determine all the variables.
12. State and apply the **voltage divider rule**.
13. State and apply the **current divider rule**.

1-1 DC Circuit Variables

We begin the treatment of electrical theory and applications by establishing the basic circuit variables and laws. We perform this process in this chapter primarily with **dc** models. Strictly speaking, the abbreviation *dc* means **direct current**, but its usage as an adjective has been extended to describe many types of circuit conditions in which all the voltage and current variables have constant values.

FIGURE 1–1
Comparison of a dc voltage and an ac voltage.

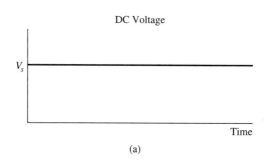

(a)

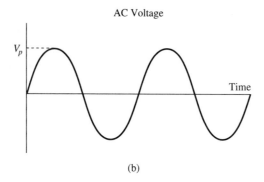

(b)

DC versus AC

The abbreviation **ac** means **alternating current**. The study of ac will constitute a major topic later in the text, but it may be necessary as we proceed to compare some of the attributes of dc with ac, and for that reason, the latter will be briefly introduced at this point. Most of the properties that we develop for dc can be extended to ac by suitable modification.

Strictly speaking, *ac* refers to **sinusoidal voltages and currents** such as encountered in the commercial power distribution system and in communications systems, but in casual usage, it often refers to any time-varying voltages and currents. (We will discuss the concepts of voltage and current shortly.) Just to ensure that the reader is clear as to the distinction, Figure 1-1 shows a comparison of a dc voltage and a true sinusoidal ac voltage. The idealized dc voltage of (a) is constant over all time as shown with a value V_s. However, the ac voltage of (b) reverses polarity in a sinusoidal manner as shown. It oscillates from a positive peak value of V_p to a negative value of $-V_p$. The various reasons and logic behind the time-varying nature of ac will be developed extensively later in the text.

Units and Prefixes

We will introduce many units throughout the text for describing the values of various electrical variables and components. Table 1-1 shows a compilation of some of the most basic

Table 1–1 Basic electric variables used in dc circuit analysis.

Variable	Symbol	Unit	Unit Abbreviation
Time	t	second	s
Charge	Q	coulomb	C
Current	I	ampere	A
Voltage	V (or E)	volt	V
Energy	W	joule	J
Power	P	watt	W
Resistance	R	ohm	Ω
Conductance	G	siemens	S

Table 1–2 Prefixes in SI units.

Value	Prefix	Abbreviation
10^{-12}	pico	p
10^{-9}	nano	n
10^{-6}	micro	μ
10^{-3}	milli	m
10^{3}	kilo	k
10^{6}	mega	M
10^{9}	giga	G

of these for circuit analysis, and these will be explained within the chapter. Others will be introduced later in the text as required.

Note that the symbols for the variables are *italicized*, but the symbols for the unit abbreviations are *not italicized*. For example, the description of a voltage of 12 volts could be expressed as $V = 12$ V. The quantity V on the left is the circuit variable and it is italicized, but the V on the right following the value of 12 is not italicized. Note also that there is a space between the number and the abbreviation. This will be the standard practice except when the number is serving as an adjective; for example, a 12-V battery.

Electrical units vary from extremely small to extremely large in magnitude. It is convenient to use prefixes to simplify the labeling system. Table 1-2 shows some of the most common prefixes used in the electrical field. For example, a current of 20 nanoamperes could be expressed as $I = 20$ nA.

As a general rule with some exceptions, we will use *lower-case* symbols for variables to represent variables that are or could be *varying with time*, whereas *upper-case* symbols will be used for variables that assume *constant values*. Therefore, because dc values are assumed to be constant, upper-case symbols will be used extensively in this chapter to represent the circuit variables.

In contrast to the rule for variables, the prefix forms apply as given for either constant or time-varying variables. For example, lower-case m refers to 10^{-3} while upper-case M refers to 10^{6}. An error here in certain situations could lead to disastrous results!

Charge

Arguably, charge is the most basic quantity in electricity, but it is rarely used as a variable of primary interest in common circuit applications. However, before any discussion of the more common circuit variables can be made, it is necessary to acknowledge and define the concept of electrical charge.

Charge is a quantity of electricity. The symbol used to represent a constant value of charge is upper-case Q. The unit of charge is the **coulomb**, which is abbreviated as C. Negatively charged electrons constitute the movement of charge within conductors, and the charge associated with one electron is $Q = 1.6 \times 10^{-19}$ C.

Voltage and Current Forms

The two most common variables that describe the behavior of electricity in the majority of applications are voltage and current. Voltage may be thought of as a type of "electrical pressure." It is measured across two points in a circuit. Current represents the rate of flow of charge and is measured through a circuit element. In casual terms, voltage is an "across variable" and current is a "through variable."

In the most general case, both voltage and current will vary with time. However, as previously discussed, the treatment in this chapter will be primarily concerned with dc circuits, in which case all voltages and currents will be considered as constant values. Whenever it is necessary to make a reference to **time** as a variable, the symbol is t and the basic unit is the **second**, which is abbreviated as lower-case **s**.

DC Current

Current is the rate of flow of electrical charge. When the flow has a constant rate in one direction, it may be considered as direct current and is abbreviated as *dc*. The symbol for dc current is the italicized symbol *I*. (The reader might wonder why *C* is not used, but it is the abbreviation for capacitance, which will be introduced in Chapter 3.) Various subscripts and/or superscripts may be added when there is more than one current under consideration. The unit of current is the **ampere**. It is abbreviated as **A**. A current flow of one ampere represents the movement of charge past a given point equal to 1 coulomb per second (C/s).

Conventional Current Flow versus Electron Flow

There are two totally opposite conventions for current flow. Within conductors, current flow consists of the movement of electrons, and this phenomenon has led to the concept of electron flow as an assumed direction for current flow. However, in the majority of engineering and engineering technology references, current flow is described in terms of so-called **conventional current flow**, which is opposite to that of electron flow. Just to ensure that the assumption is understood at the beginning, references in this book will be based on the positive direction of conventional current flow.

In theory, it wouldn't make any difference which convention is used, because no one can see the actual current flowing in the conductor anyway. However, the sign pattern for circuit analysis is much cleaner with conventional current flow when mathematical equations are involved. That is one of the major reasons why conventional current flow is so widely employed in advanced electrical circuit analysis and modeling.

The most common way of showing a current on a circuit diagram is by an arrow located adjacent to the conductor as shown in Figure 1-2. The arrow is pointed in the direction of the assumed positive direction for conventional current flow. Once that direction is assigned, the sign of the value determines the actual direction. For example, if $I = 4\,\text{A}$ in Figure 1-2, the positive conventional current flow is in the actual assumed direction. However, if the value is stated as $I = -4\,\text{A}$, the direction of conventional current flow would be opposite to that assumed.

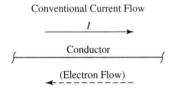

FIGURE 1–2
Conventional current flow in opposite direction to electron flow.

DC Voltage

The most widely employed symbol for dc voltage is the italicized symbol *V*, with subscripts and/or superscripts used whenever more than one voltage is being described. Many references use *E* as the symbol for voltage, but we will use *V* in this text. The unit of voltage is the **volt**.

The most common way to show a voltage on a circuit diagram is by + and − polarity symbols, which define the two points across which the voltage is measured. This concept is illustrated in Figure 1-3. The + symbol represents the most positive assumed reference point and the − symbol represents the most negative reference point. Once the direction is assigned, the sign of the value determines the actual direction of the voltage. For example, if $V = 6\,\text{V}$ in Figure 1-3, the polarity is in the direction assumed. However, if the value is stated as $V = -6\,\text{V}$, the voltage would have the opposite polarity to that assumed. The voltage between two points is also referred to *as potential difference* and in some older references as the electromotive force or emf.

FIGURE 1–3
Voltage between two points in a circuit.

Ideal Sources

There are many sources of electricity in the real world, including generators, batteries, solar cells, and other sources. Throughout the text, many of the different sources will be described, along with their realistic behavior. However, it is very desirable at the outset to describe two ideal source models: (1) the **ideal voltage source** and (2) the **ideal current source**. The models will be described here based on the dc forms.

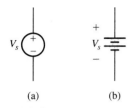

FIGURE 1–4
Schematic symbols for (a) an ideal voltage source and (b) a battery.

Ideal DC Voltage Source

The model shown in Figure 1-4(a) represents an ideal voltage source. The positive (+) and negative (−) terminals define the assumed polarity of the voltage. The (+) indicates the assumed positive terminal and the (−) indicates the assumed negative terminal.

This ideal model as considered for dc is assumed to maintain a constant voltage of value V_s independent of whatever load may be connected to it. This is an idealized concept, but there are many sources that approximate an ideal voltage source under normal operating conditions. For example, a well-charged battery may often be approximated by an ideal voltage source. Although our focus is on dc sources at the moment, we note in passing that the commercial ac power distribution system also approximates an ideal voltage source under many operating conditions, although the voltage in that case varies with time in a sinusoidal fashion, as previously noted. The fact that a source varies with time is different from its status as an assumed ideal voltage source. The latter idealization is based on the assumption that the voltage does not change when a load is connected, but the voltage may indeed be varying with time if it is not a dc source.

Whereas the model considered in Figure 1-4(a) can be used to represent virtually any voltage source, a common representation for dc sources, particularly in the case of a battery, is shown in Figure 1-4(b). Note that the positive terminal appears next to a wider line. We will use this symbol in some cases although, as previously noted, the other model can be applied to either dc or other types of sources.

Ideal DC Current Source

FIGURE 1–5
Schematic symbol for an ideal current source.

The model shown in Figure 1-5 represents an ideal dc current source. This ideal dc model is assumed to supply a constant current of value I_s independent of the load.

Even though most readers are likely familiar with sources that approximate the ideal voltage source, such as batteries, the concept of an ideal current source might seem a bit more abstract. However, many sources within active devices such as integrated circuits and transistors approximate ideal current sources. Moreover, some of the analysis methods employed in circuit models involve conversions between voltage sources and current sources, so it is necessary to treat both types of sources in circuit analysis.

Resistance

Resistance will be defined as the opposition to current flow exhibited by any device in which the resulting power is dissipated. All electrical devices exhibit resistance whether intended or not. For example, an ordinary electrical light bulb acts as a resistance. In that case, some of the energy is converted into light. The heating elements on an electric stove also act as resistances, and some of the energy in that case is converted to heat.

FIGURE 1–6
Schematic symbol for a resistance.

The symbol for resistance is R and the unit is the **ohm**. The abbreviation for ohm is the Greek upper-case omega, indicated as Ω. The **schematic** representation for a resistance is shown in Figure 1-6.

A device manufactured to exhibit a specific value of resistance is called a **resistor**. Many electronic circuits contain hundreds of resistors that are used to establish various voltage and current levels within the circuit.

Conductance

A resistance may also be described in terms of its equivalent **conductance**. Conductance will be denoted as G. It it is the reciprocal of the resistance; that is,

$$G = \frac{1}{R} \qquad (1\text{-}1)$$

The units of conductance are **siemens**, which are abbreviated as upper-case **S**. Thus, a resistance $R = 20\,\Omega$ could also be described as $G = 0.05\,\text{S}$. The use of conductance

FIGURE 1–7
Illustration of (a) a short circuit and (b) an open circuit.

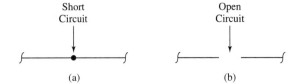

simplifies certain types of computations, but resistance units are much more common and will be employed in the majority of cases within the text.

Short Circuits and Open Circuits

Two terms that will appear very often in describing circuit configurations are the **short circuit** and the **open circuit**. A **short circuit** is a direct connection between two points as illustrated in Figure 1-7(a). Assuming zero resistance in the connection, it can be described by stating that the voltage across the short circuit is zero. However, current may be flowing through a short circuit.

An **open circuit** represents the complete lack of a connection between the two points as illustrated in Figure 1-7(b). It can be said that there is infinite resistance between the two points. It can also be described by the statement that the current flowing between the two points is zero. However, a voltage may exist across an open circuit.

Many of the common hookup and operating problems with electrical devices are a result of a short circuit between two points that should be open or an open circuit in a portion of a circuit that should be connected. Anyone reading this book who has set up a computer, a television set with components, or a stereo system may likely have encountered such a situation at some point, at least on a temporary basis until the problem was found. Searching for possible open and/or short circuits is a major part of troubleshooting. Many seemingly complex problems often reduce to a simple open or short at a point where the condition should not exist.

1-2 A Very Simple DC Circuit

Let us begin the study of circuit analysis by considering one of the most basic of all relationships and use it to establish some fundamental processes. Consider a dc voltage source of value V in volts as represented by the battery symbol shown in Figure 1-8(a). On the right-hand side is a resistor whose resistance is assumed to be of value R ohms. Although the bottom of the resistor is connected to the battery, a switch appears between the top of the battery and the top of the resistor. The switch is *open* as shown in (a) and therefore no current will flow. A circuit layout of this type is referred to as a **schematic diagram**.

In Figure 1-8(b), the switch is *closed* and current will now flow. Although electrons will flow from the negative terminal of the battery through the resistor and back to the positive terminal of the battery, that point of view will not be considered in this or later developments. Instead, the assumed conventional direction of the current flow is from the positive

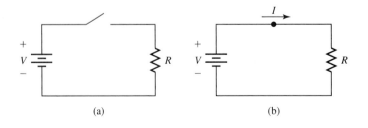

FIGURE 1–8
Simple circuit used to illustrate Ohm's law.

terminal of the battery through the load and back to the negative terminal of the battery. Learn to think that way in dealing with circuit analysis, because that is the manner widely used by engineers and technologists. The current along the top from the source to the load will be the same as the current along the bottom returning to the battery. Said differently, the circuit now constitutes a *loop* in which the current is assumed to be the same at all points around the loop.

We will assume that any resistance in the switch and the connecting wires is negligible compared with the resistance R. Moreover, we will assume an ideal voltage source with no internal resistance at this point. Thus, the voltage appearing across the resistance is the value V.

Ohm's Law

The current in amperes flowing through the resistance is determined by **Ohm's law**, which states

$$I = \frac{V}{R} \tag{1-2}$$

Thus, the current is directly proportional to the voltage across the resistance and inversely proportional to the resistance. This law is arguably the most basic circuit relationship. It will appear many times throughout the text in many different forms. Definitely commit it to memory along with the two alternate forms easily obtained from the basic equation.

$$V = RI \tag{1-3}$$

and

$$R = \frac{V}{I} \tag{1-4}$$

Power Relationships

In this simple circuit, the battery delivers power to the resistance, which absorbs the power. A lumped resistor simply dissipates the power absorbed in the form of heat. However, if the resistance is some more complex electrical component, the power may be converted to an alternate form of energy. Later in the text, the concepts of **inductance** and **capacitance** will be established and these parameters can actually store energy.

Power will be denoted for a dc circuit by the symbol P and the unit is the **watt**, which will be abbreviated as W. For a dc voltage V and a dc current I expressed in their basic units, the power P in watts is given by

$$P = VI \tag{1-5}$$

In the simple dc circuit under consideration, the battery delivers this power to the resistance R, which absorbs it. A positive resistance (the only type considered in this text) will always absorb power when current is flowing through it. When there is only one source, which is the case here, the source must deliver the power. When there is more than one source, it is actually possible for a particular source to absorb some of the power delivered by other sources. For example, consider the process of recharging a battery, in which case the battery receives power supplied by the charger. We will consider the concepts of power delivered and power absorbed in the next section.

Alternate Expressions for Power

The product of voltage and current as given by Equation 1-5 is always valid for either a dc source or a resistance. However, two alternate forms are quite useful for dealing with the power dissipated in a resistance. In one form, the expression $V = RI$ as obtained from Ohm's law is substituted in Equation 1-5 and the power expressed in terms of current and resistance is

$$P = I^2 R \qquad (1\text{-}6)$$

In still another form, the expression $I = V/R$ as obtained from Ohm's law is substituted in Equation 1-5 and the power expressed in terms of voltage and resistance is

$$P = \frac{V^2}{R} \qquad (1\text{-}7)$$

The three power relationships occur so frequently that it is recommended that they be committed to memory.

Ground

The term **ground** appears extensively in electrical and electronic applications. It can have different meanings according to the particular situation. In general, the term refers to a specific reference point from which various voltages are measured. In commercial power distribution systems, the actual earth ground may serve as a point of reference. In most electronic equipment, a so-called **common ground** point may be established by a metal frame or conducting plane within the equipment, which may or may not be connected to the power system ground.

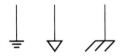

FIGURE 1–9
Different types of ground symbols.

Figure 1-9 displays several symbols that can be used to represent a ground. The one on the left is the most common one, and we will use it throughout the text. However, the other two may appear on various schematic diagrams. In fact, to minimize noise and pickup, some circuits employ more than one ground, in which case more than one of these symbols may appear in the same circuit.

To consider a simple application of a ground concept, consider the dc circuit of Figure 1-8(b) redrawn in the form of Figure 1-10. Although there is no physical connection shown between the negative terminal of the battery and the lower terminal of the resistance, the presence of the two ground symbols establishes a connection between them. In other words, all points on a schematic diagram with a ground symbol can be assumed to be connected to the same point, as least as far as circuit analysis is concerned. In many practical applications, there might exist small voltage differences between ground points that are supposedly at the same potential. Therefore, grounding tends to be somewhat of an "art" form for many circuits, particularly those with small signal levels, when unwanted pickup and noise are to be minimized. Moreover, there are serious safety considerations involved with grounding of power circuits.

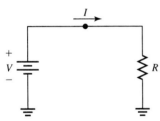

FIGURE 1–10
Simple circuit redrawn with ground symbols.

Effective Values of AC Voltage and Current

The major emphasis throughout this chapter is on dc circuits, but to get the reader to think about some concepts of ac, we introduce the term **effective value**. All you need to know for now is that the **effective value** of an ac voltage or current is the value that is used in computing ac power in a resistance. Thus, Equations 1-5, 1-6, and 1-7 can be applied to ac power dissipated in a resistance, provided it is the *effective value* of the voltage and/or the current used. When reference is made to 120 V ac, that value is the *effective value* of the time-varying ac sinusoidal voltage (also called the **root-mean-square** or **rms value**). Moreover, ac power represents the **average power** as determined over one complete cycle of the voltage and current. We don't want to confuse the issue at this point, but this extension is quite simple and easy to apply. It will be illustrated in Example 1-4.

EXAMPLE 1-1

A simple circuit of the form shown in Figure 1-8(b) has $V = 12$ V and $R = 4.7$ kΩ. Determine the current I.

SOLUTION Applying Ohm's law, we have

$$I = \frac{V}{R} = \frac{12}{4.7 \times 10^3} = 2.553 \times 10^{-3} \text{A} = 2.553 \text{ mA} \tag{1-8}$$

About the only challenge here was rounding off the answer and changing the units.

EXAMPLE 1-2

For the simple circuit of Example 1-1, determine the power delivered by the source and the power dissipated by the resistance.

SOLUTION Don't let the extra wording here cause any confusion, because the power delivered by the source is exactly equal to the power dissipated by the resistance. It can be determined as

$$P = VI = 12 \times 2.553 \times 10^{-3} = 30.64 \times 10^{-3} \text{ W} = 30.64 \text{ mW} \tag{1-9}$$

Even though the procedure was quite simple, let's use this opportunity to make a point. In the computation of Equation 1-9, the calculated result obtained from Example 1-1 (I) was used. With the additional calculation, this provided two possible points for error and round-off. However, it is possible to calculate power directly from voltage and resistance, and this approach means that we don't have to deal with the intermediate step to obtain the desired result. Hence, we can express the power as

$$P = \frac{V^2}{R} = \frac{(12)^2}{4.7 \times 10^{-3}} = 30.64 \times 10^{-3} \text{ W} = 30.64 \text{ mW} \tag{1-10}$$

In this simple example, the results are exactly the same. The bottom line is that when you have a choice of more than one path to take, the one involving fewer results of other calculations might be the wiser choice if the number of calculations is about the same. Take this as a wise suggestion to consider rather than a rigid rule, because it is very likely that the authors don't necessarily follow this practice in all cases!

EXAMPLE 1-3

The output current of a certain integrated circuit is 6 mA and it is flowing into a resistance of value 5 kΩ. Determine the voltage across the resistance.

SOLUTION The voltage is determined by Ohm's law. It is

$$V = RI = 5 \times 10^3 \times 6 \times 10^{-3} = 30 \text{ V} \tag{1-11}$$

EXAMPLE 1-4

Determine the "hot" resistance of a 60-W bulb operated from an ac effective voltage of 120 V.

SOLUTION The effective value of 120 V is the value used in determining the ac or average power. We begin with

$$P = \frac{V^2}{R} \tag{1-12}$$

Solve for R.

$$R = \frac{V^2}{P} = \frac{(120)^2}{60} = 240 \; \Omega \tag{1-13}$$

EXAMPLE 1-5

The power dissipated in a certain resistance is 100 W and the current is 4 A. Determine the resistance.

SOLUTION We begin with
$$P = I^2 R \tag{1-14}$$

Solve for R.
$$R = \frac{P}{I^2} = \frac{100}{(4)^2} = 6.250 \ \Omega \tag{1-15}$$

1-3 Measuring DC Circuit Variables

The basic instrument for measuring voltage is the **voltmeter** and the basic instrument for measuring current is the **ammeter**. One instrument for measuring resistance is the **ohmmeter**. Instruments called **multimeters** can perform all three of these functions for dc and ac voltage and current measurements by changing the mode and scales provided. Inexpensive versions of these instruments are readily available at hardware and consumer electronics stores. They are adequate for routine applications. However, the more accurate and rugged (and more expensive) laboratory-grade instruments usually are obtained through specialized electrical and electronic distributors. We will not pursue the internal details of such instruments, but instead will focus on the ideal models.

Ideal DC Voltmeter

A dc voltmeter is connected *across* the two points for which the measurement is desired, as illustrated in Figure 1-11. The ideal voltmeter is characterized by an *infinite* input resistance. The practical implication of this condition is that the voltmeter will draw no current from the circuit to which it is connected and therefore will not disturb the circuit. Practical voltmeters have resistances that can be as high as many millions of ohms.

One important consideration in making a voltmeter measurement is to ensure that the polarity of the instrument terminals is correct for a dc voltage and to ensure that the range is set to a value greater than that of the voltage being measured. For many digital instruments having input isolation, a mistake in the initial connection or range might not do any damage. However, with some of the more classical analog voltmeters using a needle movement, an instrument may be seriously damaged by connecting with the wrong polarity, or with a scale set too low. When such an instrument must be employed, determine as much information as possible about the expected polarity and the range of the voltage being measured. If in doubt, start with the highest possible range of the instrument and note the direction of movement. If the deflection is negative, reverse the connections. Then, reduce the range until the deflection is as high as possible without exceeding the maximum level.

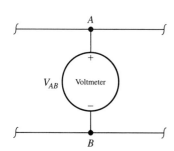

FIGURE 1–11
A voltmeter is connected between two points in a circuit across which the voltage is to be measured.

The voltage V_{AB} in Figure 1-11 utilizes **double-subscript notation**, which is sometimes convenient to use. The first subscript (A) is the assumed positive terminal and the second (B) is the assumed negative terminal. Thus, $V_{BA} = -V_{AB}$.

Ideal DC Ammeter

A dc ammeter must be connected within the branch for which the current is to be measured as shown in Figure 1-12. The positive terminal of a dc ammeter must be connected to the point at which the positive current enters as illustrated. The ideal ammeter should have *zero* resistance so that it does not disturb the circuit in which it is connected. This is opposite to that of a voltmeter. However, remember that voltage is measured *across* two points within a circuit, whereas current is measured through a branch.

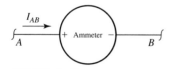

FIGURE 1–12
An ammeter is connected within the branch for which current is to be measured.

The current I_{AB} in Figure 1-12 utilizes **double-subscript notation**, which is sometimes convenient to use. The first subscript (A) is the terminal in which the current is assumed to enter and the second (B) is the terminal in which the current is assumed to exit. Thus, $I_{BA} = -I_{AB}$.

Comparison of Voltage and Current Measurement

Comparing the procedures for both a voltmeter and an ammeter, voltage is usually easier to measure. This is mostly because the circuit normally does not have to be reconfigured to measure voltage across two points. However, unless the ammeter is built into the circuit configuration, it is usually necessary to disconnect part of a circuit to insert an ammeter. For that reason, much routine troubleshooting is performed with voltage measurements to the extent possible.

Resistance Measurements

Although there are special instruments to measure resistance (including the classical **Wheatstone bridge**), most routine dc resistance measurements are made with multimeters that have one or more ohmmeter scales. The ohmmeter scales of these multimeters excite the circuit with an internal source (usually a battery). They have calibrated scales that employ Ohm's law in some fashion to provide resistance values in ohms. Because power is provided to the circuit to perform the measurement, it is essential that the circuit being measured be de-energized or turned off to make the measurement. It might also be necessary to disconnect the part of the circuit for which the measurement is desired from the remainder to avoid shunting effects. This situation will be clearer after we study equivalent resistance.

Certain measurements with an ohmmeter or a bridge can be misleading because some resistive devices are nonlinear, and the resistance might vary with the current through the device. In this case, it might be necessary to establish operation at the desired level and employ both a voltmeter and an ammeter for measurement. Ohm's law can then be used to determine the resistance under the specified operating conditions.

Measurement Summary

Although it is certainly possible to measure voltage, current, and resistance separately, voltage is usually the simplest to measure. Indeed, a common approach with many circuits is to connect one terminal to the common ground and move around the circuit with the other terminal, providing care with regard to possible safety hazards, polarities, and scale settings as necessary. It is always prudent to turn off the equipment before moving the terminals of the voltmeter. Moreover, if the voltmeter is a classical analog type, it will be necessary to reverse the terminals if the voltage at a given point has the opposite polarity to that of the preceding point.

One particular useful feature of the ohmmeter function is to measure circuit **continuity**. Assuming that there is no power within a circuit, as discussed earlier, for ohmmeter measurements, the continuity between two points may be checked by a measurement of zero or a very small resistance between the two points. You may use this procedure to test for open circuits and breakage between two points that should be connected, and points where shorts might exist.

Finally, although we are considering only dc measurements at this point, we should at least mention the ubiquitous **oscilloscope** here. This device provides an instantaneous plot of a voltage versus time. It is essential in dealing with many time-varying voltages. We can also use an oscilloscope to measure dc voltages if it is set on **direct coupling** (also denoted as **dc**). In that case, a dc voltage is simply a vertical displacement of a horizontal line.

EXAMPLE 1-6

We determine the resistance of a device under operating conditions by measuring the voltage across the device and the current through it. The voltmeter reads 120.0 V and the ammeter reads 2.012 A. Determine the resistance.

SOLUTION The resistance is determined by Ohm's law.

$$R = \frac{120.0}{2.012} = 59.64 \ \Omega \qquad (1\text{-}16)$$

1-4 Power and Energy

For the simple example considered earlier, it was obvious that the dc source delivered the power and the resistance absorbed it. However, in more complex circuits, it is necessary to investigate the concept of **power delivered** versus **power absorbed**.

Power Delivered

Consider the situation depicted in Figure 1-13. When the current I is leaving the positive terminal of a device, that device is delivering power.

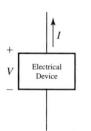

FIGURE 1–13
Reference directions for power **delivered**.

Power Absorbed

Next, consider the situation shown in Figure 1-14. When the current I is entering the positive terminal of a device, that device is absorbing power.

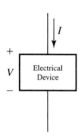

FIGURE 1–14
Reference directions for power **absorbed**.

Discussion

In the simple circuit considered earlier, the current is flowing out of the positive terminal of the dc source and into the positive terminal of the resistance. Therefore, based on the power conventions, the source is delivering the power and the resistance is absorbing this power, which is certainly the case.

As mentioned earlier, we will consider only positive resistance in this text. (There is a concept called negative resistance that arises in certain specialized circuits.) Positive resistance always absorbs power. This means that the polarity of the voltage across a resistance will have its positive terminal at the point where the current enters. This property will be important later when establishing reference polarities for voltage drops in more complex circuits.

The same conventions indicated here for dc also apply to ac and other situations. However, whereas the power absorbed by a resistance is always dissipated in some form or another, the energy associated with other circuit parameters may be stored and released later. The two energy storage parameters are capacitance and inductance. They will be introduced later.

Negative Signs

Consider that the current is assumed to be leaving the assumed positive terminal of a device and the resulting power delivered by the device is formulated. However, suppose that either the voltage or the current (but not both) have a negative value. This will cause the calculated power *delivered* to be *negative*. This means only that the power is actually *absorbed*. The converse is also true; that is, if an assumed power *absorbed* turns out to be *negative*, this means that the power is actually *delivered*.

Energy

Power is more often used in electrical system specifications and circuit analysis, but energy is arguably more basic. Energy is the actual work performed or the capacity to perform work. It will be represented for dc by the symbol W. The basic unit is the **joule**, which is abbreviated as **J**. There are several variations on units, including the widely employed unit of **kilowatt-hour (kWh)**. Note that W is the symbol for energy, but **W** is the abbreviation for the units of power (watts). The use of the same symbol should not cause any problem because the italicized form will always be on the left-hand side of an equation and the nonitalicized form will follow a specific value.

Power is the rate of change of energy, and later, when we use calculus, we will develop a more rigorous relationship between power and energy. At this point, accept the fact that if work is being performed at a rate of 1 joule/second, the power generated is 1 watt; that is, 1 W = 1 J/s.

For a dc circuit with constant voltage V and constant current I, the power $P = VI$ is also constant. The work W performed over a period of time T based on this constant power level is

$$W = PT \tag{1-17}$$

Although the focus here is on dc, it should be noted that the relationship of Equation 1-17 applies for ac provided that the power P is interpreted as the *average power* as determined over the interval of T.

The conventions of power delivered and power absorbed hold for energy as well. Thus, energy may be delivered or absorbed.

To convert between joules and kilowatt-hours, it is necessary to change from seconds to hours and from watts to kilowatts. The following conversion formula is useful:

$$W(\text{J}) = 3.6 \times 10^6 \times W(\text{kWh}) \tag{1-18}$$

The conversion formula comes from the fact that one hour = 3600 seconds and 1 kilowatt = 1000 watts.

EXAMPLE 1-7

Assume that a family leaves a 60-W light bulb on for the duration of a two-week trip. If electricity costs 8 cents per kilowatt-hour, determine the cost incurred.

SOLUTION The total time in hours is 2 weeks × 7 days/week × 24 hours/day = 336 hours. The power in kilowatts is 0.06 kW. Hence, the energy expended is

$$W = 0.06 \times 336 = 20.16 \text{ kWh} \tag{1-19}$$

The cost is

$$\text{Cost} = 20.16 \text{ kWh} \times 0.08 \text{ dollars/kWh} = \$1.61 \tag{1-20}$$

1-5 Kirchhoff's Laws

Along with Ohm's law, two other basic laws for dealing with circuits are Kirchhoff's voltage law and Kirchhoff's current law. The former will be abbreviated throughout the text as **KVL** and the latter will be abbreviated as **KCL**.

Kirchhoff's Voltage Law

Kirchhoff's voltage law (KVL) may be stated as follows: The algebraic sum of the voltages around any closed loop is zero. For a dc circuit, this statement may be expressed mathematically as

$$\sum_n V_n = 0 \tag{1-21}$$

In the interpretation of this law, it is necessary to establish a sign convention. The convention that will be used in most cases in this text is that **a** voltage drop will be treated as a positive value and a voltage rise will be treated as a negative value. The opposite convention would be equally valid, but in later developments, the convention used will typically result in fewer negative signs. An alternate viewpoint is that the sum of the voltage rises is equal to the sum of the voltage drops.

To illustrate KVL, consider the loop shown in Figure 1-15. None of the components shown here are identified, but assume that we begin below the voltage V_1 and "walk"

FIGURE 1–15
A loop used to illustrate Kirchhoff's voltage law.

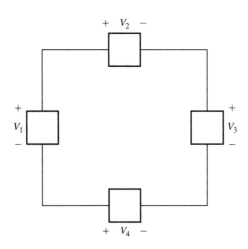

around the loop clockwise. We note that we will encounter two rises (V_1 and V_4) and two drops (V_2 and V_3). Application of KVL results in

$$-V_1 + V_2 + V_3 - V_4 = 0 \tag{1-22}$$

Kirchhoff's Current Law

Kirchhoff's current law (KCL) may be stated as follows: The algebraic sum of the currents at any node is zero. A node is a junction point between two or more branches. For a dc circuit, KCL may be expressed mathematically as

$$\sum_n I_n = 0 \tag{1-23}$$

As in the case of KVL, it is necessary to establish a sign convention. The convention that we will use in most cases is that current leaving the node will be treated as a positive value and current entering the node will be treated as a negative value. The opposite convention also would work.

To illustrate KCL, consider the node shown in Figure 1-16. Two currents (I_1 and I_2) are leaving and two currents (I_3 and I_4) are entering. KCL applied to the node reads

$$I_1 + I_2 - I_3 - I_4 = 0 \tag{1-24}$$

An alternate viewpoint is that *the net current entering the node is equal to the net current leaving the node.*

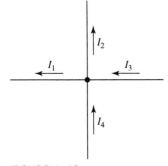

FIGURE 1–16
Illustration of Kirchhoff's current law at a node.

▌▌ EXAMPLE 1-8

For the circuit of Figure 1-17, determine the value of the voltage V_x.

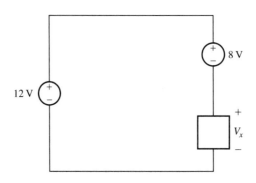

FIGURE 1–17
Circuit of Example 1-8.

1-6 • Equivalent Resistance

SOLUTION We begin below the 12-V source and move clockwise around the loop applying KVL.

$$-12 + 8 + V_x = 0 \tag{1-25}$$

Solve for V_x.

$$V_x = 4 \text{ V} \tag{1-26}$$

In this case, the unknown voltage had a positive value, so the polarity is in the direction assumed. If the value had been negative, the polarity would be opposite to that assumed.

EXAMPLE 1-9

For the circuit of Figure 1-18, determine the value of the current I_x.

SOLUTION Assuming currents leaving as positive, we have

$$-8 + 3 + I_x = 0 \tag{1-27}$$

Solve for I_x.

$$I_x = 5 \text{ A} \tag{1-28}$$

Because the value is positive, the current is flowing in the direction assumed.

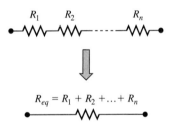

FIGURE 1–18
Circuit of Example 1-9.

1-6 Equivalent Resistance

Frequently, circuits will contain two or more resistances connected so that the external terminal behavior can be simplified and represented for analysis purposes by a single resistance. The most common configurations of this type are the series and parallel configurations. Moreover, often we can simplify more complex configurations by successive application of the series and parallel operations.

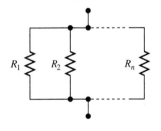

FIGURE 1–19
Resistances in series and the equivalent single resistance.

Series Resistance

A **series** combination of several resistances is illustrated in Figure 1-19. To be a true series connection, there must not be any branches connected to any of the intermediate nodes. We provide a guided exercise for the reader (Problem 1-33) to show that the equivalent resistance R_{eq} seen at the two outside terminals is given by

$$R_{eq} = R_1 + R_2 + \cdots + R_n \tag{1-29}$$

Stated in words, resistances in series add.

Parallel Resistance

A **parallel** combination of several resistances is illustrated in Figure 1-20. The connection along the top is referred to as a **bus**, as is the one on the bottom, and it is assumed that there is negligible resistance in the buses. We provide a guided exercise for the reader (Problem 1-34) to show that the equivalent resistance R_{eq} seen at the two terminals can be determined by the formula

$$\frac{1}{R_{eq}} = \frac{1}{R_1} + \frac{1}{R_2} + \cdots + \frac{1}{R_n} \tag{1-30}$$

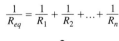

FIGURE 1–20
Resistances in parallel and the equivalent single resistance.

Stated in words, the reciprocals of resistances in parallel add to form the reciprocal of the equivalent resistance.

An alternate formulation for the parallel configuration, which is often easier to manipulate, is first to convert all the resistances to their equivalent conductances. Let

$G_1 = 1/R_1$, $G_2 = 1/R_2$, and so on. The equivalent conductance G_{eq} is then

$$G_{eq} = G_1 + G_2 + \cdots + G_n \tag{1-31}$$

Stated in words, conductances in parallel add. After the equivalent conductance is determined, its reciprocal can then be calculated and it will be the equivalent resistance.

Two Resistances in Parallel

A common situation arising in many circuits is that of two resistances in parallel. Even though either of the two preceding formulas can be used, there is a form that is often easier to apply when there are only two values. It can be readily shown that

$$R_{eq} = \frac{R_1 R_2}{R_1 + R_2} \tag{1-32}$$

Thus for two resistances in parallel, the equivalent resistance is the product of the resistances divided by the sum. Sometimes, this may even be appropriate for dealing with more than two. First, the parallel combination of any two is determined. Then that equivalent value is parallel combined with the third, and so on.

Some Useful Properties

Although the formulas have been covered, it is useful to point out a few properties, some of which are obvious, and others of which may not be so obvious.

- The net resistance of a number of resistances in series is always greater than the largest one.
- The net resistance of N equal resistances of value R in series is NR.
- The net resistance of a number of resistances in parallel is always smaller than the smallest one.
- The net resistance of N equal resistances of value R in parallel is R/N.
- If two resistances whose values are widely separated are connected in parallel, the net resistance is only slightly less than the smaller one.

This last property might seem a bit vague, and one can always work out the value. However, it has practical implications when performing voltage measurements in a circuit. Any voltmeter will have some loading effect, so ideally the input resistance to the voltmeter should be much greater than the resistance level of the circuit being measured, in which case the loading effect will be minimal.

■ **EXAMPLE 1-10** For the circuit of Figure 1-21, determine an equivalent resistance as "seen" at the input terminals.

SOLUTION The equivalent resistance can be determined by successive applications of the series and parallel combination processes. Refer to Figure 1-22 for the development that follows. All the computations involve applications of the formulas considered earlier to simple numbers, so the steps will be delineated and the reader may verify the results.

We begin with the two parallel resistances shown on the right of Figure 1-21 and on the top left of Figure 1-22. We can show that the equivalent resistance of 6 Ω in parallel with 12 Ω is 4 Ω. Next, this equivalent resistance appears in series with the resistance of 2 Ω and the net series equivalent is 6 Ω. This resistance appears in parallel with the resistance of 12 Ω in the middle of the circuit. The net resistance is 4 Ω. Finally, this resistance appears in series with 3 Ω and the net resistance seen at the terminals is 7 Ω.

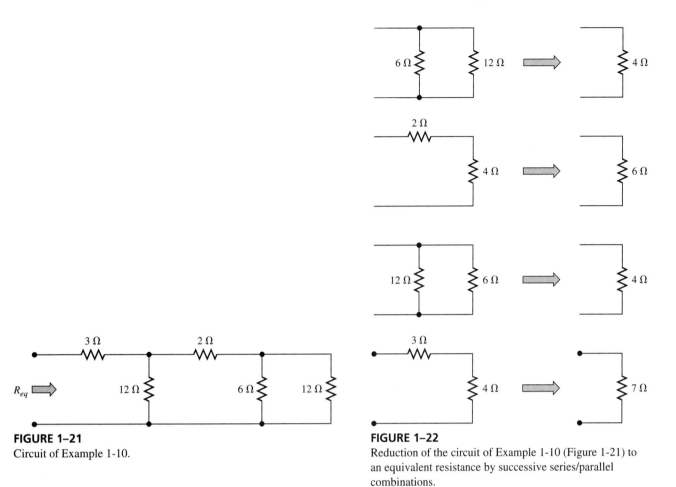

FIGURE 1–21
Circuit of Example 1-10.

FIGURE 1–22
Reduction of the circuit of Example 1-10 (Figure 1-21) to an equivalent resistance by successive series/parallel combinations.

Note that these values are good "teaching values" because they yield simple results. Seldom do any real-life circuits have such nice numbers.

1-7 Single-Loop or Series Circuit

One of the simplest circuits to analyze is the **single-loop** or **series circuit**. An example of this type is shown in Figure 1-23. In general, the loop may contain any arbitrary number of voltage sources and resistances. For the moment, however, we will assume that there are no current sources in the loop.

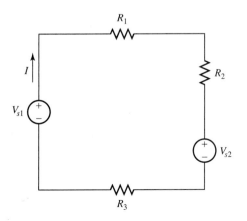

FIGURE 1–23
Representative single-loop (series) circuit containing voltage sources and resistances.

Constraint for Single-Loop Circuit

The key to analyzing a single loop circuit is the basic constraint that the current at any point in the loop must be the same. This fact may be verified by considering any node and any component. KCL requires that the current entering a node must equal the current leaving the node and the current entering a component must equal the current leaving it. The only conclusion is that the current must be the same at all points in the loop.

Depending on what is given, one or more unknowns may be determined by applying KVL to the loop and Ohm's law to each resistance. A common situation is when the voltage sources and resistances are known, but in which the current must be determined. To solve for the current, perform the following steps:

(1) Assume an unknown current I. It really doesn't matter which direction is assumed because the sign of the result will determine whether the assumed direction is correct or not. However, with a little practice, it will be feasible in many cases to assume the correct direction, which will eliminate the nuance of dealing with a negative sign for the result.

(2) Start at some point and "walk" all the way around the loop applying KVL. Although the direction is arbitrary, we recommend that the movement be in the direction of the assumed current.

(3) Choose a sign pattern concerning drops and rises. As mentioned earlier, the one that we will follow extensively is that a voltage rise will be considered as negative ($-$) and a voltage drop will be considered as positive ($+$).

(4) For each resistance, the positive terminal will be considered as the point where the current enters the resistance. Thus if the sign pattern of step (3) is chosen, this will mean that all resistive voltage drops will be considered as positive.

(5) Solve for the unknown current using algebra. All voltages across resistances may then be easily determined using Ohm's law and the power associated with each component may also be determined.

Presence of a Current Source

In a single-loop circuit, it is meaningful to consider the possibility of only one current source. If there were more than one with different values, it would create a physical contradiction, and if they were equal, it would be equivalent to a single source. Thus, we will consider the possible presence of only one current source.

If there is an ideal current source in a single-loop circuit, the current is forced to be equal to the value of the current source. We use Ohm's law to determine the voltage across each resistance. One other variable to be determined is the voltage across the current source. That may be determined by an application of KVL for the loop. We illustrate this process in Example 1-12.

Application to the Representative Circuit

Consider the application of the preceding strategy to the representative circuit of Figure 1-23. The current I is assumed to be clockwise as shown. We will begin at a point just below V_{s1}, move upward through the source, and continue around the loop in the direction of the current. Because the movement through V_{s1} represents a *rise* in potential (movement from negative terminal to positive terminal), it will carry a negative sign. However, for the movement through V_{s2}, we enter at the positive terminal and leave at the negative terminal. This represents a drop in potential, and it will carry a positive sign. Because we are moving in the direction of the current, we assume that the voltage across each resistance has its positive terminal at the point of entry, so all resistive voltages will be drops and will carry positive signs. The equation follows.

$$-V_{s1} + R_1 I + R_2 I + V_{s2} + R_3 I = 0 \tag{1-33}$$

This equation applies irrespective of which variables are known or unknown. For the sake of discussion, let us solve for the current in terms of the other quantities in the equation. Algebraic manipulation yields

$$I = \frac{V_{s1} - V_{s2}}{R_1 + R_2 + R_3} \quad (1\text{-}34)$$

There is an intuitive pattern emerging from this expression that is worthy of discussion. First, the source V_{s1} would tend to force the current to flow in the direction assumed, and it appears as a positive term in the numerator. However, the source V_{s2} would tend to force the current to flow opposite to the direction assumed, and it appears as a negative term in the numerator. In other words, the two voltage sources act in opposition, and the net voltage that would force current to flow in the direction assumed is $V_{s1} - V_{s2}$. The denominator is simply the equivalent series resistance around the loop. The final form of the expression for the current reduces to a version of Ohm's law for the entire loop.

Although the interpretation of the preceding paragraph is valuable, and we encourage readers to use it as a check, for complex loops we recommend that the basic approach of applying KVL and "walking" around the loop be employed. The chances of a sign error are less when this methodology is applied consistently.

■ **EXAMPLE 1-11**

For the circuit of Figure 1-24, determine (a) loop current, (b) voltages across all resistances, and (c) power delivered or absorbed by each component in the circuit.

SOLUTION This circuit has the same form as used in this section for illustration, but we will go through the full procedure again. Although the reader's intuition might not have reached this level yet, because the 32-V source is greater than the 12-V source, the net current will have a clockwise direction. We will assume that is the case. However, it really wouldn't matter which direction is assumed, because the sign pattern would take care of that nuance. Note that the voltage across each of the resistances is labeled and is consistent with the assumed direction of current.

(a) We begin by walking around the loop again and we have

$$-32 + 5I + 3I + 12 + 2I = 0 \quad (1\text{-}35)$$

Solve for I.

$$I = \frac{32 - 12}{5 + 3 + 2} = \frac{20}{10} = 2 \text{ A} \quad (1\text{-}36)$$

Anyone having some background probably could have deduced this result in one step, but we will continue to defend the position of writing out the loop equation unless you have a lot of experience with this process.

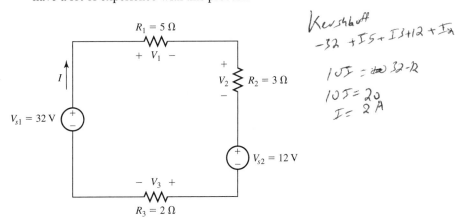

FIGURE 1–24
Circuit of Example 1-11.

(b) The voltage drops across the three resistances are determined as follows:

$$V_1 = R_1 I = 5 \times 2 = 10 \text{ V} \tag{1-37}$$
$$V_2 = R_2 I = 3 \times 2 = 6 \text{ V} \tag{1-38}$$
$$V_3 = R_3 I = 2 \times 2 = 4 \text{ V} \tag{1-39}$$

(c) The voltage source V_{s1} *delivers* power because current is flowing out of its positive terminal and the value is

$$P_{s1} = V_1 I = 32 \times 2 = 64 \text{ W} \tag{1-40}$$

On the other hand, V_{s2} *absorbs* power because current is flowing into its positive terminal and the value is

$$P_{s2} = V_2 I = 12 \times 2 = 24 \text{ W} \tag{1-41}$$

The three resistances each *absorb* (and dissipate) power and the three values are

$$P_{R1} = I^2 R_1 = (2)^2 \times 5 = 20 \text{ W} \tag{1-42}$$
$$P_{R2} = I^2 R_2 = (2)^2 \times 3 = 12 \text{ W} \tag{1-43}$$
$$P_{R3} = I^2 R_3 = (2)^2 \times 2 = 8 \text{ W} \tag{1-44}$$

We could just as easily have used one of the alternate formulas for the power associated with the resistors.

It is helpful to construct a table to display the power values.

Variable	Power Delivered	Power Absorbed
V_{s1}	64 W	
V_{s2}		24 W
R_1		20 W
R_2		12 W
R_3		8 W
Totals	64 W	64 W

The final result is that power delivered = power absorbed, a result that must be satisfied.

EXAMPLE 1-12

For the circuit of Figure 1-25, determine (a) loop current, (b) voltages across all resistances, (c) voltage across the current source, and (d) power delivered or absorbed by each component in circuit.

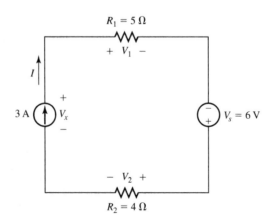

FIGURE 1–25
Circuit of Example 1-12.

SOLUTION

(a) The loop contains one current source of value 3 A so this forces the loop current I in the direction shown to be

$$I = 3 \text{ A} \tag{1-45}$$

(b) The voltage drops across the two resistances are

$$V_1 = R_1 I = 5 \times 3 = 15 \text{ V} \tag{1-46}$$
$$V_2 = R_2 I = 4 \times 3 = 12 \text{ V} \tag{1-47}$$

(c) Before dealing with power, we must determine the voltage across the current source. In general, the voltage across a current source may be in either direction. In this case, we will assume that it is positive at the top and denote this voltage as V_x. Starting below the current source and forming a KVL loop, we have

$$-V_x + 15 - 6 + 12 = 0 \tag{1-48}$$

Solve for V_x.

$$V_x = 21 \text{ V} \tag{1-49}$$

Thus, we "guessed" correctly on the direction for the voltage across the current source and it is positive at the top.

(d) Because the current is flowing out of the positive terminal of the current source, it is *delivering* power given by

$$P_I = V_x I = 21 \times 3 = 63 \text{ W} \tag{1-50}$$

The voltage source V_s is also delivering power because the current is flowing out of its positive terminal and it is

$$P_V = VI = 6 \times 3 = 18 \text{ W} \tag{1-51}$$

The two resistances are absorbing power and the values are

$$P_{R1} = I^2 R_1 = (3)^2 \times 5 = 45 \text{ W} \tag{1-52}$$
$$P_{R2} = I^2 R_1 = (3)^2 \times 4 = 36 \text{ W} \tag{1-53}$$

As in the preceding example, it is convenient to construct a table.

Variable	Power Delivered	Power Absorbed
Current Source	63 W	
Voltage Source	18 W	
R_1		45 W
R_2		36 W
Totals	81 W	81 W

1-8 Single Node-Pair or Parallel Circuit

Another simple circuit that arises in many practical problems is that of a **single node-pair** or **parallel circuit**. The most common way in which this circuit appears is one in which there is a single voltage source. We show an example of this in Figure 1-26. We will consider the presence of only one voltage source in this development, because the presence of two or more sources of different values in parallel would represent a physical contradiction. If there were two or more voltage sources in parallel with the same value, it would be

FIGURE 1-26

Single node-pair (parallel) circuit with voltage source and resistances.

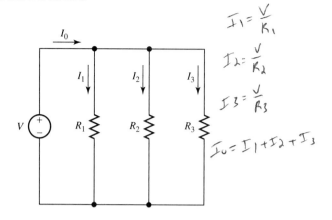

$$I_1 = \frac{V}{R_1}$$
$$I_2 = \frac{V}{R_2}$$
$$I_3 = \frac{V}{R_3}$$
$$I_0 = I_1 + I_2 + I_3$$

equivalent to a single voltage source, so we need consider only one source for this purpose. We will discuss the presence of current sources later in this section.

Constraint for Single Node-Pair Circuit

The key to analyzing a single node-pair circuit is that the voltage across the two nodes must be the same for all elements connected in parallel. We can verify this assumption readily by expressing KVL around any of the loops within the circuit. Because there are only two voltages around any loop (a rise and a drop), the voltages must be equal. Based on the assumption that the voltage source is ideal, the voltage across each component is the value of the source voltage.

The circuit of Figure 1-26 utilizes two **buses**, which are the connection along the top and the connection along the bottom. This manner of drawing the circuit is very convenient, and many circuit layouts utilize buses. However, the key to the single node-pair property is that all points along the top have the same potential, and all points along the bottom have the same potential. The connecting wires are therefore assumed to have negligible resistance. In fact, it would be possible to redraw the circuit with all connections on the top coming to a single point and all connections on the bottom coming to a single point. Thus, from an analysis viewpoint, an ideal bus is equivalent to a single node.

Procedure

The approach that follows for the representative circuit is essentially the same for any parallel circuit having an ideal voltage source in parallel with an arbitrary number of resistances. We will assume the common situation in which the voltage source and the resistance values are known, but in which each of the resistive currents and the source current must be determined. We can modify the approach to determine other variables if the need arises. The procedure follows.

(1) Assume a branch current for each of the resistances. The direction should be such that the current is entering each branch at the positive voltage terminal as illustrated in Figure 1-26.

(2) Assume a current flowing out of the voltage source. The source will be delivering the power, so the direction will be out of the positive terminal of the source.

(3) Determine the current for each of the branches using Ohm's law.

(4) Determine the source current by the application of KCL at the node or bus at which the current is leaving the source. (Actually, it wouldn't matter which node is used because the equations at the two nodes are mathematically equivalent, but the sign pattern is cleaner at the positive node.)

(5) Additional computations involving power delivered by the source and power dissipated by the resistances can be made as needed.

Application to the Representative Circuit

The circuit of Figure 1-26 has been labeled with the three individual resistive currents and the net current. First, the three resistive currents are calculated.

$$I_1 = \frac{V}{R_1} \tag{1-54}$$

$$I_2 = \frac{V}{R_2} \tag{1-55}$$

$$I_3 = \frac{V}{R_3} \tag{1-56}$$

We next apply KCL at the upper node, assuming that current entering is negative and current leaving is positive. We have

$$-I_0 + I_1 + I_2 + I_3 = 0 \tag{1-57}$$

This leads to

$$I_0 = I_1 + I_2 + I_3 \tag{1-58}$$

Most readers probably could have written down the last form of the equation directly from the concept that the current entering the node must equal the current leaving the node. The three values of current expressed in terms of the voltage source could now be substituted in the last equation and power values could be determined. The examples that follow the section will provide some illustrations of that nature.

Single Node-Pair with Current Sources and Resistances

Next, consider the situation in which there is no parallel voltage source to establish the voltage, but in which there are an arbitrary number of current sources and resistances. The representative circuit of Figure 1-27 should suffice to establish the general procedure for analysis. The procedure follows.

(1) Assume an unknown voltage V. It really doesn't matter which direction is assumed because the sign of the result will determine whether the assumed direction is correct or not. However, with a little practice, it will be feasible in many cases to assume the correct direction, which will eliminate some of the nuance of dealing with negative signs.

(2) Apply KCL to one of the two nodes. It is best to apply it at the node assumed to be the most positive, although the equation at the other node will be mathematically equivalent.

(3) Choose a sign pattern concerning current leaving and current entering. The one that will be followed extensively is that current entering will be considered as negative ($-$) and current leaving will be considered positive ($+$).

(4) For each resistance, the positive terminal will be considered as the point where the current enters the resistance. Thus, if the sign pattern of step (3) is chosen, this will mean that all currents through resistances will be considered as positive.

(5) Solve for the unknown voltage using algebra. All currents through resistances may then be determined easily using Ohm's law, and the power associated with each component may be determined.

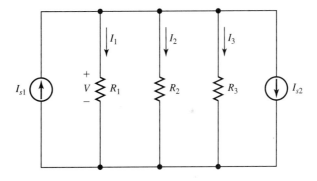

FIGURE 1–27
Representative single node-pair circuit containing current sources and resistances.

Application to the Representative Circuit

Consider the application of the preceding procedure to the circuit of Figure 1-27. A voltage V is assumed across the circuit and KCL is applied to the upper node. The equation reads

$$-I_{s1} + \frac{V}{R_1} + \frac{V}{R_2} + \frac{V}{R_3} + I_{s1} = 0 \tag{1-59}$$

Assuming that the current source and resistance values are known, we can determine the voltage V. We can then calculate the currents through the resistances and various power values.

▌ EXAMPLE 1-13

For the circuit of Figure 1-28, determine (a) voltage across the circuit, (b) currents through the three resistances, (c) current through voltage source, and (d) all values of power delivered and power absorbed.

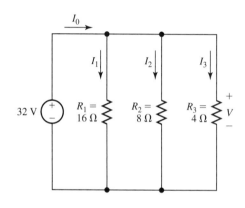

FIGURE 1–28
Circuit of Example 1-13.

SOLUTION Note the labels provided for the circuit elements.

(a) The voltage across the circuit is immediately obvious. It is

$$V = 32 \text{ V} \tag{1-60}$$

(b) This voltage appears across the three resistances, and the three currents are then determined as

$$I_1 = \frac{V}{R_1} = \frac{32}{16} = 2 \text{ A} \tag{1-61}$$

$$I_2 = \frac{V}{R_1} = \frac{32}{8} = 4 \text{ A} \tag{1-62}$$

$$I_3 = \frac{V}{R_3} = \frac{32}{4} = 8 \text{ A} \tag{1-63}$$

(c) Application of KCL at the upper node and solution for I_0 result in

$$I_0 = I_1 + I_2 + I_3 = 2 + 4 + 8 = 14 \text{ A} \tag{1-64}$$

(d) Because the voltage source is the only source, it must deliver power. It is

$$P_V = VI_0 = 32 \times 14 = 448 \text{ W} \tag{1-65}$$

The three values of power dissipated in the resistances are

$$P_{R1} = \frac{V^2}{R_1} = \frac{(32)^2}{16} = 64 \text{ W} \tag{1-66}$$

$$P_{R2} = \frac{V^2}{R_2} = \frac{(32)^2}{8} = 128 \text{ W} \tag{1-67}$$

$$P_{R1} = \frac{V^2}{R_1} = \frac{(32)^2}{4} = 256 \text{ W} \tag{1-68}$$

A table providing a compilation of the power values follows.

Variable	Power Delivered	Power Absorbed
Voltage Source	448 W	
R_1		64 W
R_2		128 W
R_3		256 W
Totals	448 W	448 W

■ EXAMPLE 1-14

For the circuit of Figure 1-29, determine (a) voltage across the node pair, (b) current through each of the resistances, and (c) all values of power delivered and absorbed.

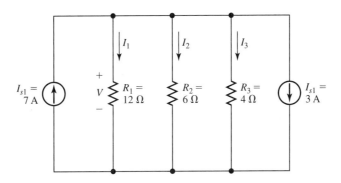

FIGURE 1–29
Circuit of Example 1-14.

SOLUTION

(a) The 7-A source acting alone would make the upper node the positive node and the 3-A source acting alone would make the lower node the positive node. Because the former is larger than the latter, we will assume that the upper node will be positive and the direction of the voltage is set accordingly. Assuming currents entering as negative and currents leaving as positive, KCL for the upper node reads

$$-7 + \frac{V}{12} + \frac{V}{6} + \frac{V}{4} + 3 = 0 \tag{1-69}$$

Solution of this equation leads to

$$V = 8 \text{ V} \tag{1-70}$$

(b) Ohm's law may now be used to determine the three resistive currents.

$$I_1 = \frac{V}{12} = \frac{8}{12} = 0.6667 \text{ A} \tag{1-71}$$

$$I_2 = \frac{V}{6} = \frac{8}{6} = 1.333 \text{ A} \tag{1-72}$$

$$I_3 = \frac{V}{4} = \frac{8}{4} = 2 \text{ A} \tag{1-73}$$

(c) Because current is leaving the source on the left at its most positive terminal, it is *delivering* power as given by

$$P_{s1} = VI_{s1} = 8 \times 7 = 56 \text{ W} \tag{1-74}$$

Conversely, current is entering the source on the left at its most positive terminal, so it is *absorbing* power as given by

$$P_{s2} = VI_{s2} = 8 \times 3 = 24 \text{ W} \tag{1-75}$$

The three resistances are *absorbing* power and the values are

$$P_{R1} = \frac{V^2}{R_1} = \frac{(8)^2}{12} = 5.333 \text{ W} \tag{1-76}$$

$$P_{R2} = \frac{V^2}{R_2} = \frac{(8)^2}{6} = 10.667 \text{ W} \tag{1-77}$$

$$P_{R3} = \frac{V^2}{R_3} = \frac{(8)^2}{4} = 16 \text{ W} \tag{1-78}$$

A compilation of the power values is provided in the table that follows.

Variable	Power Delivered	Power Absorbed
I_{s1}	56 W	
I_{s2}		24
R_1		5.333 W
R_2		10.667 W
R_3		16 W
Totals	56 W	56 W

1-9 Voltage and Current Divider Rules

We now consider two simple rules that are very convenient in analyzing certain circuit configurations. These rules could be considered as "shortcuts" in many situations. They are applicable both to simple circuits and as a portion of many more complex circuits. They will be denoted as the **voltage divider rule** and the **current divider rule**, respectively.

Voltage Divider Rule

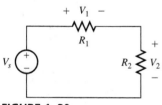

FIGURE 1–30
Circuit used to illustrate the voltage divider.

The voltage divider rule will be illustrated by the circuit of Figure 1-30. A voltage source and two resistors in series form a single-loop circuit. The circuit could indeed be analyzed by the single-loop concept considered earlier. However, suppose we are interested in only the voltages across the two resistances. A guided exercise for the reader (Problem 1-35) will show that we can determine the two voltages with the following equations:

$$V_1 = \frac{R_1}{R_1 + R_2} V_s \tag{1-79}$$

$$V_2 = \frac{R_2}{R_1 + R_2} V_s \tag{1-80}$$

The thought process underlying these equations is as follows: The voltage across an individual resistance is determined by forming the ratio of the particular resistance to the total series resistance, multiplied by the voltage at the input to the divider.

Two comments are appropriate:

(1) The rule may be extended to any number of resistances in a single loop. The numerator resistance for each computation is the particular resistance across which the voltage is desired and the denominator is the sum of all the resistances in the loop.

(2) The input voltage does not need to be a source. It can be any voltage that is known to exist across the series combination.

Note that the larger the resistance, the larger the value of the voltage across it. Moreover, all of the voltage drops across the series resistances must equal the net voltage across the combination.

Current Divider Rule

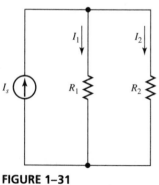

FIGURE 1–31
Circuit used to illustrate the current divider.

The current divider rule is illustrated by the circuit of Figure 1-31. A current source and two parallel resistances form a single node-pair circuit that could be analyzed by the procedure developed earlier for this type of configuration. However, suppose we are interested only in the currents through the two resistances. A guided exercise for the reader (Problem 1-36) will show that we can determine the two currents with the following equations:

$$I_1 = \frac{R_2}{R_1 + R_2} I_s \tag{1-81}$$

$$I_2 = \frac{R_1}{R_1 + R_2} I_s \tag{1-82}$$

At first glance, the current divider equations may appear similar to the voltage divider equations, but there is a major difference. In the voltage divider equations, the numerator is the resistance across which the voltage is desired, but in the current divider equations, the other resistance appears in the numerator. Because of the difference in the numerator form, the current divider rule is a bit more tricky to extend to more than two resistances and will not be considered at this point.

In a similar fashion to the voltage divider rule, the current flowing into the parallel combination need not be a source. It can be any current that is known, and the current divider rule simply shows how the current divides into two parts.

▮▮ EXAMPLE 1-15

Use the voltage divider rule to determine the voltages across the two resistors in Figure 1-32.

SOLUTION The voltage V_1 is given by

$$V_1 = \frac{12{,}000}{12{,}000 + 33{,}000} \times 45 = \frac{12}{45} \times 45 = 12 \text{ V} \tag{1-83}$$

The voltage V_2 is

$$V_2 = \frac{33{,}000}{12{,}000 + 33{,}000} \times 45 = \frac{33}{45} \times 45 = 33 \text{ V} \tag{1-84}$$

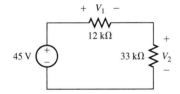

FIGURE 1–32
Circuit of Example 1-15.

The reader is invited to check these results by using the procedure for a single-loop circuit. Incidentally, we have chosen to keep the resistances in their basic units, but because the voltage divider ratio involves ohms divided by ohms, it would have been feasible in this case to express both the numerator and denominator of these ratios in thousands of ohms to simplify the expressions somewhat.

▮▮ EXAMPLE 1-16

Use the current divider rule to determine the two resistive currents in Figure 1-33.

SOLUTION The current I_1 is given by

$$I_1 = \frac{8200}{1200 + 8200} \times 0.1 = \frac{8200}{9400} \times 0.1 = 0.08723 \text{ A} = 87.23 \text{ mA} \tag{1-85}$$

The current I_2 is

$$I_2 = \frac{1200}{1200 + 8200} \times 0.1 = \frac{1200}{9400} \times 0.1 = 0.01277 \text{ A} = 12.77 \text{ mA} \tag{1-86}$$

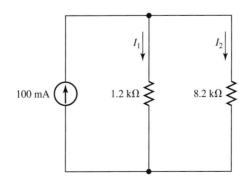

FIGURE 1–33
Circuit of Example 1-16.

As in the previous example, it would have been feasible to express the resistive values for both the numerator and denominator in thousands of ohms, and we also could have retained the current directly in milliamperes. However, until the reader develops a certain level of intuition concerning electrical units, it will be the practice in most cases in this book to use basic units until the final answer is obtained. Even when there are more steps and messy numbers, the likelihood of a mistake is less when the computations are performed with the basic units.

Note that the smallest value of resistance has the largest value of current, and vice versa. We can verify readily that the voltages across the two resistances are the same, which is a requirement for a parallel configuration.

PROBLEMS

1-1 Determine the current I in the circuit of Figure P1-1.

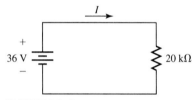

FIGURE P1–1

1-2 Determine the current I in the circuit of Figure P1-2.

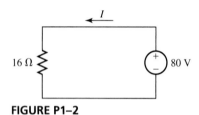

FIGURE P1–2

1-3 Determine the current I in the circuit of Figure P1-3.

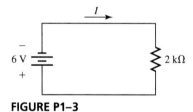

FIGURE P1–3

1-4 Determine the current I in the circuit of Figure P1-4.

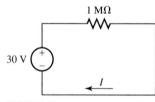

FIGURE P1–4

1-5 Determine the power delivered by the source in Problem 1-1.

1-6 Determine the power delivered by the source in Problem 1-2.

1-7 Determine the power delivered by the source in Problem 1-3.

1-8 Determine the power delivered by the source in Problem 1-4.

1-9 Determine the voltage V in the circuit of Figure P1-9.

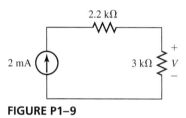

FIGURE P1–9

1-10 Determine the voltage V in the circuit of Figure P1-10.

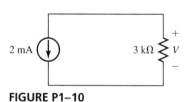

FIGURE P1–10

1-11 Determine the resistance of a 20-W bulb operating from a dc voltage of 12 V.

1-12 Determine the resistance of a 500-W heating element operating from an ac effective voltage of 120 V.

1-13 A radio transmitter has an ac average power input of 1 kW with no modulation, and it acts effectively as a resistance of value 50 Ω. Determine the ac effective current and the ac effective voltage.

1-14 The power dissipated in a certain resistance of 20 Ω is 500 W. Determine the current and the voltage.

1-15 Various voltages in a circuit are measured with respect to a common ground reference. The voltage at point A is 40 V and the voltage at point B is 22 V. Determine the voltage V_{AB}.

1-16 Repeat the analysis of Problem 1-15 if the voltage at point A is 12 V and the voltage at point B is −20 V.

1-17 The voltage across a resistance is 15 V and the current is 0.5 mA. Determine the resistance.

1-18 The voltage across a resistance is 10 V and the current is 2 μA. Determine the resistance.

1-19 Determine the cost of leaving a 10-W bulb on for one year if electricity costs $0.09 per kilowatt-hour. (Assume that it is not a leap year!)

1-20 Determine the energy in joules dissipated in the bulb of Problem 1-19 in one year.

1-21 Determine the voltages V_1 and V_2 in Figure P1-21.

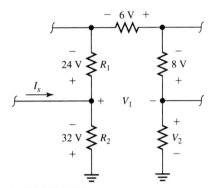

FIGURE P1–21

1-22 Determine the voltage V_x in Figure P1-22.

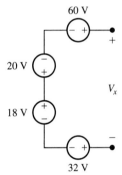

FIGURE P1–22

1-23 Assume in Figure P1-21 that $R_1 = 1\ \text{k}\Omega$ and $R_2 = 2\ \text{k}\Omega$. Determine the current I_x.

1-24 Assume for the circuit of Figure P1-22 that a resistive load is connected that draws 2 A. (a) Determine the power delivered or absorbed by each of the sources. (b) Determine the power dissipated in the resistive load.

1-25 Determine the equivalent resistance seen at the input terminals in Figure P1-25.

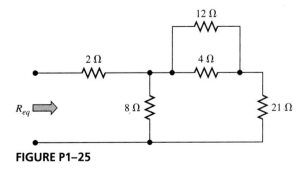

FIGURE P1–25

1-26 Determine the equivalent resistance seen at the input terminals in Figure P1-26.

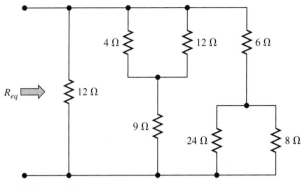

FIGURE P1–26

1-27 For the circuit of Figure P1-27, determine (a) loop current, (b) voltages across all resistances, and (c) power delivered or absorbed by each component in circuit.

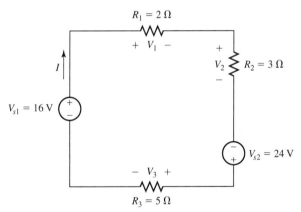

FIGURE P1–27

1-28 For the circuit of Figure P1-28, determine (a) loop current, (b) voltages across all resistances, and (c) power delivered or absorbed by each component in circuit.

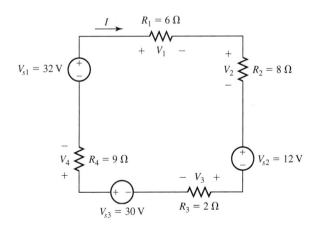

FIGURE P1–28

1-29 For the circuit of Figure P1-29, determine (a) loop current, (b) voltages across all resistances, (c) voltage across the current source, and (d) power delivered or absorbed by each component in the circuit.

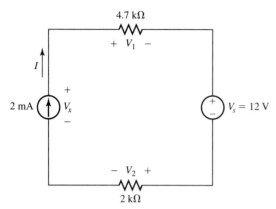

FIGURE P1–29

1-30 For the circuit of Figure P1-30, determine (a) loop current, (b) voltages across all resistances, (c) voltage across the current source, and (d) power delivered or absorbed by each component in the circuit.

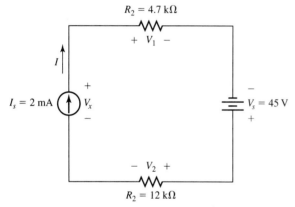

FIGURE P1–30

1-31 For the circuit of Figure 1-31, determine (a) voltage across the circuit, (b) currents through the three resistances, (c) current through the voltage source, and (d) all values of power delivered and power absorbed.

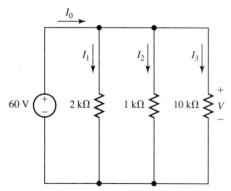

FIGURE P1–31

1-32 For the circuit of Figure P1-32, determine (a) voltage across the circuit, (b) currents through the two resistances, (c) current through the voltage source if $I = 30$ mA, and (d) the value of I such that $I_0 = 0$.

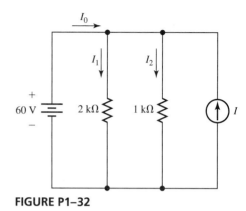

FIGURE P1–32

1-33 The equivalent resistance formula for resistances in series will be derived in this problem. Consider the series connection of n resistances as shown in Figure 1-19. Assume that a voltage source V is connected across the series combination and assume a current I flowing through the combination. Write a KVL equation for the loop with all voltage drops expressed in terms of the current. Then solve for the equivalent resistance as $R_{eq} = V/I$ and show that the series equivalent formula is obtained.

1-34 The equivalent resistance formula for resistances in parallel will be derived in this problem. Consider the parallel connection of n resistances as shown in Figure 1-20. Assume that a voltage source V is connected across the series combination and assume a current I flowing from the source. Express the current through each resistance in terms of the voltage and determine the net current I_0 flowing from the source. Then solve for the reciprocal of the resistance as $1/R_{eq} = I_0/V$ and show that the parallel equivalent formula is obtained.

1-35 The voltage divider rule will be derived in this example. Consider the circuit of Figure 1-30 and assume a current I flowing from the positive terminal of the source. Write a loop equation and solve for I. Use this result to determine the voltage drop across each of the resistances, and show that the voltage divider rule is obtained for each resistance.

1-36 The current divider rule will be derived in this example. Consider the circuit of Figure 1-31 and assume a voltage V across the circuit. Write a node-pair equation at the upper node and solve for V. Use this result to determine the currents through the two resistances and show that the current divider rule is obtained for each resistance.

General DC Circuit Analysis 2

OVERVIEW AND OBJECTIVES

The preceding chapter covered the presentation of the basic circuit laws and the models used in circuit analysis. All circuits were fairly simple in the configurations involved, and could be analyzed either by a single equation or by a series of step-by-step procedures. We are now ready to study the approaches that can be used for more complex circuits. However, we will continue to emphasize dc circuits until these procedures are thoroughly mastered.

Objectives

After completing this chapter, the reader should be able to

1. Write the **mesh current equations** for a given circuit and solve them.
2. Write the **node voltage equations** for a given circuit and solve them.
3. Transform a voltage source in series with a resistance to a current source in parallel with the resistance.
4. Transform a current source in parallel with a resistance to a voltage source in series with the resistance.
5. State **Thevenin's** and **Norton's** theorems.
6. Apply Thevenin's and Norton's theorems for a dc resistive circuit.
7. Explain how Thevenin and Norton equivalent circuits can be used to analyze the behavior of a given circuit with respect to an external load.

2-1 Mesh Current Analysis

The first general method that we will cover is that of **mesh current analysis**. In general, a **mesh** is an open area in a circuit in which a complete loop may be formed without crossing any wires. For example, the circuit form shown in Figure 2-1 contains four meshes.

Mesh current analysis is easier to apply when all the sources involved are voltage sources. That will be the initial assumption. We will assume that all of the voltage source values and the resistances are known. The procedure for writing mesh equations with voltage sources and resistances is as follows:

(1) Assume that each mesh contains a circulating unknown loop current and label the currents in a systematic manner such as I_1, I_2, and so on, or as I_a, I_b, and so on. Although

FIGURE 2–1
Form of the mesh current variables in a representative circuit layout.

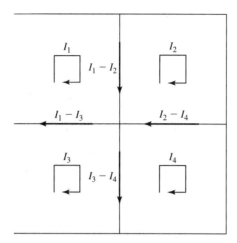

the directions are arbitrary, and we will take more liberty later in the text, initially we recommend that all currents be assumed in the clockwise direction as shown in Figure 2-1.

(2) Recognize that each current in any outer branch will be equal to a particular mesh current, but we can express each current in an inner branch as the difference between two mesh currents if all the mesh currents are assumed in the same direction. Note the four inner branch currents that have been labeled on Figure 2-1. Any branch current could be interpreted as flowing in either direction by redefining it. For example, $I_1 - I_2$ flowing downward could be redefined as $I_2 - I_1$ flowing upward. These dual interpretations will prove useful in writing the equations.

(3) Write a KVL equation for each mesh or loop while moving in the direction of the particular mesh current. We will follow the convention established in the preceding chapter that rises will be considered as negative and drops as positive. Voltage sources will be expressed in accordance with their signs and values, but voltages across resistances will be treated as drops based on the assumption that the applicable current is entering the positive terminal. This procedure will become clear in the examples that follow.

Current Source

The presence of an ideal current source can result in one or more mesh currents assuming a particular value. The procedure of working directly with current sources in mesh equations can be a little tricky, so we will sidestep that approach. Instead, it will be recommended that either node voltage analysis be employed (to be covered in Section 2-2) or a source transformation be employed (to be covered in Section 2-3).

▍ EXAMPLE 2-1

Write the mesh equations for the circuit of Figure 2-2 and solve for I_1 and I_2.

SOLUTION We begin below the 25-V source and move around the mesh in a clockwise direction assigning positive signs to voltage drops and negative signs to voltage rises. Recognize that in the direction we are moving, the current through the 4-Ω resistance is I_1 and the current through the 5-Ω resistance is $I_1 - I_2$. The equation for mesh 1 follows.

$$-25 + 4I_1 + 5(I_1 - I_2) - 10 = 0 \qquad (2\text{-}1)$$

For mesh 2, we will begin below the 10-V source and move upward. Note in this direction, the 10-V source represents a drop. Note also, and this is very important, we will now interpret the current through that branch to be upward, and the value will be $I_2 - I_1$ in that direction. Thus, we interpret that the assumed current is always moving in the same

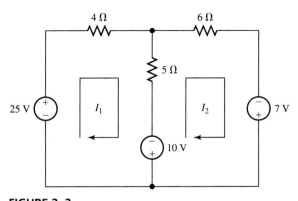

FIGURE 2–2
Circuit of Example 2-1.

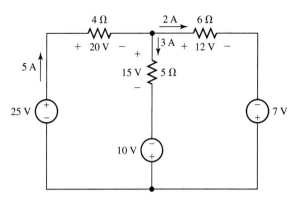

FIGURE 2–3
Circuit of Example 2-1 with variables labeled.

direction as the particular mesh current. That will result in an assumed voltage drop across any resistance in the branch. The equation for mesh 2 is

$$10 + 5(I_2 - I_1) + 6I_3 - 7 = 0 \tag{2-2}$$

These equations can now be rearranged in the standard form for simultaneous solution.

$$9I_1 - 5I_2 = 35 \tag{2-3}$$
$$-5I_1 + 11I_2 = -3 \tag{2-4}$$

Readers who need to review the process for solving simultaneous equations may refer to Appendix A. Simultaneous solution of the preceding two equations results in

$$I_1 = 5 \text{ A} \tag{2-5}$$
$$I_2 = 2 \text{ A} \tag{2-6}$$

Once the currents are determined, we can determine other variables in the circuit with the methods of Chapter 1. To illustrate the process, consider Figure 2-3. The current through the middle branch is $I_1 - I_2 = 5 - 2 = 3$ A. The voltages across the various resistances are readily computed with Ohm's law and are labeled on the circuit. It can be readily shown that KVL is satisfied around each mesh, and also around the outside of the circuit.

■ **EXAMPLE 2-2**   Write the mesh equations for the circuit of Figure 2-4 and solve for the three mesh currents.

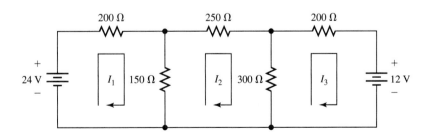

FIGURE 2–4
Circuit of Example 2-2.

SOLUTION   We will begin below the 24-V battery and walk clockwise around the first loop. The equation reads

$$-24 + 200I_1 + 150(I_1 - I_2) = 0 \tag{2-7}$$

For mesh 2, we begin below the 150-Ω resistance. The equation in this case is

$$150(I_2 - I_1) + 250I_2 + 300(I_2 - I_3) = 0 \tag{2-8}$$

Finally, the equation for mesh 3 is

$$300(I_3 - I_2) + 200I_3 + 12 = 0 \tag{2-9}$$

The three mesh equations need to be rearranged in standard form. The results are

$$350I_1 - 150I_2 = 24 \tag{2-10}$$
$$-150I_1 + 700I_2 - 300I_3 = 0 \tag{2-11}$$
$$-300I_2 + 500I_3 = -12 \tag{2-12}$$

Note that there is no I_3 term in the first equation and no I_1 term in the third equation. This means that the coefficients of those currents in the respective equations are zero. This point is important if a matrix solution of the equations is used.

Simultaneous solution of the three equations yields

$$I_1 = 0.07147 \text{ A} = 71.47 \text{ mA} \tag{2-13}$$
$$I_2 = 0.006777 \text{ A} = 6.777 \text{ mA} \tag{2-14}$$
$$I_3 = -0.01994 \text{ A} = -19.94 \text{ mA} \tag{2-15}$$

Other variables could now be determined for the circuit if needed.

2-2 Node Voltage Analysis

The second general method that we will cover is that of **node voltage analysis**. Between mesh current analysis of the preceding section and node voltage analysis to be covered in this section, virtually any practical circuit can be analyzed. In many cases, it is an arbitrary choice as to which method is used. However, there are circuits in which one of the two methods might lead to a more direct solution, so both methods should be studied carefully.

Node voltage analysis is easier to apply if either of the following two conditions are met: (1) Any voltage sources present are connected to the common ground (to be discussed shortly) or (2) all sources are current sources. The first condition is much more common and we will consider it first. If the circuit contains any ungrounded voltage sources, it is recommended that either mesh current analysis be employed or a source transformation (to be discussed in Section 2-3) be used.

Common Ground

In node voltage analysis, a common ground point must be selected and all voltages are measured with respect to this common ground. The ground might or might not correspond to a possible ground point in the actual circuit, although we recommend highly that any actual ground be used as the common ground whenever practicable. The general form is illustrated in Figure 2-5, in which the lower bus is considered as a ground node and there are three voltages that we can measure with respect to this ground.

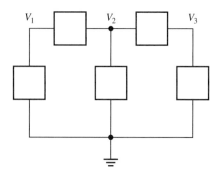

FIGURE 2–5
Form of the node voltage variables in a representative circuit.

Procedure

(1) With a common ground assigned, assume a node voltage at all other nodes and identify the unknown voltages in a systematic form such as V_1, V_2, and so on, or as V_a, V_b, and so on.

(2) At all nodes at which known ideal voltage sources with respect to ground are present, note the values of the voltages for reference in the equations that follow.

(3) Write a KCL equation for each node containing an unknown voltage. We will follow the convention established in the preceding chapter that current leaving will be considered as positive and current entering as negative. We will express current sources in accordance with their signs and values, but currents through resistances will be treated as leaving a given node, based on the assumption that the voltage at the given node is the more positive one. This procedure will become clear in the examples that follow.

▌ EXAMPLE 2-3

Rework the circuit of Example 2-1 using node voltage analysis.

SOLUTION The circuit of Figure 2-2 is redrawn in Figure 2-6 and the bus along the bottom is selected for the common ground. There are no current sources, and the three voltage sources are all connected to the common ground, so the voltages at those three nodes are known. From left to right, they are 25 V, -10 V, and -7 V.

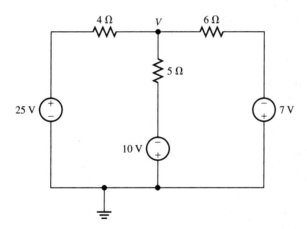

FIGURE 2–6
Circuit of Example 2-3.

The only unknown voltage is the one at the upper middle node, which is indicated simply as V. Although we don't yet know the sign of this voltage, we assume for the purpose of writing the node voltage equation that it is more positive than the voltages on the other sides of the resistances. Thus, the voltage across the 4-Ω resistance is considered as $V - 25$; the voltage across the 5-Ω resistance is considered as $V - (-10) = V + 10$; and the voltage across the 6-Ω resistance is $V - (-7) = V + 7$. The three currents at this node are then considered as leaving and we have

$$\frac{V-25}{4} + \frac{V+10}{5} + \frac{V+7}{6} = 0 \tag{2-16}$$

Simplification results in

$$0.6167V = 3.0833 \tag{2-17}$$

and

$$V = 5.000 \text{ V} \tag{2-18}$$

This result may be verified by the work of Example 2-1. Note that this voltage is actually less than the voltage on the left (25 V) but greater than the other two voltages (-10 V and -7 V), but in writing the equations, it was correct to assume that it was larger than all three insofar as expressing currents leaving was concerned. The signs take care of themselves if the correct methodology is employed.

EXAMPLE 2-4

Rework the circuit of Example 2-2 using node voltage analysis.

SOLUTION The circuit of Figure 2-3 is redrawn in Figure 2-7 and the bus along the bottom is selected for the common ground. As in the previous example, there are no current sources and the two voltage sources are both connected to the common ground, so the voltages at those two nodes are known. They are 24 V and 12 V, respectively.

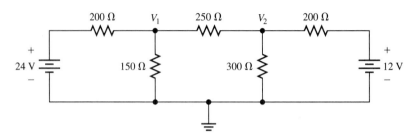

FIGURE 2–7
Circuit of Example 2-4.

The two unknown voltages are labeled as V_1 and V_2 as shown. We begin at node 1 and express the currents leaving that node. Note that because the 150-Ω resistance is grounded, the voltage across it is simply $V_1 - 0 = V_1$. The node voltage equation at node 1 is then

$$\frac{V_1 - 24}{200} + \frac{V_1}{150} + \frac{V_1 - V_2}{250} = 0 \tag{2-19}$$

When we move to node 2, we assume that it is the most positive, and as at node 1, we consider currents leaving. The equation at node 2 is

$$\frac{V_2 - V_1}{250} + \frac{V_2}{300} + \frac{V_2 - 12}{200} = 0 \tag{2-20}$$

These equations need to be rearranged in standard form for simultaneous solution. The result, after a bit of arithmetic, is

$$15.667 \times 10^{-3} V_1 - 4 \times 10^{-3} V_2 = 0.120 \tag{2-21}$$

$$-4 \times 10^{-3} V_1 + 12.333 \times 10^{-3} V_2 = 0.06 \tag{2-22}$$

Simultaneous solution yields

$$V_1 = 9.705 \text{ V} \tag{2-23}$$

$$V_2 = 8.013 \text{ V} \tag{2-24}$$

To compare these results with those obtained in Example 2-2, it would be necessary to calculate the voltages at the same points with respect to the lower bus using the results of Example 2-2, or to calculate the currents of Example 2-2 using the results of this example. The reader is invited to show that either process results in agreement, although slight roundoff errors might occur based on the numbers involved.

EXAMPLE 2-5

Determine the voltages V_1 and V_2 in Figure 2-8 using node voltage analysis.

SOLUTION There are no voltage sources in this circuit, so there are no known voltages at the outset. In this case, we apply KCL at each of the upper nodes, expressing positive currents as leaving. Of course, current sources are expressed in terms of their actual values.

We will begin in this example to take a small liberty to simplify the numbers somewhat. We know that volts/ohms = amperes. Well, it turns out that volts/kilohms = milliamperes. In other words, if the 10^3 factor associated with kilohms is ignored, the

FIGURE 2-8
Circuit of Example 2-5.

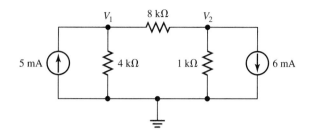

currents resulting from the division are expressed in milliamperes. Thus, we will express the currents in mA and the resistances in kΩ. If this process is not clear, you may reformulate the equations in the basic units. This is recommended if there is any doubt.

Using this shortened process, KCL at node 1 reads

$$-5 + \frac{V_1}{4} + \frac{V_1 - V_2}{8} = 0 \tag{2-25}$$

The corresponding KCL equation at node 2 is

$$\frac{V_2 - V_1}{8} + \frac{V_2}{1} + 6 = 0 \tag{2-26}$$

Simplification yields

$$0.375V_1 - 0.125V_2 = 5 \tag{2-27}$$
$$-0.125V_1 + 1.125V_2 = -6 \tag{2-28}$$

Simultaneous solution yields

$$V_1 = 12 \text{ V} \tag{2-29}$$
$$V_2 = -4 \text{ V} \tag{2-30}$$

2-3 Source Transformations

We made a few references to **source transformations** earlier in the chapter. These are analysis tools that permit one to reconfigure a circuit to simplify the analysis. They might have occasional application in the actual design or implementation of a real circuit, but they are mostly analysis procedures to aid in analyzing circuit equations. The two transformations shown in Figure 2-9 involve (a) converting a voltage source to a current source and

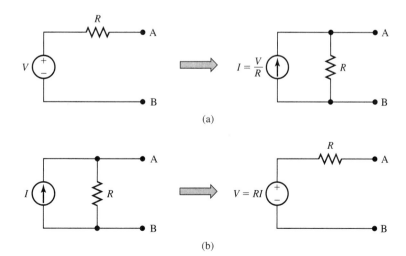

FIGURE 2-9
Source transformations.

(b) converting a current source to a voltage source. It should be stressed at the outset that a resistance must be present in both cases to permit the transformation. It should also be stressed that these transformations are valid only for analysis *external to the terminals involved*.

Conversion of Voltage Source to Current Source

The transformation of Figure 2-9(a) permits a voltage source in series with a resistance R to be converted to a current source in parallel with the resistance, both of which produce the same *external effects* when connected to additional circuitry. The value of the current source is the current that would flow in a short connected across the points A and B. It is

$$I = \frac{V}{R} \tag{2-31}$$

Conversion of Current Source to Voltage Source

The transformation of Figure 2-9(b) permits a current source in parallel with a resistance R to be converted to a voltage source in series with the resistance, both of which produce the same *external effects* when connected to additional circuitry. The value of the voltage source is the voltage that would appear across the open circuit terminals A and B. It is

$$V = RI \tag{2-32}$$

Application to Mesh and Node Analysis

We have seen that mesh current analysis involves summing voltages around loops. Therefore, if a current source is present, it can often be converted to a voltage source by the second transformation. It is necessary that there be a resistance in parallel with the source for this to be possible.

We have seen that node voltage analysis involves summing currents at nodes. We have also seen that a grounded voltage source represents a node for which the voltage is already known and poses no problem. However, an ungrounded voltage source can often be converted to a current source by the first transformation. It is necessary that there be a resistance in series with the voltage source for this to be possible.

Remember that when these transformations are used, only results *external* to the terminals of the sources will be valid.

▌▌ EXAMPLE 2-6

This example will demonstrate with a simple circuit how a source transformation may be used and that it will produce the same external result. Consider the parallel circuit of Figure 2-10(a). (a) Determine the voltage V directly without changing the configuration. (b) Perform a source transformation to the left of the terminals A and B and determine the voltage V with the new circuit.

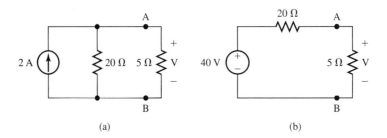

FIGURE 2–10
Circuit of Example 2-6.

SOLUTION

(a) As given, the circuit is a single node-pair circuit and an equation at the upper node is

$$-2 + \frac{V}{20} + \frac{V}{5} = 0 \tag{2-33}$$

This result leads to

$$0.25V = 2 \tag{2-34}$$

or

$$V = 8\,\text{V} \tag{2-35}$$

At least two other methods could have been used without changing the configuration. One way is to determine the equivalent resistance for the two resistors in parallel, and to apply Ohm's law to determine the voltage across the parallel combination. Another way is to apply the current divider rule to determine the current through the 5-Ω resistance and then apply Ohm's law for that particular resistance. The reader is invited to make those computations, which should result in the same value of voltage.

(b) The source transformation is applied to the 2-A current source and the 20-Ω resistance. The result is a 40-V voltage source in series with the 20-Ω resistance as shown in Figure 2-10(b). Although there are several approaches that could be used for this series configuration, we will employ the voltage divider rule to obtain V.

$$V = \frac{5}{5 + 20} \times 40 = 8\,\text{V} \tag{2-36}$$

Perhaps we have belabored this simple circuit too much, but remember that more than one circuit analysis tool can often be applied. It is a good check to use different methods when appropriate. The major intent was to demonstrate the validity of a source transformation and show how it can be used to change the configuration of a circuit completely for analysis purposes.

In general, the results obtained after making a transformation apply only to the external circuit, which in this case is the resistance to the right of A and B. The circuit to the left is different, but it produces the same external behavior.

2-4 Thevenin's and Norton's Theorems

In the preceding section, we learned how to convert a voltage source in series with a resistance to a current source in parallel with the same resistance, and we learned the corresponding inverse transformation. In many situations, these transformations are convenient to simplify a circuit configuration for easier analysis and interpretation.

We will now take a much larger step in showing how to represent a more complex circuit at a set of terminals as far as the external behavior is concerned. The theorems that are used to describe these operations are known as **Thevenin's theorem** and **Norton's theorem**. The theorems will be stated first, and then we will discuss the processes involved.

Thevenin's Theorem

Consider the block shown on the left in Figure 2-11. As applied to dc resistive circuits, Thevenin's theorem states that as far as any external behavior is concerned, the circuit can

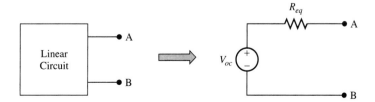

FIGURE 2–11
General circuit and its Thevenin equivalent circuit.

FIGURE 2–12
Linear circuit and its Norton equivalent circuit.

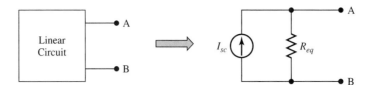

be replaced by a single voltage source in series with a single resistance as shown on the right.

The value of the voltage is the **open-circuit voltage** that would be measured across the terminals A and B.

Norton's Theorem

Consider the block shown on the left in Figure 2-12. As applied to dc resistive circuits, Norton's theorem states as far as any external behavior is concerned, the circuit can be replaced by a single current source in parallel with a single resistance as shown on the right.

The value of the current is the **short-circuit current** that would be measured through a short-circuit connected to the output terminals A and B. Understand that this is primarily an analysis tool rather than one that would normally be used in a laboratory. Attempting to measure the short-circuit current for many real circuits is not practical and could even be hazardous in some cases!

Resistance Value

The resistance value is the same for either the Thevenin representation or the Norton representation. Therefore, it should be clear from the work of the preceding section that once either form is known, we can readily determine the other with a source transformation.

Non-Feedback Circuits

In this text, we will limit the application of these theorems to those circuits that do not contain electronic feedback. This will simplify greatly the development that follows and it will suffice for the objectives of the text.

Determining the Resistance

To determine the equivalent resistance, first it is necessary to de-energize all internal sources. The equivalent resistance is then the value "seen" looking back into the circuit from the output terminals.

To de-energize an ideal voltage source, replace it (on paper) by a short circuit. To de-energize an ideal current source, replace it by an open circuit. For a dc circuit without feedback, the circuit will then reduce to a purely resistive form, for which we can often determine the equivalent resistance by successive series and/or parallel combinations.

Use of Source Transformations

We can determine the Thevenin or Norton forms of some circuits more easily by the use of successive source transformations. We will illustrate this process in Example 2-9.

III EXAMPLE 2-7

Determine the Thevenin equivalent circuits for the circuit of Figure 2-13(a) at the terminals A-B.

SOLUTION This circuit is about as easy as it gets for determining the Thevenin equivalent form, but this layout, sometimes referred to as an L-form, arises in many practical applications (usually with more realistic values). The first step is to determine the

FIGURE 2–13
Determining the Thevenin equivalent circuit in Example 2-7.

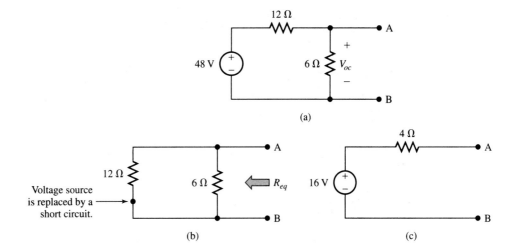

open-circuit voltage V_{oc} at the terminals A and B. There are several ways to do this. We will choose the voltage divider rule. The value of this voltage is

$$V_{oc} = \frac{6}{6 + 12} \times 48 = 16 \text{ V} \tag{2-37}$$

This voltage will be the value of the voltage source in the Thevenin equivalent circuit.

Next, the voltage source is de-energized by replacing it with a short circuit, as shown in Figure 2-13(b). We then look back from the output terminals and determine R_{eq}.

Based on many years of experience observing typical student errors, a comment is in order at this point. In the original circuit with no load on the output, it is correct to say that the two resistors are connected in series as far as the source is concerned. The series resistance is, of course, 9 Ω and a common error is to assume this value for the Thevenin resistance. However, when we determine the Thevenin or Norton equivalent circuit, we *look back* from the output terminals and the two resistors are in *parallel* based on this interpretation. Thus, the equivalent resistance is

$$R_{eq} = \frac{12 \times 6}{12 + 6} = 4 \text{ Ω} \tag{2-38}$$

The resulting Thevenin equivalent circuit is shown in Figure 2-13(c).

While not shown, the corresponding Norton equivalent circuit can now be readily determined with a source transformation. The value of the current source would be $16/4 = 4$ A directed upward, and it would be in parallel with the resistance of 4 Ω.

It also might help to consider how to determine the Norton equivalent circuit directly from the original circuit. If a short were placed across the output terminals, no current would flow through the 6-Ω resistance and the actual current would be 48 V/12 Ω = 4 A downward. This would mean that the corresponding current source would have to point upward. Determination of the equivalent resistance is the same as for the Thevenin circuit.

■ EXAMPLE 2-8

Determine the Thevenin equivalent circuit for the circuit of Figure 2-14(a) at the terminals A-B. The circuit to the left of the 10-Ω resistance is the same as that in Example 2-7.

SOLUTION This circuit has been created as an extension of the preceding example to make an important point. It can be referred to as a T-configuration, which also arises in many practical applications.

Now for an important point. The open-circuit voltage at the terminals A and B is the same as the voltage across the 6-Ω resistance. That is because no current is assumed to flow

FIGURE 2–14
Determining the Thevenin equivalent circuit in Example 2-8.

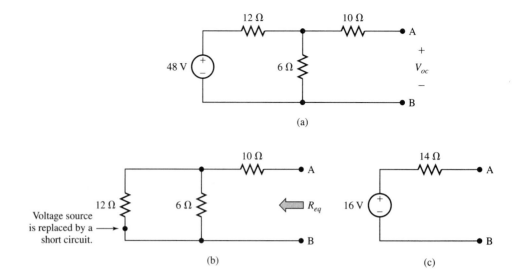

in an ideal voltmeter, and therefore there will be no drop across the 10-Ω resistance. Thus, the open-circuit voltage is again

$$V_{oc} = \frac{6}{6+12} \times 48 = 16 \text{ V} \tag{2-39}$$

Be aware of this type of situation, because it tends to arise in many practical circuits.

Of course, the equivalent resistance will be different in this case. It can be determined from Figure 2-14(b). The value is

$$R_{eq} = \frac{12 \times 6}{12 + 6} + 10 = 4 + 10 = 14 \text{ Ω} \tag{2-40}$$

The final Thevenin equivalent circuit is shown in Figure 2-14(c). The Norton equivalent circuit would be a current source of value $16/14 = 1.143$ A in parallel with 14 Ω.

■ EXAMPLE 2-9

Determine the Thevenin equivalent circuit for the circuit of Figure 2-15.

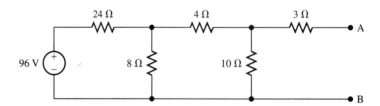

FIGURE 2–15
Circuit of Example 2-9.

SOLUTION The form of this circuit is known as a **ladder network**. It arises in many practical applications. To determine the Thevenin equivalent circuit, one approach would be to determine the open-circuit voltage and subsequently determine the resistance, looking back from the output with the one source de-energized. The latter step could be easily achieved with successive parallel and series combinations, but the determination of the open-circuit voltage could require solving a set of simultaneous equations.

Let's see if we can't use a simpler step-by-step approach using the techniques considered in the past two examples. Refer to Figure 2-16 for reference in the steps that follow. The procedure used will be to begin on the left-hand side and successively determine Thevenin and/or Norton forms as we move to the right. Alternately, source transformations could be used for some of the steps if desired.

We begin the process with the left-hand L-configuration as shown in Figure 2-16(a). This portion of the circuit has the form of the simpler circuit of Example 2-7. Without

FIGURE 2–16
Steps involved in determining the Thevenin equivalent circuit of Example 2-9.

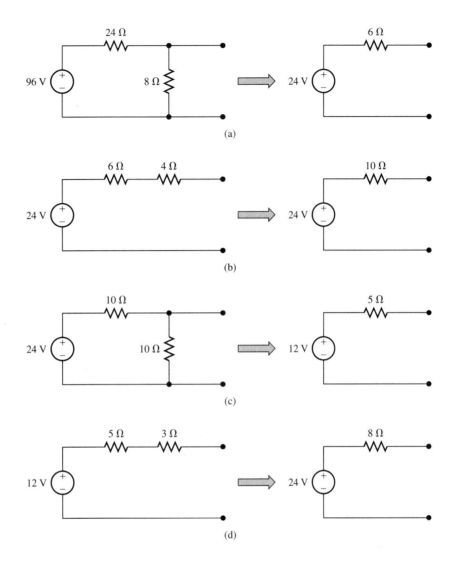

showing the calculations, the reader can readily verify using the voltage divider rule that the open-circuit voltage is 24 V. The equivalent resistance when the voltage source is replaced with a short circuit is the parallel combination of 24 Ω and 8 Ω, which results in 6 Ω. The corresponding Thevenin equivalent circuit is shown on the right.

The next step is to add the 4-Ω resistance as shown in Figure 2-16(b). The additional series resistance does not change the open-circuit voltage, but the resistance is now $6 + 4 = 10$ Ω as shown on the right.

Next, the 10-Ω shunt resistance is placed across the preceding part of the circuit as shown in Figure 2-16(c). Again, we have an L-configuration and the open-circuit voltage in this case is half of the source voltage; that is, 12 V. Likewise, the parallel combination of two 10-Ω resistances is half the value or 5 Ω. The equivalent circuit is shown on the right.

The final step involves adding the 3-Ω resistance in series as shown in Figure 2-16(d). The net Thevenin equivalent circuit is 12 V in series with 8 Ω as shown on the lower right.

This problem has illustrated how a more complex circuit can be analyzed by using a series of simpler steps. Not every circuit lends itself to this type of approach, but many do, and this approach provides a great deal of insight as to how each component affects the overall circuit.

Once again, we will emphasize that the equivalent circuit can be used to predict the external behavior, but the original circuit to the left of A and B has now been lost, and it is not possible to predict the internal behavior from the Thevenin equivalent circuit.

EXAMPLE 2-10

The circuit to the left of A-B in Figure 2-17(a) is the same as in Example 2-9. The circuit is to be connected to an external load resistance as shown. Determine and plot equations for (a) voltage, (b) current, and (c) power as functions of the load resistance R_L.

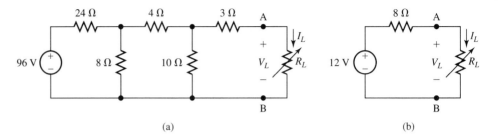

FIGURE 2–17
Circuit of Example 2-10.

SOLUTION This example will demonstrate some of the major applications of why a Thevenin or Norton equivalent circuit is such a powerful tool. Without the use of an equivalent circuit, it would be necessary to perform a complete circuit solution for each value of load resistance. By using an equivalent circuit to the left of A-B, the analysis can be reduced to that of a simple circuit.

The Thevenin equivalent circuit of Example 2-9 (Figure 2-16) is shown in Figure 2-17(b) with the load resistance connected. In each of the steps that follow, we will calculate and plot the variable as a function of the independent variable R_L.

(a) The load voltage V_L is readily calculated using the voltage divider rule.

$$V_L = \frac{R_L}{R_L + 8} \times 12 = \frac{12R_L}{R_L + 8} \tag{2-41}$$

A plot of the voltage versus the load resistance over a reasonable range is shown in Figure 2-18. Although the final value is not shown on the graph, the voltage rises toward a final value of 8 V as the load resistance approaches infinity (an open circuit).

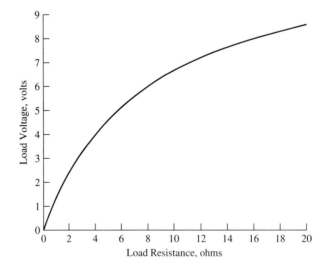

FIGURE 2–18
Load voltage as a function of load resistance.

(b) Once the load voltage is known, the current is readily determined by Ohm's law.

$$I_L = \frac{V_L}{R_L} = \frac{12R_L}{(R_L + 8)R_L} = \frac{12}{R_L + 8} \tag{2-42}$$

A plot of the current versus the load resistance is shown in Figure 2-19. The current begins at its maximum value of $12\,\text{V}/8\,\Omega = 1.5$ A and eventually approaches zero as the load resistance approaches infinity.

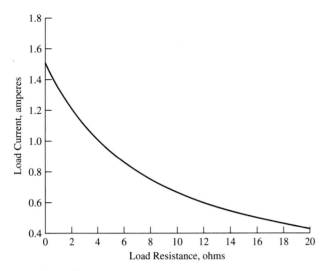

FIGURE 2–19
Load current as a function of load resistance.

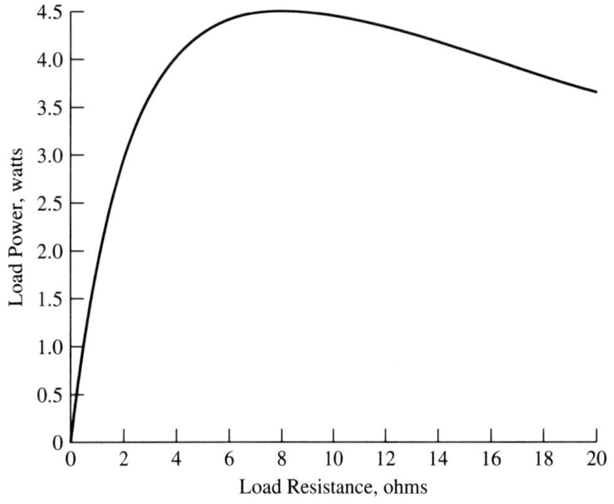

FIGURE 2–20
Load power as a function of load resistance.

(c) The load power P_L is given by

$$P_L = V_L I_L = \frac{12 R_L}{R_L + 8} \times \frac{12}{R_L + 8} = \frac{144 R_L}{(R_L + 8)^2} \tag{2-43}$$

A plot of the load power versus the load resistance is shown in Figure 2-20. This graph illustrates a classical result in circuit theory known as the **maximum power transfer theorem**: When a source has an internal resistance that cannot be changed, maximum power is delivered to an external load when the load resistance is equal to the internal resistance. In this case, the internal resistance is the equivalent Thevenin resistance of 8 Ω.

PROBLEMS

2-1 Write the mesh current equations for the circuit of Figure P2-1 and solve for I_1 and I_2.

2-2 Write the mesh current equations for the circuit of Figure P2-2 and solve for I_1 and I_2.

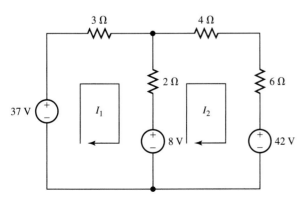

FIGURE P2–1

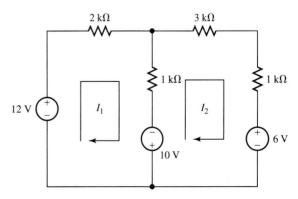

FIGURE P2–2

2-3 Write the mesh current equations for the circuit of Figure P2-3 and solve for the three mesh currents.

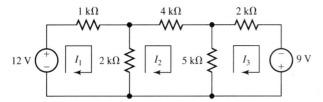

FIGURE P2–3

2-4 Write the mesh current equations for the circuit of Figure P2-4 and solve for the three mesh currents.

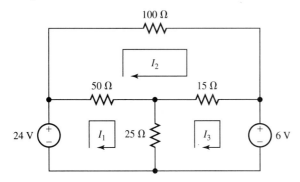

FIGURE P2–4

2-5 Write the mesh current equations for the circuit of Figure P2-5 and solve for the three mesh currents.

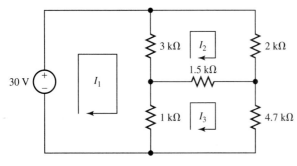

FIGURE P2–5

2-6 Write the mesh current equations for the circuit of Figure P2-6 and solve for the three mesh currents.

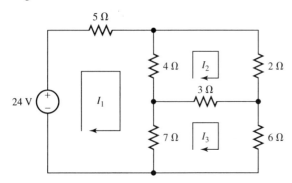

FIGURE P2–6

2-7 Establish a ground node on the lower bus in the circuit of Problem 2-1 (Figure P2-1) and solve for the voltage in the upper middle node using node voltage analysis.

2-8 Establish a ground node on the lower bus in the circuit of Problem 2-2 (Figure P2-2) and solve for the voltage in the upper middle node using node voltage analysis.

2-9 Establish a ground node on the lower bus in the circuit of Problem 2-3 (Figure P2-3) and solve for the two unknown node voltages using node voltage analysis.

2-10 Establish a ground node on the lower bus in the circuit of Problem 2-4 (Figure P2-4) and solve for the voltage in the upper middle node using node voltage analysis.

2-11 Establish a ground node on the lower bus in the circuit of Problem 2-5 (Figure P2-5) and solve for the voltages in the middle two nodes using node voltage analysis.

2-12 Establish a ground node on the lower bus in the circuit of Problem 2-6 (Figure P2-6) and solve for the voltages at the upper right and middle two nodes using node voltage analysis.

2-13 The circuit of Problem 2-5 is an example of an unbalanced bridge circuit. It does not lend itself to the simple series and parallel combinations considered in the last chapter. One way to determine the equivalent resistance as viewed by the source is to solve for the ratio of V_s/I_1, where $V_s = 30$ V in this case. Use the results of Problem 2-5 to determine the equivalent resistance.

2-14 Determine the equivalent resistance as viewed by the source for the circuit of Problem 2-6. Refer to the procedure discussed in Problem 2-13.

2-15 Determine the Thevenin and Norton equivalent circuits for the circuit of Figure P2-15 at the terminals A-B.

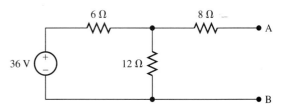

FIGURE P2–15

2-16 Determine the Thevenin and Norton equivalent circuits for the circuit of Figure P2-16 at the terminals A-B.

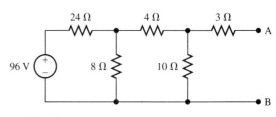

FIGURE P2–16

2-17 Determine the Thevenin and Norton equivalent circuits for the circuit of Figure P2-17 at the terminals A-B.

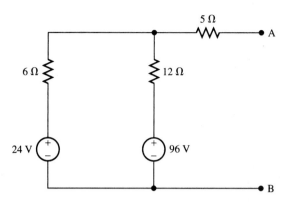

FIGURE P2–17

2-18 Determine the Thevenin and Norton equivalent circuits for the circuit of Figure P2-18 at the terminals A-B.

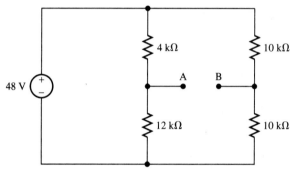

FIGURE P2–18

2-19 The circuit of Figure P2-19 is a Wheatstone bridge circuit with arbitrary values of the resistances. Using mesh current analysis, prove that the current $I = 0$ when the following equality is satisfied: $R_1/R_2 = R_3/R_4$.

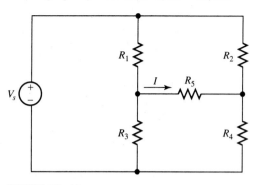

FIGURE P2–19

2-20 Perform the proof of Problem 2-19 using node voltage analysis.

2-21 Two batteries are connected through resistances to a load R_L as shown in Figure P2-21. Determine a relationship between the various circuit parameters so that all the power in R_L is delivered by V_{s1} and no power is either absorbed or delivered by V_{s2}.

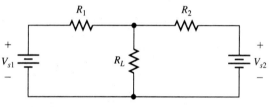

FIGURE P2–21

2-22 For the circuit of Problem 2-21, determine a relationship between the various circuit parameters so that half the power in R_L is delivered by each of the two sources.

Transient Circuits

OVERVIEW AND OBJECTIVES

The first two chapters were devoted to basic circuit laws and the analysis of dc circuits with resistance as the only passive circuit parameter. In this chapter, we introduce the additional passive parameters of inductance and capacitance. Whereas the voltage-current relationship for a resistance is algebraic in form, we will see that the voltage-current relationships for inductance and capacitance involve the calculus operations of differentiation and integration.

Objectives

After completing this chapter, the reader should be able to

1. Discuss the form and notation for a time-varying voltage and/or current.
2. State Ohm's law for a resistance in terms of time-varying voltage and current.
3. Define the three passive circuit parameters and discuss each one in terms of **energy dissipation** or **energy storage**.
4. State and apply the different forms of the **voltage-current relationships** for an inductance and a capacitance.
5. For piecewise linear voltage or current functions, calculate and plot the waveforms of voltage, current, and power for inductance and capacitance.
6. Determine the energy stored in an inductance and a capacitance.
7. Draw the initial equivalent circuits for an inductance and a capacitance for the case of no initial energy and for when there is initial energy stored.
8. Solve for the initial values of voltage and current in a circuit with inductance and capacitance.
9. Draw the final steady-state equivalent circuits for inductance and capacitance when all excitations are dc.
10. Solve for the final values of voltage and current in a circuit with inductance and capacitance and when all excitations are dc.
11. Determine the voltage or current response of a **first-order circuit** with dc sources and sketch the waveforms.

3-1 Basic Circuit Parameters

There are three types of **passive circuit parameters**. By passive, we mean they are incapable of creating energy. The three parameters are resistance, inductance, and capacitance. We have already studied resistance, and we know that it **dissipates** power and energy and satisfies Ohm's law. The other two parameters are inductance and capacitance. Both of these parameters in their ideal pure form do not dissipate power and energy, but instead store energy and are capable of releasing it at a later time. Inductance and capacitance are **energy storage parameters**. For reasons that will not be clear until we study ac circuits, inductance and capacitance are also referred to as **reactive parameters**.

There are two forms of inductance: **self-inductance** and **mutual inductance**. We will study mutual inductance in Chapter 14, where we also will study the major application of mutual inductance, the *transformer*. Here we will emphasize the circuit models of inductance, and all inductance that we consider will be self-inductance, in which case we will drop the prefix "self" for the sake of brevity.

Time-Varying Variables

As we begin the study of energy storage parameters, we must consider the circuit variables of voltage, current, power, and energy to be functions of the time variable t. To that end, we will use lower-case symbols such as $v(t), i(t), p(t),$ and $w(t)$. In some cases, we will retain the (t) to emphasize the time-varying nature of the variable. However, to simplify many of the expressions, we will often employ $v, i, p,$ and w without explicitly showing the time dependency. Just remember that the use of lower-case symbols means that a function *could* vary with time.

Resistance

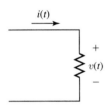

FIGURE 3–1
Resistance with instantaneous time-varying voltage and current.

We have already studied resistance extensively in the first two chapters. Referring to Figure 3-1, it is meaningful now to extend Ohm's law to the case where both the voltage and current are varying with time. Hence, we have

$$i(t) = \frac{v(t)}{R} \tag{3-1}$$

$$v(t) = Ri(t) \tag{3-2}$$

We have retained the time-varying symbols here to make a point. Ohm's law for time-varying quantities reads just as it does for dc, except that both the voltage and current are now functions of time. Because the constant of proportionality between them is either $1/R$ or R, which in either case is a positive real constant, *waveforms of the voltage and current associated with an ideal resistance have the same form.*

The preceding property can be exploited in practical measurement systems. As mentioned in Chapter 1, it usually is more convenient to measure voltage than current. Therefore, the nature of a current waveform can often be deduced from the observation of the voltage across a known resistance. In fact, even if there is no resistance available for this purpose, it might be convenient in some cases to insert a small sampling resistor in a circuit so that an oscilloscopic measurement of current can be made indirectly. Take care when doing this, because most oscilloscopic measurements must be made with respect to a common circuit ground. Moreover, the resistance must be sufficiently small that it does not change the nature of the circuit.

Inductance

Inductance is a property that results from magnetic flux generated in a circuit by current flow. Although all circuit configurations possess stray inductance, here we will emphasize a lumped quantity of inductance, which is called an **inductor**. A lumped inductor typically is composed of many turns of wire, often placed on a **ferromagnetic** core (to be discussed

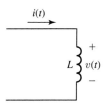

FIGURE 3–2
Inductance with instantaneous time-varying voltage and current.

later) that strengthens the magnetic flux. A descriptive word for some inductors is **coil**. Some older references use the term **choke** for large inductors used in power supplies.

The schematic symbol for inductance is shown in Figure 3-2. It displays a shape that is descriptive of its nature. The symbol for inductance is L. The unit of inductance is the **henry**. Its abbreviation is H.

Inductive Voltage-Current Relationships

The voltage across an inductance is proportional to the rate of change of current with respect to time. It can be expressed as

$$v(t) = L\frac{di(t)}{dt} \tag{3-3}$$

The inverse relationship for current expressed in terms of voltage is

$$i(t) = \frac{1}{L}\int_{-\infty}^{t} v(t)\, dt \tag{3-4}$$

In many analysis situations, the time $t = 0$ will be used as a starting point for analysis. In that case, an alternate expression for the current is

$$i(t) = \frac{1}{L}\int_{0}^{t} v(t)\, dt + I_0 \tag{3-5}$$

where I_0 is the initial current flowing in the inductance at $t = 0$.

Capacitance

Capacitance is a property based on the presence of an electric field between conducting media resulting from charge. Although all circuit configurations possess stray capacitance, our emphasis here will be on a lumped quantity of capacitance, which is called a **capacitor**. An older term that might be found in some references is that of a **condenser**.

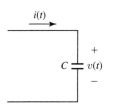

FIGURE 3–3
Capacitance with instantaneous time-varying voltage and current.

A popular schematic symbol for a capacitor, and the one that we will use in this text, is shown in Figure 3-3. In an alternate symbol, one of the sides is curved like an arc of a circle. The symbol with the parallel lines is the same as that of the contact points in a programmable controller (which we will consider in Chapter 17) but there is little likelihood of confusion.

The "gap" between the two lines or conductors in the capacitor schematic symbolizes that there is no conducting medium between them. The unit of capacitance is the **farad**, which is abbreviated as F. A farad is a very large unit. Most practical capacitors have capacitance values in the range of microfarads, nanofarads, and even picofarads.

Capacitive Voltage-Current Relationships

The current flow in a capacitance is proportional to the rate of change of voltage with respect to time. It can be expressed as

$$i(t) = C\frac{dv(t)}{dt} \tag{3-6}$$

The inverse relationship for voltage expressed in terms of current is

$$v(t) = \frac{1}{C}\int_{-\infty}^{t} i(t)\, dt \tag{3-7}$$

In many analysis situations, the time $t = 0$ will be used as a starting point for analysis. In that case, an alternate expression for the voltage is

$$v(t) = \frac{1}{C}\int_{0}^{t} i(t)\, dt + V_0 \tag{3-8}$$

where V_0 is the initial voltage across the capacitor at $t = 0$.

Energy Storage

It has been stated that both inductance and capacitance *store energy* rather than dissipate it. The manner in which the energy is stored is different for the two components. Energy stored in an inductance results from current flowing in the inductance. It is a form of **kinetic energy**. Conversely, energy stored in a capacitance results from a voltage existing across the capacitance. It is a form of **potential energy**. A different point of view is that energy in an inductance results from *charge in motion* and energy in a capacitance results from *charge at rest*.

Inductive Energy

Let w_L represent the energy stored in an inductance. It is a function of the inductance L and the current i flowing in it at a given time. It is given by

$$w_L = \tfrac{1}{2} L i^2 \tag{3-9}$$

Capacitive Energy

Let w_C represent the energy stored in a capacitance. It is a function of the capacitance C and the voltage v across it at a given time. It is given by

$$w_C = \tfrac{1}{2} C v^2 \tag{3-10}$$

Instantaneous Power

The instantaneous power will be denoted as $p(t)$. It is the product of the instantaneous voltage and the instantaneous current.

$$p(t) = v(t) i(t) \tag{3-11}$$

Based on the assumed power conventions established in Chapter 1, a positive value of the power means that the power is being absorbed and a negative value means that the power is being delivered. As we will see shortly, power can be delivered by an energy storage element based on energy stored in the device at an earlier time.

Energy

The total energy absorbed by a device at any time t will be denoted as $w(t)$. It is the integral over time of the power. It can be stated analytically as

$$w(t) = \int_{-\infty}^{t} p(t)\, dt = \int_{-\infty}^{t} v(t) i(t)\, dt \tag{3-12}$$

Conversely, the instantaneous power is the rate of change of energy with respect to time.

$$p(t) = \frac{dw(t)}{dt} \tag{3-13}$$

EXAMPLE 3-1

The 0.5-µF capacitor of Figure 3-4 has the waveform of voltage shown across it. Calculate and plot as instantaneous functions of time (a) the *current* and (b) the *power*.

SOLUTION The current flow in a capacitance is proportional to the rate of change of voltage as given by Equation 3-6. The waveform of voltage is a **piecewise linear (PWL)**

3-1 • Basic Circuit Parameters   55

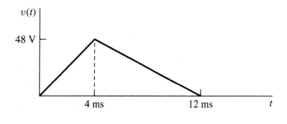

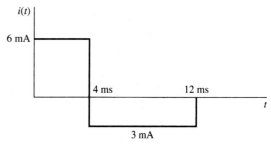

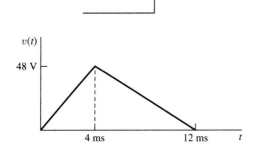

FIGURE 3–4
Capacitor and voltage of Example 3-1.

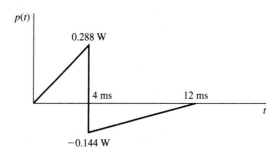

FIGURE 3–5
Waveforms of Example 3-1.

function, which is one composed of straight-line segments. The derivative or rate of change of a particular segment can be determined by evaluating its slope, which can be performed with simple arithmetic. The slope is obtained by dividing the change in the dependent variable by the change in the independent variable, and the result may either be positive or negative.

Refer to the waveforms of Figure 3-5 for the analysis that follows. The waveform of voltage is repeated at the top of the figure for convenience. This PWL function is composed of two straight-line segments, one with a positive slope and one with a negative slope. For each segment, we need to calculate the derivative or slope and multiply that value by the capacitance of $0.5 \ \mu F = 0.5 \times 10^{-6}$ F.

(a) $\qquad\qquad\qquad\qquad 0 \leq t \leq 4$ ms

The derivative (or slope) in this interval is positive. It is calculated by forming the change in voltage divided by the change in current.

$$\frac{dv}{dt} = \frac{48 \text{ V}}{4 \times 10^{-3} \text{ s}} = 12 \times 10^3 \text{ V/s} \qquad (3\text{-}14)$$

The current is

$$i = C\frac{dv}{dt} = 0.5 \times 10^{-6} \times 12 \times 10^3 = 6 \times 10^{-3} \text{A} = 6 \text{ mA} \qquad (3\text{-}15)$$

$$4 \leq t \leq 12 \text{ ms}$$

In this interval, the change in voltage is negative and the slope is

$$\frac{dv}{dt} = \frac{-48 \text{ V}}{8 \times 10^{-3} \text{ s}} = -6 \times 10^3 \text{ V/s} \tag{3-16}$$

$$i = C\frac{dv}{dt} = -0.5 \times 10^{-6} \times 6 \times 10^3 = -3 \times 10^{-3} \text{ A} = -3 \text{ mA} \tag{3-17}$$

The current waveform is shown below the voltage waveform on Figure 3-5. Note that as long as the voltage is increasing, the current is positive, but when the voltage begins to decrease, the current reverses its direction.

Nothing has been said thus far as to whether the voltage is causing the current or vice versa. It doesn't make any difference as far as the waveforms are concerned. A voltage source having the behavior of the top waveform of Figure 3-5 would cause a current to flow, as shown by the middle waveform. Likewise, a current source of the form shown in the middle would produce the voltage at the top.

(b) While the instantaneous power could be difficult to determine graphically for complex waveforms, the functions given here lend themselves to a simplified analysis. Consider first the interval between 0 and 4 ms. The current has a constant positive value in this interval, so the power must necessarily have the shape of the voltage waveform. This means that the power will be a triangular function. A key point on the curve is the value of the power immediately to the left of the discontinuity in the current. It is

$$p = (48 \text{ V}) \times (6 \times 10^{-3} \text{ A}) = 0.288 \text{ W} \tag{3-18}$$

For the interval between 4 ms and 12 ms, the current is a negative constant while the voltage is a decreasing positive triangular function. Therefore, the power is an inverted or negative triangular function. A key point on the curve is the value of the power immediately to the right of the discontinuity in the current. It is

$$p = (48 \text{ V}) \times (-3 \times 10^{-3} \text{ A}) = -0.144 \text{ W} \tag{3-19}$$

Note that the power during the first interval is positive, meaning that the capacitance is absorbing the energy. However, during the second interval, the power is negative, which means that the capacitor is delivering the energy back to some external circuit not shown.

EXAMPLE 3-2

Determine the peak value of the energy stored in the capacitance of Example 3-1.

SOLUTION First, note that we could obtain the energy as a function of time by integrating the power over time, but our interest is the peak energy. From Equation 3-10, we observe that the peak energy occurs when the value of voltage is at its peak. Let V_p represent the peak voltage, which is 48 V at the time of 4 ms. We denote the value of this energy as W_p. It is

$$W_p = \tfrac{1}{2}CV_p^2 = \tfrac{1}{2} \times 0.5 \times 10^{-6} \times (48)^2 = 576 \times 10^{-6} \text{ J} = 576 \text{ μJ} \tag{3-20}$$

The voltage returns to a value of zero, so all of this stored energy is released back to the circuit.

EXAMPLE 3-3

The current waveform shown at the top of Figure 3-6 flows into an initially uncharged 0.5-μF capacitance. Calculate and plot the voltage as a function of time.

SOLUTION The pertinent relationship is that of Equation 3-8 with $V_0 = 0$ because the capacitor is initially uncharged.

$$v(t) = \frac{1}{C}\int_0^t i(t)\,dt = \frac{1}{0.5 \times 10^{-6}}\int_0^t i(t)\,dt = 2 \times 10^6 \int_0^t i(t)\,dt \tag{3-21}$$

FIGURE 3–6
Waveforms of Example 3-3.

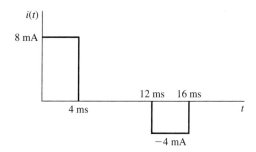

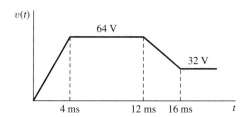

The integration involves determining the area under the curve of current as time increases and multiplying that area by the factor in front of the integral. In contrast to differentiation, integration is an accumulative process. Therefore, as the area is summed, the resulting function cannot change abruptly. It will remain at any level reached until further area occurs. Positive area will cause the integral to increase and negative area will cause the integral to decrease. Let us consider four different intervals for this purpose.

$$0 \leq t \leq 4 \text{ ms}$$

In this first interval, the current is a positive constant, so the area must increase at a linear rate. A key value of time is 4 ms. The net area accumulated between 0 and 4 ms is the area of the rectangle shown. The area is $(8 \times 10^{-3} \text{ A}) \times (4 \times 10^{-3} \text{ s}) = 32 \times 10^{-6} \text{ A} \cdot \text{s} = 32 \times 10^{-6}$ C. (Note that 1 ampere · second = 1 coulomb.) The voltage at this point is

$$v(4 \text{ ms}) = 2 \times 10^6 \times 32 \times 10^{-6} = 64 \text{ V} \qquad (3\text{-}22)$$

Thus, the voltage ramps from 0 to 64 V as shown in the bottom waveform of Figure 3-6.

$$4 \leq t \leq 12 \text{ ms}$$

Because the current is zero in this interval, the voltage remains at a level of 64 V.

$$12 \leq t \leq 16 \text{ ms}$$

The current is a negative constant in this interval, so the area will decrease at a linear rate. The net change in area is $(-4 \times 10^{-3} \text{ A}) \times (4 \times 10^{-3} \text{ s}) = -16 \times 10^{-6}$ C. The voltage change is obtained by multiplying this value by the constant in front of the integral. This change is then added algebraically to the initial value at 12 ms and the voltage at 16 ms is

$$v(16 \text{ ms}) = 64 - 2 \times 10^6 \times 16 \times 10^{-6} = 32 \text{ V} \qquad (3\text{-}23)$$

$$t \geq 16 \text{ ms}$$

The capacitor is charged to 32 V and will remain at that level until further current flows.

EXAMPLE 3-4

The upper waveform of current shown in Figure 3-7 is flowing in an inductance of value L. It can be described by the equation

$$i(t) = I_p \sin \omega t \qquad (3\text{-}24)$$

Determine the voltage across the inductance as a function of time.

FIGURE 3–7
Waveforms of Example 3-4.

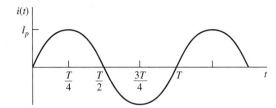
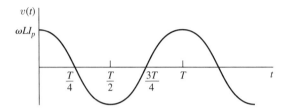

SOLUTION The voltage across an inductance as a function of time is given by Equation 3-3.

$$v = L\frac{di}{dt} \tag{3-25}$$

The derivative of the current is given by

$$\frac{di}{dt} = (I_p \cos \omega t) \times \omega = \omega I_p \cos \omega t \tag{3-26}$$

The voltage across the inductance is

$$v = L(\omega I_p \cos \omega t) = \omega L I_p \cos \omega t \tag{3-27}$$

The waveform is shown as the bottom curve of Figure 3-7. This example represents sort of a preview of things to come in later chapters. In previous examples, we have seen that the voltage and current associated with energy storage elements can have drastically different shapes. However, in the case of sinusoidal voltage or current, the shapes are the same, but they have different phase angles. In the case of the inductance, the voltage *leads* the current by 90°. In the case of a capacitance, we will show later that the voltage *lags* the current by 90°.

3-2 Boundary Conditions for Energy Storage Elements

We have seen that the nature of the voltage and current relationships associated with energy storage elements follow the laws of calculus rather than algebra. Consequently, the instantaneous waveforms of voltage and current associated with these parameters might be quite different and, in some cases, might require the solution of differential equations to determine the nature of the response functions.

It may be somewhat involved to determine the instantaneous behavior of voltage and current for all time in many circuits, but we may predict certain conditions easily with dc equivalent circuits. They are the **initial conditions** and the **steady-state dc conditions**. For either of these types of conditions, the circuit models reduce to simple dc forms that we may analyze using the methods of earlier chapters.

Inductive Current and Capacitive Voltage Boundary Behavior

Let i_L represent the instantaneous current flow in an inductance and let v_C represent the instantaneous voltage across a capacitance. Let $t = 0$ represent the beginning time for a circuit analysis to be performed. Let $t = 0^-$ represent an infinitesimally small time before the beginning and $t = 0^+$ represent a similar time immediately after the beginning of the analysis. The following boundary conditions then describe the behavior of these elements:

$$i_L(0^+) = i_L(0^-) \tag{3-28}$$

$$v_C(0^+) = v_C(0^-) \tag{3-29}$$

Stated in words, these equations state that the current flow in an inductance cannot change instantaneously and the voltage across a capacitance cannot change instantaneously. The variables involved are those that establish the energy stored in the respective components, so from a different perspective, it takes some time to establish or reestablish a certain energy level in the components.

Don't let the − and + confuse you as far as substitution in any equation is concerned, because both 0^- and 0^+ would be treated as 0. However, the signs convey the message of "just before" and "just after."

As is often the case, there are possible exceptions to the rule (at least in theory). Situations known as impulsive conditions theoretically can result in abrupt changes in the variables indicated, but we will not pursue them in this text.

Note that the opposite terminal variables can experience sudden changes. The voltage across an inductance can change instantaneously and the current flow in a capacitance can change instantaneously.

Initial Equivalent Circuits with No Energy Storage

We can use the preceding boundary conditions to infer some equivalent circuits that describe the behavior of inductance and capacitance immediately after any type of switching action occurs. Consider first the situation in which there is no initial energy stored in the elements, meaning that an inductor has no current flowing in it and a capacitor has no voltage across it. An inductor with no initial current flowing in it is called an **unfluxed inductor**, and a capacitor with no initial voltage across it is referred to as an **uncharged capacitor**.

The equivalent circuit immediately following a switching action of an initially unfluxed inductor is shown in Figure 3-8(a). This equivalent circuit applies only for an infinitesimally short time, but the model is that of an open circuit.

The corresponding equivalent circuit of an initially uncharged capacitor is shown in Figure 3-8(b). In the case of the capacitor, the model is that of a short circuit.

Initial Equivalent Circuits with Energy Storage

If an inductor has initial energy stored in it, then there is current flowing at the moment that switching occurs. The inductor will tend to resist any change in current initially, so it will act as a source of constant current. The equivalent circuit model at the instant of switching is shown in Figure 3-9(a).

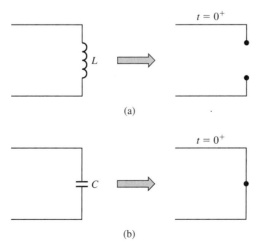

FIGURE 3-8
Initial equivalent circuits for unfluxed inductance and uncharged capacitance.

FIGURE 3–9
Initial equivalent circuits for fluxed inductance and charged capacitance.

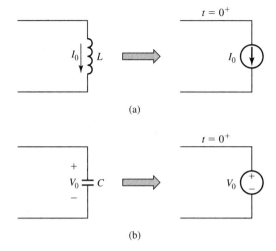

If a capacitor has initial energy stored in it, then there is a voltage across it when switching occurs. The capacitor will tend to resist any change in voltage initially, so it will act as a source of constant voltage. The equivalent circuit model in this case is shown in Figure 3-9(b).

Steady-State DC Conditions

Consider next the situation in which a circuit contains inductance and/or capacitance along with resistance, but in which all the excitations in the circuit are dc sources. As long as there is any resistance at all in the circuit, eventually the circuit will settle into a so-called steady-state behavior in which all voltages and currents assume constant values. This situation will be referred to as a **steady-state dc condition**. It is customary to use the symbol $t = \infty$ to represent the time associated with this condition even though, in many circuits, the time involved might represent only a small fraction of a second.

In the dc steady state, derivatives with respect to time of all currents and all voltages have essentially become zero. A review of the terminal equations then leads to the conclusion that inductive voltages and capacitive currents become zero in the dc steady state.

Steady-State DC Models

Based on the logic of the preceding discussion, the equivalent circuits in the dc steady state are shown in Figure 3-10. Assuming an ideal inductor, it acts as a short circuit, as shown in (a). (In practice, all real inductors contain resistance, but we are ignoring it.)

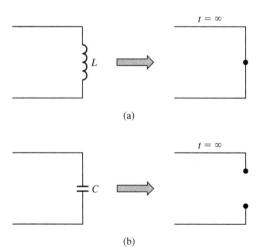

FIGURE 3–10
Steady-state dc equivalent circuits for inductance and capacitance.

The idea of an ideal inductor becoming a short circuit to dc is reasonable because it is actually composed of a coil of wire. For a voltage to exist across it, there must be a changing current, and the process reduces to the notion of changing magnetic flux inducing a voltage, a phenomenon that will be studied later in the book concerning transformers, generators, and motors.

The equivalent circuit of an ideal capacitor in the dc steady state is shown in Figure 3-10(b). It acts as an open circuit under those conditions.

The idea of an ideal capacitor becoming an open circuit to dc is reasonable, because the two conducting media are insulated from each other by a **dielectric** material. Current flow is actually the movement of charge from one side to the other via the external circuit, and for this charge to flow, the voltage must be changing with time. In casual conversation, it is customary to speak of current flow *through* a capacitor, but the actual movement is from one side to the other via the external circuit. The effect from an external point of view is the same as if the current were flowing through the capacitor.

As a general rule with some exceptions, capacitors can be made to be much more ideal than inductors. There might be some very small leakage through the dielectric material in a capacitor, but except for some of the larger electrolytic types, they may be assumed for most practical purposes to be ideal. In contrast, most real inductors have some resistance, and some of the larger inductors may have fairly large resistance values.

Skin Effect

The resistance of most inductors cannot be determined from a dc resistance measurement because the effective resistance for ac might be much greater than for dc because of a phenomenon called the **skin effect**. Because of the inductance within a wire, current flow tends to move toward the outer portion of a conductor as the frequency increases, which results in an effective increase in resistance.

Some Steady-State Applications

Several applications for capacitance and inductance will appear throughout the text, but we will consider a few based on the dc steady-state models now. Because a capacitance appears as an open circuit for steady-state dc, capacitors are widely used in numerous circuits to block dc while passing the time-varying part of a signal (referred to casually as the "ac component"). In a contrasting sense, an inductance can be used to represent an open circuit for an ac component while allowing steady-state dc current to flow through it.

▌ EXAMPLE 3-5

The switch in the circuit of Figure 3-11(a) is closed at $t = 0$, and there is no initial energy storage. Construct an equivalent circuit at $t = 0^+$ and determine the initial values of the variables shown on the diagram.

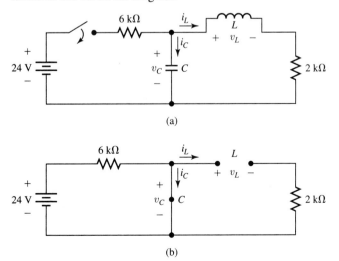

FIGURE 3–11
Circuit of Example 3-5 and its initial equivalent circuit.

SOLUTION There is no initial energy stored in the circuit; therefore the capacitor will be initially uncharged and the inductor will be initially unfluxed. Thus, the capacitor will act like a short circuit and the inductor will act like an open circuit. The initial equivalent circuit is in Figure 3-11(b).

Because the capacitor acts as a short circuit and the inductor acts an open circuit, we can immediately say that

$$v_C(0^+) = 0 \tag{3-30}$$

and

$$i_L(0^+) = 0 \tag{3-31}$$

As far as the other variables specified, don't let the dangling wires confuse you. The circuit reduces to a simple Ohm's law situation because the entire source voltage appears across the left-hand resistance and the initial current flow in the capacitor is

$$i_C(0^+) = \frac{24}{6 \times 10^3} = 4 \text{ mA} \tag{3-32}$$

Although it is possible in general to have an initial voltage appearing across the open circuit represented by an inductance, in this case, it is zero; that is,

$$v_L(0^+) = 0 \tag{3-33}$$

▮ EXAMPLE 3-6

For the circuit of Example 3-5, determine the steady-state values of the variables identified in that example.

SOLUTION For convenience, the circuit is repeated in Figure 3-12(a). Because the circuit excitation is a dc voltage, the steady-state circuit model can be created, and it is shown in Figure 3-12(b). Note that the capacitor is now represented as an open circuit and the inductor is represented as a short circuit. Thus, we can immediately say that

$$i_C(\infty) = 0 \tag{3-34}$$

and

$$v_L(\infty) = 0 \tag{3-35}$$

As in the preceding example, we have some dangling connectors, but the circuit reduces essentially to a single-loop circuit. The steady-state current $i_L(\infty)$ is

$$i_L(\infty) = \frac{24}{6 \times 10^3 + 2 \times 10^3} = \frac{24}{8 \times 10^3} = 3 \text{ mA} \tag{3-36}$$

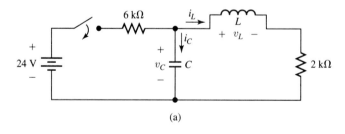

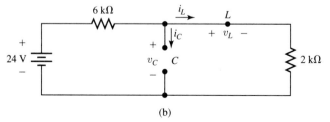

FIGURE 3–12
Circuit of Example 3-6 and its steady-state dc equivalent circuit.

The steady-state voltage across the capacitor is the same as the voltage across the right-hand resistance. It is determined by Ohm's law as

$$v_C(\infty) = 2 \times 10^3 \times 3 \times 10^{-3} = 6 \text{ V} \tag{3-37}$$

These results infer that after the circuit is settled into a steady-state dc condition, the inductance will be fluxed to a current of 3 mA and the capacitor will be charged to a voltage of 6 V.

EXAMPLE 3-7

The circuit of Figure 3-13(a) is the same as that considered in the previous two examples except that the capacitor is initially charged to 8 V and the inductor is initially fluxed to 1 mA with the directions shown. At $t = 0$, the same 24-V dc source of the previous two examples is switched into the circuit. Determine the initial value of the variables identified on the circuit diagram.

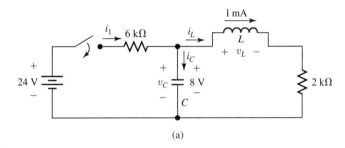

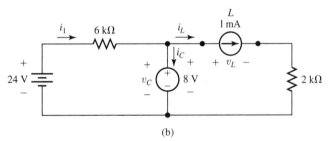

FIGURE 3–13
Circuit of Example 3-7 and its initial equivalent circuit.

SOLUTION The initial equivalent circuit is shown in Figure 3-13(b). In contrast to Example 3-5, the capacitor will initially act as a constant voltage of 8 V and the inductor will act as a constant current of 1 mA. We can immediately state that

$$v_C(0^+) = 8 \text{ V} \tag{3-38}$$

and

$$i_L(0^+) = 1 \text{ mA} \tag{3-39}$$

The circuit in this case contains two loops, but simultaneous equations are not required. The easiest way to solve the circuit is to recognize that the net voltage across the 6-kΩ resistance is $24 - 8 = 16$ V, with the positive terminal on the left. Thus, the initial current $i_1(0^+)$ is

$$i_1(0^+) = \frac{16}{6 \times 10^3} = 2.667 \text{ mA} \tag{3-40}$$

At the node in the middle, a current of 2.667 mA is entering from the left, and a current of 1 mA is leaving on the right. The initial current flow into the capacitor is then

$$i_C(0^+) = 2.667 - 1 = 1.667 \text{ mA} \tag{3-41}$$

The initial voltage across the 2-kΩ resistance is

$$v_2(0^+) = 2 \times 10^3 \times 1 \times 10^{-3} = 2 \text{ V} \tag{3-42}$$

The initial inductive voltage is then determined from KVL applied to the right-hand loop.

$$v_L(0^+) = 8 - 2 = 6 \text{ V} \tag{3-43}$$

Although the circuit begins with different initial conditions than in the preceding two examples, the final steady-state circuit will be the same as in Example 3-5, which was given in Figure 3-12(b). The initial voltage on the capacitor and the initial current in the inductance affects the initial values and the magnitude of the transient behavior, but the final response is dependent only on the 24-V actual source.

3-3 First-Order Circuits with DC Excitations

We have introduced the voltage-current relationships for the three basic circuit parameters and have covered exercises for dealing with the terminal relationships. We are now ready to consider combinations of two or more of these elements and how the voltage and current will behave with these combinations.

The simplest type of circuit containing two or more types of elements is a **first-order circuit** form. A first-order form is one in which the voltage and current relationships involve a **first-order differential equation**. As it turns out, all circuits of this type are reducible to either one of the following forms: (1) a single resistance and a single inductance or (2) a single resistance and a single capacitance. Circuits in the first category are called **RL circuits** and circuits in the second category are called **RC circuits**. A combination that will *not* be considered in this section is one containing both L and C.

We will further restrict our consideration here to circuits having only dc sources present. This could include initial currents in inductors or initial voltages on capacitors. Whereas all of the preceding statements may seem rather restrictive, many applications involve first-order circuits with dc or "dc-like" sources. The latter description includes initial conditions and pulse waveforms.

First-Order Differential Equation with DC Excitation

Let y represent an arbitrary variable to be determined from physical consideration, which for the present discussion will be either a voltage or a current. Assuming that the independent variable is time t, the first-order differential equation with dc sources will be of the form

$$\frac{dy}{dt} + \frac{1}{\tau}y = K \tag{3-44}$$

The value K is a constant that is dependent on the dc voltage or current excitation. The quantity τ is called the **time constant**. It plays a significant role in how long a circuit response reaches a steady-state condition. Without showing the details, it turns out that the solution of this differential equation is always of the form

$$y = A + Be^{-t/\tau} \tag{3-45}$$

Thus, first-order circuits with dc sources have voltages and currents that display this form. Therefore, we can determine any response in the circuit from a knowledge of the constants A, B, and τ.

Determining the Time Constant

The time constant for the circuit depends on whether it is an *RL* circuit or an *RC* circuit. Assuming that the circuit has been reduced to one containing a single value of R and a single value of C, or a single value of R and a single value of L, the time constant is evaluated by one or the other of the following two equations:

$$\tau = RC \tag{3-46}$$

$$\tau = \frac{L}{R} \tag{3-47}$$

Determining A and B

It turns out that A and B can be determined from the equivalent circuit models provided in the last section. First, the initial equivalent circuit model is constructed and the initial value $y(0^+)$ is determined. Next, the dc steady-state equivalent circuit model is constructed and the final value $y(\infty)$ is determined. From these two values, we can determine the constants A and B. Noting that $y(\infty) = A$, the constants may be expressed as

$$A = y(\infty) \tag{3-48}$$

and

$$B = y(0) - y(\infty) \tag{3-49}$$

An alternate form for the general solution can then be expressed

$$y(t) = y(\infty) + [y(0^+) - y(\infty)]e^{-t/\tau} \tag{3-50}$$

This latter form has the advantage that one need only "plug in" the initial value, the final value, and the time constant to determine the net response as a function of time.

Sketching the Response

Any response of the form considered can be sketched easily. The function begins at the initial value $y(0^+)$ and eventually approaches the final value $y(\infty)$. It is concave toward the final value, as we will demonstrate in the examples that follow this section.

Effect of Time Constant

The larger the value of the time constant, the longer it will take for a given response to settle into its final steady state. Some values of the exponential function at different multiples of the time constant are shown in the table that follows.

t	$e^{-t/\tau}$
τ	0.368
2τ	0.135
3τ	0.0498
4τ	0.0183
5τ	0.00674

We can interpret these results in terms of the fraction between the initial and final values at different multiples of the time constant. For example, at one time constant, the remaining difference between initial and final values is about 36.8 percent of the difference between initial and final values.

Note that at five time constants, the difference is less than 1 percent. This serves as a useful rule of thumb for many applications. In other words, the circuit may be assumed to have reached a steady-state dc condition in about five time constants. Remember, however, that this is only a reasonable estimate that must be tempered with other factors.

■ EXAMPLE 3-8

The switch in the circuit of Figure 3-14(a) is closed at $t = 0$ and there is no initial energy stored in the inductance. Determine the equations for the instantaneous voltage across the inductance and the current in the loop, and sketch the results.

SOLUTION The initial equivalent circuit ($t = 0^+$) is shown in Figure 3-14(b). Because both $i(t)$ and $v_L(t)$ are of interest, we need to evaluate both values at this time. We can see readily that

$$v_L(0^+) = 12 \text{ V} \tag{3-51}$$

$$i(0^+) = 0 \tag{3-52}$$

FIGURE 3–14
Circuit forms of Example 3-8.

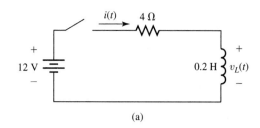

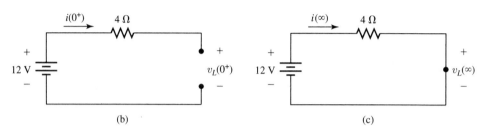

The dc steady-state equivalent circuit ($t = \infty$) is shown in Figure 3-14(c). The voltage and current are

$$v_L(\infty) = 12 \text{ V} \tag{3-53}$$

$$i_L(\infty) = \frac{12 \text{ V}}{4 \, \Omega} = 3 \text{ A} \tag{3-54}$$

The time constant is

$$\tau = \frac{L}{R} = \frac{0.2 \text{ H}}{4 \, \Omega} = 0.05 \text{ s} \tag{3-55}$$

The voltage is given by

$$v_L(t) = v_L(\infty) + [v(0^+) - v(\infty)]e^{-t/\tau} = 0 + (12 - 0)e^{-t/0.05} = 12e^{-20t} \tag{3-56}$$

The current is

$$i(t) = i(\infty) + [i(0^+) - i(\infty)]e^{-t/\tau} = 3 + (0 - 3)e^{-t/0.05} = 3(1 - e^{-20t}) \tag{3-57}$$

We might sketch the two functions by placing the initial and final values on a graph and letting the variable change from the initial to the final value in an exponential fashion.

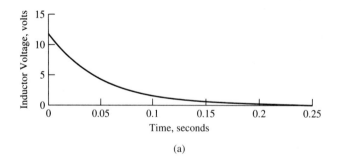

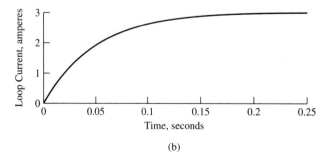

FIGURE 3–15
(a) Inductor voltage of Example 3-8. (b) Loop current of Example 3-8.

To obtain an intuitive sense about the effect of the time constant, the inductive voltage decreases to 36.8 percent of its initial value in 0.05 s, and the current increases to 63.2 percent of its final value in the same time interval. Using the rule of thumb mentioned in this section, final steady-state conditions will be reached for most practical purposes in about $5 \times 0.05 = 0.25$ s.

EXAMPLE 3-9

The switch in the circuit of Figure 3-16(a) is closed at $t = 0$ and there is no initial energy stored in the capacitance. Determine the equations for the instantaneous current in the loop and the voltage across the capacitance and sketch the results.

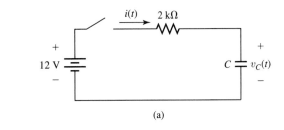

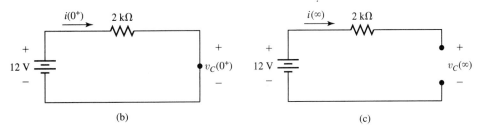

FIGURE 3–16
Circuit forms of Example 3-9.

SOLUTION The initial equivalent circuit is shown in Figure 3-16(b). The initial values of the two variables of interest are

$$i(0^+) = \frac{12 \text{ V}}{2 \times 10^3} = 6 \times 10^{-3} \text{ A} = 6 \text{ mA} \tag{3-58}$$

$$v_C(0^+) = 0 \tag{3-59}$$

The dc steady-state equivalent circuit is shown in Figure 3-16(c). The final values of the current and capacitive voltage are

$$i(\infty) = 0 \tag{3-60}$$

$$v_C(\infty) = 12 \text{ V} \tag{3-61}$$

The time constant is

$$\tau = RC = 2 \times 10^3 \times 1 \times 10^{-6} = 2 \times 10^{-3} \text{ s} = 2 \text{ ms} \tag{3-62}$$

The current is

$$i(t) = i(\infty) + [i(0^+) - i(\infty)]e^{-t/\tau} = 0 + (6 \times 10^{-3} - 0)e^{-t/2 \times 10^{-3}} = 6 \times 10^{-3}e^{-500t} \tag{3-63}$$

The capacitive voltage is

$$v_C(t) = v_C(\infty) + [v_C(0^+) - v_C(\infty)]e^{-t/\tau} = 12 + (0 - 12)e^{-t/2 \times 10^{-3}} = 12(1 - e^{-500t}) \tag{3-64}$$

Plots of the current and capacitive voltage are shown in Figure 3-17. Based on the rule of thumb, this circuit will have settled into a dc steady state in about 5×2 ms $= 10$ ms.

FIGURE 3–17
(a) Loop current of Example 3-9. (b) Capacitor voltage of Example 3-9.

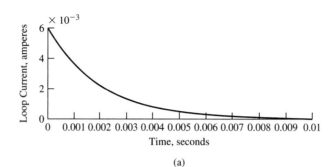

(a)

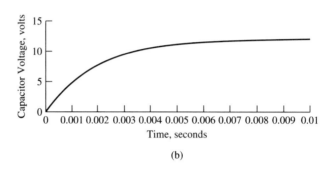

(b)

■ EXAMPLE 3-10

The circuit of Figure 3-18(a) is the same as that of Example 3-9 except for one difference. In this case, the capacitor is initially charged to 8 V with the positive terminal at the bottom as shown. Determine the equations for the instantaneous current in the loop and the voltage across the capacitance, and sketch the results.

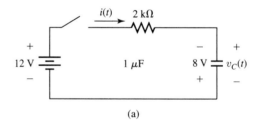

(a)

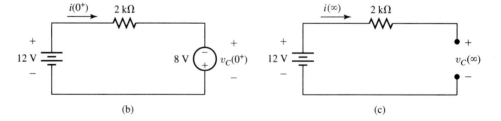

FIGURE 3–18
Circuit forms of Example 3-10.

(b)　　　　　　　　　　　　　　　(c)

SOLUTION The initial equivalent circuit is shown in Figure 3-18(b). The capacitor momentarily acts as a constant voltage of 8 V with its positive terminal at the bottom. The result is a single-loop circuit with the two voltages aiding each other as far as current flow is concerned. The initial values of the two variables of interest are

$$i(0^+) = \frac{12\text{ V} + 8\text{ V}}{2\text{ k}\Omega} = 10 \times 10^{-3} \text{A} = 10 \text{ mA} \tag{3-65}$$

$$v_C(0^+) = -8 \text{ V} \tag{3-66}$$

Note that the two voltages aid each other to produce a larger value of current than in the previous problem, but the capacitive voltage is negative in terms of the assumed reference.

The steady-state dc equivalent circuit is shown in Figure 3-18(c). We can assume that the initial voltage across the capacitor has now been dissipated in the circuit resistance. The

FIGURE 3–19
(a) Loop current of Example 3-10. (b) Capacitor voltage of Example 3-10.

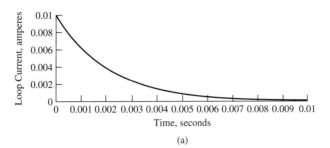

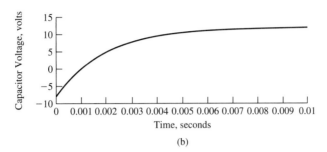

final values are therefore the same as in Example 3-8.

$$i(\infty) = 0 \tag{3-67}$$

$$v_C(\infty) = 12 \text{ V} \tag{3-68}$$

The time constant is unchanged.

$$\tau = RC = 2 \times 10^3 \times 1 \times 10^{-6} = 2 \times 10^{-3} \text{ s} = 2 \text{ ms} \tag{3-69}$$

The current is

$$i(t) = i(\infty) + [i(0^+) - i(\infty)]e^{-t/\tau} = 0 + (10 \times 10^{-3} - 0)e^{-t/2 \times 10^{-3}} = 10 \times 10^{-3} e^{-500t} \tag{3-70}$$

The capacitive voltage is

$$v_C(t) = v_C(\infty) + [v_C(0^+) - v_C(\infty)]e^{-t/\tau} = 12 + (-8 - 12)e^{-t/2 \times 10^{-3}} = 12 - 20e^{-500t} \tag{3-71}$$

The two time functions are shown in Figure 3-19. The presence of the initial voltage on the capacitor does not affect the final result, but it does have a significant effect on both the initial and intermediate behavior.

3-4 Inductance and Capacitance Combinations

In all the developments considered thus far in the chapter, we have not considered more than one inductor or more than one capacitor in a given circuit. In general, the presence of more than one inductor or capacitor could mean that the circuit is of a higher-order form than considered in this chapter. However, in some cases, the combination could reduce to a simple series or parallel configuration.

In Chapter 1, we introduced techniques for determining the equivalent resistance of a series or parallel combination of more than one resistor. We can develop similar procedures for inductors or capacitors in series or parallel combinations.

We emphasize that the rules for combining inductors assume that there is no magnetic coupling between the different inductors.

Inductors in Series

Refer to the circuit of Figure 3-20. Inductances without mutual coupling in *series* combine exactly like resistances in *series*. The net inductance is

$$L_{eq} = L_1 + L_2 + \cdots + L_n \tag{3-72}$$

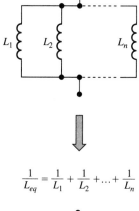

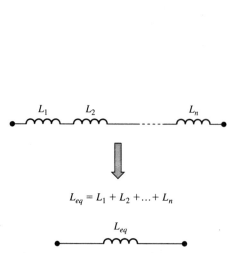

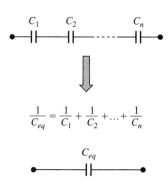

FIGURE 3–20
Inductances in series and the equivalent single inductance when there is no mutual inductance between them.

FIGURE 3–21
Inductances in parallel and the equivalent single inductance when there is no mutual inductance between them.

FIGURE 3–22
Capacitances in series and the equivalent single capacitance.

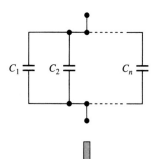

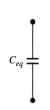

FIGURE 3–23
Capacitances in parallel and the equivalent single capacitance.

Inductors in Parallel

Refer to the circuit of Figure 3-21. Inductances without mutual coupling in parallel combine exactly like resistances in parallel. The net inductance is determined from the formula

$$\frac{1}{L_{eq}} = \frac{1}{L_1} + \frac{1}{L_2} + \cdots + \frac{1}{L_n} \tag{3-73}$$

Capacitors in Series

Refer to the circuit of Figure 3-22. Capacitances in series combine like resistances in parallel, and we determine the net capacitance from the formula

$$\frac{1}{C_{eq}} = \frac{1}{C_1} + \frac{1}{C_2} + \cdots + \frac{1}{C_n} \tag{3-74}$$

Capacitors in Parallel

Refer to the circuit of Figure 3-23. Capacitances in parallel combine like resistances in series. The net capacitance is

$$C_{eq} = C_1 + C_2 + \cdots + C_n \tag{3-75}$$

To summarize, inductors without mutual coupling combine in the same sense as resistors, but capacitors combine in the opposite sense.

▌ EXAMPLE 3-11

Determine the equivalent capacitance of the parallel combination of two capacitors with the following values: 1 μF and 0.02 μF.

SOLUTION Because the capacitances are given in microfarads, it is simpler to work directly with those units. The net capacitance is

$$C_{eq} = C_1 + C_2 = 1 + 0.02 = 1.02 \text{ μF} \tag{3-76}$$

EXAMPLE 3-12

Determine the equivalent capacitance of the series combination of the two capacitors of Example 3-11.

SOLUTION Working again with microfarads, the capacitance is determined as follows:

$$\frac{1}{C_{eq}} = \frac{1}{C_1} + \frac{1}{C_2} = \frac{1}{1} + \frac{1}{0.02} = 1 + 50 = 51 \qquad (3\text{-}77)$$

$$C_{eq} = \frac{1}{51} = 0.0196 \ \mu\text{F} \qquad (3\text{-}78)$$

Note that the equivalent capacitance is not much smaller than the smallest of the two capacitances, a result of the widely different values.

PROBLEMS

3-1 The 0.2-μF capacitor of Figure P3-1 has the waveform of voltage shown across it. Calculate and plot as instantaneous functions of time (a) the current and (b) the power.

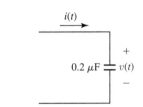

FIGURE P3–1

3-2 The 20-mH inductor of Figure P3-2 has the waveform of current shown flowing in it. Calculate and plot as instantaneous functions of time (a) the voltage and (b) the power.

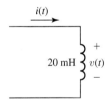

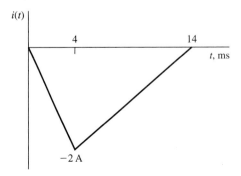

FIGURE P3–2

3-3 The 2-μF capacitor of Figure P3-3 has the waveform of current shown flowing in it and the capacitor is initially uncharged. Calculate and plot as an instantaneous function of time the voltage across the capacitor.

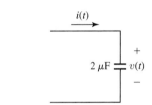

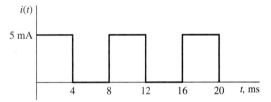

FIGURE P3–3

3-4 The 4-H inductor of Figure P3-4 has the waveform of voltage shown across it and the inductor is initially

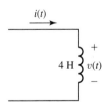

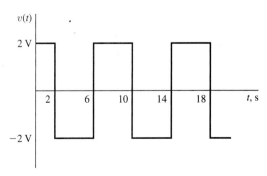

FIGURE P3–4

unfluxed. Calculate and plot as an instantaneous function of time the current flowing into the inductor.

3-5 For the capacitor and waveform of Problem 3-1, determine the peak energy stored in the capacitor.

3-6 For the inductor and waveform of Problem 3-2, determine the peak energy stored in the inductor.

3-7 For the capacitor and waveform of Problem 3-3, determine the peak energy stored in the capacitor.

3-8 For the inductor and waveform of Problem 3-4, determine the peak energy stored in the inductor.

3-9 The voltage across a capacitor with capacitance C is given by

$$v(t) = V_p \sin \omega t$$

Determine an expression for the current flowing into the capacitor.

3-10 The current flow into an initially uncharged capacitor with capacitance C is given by

$$i(t) = I_p \cos \omega t$$

Determine an expression for the voltage across the capacitor.

3-11 The switch in the circuit of Figure P3-11 is closed at $t = 0$, and there is no initial energy storage. Construct an equivalent circuit at $t = 0^+$ and determine the initial values of the variables shown on the diagram.

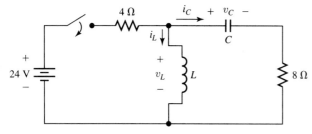

FIGURE P3-11

3-12 The switch in the circuit of Figure P3-12 is closed at $t = 0$, and there is no initial energy storage. Construct an equivalent circuit at $t = 0^+$ and determine the initial values of the variables shown on the diagram.

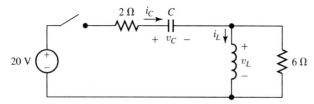

FIGURE P3-12

3-13 For the circuit of Problem 3-11 (Figure P3-11), determine the steady-state values of the variables identified in that problem.

3-14 For the circuit of Problem 3-12 (Figure P3-12), determine the steady-state values of the variables identified in that problem.

3-15 The switch in the circuit of Figure P3-15 is closed at $t = 0$, and the initial value of the capacitor voltage is shown. Construct an equivalent circuit at $t = 0^+$ and determine the initial values of the variables shown on the diagram.

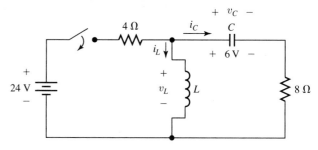

FIGURE P3-15

3-16 The switch in the circuit of Figure P3-16 is closed at $t = 0$, and the initial value of the capacitor voltage and inductor current are shown. Construct an equivalent circuit at $t = 0^+$ and determine the initial values of the variables shown on the diagram.

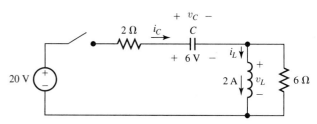

FIGURE P3-16

3-17 The switch in the circuit of Figure P3-17 is closed at $t = 0$ and there is no initial energy stored in the inductance. Determine the equations for the instantaneous voltage across the inductance and the current in the loop and sketch the results.

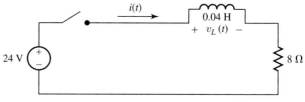

FIGURE P3-17

3-18 The switch in the circuit of Figure P3-18 is closed at $t = 0$ and there is an initial current in the inductance as

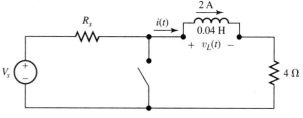

FIGURE P3-18

shown. Determine the equations for the instantaneous voltage across the inductance and the current in the loop and sketch the results.

3-19 The switch in the circuit of Figure P3-19 is closed at $t = 0$ and there is no initial energy stored in the capacitance. Determine the equations for the instantaneous voltage across the capacitance and the current in the loop and sketch the results.

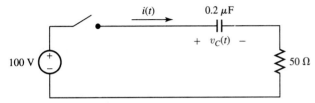

FIGURE P3–19

3-20 The switch in the circuit of Figure P3-20 is closed at $t = 0$ and there is an initial voltage stored on the capacitance as shown. Determine the equations for the

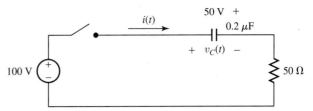

FIGURE P3–20

instantaneous voltage across the capacitance and the current in the loop and sketch the results.

3-21 Determine the equivalent capacitance of the parallel combination of three capacitors with the following values: 2 µF and 0.1 µF, and 0.005 µF.

3-22 Determine the equivalent capacitance of the parallel combination of five capacitors with equal values of 0.2 µF.

3-23 Determine the equivalent capacitance of the series combination of the three capacitors of Problem 3-21.

3-24 Determine the equivalent capacitance of the series combination of the five capacitors of Problem 3-22.

ALTERNATING CURRENT CIRCUITS

4 AC Circuits I

5 AC Circuits II

AC Circuits I

OVERVIEW AND OBJECTIVES

Chapters 1 and 2 were devoted to basic circuit laws and dc circuit analysis, with resistance as the only passive parameter. Chapter 3 introduced inductance and capacitance and the instantaneous behavior associated with their voltages and currents.

We are now ready to develop the techniques for analyzing **steady-state sinusoidal circuits**, which are also commonly called **ac circuits**. The description will include all three of the basic circuit parameters, and the excitation will be assumed to be a single frequency sinusoid. Moreover, we will assume that all transients have settled and that the circuit is operating in *steady-state ac*. This means that with a single-frequency sinusoidal excitation, all voltages and currents in the circuit will also be sinusoidal functions having the same frequency, but with different magnitudes and phase angles. It is known that all voltages and currents are sinusoidal functions with the same frequency, so it is necessary to determine only their peak values or magnitudes and their phase angles.

Objectives

After completing this chapter, the reader should be able to

1. Write the equation for both an ac **sinusoidal voltage** and a **sinusoidal current**.
2. For a sinusoidal function, determine the (a) **peak value**, (b) **angular frequency**, (c) **cyclic frequency**, (d) **period**, and (e) **phase angle**.
3. State **Euler's equation** and discuss its significance in establishing ac **phasors**.
4. Discuss the properties of the complex plane and show how a phasor is displayed.
5. State the **rectangular, exponential,** and **polar** forms of a phasor and convert among the different forms.
6. State the ac steady-state voltage-current relationships for the three passive parameters.
7. Determine the steady-state ac **impedances** for the three passive parameters.
8. Convert a circuit to its ac phasor form with sources replaced by phasors and passive parameters replaced by their real, imaginary, or complex impedances.
9. Solve for phasor voltages and currents in a steady-state ac circuit and determine the instantaneous forms from the phasors.
10. Show how phasors add graphically in the complex plane.

4-1 Sinusoidal Functions

Why are several chapters in this text devoted to ac sinusoidal circuits? It is because the sinusoidal function is arguably the most important type of single waveform in the application of electrical and electronic systems. Some digital enthusiasts might question this statement, but even digital data transmission often requires the use of sine waves.

First, commercial ac power systems utilize sinusoidal waveforms. Second, communication systems use sinusoidal waveforms as the so-called carrier functions. Third, numerous digital techniques use sinusoids to encode the data for transmission through electromagnetic propagation or over wire links. Thus, the study of electrical theory mandates a comprehensive treatment of sinusoidal ac circuit analysis methods.

Sine Function

The first topic to be considered is the nature of a sinusoidal function and its parameters. Readers likely are familiar with the function from a study of trigonometry, but the form used in circuit theory has its own special terminology.

Either a current or a voltage could be used in the development, but we will arbitrarily select a current function for this early discussion. Just remember that a voltage function will follow the same format, but with a change in the notation for the peak value.

Consider then a current $i(t)$ expressed as a sine function in the form

$$i(t) = I_p \sin \omega t \tag{4-1}$$

This function is shown in Figure 4-1. It passes through the origin with a positive slope as shown. The parameters in the equation are

I_p = peak value of current in amperes (A)

ω = angular frequency in radians/second (rad/s)

The **angular frequency** is related to the cyclic frequency f in hertz (Hz) by

$$\omega = 2\pi f \tag{4-2}$$

The **cyclic frequency** in Hz is related to the period T in seconds (s) by

$$T = \frac{1}{f} \tag{4-3}$$

To further enhance these definitions, the period T is the time that the sinusoid takes to complete *one cycle*. It is shown in Figure 4-1. The frequency is the number of cycles per second, with 1 hertz = 1 cycle/second. For each cycle, the angle undergoes a change of 2π radians, so the *angular frequency* is 2π times the number of cycles per second. In mechanics, the *angular frequency* is called the *angular velocity*. It is a very important variable in the design and operation of rotating machinery.

Cosine Function

A current expressed as a cosine function is given by

$$i(t) = I_p \cos \omega t \tag{4-4}$$

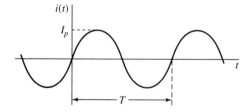

FIGURE 4–1
Sine current function.

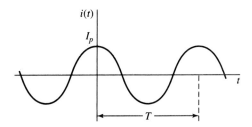

FIGURE 4–2
Cosine current function.

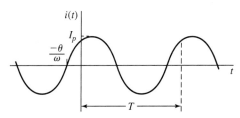

FIGURE 4–3
General sinusoidal current with an arbitrary phase angle.

This function is shown in Figure 4-2 and differs from the sine function only by the fact that it has a peak value at the origin. The parameters in the equation are the same as for the sine function. In angular units, the cosine *leads* the sine by 90° or $\pi/2$ radians. In terms of time, the cosine leads the sine by one-quarter of a cycle or $T/4$ seconds.

General Sinusoidal Function

Both the sine and the cosine are considered as **sinusoidal functions** or **sinusoids**. A more general form can be described by

$$i(t) = I_p \sin(\omega t + \theta) \tag{4-5}$$

where θ is some arbitrary phase angle. The form of this function for a positive angle of about 45° is shown in Figure 4-3. Note that the positive phase angle shifts the function to the left of the origin.

Units for Angles

It is very common throughout the engineering and scientific world to use degrees for angular measurement. That practice will be followed in this book for additive phase angles such as θ in the preceding equation. However, degrees are artificial units. The most natural units are **radians (rad)**. In fact, the product ωt has the dimensions of radians, and when it is necessary to actually add a constant phase angle to this value, it must be converted to radians. This will be illustrated in Example 4-1.

An angle of 2π radians = 360° and the following conversion formulas are appropriate:

$$\text{Angle(radians)} = \frac{\pi}{180} \times \text{Angle(degrees)} \tag{4-6}$$

$$\text{Angle(degrees)} = \frac{180}{\pi} \times \text{Angle(radians)} \tag{4-7}$$

An easy way to remember which one is used is that there will be more degrees than radians, so the ratio that results in that inequality will be the correct one.

Relative Phase Sequence

The diagram of Figure 4-4 provides a convenient means for determining the relative phase sequence of the sine and cosine functions. The positive direction of rotation is counterclockwise. Thus, the +cosine function *leads* the +sine function by 90°. An equivalent statement is that the +sine function *lags* the +cosine function by 90°.

FIGURE 4–4
Relative phase sequence of sine and cosine functions.

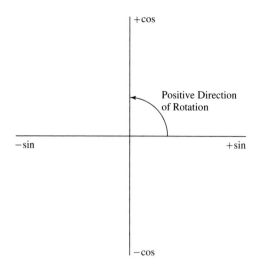

▎ EXAMPLE 4-1

A voltage $v(t)$ in volts is given by

$$v(t) = 50 \sin(2\pi \times 100t + 45°) \qquad (4\text{-}8)$$

Determine the (a) peak value, (b) angular frequency, (c) cyclic frequency, (d) period, and (e) value of the voltage when $t = 2$ ms.

SOLUTION

(a) The peak value is the multiplier of the sinusoidal function and is

$$V_p = 50 \text{ V} \qquad (4\text{-}9)$$

(b) The angular frequency is the multiplier of t in the sinusoidal function argument. It is

$$\omega = 2\pi \times 100 = 628.3 \text{ rad/s} \qquad (4\text{-}10)$$

(c) The cyclic frequency is obtained by dividing the angular frequency by 2π.

$$f = \frac{\omega}{2\pi} = \frac{2\pi \times 100}{2\pi} = 100 \text{ Hz} \qquad (4\text{-}11)$$

(d) The period is the reciprocal of the frequency.

$$T = \frac{1}{100} = 0.01 \text{ s} = 10 \text{ ms} \qquad (4\text{-}12)$$

(e) As stated earlier, the product ωt has the dimensions of radians and θ is expressed in degrees. To evaluate the function at any value of t, we must first convert the phase angle to radians and rewrite the voltage as

$$v(t) = 50 \sin(2\pi \times 100t + \pi/4) \qquad (4\text{-}13)$$

To perform the analysis, the calculator should be set to the radian mode. We can then perform the evaluation as follows:

$$v(2 \text{ ms}) = 50 \sin\left(2\pi \times 100 \times 2 \times 10^{-3} + \frac{\pi}{4}\right) = 50 \sin(0.65\pi) = 44.55 \text{ V} \qquad (4\text{-}14)$$

4-2 Phasors

Very early in the history of electrical engineering, a simplified method for determining the steady-state response in ac circuits without completing the complete solution of the differential equation was developed. This technique uses complex numbers to represent the voltages and currents in circuits. In making this representation, the steady-state behavior can be completely determined using the algebra of complex numbers.

Euler's Formula

The key to ac circuit analysis is to represent sinusoidal functions as so-called **phasors**, which are complex numbers. Complex numbers are based on using $\sqrt{-1}$ as a multiplier for the *imaginary part*. In most mathematics texts, this quantity is denoted as *i*. However, because *i* is universally used in circuit analysis for current, electrical engineers and technologists use $j = \sqrt{-1}$.

A basic relationship used in establishing complex numbers is **Euler's formula**, which states

$$e^{j\theta} = \cos\theta + j\sin\theta \qquad (4\text{-}15)$$

The significance of this relationship will become clearer as we progress through the development that follows.

Phasor Definitions

Consider the rectangular coordinate system shown in Figure 4-5. The *x*-axis is called the **real axis** and the *y*-axis is called the **imaginary axis**. The entire coordinate system defined in this manner is called the **complex plane**.

Assume that a complex number is represented as a vector drawn from the origin to some arbitrary point with coordinates (A_x, A_y) as shown. We will use the boldface notation **A** to represent this complex number which, in ac circuit theory, is called a *phasor*. The length of the phasor is A and the angle with respect to the real axis is θ. The projection on the *x*-axis or **real part** will be denoted as A_x and the projection on the *y*-axis or **imaginary part** will be denoted as A_y. We have

$$A_x = A\cos\theta \qquad (4\text{-}16)$$
$$A_y = A\sin\theta \qquad (4\text{-}17)$$

Next, we multiply both sides of Equation 4-15 by A and we obtain

$$Ae^{j\theta} = A\cos\theta + jA\sin\theta \qquad (4\text{-}18)$$

The result of this operation is the form of the complex number or phasor expressed in two ways. Starting on the right-hand side of Equation 4-18, the first term is the projection on the *x*-axis, the real part of the complex number. The multiplier of *j* in the second term on the right is the projection on the *y*-axis, the imaginary part of the complex number. This complex sum represents the **rectangular form** of the complex number.

The left-hand side of the equation is the **exponential form** of the complex number. Although the form stated is the mathematically correct one, there is an engineering form

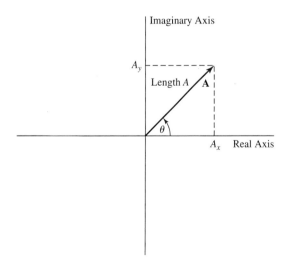

FIGURE 4–5
Complex plane with phasor representation.

widely used that will be provided in the statement that follows. The complex phasor will be denoted as **A**. It can be stated in three separate forms.

$$\mathbf{A} = Ae^{j\theta} = A_x + jA_y \stackrel{\Delta}{=} A\angle\theta \qquad (4\text{-}19)$$

The last form will be referred to as the **polar form** of the complex number, but remember that this really refers to the complex exponential definition, so that the arithmetic operations that follow will be justified. The real part of a complex number **A** is often denoted as Re[**A**] and the imaginary part is denoted as Im[**A**]. Because usage of the polar form is widespread in engineering applications, it will be used extensively in this text.

Equations 4-16 and 4-17 provide the necessary relationships for polar to rectangular conversion. If the rectangular form is given, the polar form can be determined by the right-triangle relationships

$$A = \sqrt{A_x^2 + A_y^2} \qquad (4\text{-}20)$$

$$\theta = \tan^{-1}\frac{A_y}{A_x} \qquad (4\text{-}21)$$

The angle may be in any one of the four quadrants, so it is necessary to consider the signs of both A_x and A_y in determining the inverse tangent.

Most modern scientific calculators have automatic rectangular-to-polar and polar-to-rectangular functions provided. To use these functions, the signs of the separate real and imaginary parts are entered with the values, and the angle is determined automatically. The reader is encouraged to use these operations to support the work of this chapter. However, it is also recommended that you try to check a few of the operations manually so that a better understanding of the angle location can be achieved.

It is assumed that the reader has likely encountered either the process of arithmetic operations with complex numbers or the somewhat equivalent process of combining vectors having magnitudes and angles. For that reason, the arithmetic operations are treated in Appendix B and the reader may refer to that appendix for further information on these operations.

Representation of Current and Voltage as Phasors

Let us now reconsider the sinusoidal function considered earlier and repeated here for convenience.

$$i(t) = I_p \sin(\omega t + \theta) \qquad (4\text{-}22)$$

Using the definitions of complex phasor form, we can express this function as

$$i(t) = \text{Im}\big[I_p e^{(j\omega t + \theta)}\big] = \text{Im}\big[I_p e^{j\theta} e^{j\omega t}\big] \qquad (4\text{-}23)$$

One way to visualize this process is shown in Figure 4-6. Consider the phasor on the left that assumes a value θ at $t = 0$. Consider that it rotates in the counterclockwise direction with angular velocity ω. The imaginary part is the projection on the vertical axis as shown on the right and defined by Equation 4-22. The *counterclockwise* direction of rotation is defined as the *positive* direction of rotation.

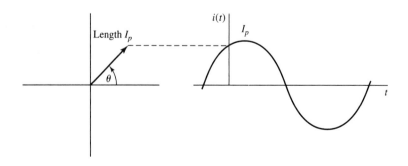

FIGURE 4–6

Generation of sinusoidal function by a rotating phasor.

Within the brackets of the last part of Equation 4-23, the first two factors represent the complex exponential form of a phasor. We can express this product as

$$\mathbf{I} = I_p e^{j\theta} \triangleq I_p \angle \theta \qquad (4\text{-}24)$$

The quantity $\mathbf{I}$ is the phasor representation of the time-varying current $i(t)$. It is easily determined by noting the peak value and the angle of the current. Some references refer to it as the **phasor transform** of the time-varying function.

Therefore, the instantaneous sinusoidal function can be expressed as

$$i(t) = \text{Im}[\mathbf{I}e^{j\omega t}] \qquad (4\text{-}25)$$

In subsequent developments, we will drop the imaginary part designation with the understanding that we are using the complex form for mathematical convenience.

Let us now try to explain what is going on here. The sinusoidal function has been replaced by a complex number multiplied by an exponential time-varying function that can represent either a sine or a cosine. The exponential function lends itself to differentiation and integration in a simpler fashion than the sine and cosine, as will be seen in the next section. Moreover, if all functions in the circuit are sinusoids with the same frequency, the complex exponential factor can be suppressed and all of the analysis can be performed with fixed phasors. These operations will be illustrated in the examples that follow.

It should be noted that either the real part (cosine) or the imaginary part (sine) could be used and that many books use the former. However, the choice made here was to use the sine function, because the authors prefer it as the reference for basic circuit analysis.

EXAMPLE 4-2   A sinusoidal voltage in volts is given by

$$v_1(t) = 8\sin(1000t + 60°) \qquad (4\text{-}26)$$

Determine the phasor form of the voltage.

SOLUTION   By inspection, the phasor is

$$\mathbf{V_1} = 8 \text{ V} \angle 60° \qquad (4\text{-}27)$$

Although the reader might not appreciate all the mathematical gyrations involved in establishing the process, the conversion of the time function to a phasor is a very simple operation. Based on the sine reference, the phasor magnitude is the peak value and the angle is the angle with respect to the real axis.

EXAMPLE 4-3   A different sinusoidal voltage in volts is given by

$$v_2(t) = 6\cos(1000t - 120°) \qquad (4\text{-}28)$$

Convert the voltage to a phasor.

SOLUTION   Because the phase angle reference in this text is based on the sine function, by using the relative phase sequence of Figure 4-4, we can rewrite this function as

$$v_2(t) = 6\sin(1000t - 30°) \qquad (4\text{-}29)$$

The phasor is

$$\mathbf{V_2} = 6 \text{ V} \angle -30° \qquad (4\text{-}30)$$

EXAMPLE 4-4   A current phasor is given by

$$\mathbf{I} = 2 \text{ A} \angle 150° \qquad (4\text{-}31)$$

Determine the instantaneous current if the frequency is 60 Hz.

SOLUTION The magnitude of the phasor is 2 A and the angle is 150°. The cyclic frequency is 60 Hz and the corresponding angular frequency is

$$\omega = 2\pi f = 2\pi \times 60 = 376.99 \text{ rad/s} \quad (4\text{-}32)$$

This radian frequency will appear frequently later in the book when power systems are studied, so we will express it simply as 377 rad/s.

The instantaneous form in amperes may then be expressed as

$$i(t) = 2\sin(377t + 150°) \quad (4\text{-}33)$$

We can also express this current in terms of the cosine function by noting from Figure 4-4 that an angle of $+150°$ with respect to the positive sine axis is an angle of $150° - 90° = 60°$ with respect to the positive cosine axis. Thus, an alternate form is

$$i(t) = 2\cos(377t + 60°) \quad (4\text{-}34)$$

■ EXAMPLE 4-5

Use phasors to form the sum $v_0(t)$ of the sinusoidal functions of Examples 4-2 and 4-3. Draw the corresponding phasor diagram.

SOLUTION The desired sum is

$$v_0(t) = v_1(t) + v_2(t) \quad (4\text{-}35)$$

Even though various trigonometric identities could be manipulated to perform the combination, the process is quite straightforward with phasors. Understand that this process requires that both sinusoids have the *same frequency* and the resulting sinusoid will also have that same frequency. The phasor sum is then

$$\mathbf{V_0} = \mathbf{V_1} + \mathbf{V_2} = 8\angle 60° + 6\angle -30° \quad (4\text{-}36)$$

To perform this addition each of the phasors must be converted to rectangular form. We have

$$\mathbf{V_0} = (4 + j6.9282) + (5.1962 - j3) = 9.1962 + j3.9282 = 10 \text{ V}\angle 23.13° \quad (4\text{-}37)$$

A graphical interpretation of this process is illustrated in Figure 4-7.

The resulting sinusoidal voltage is obtained by expressing the peak value of the voltage as the phasor magnitude and the angle with respect to the sine reference as the angle of the phasor. The result in volts is

$$v_0(t) = 10\sin(1000t + 23.13°) \quad (4\text{-}38)$$

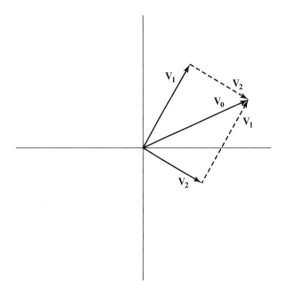

FIGURE 4–7
Phasor diagram of Example 4-5.

4-3 AC Voltage-Current Relationships

Using the concepts of phasors, we will now investigate the relationships between the voltage and current for each of the three passive circuit parameters. To enhance the presentation, both the instantaneous time form and the phasor form will be used in each case. Once again, we will begin with a current and use it as the reference. For convenience, we will use a basic sine function. The instantaneous form of the current is

$$i(t) = I_p \sin \omega t = \text{Im}[\mathbf{I}e^{j\omega t}] \tag{4-39}$$

where

$$\mathbf{I} = I_p \angle 0° \tag{4-40}$$

To simplify the developments that follow, the imaginary part designation will be dropped, but it should be understood to justify fully the mathematical steps. Thus the instantaneous current representation with the time variation retained will be assumed as

$$i(t) = \mathbf{I}e^{j\omega t} \tag{4-41}$$

Resistance

For a resistance, Ohm's law applied to the preceding form is simply

$$v(t) = Ri(t) = R\mathbf{I}e^{j\omega t} = \mathbf{V}e^{jwt} \tag{4-42}$$

Suppressing or canceling the time variation factor, the fixed phasor representation is

$$\mathbf{V} = R\mathbf{I} \tag{4-43}$$

Because R is a positive real constant, the angle of the voltage is the same as the angle of the current and the peak value of the voltage is R times the peak value of the current as dictated by Ohm's law. The voltage and current are said to be *in phase*.

The phasor diagram showing the voltage and current phasors along with the corresponding instantaneous forms are shown in Figure 4-8. The phasor diagram provides a direct interpretation of the phase relationships without displaying the time behavior. Of course, for a resistance, the situation is quite simple, but as will be seen shortly, the situation is more complex for either an inductance or a capacitance.

Differentiation and Integration of the Complex Exponential Function

Before considering the relationships for an inductance and a capacitance, let us consider the process of applying differentiation and integration to a complex exponential function. Assume the exponential current function of Equation 4-41 and determine its derivative with respect to time.

$$\frac{di}{dt} = j\omega \mathbf{I} e^{j\omega t} \tag{4-44}$$

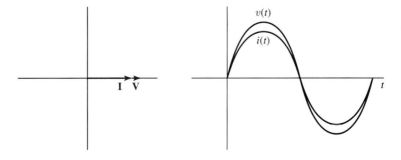

FIGURE 4-8
Phasor diagram for resistance and the instantaneous sinusoidal forms.

The definite integral based on the lower limit of the integral assuming a value of 0 is

$$\int_{-\infty}^{t} i \, dt = \frac{\mathbf{I}e^{j\omega t}}{j\omega} \tag{4-45}$$

The assumption of the lower limit value being zero is reasonable in the application of the integral to ac circuit forms.

Now let's see what the preceding equations tell us. Equation 4-44 says that to differentiate a complex exponential time function, we multiply the phasor form by $j\omega$. Conversely, Equation 4-45 says that to integrate a complex exponential time function, we divide the phasor by $j\omega$. Thus, the processes of differentiation and integration have been reduced to multiplication and division of the phasor forms.

We will next consider inductance and capacitance and will show how the voltage-current relationships for these parameters change from derivative and integral forms to multiplication and division using the phasor definitions.

Inductance

The instantaneous voltage across an inductance in terms of the current is

$$v = L\frac{di}{dt} = L\frac{d}{dt}[\mathbf{I}e^{j\omega t}] = j\omega L \mathbf{I} e^{j\omega t} = \mathbf{V}e^{jwt} \tag{4-46}$$

where $\mathbf{V}$ is the phasor representing the voltage.

After suppressing the time variation based on the last two terms, we have

$$\mathbf{V} = j\omega L \mathbf{I} \tag{4-47}$$

The derivative relationship has been reduced to an algebraic relationship. Let's see what the factor j in front means. We know that j has a magnitude of 1 and an angle of 90°, that is,

$$j = 1 \angle 90° \tag{4-48}$$

When phasors are multiplied, the angles add. Thus, the angle of the voltage will be the angle of the current plus 90°.

The effect of the factor ωL will be studied in more detail later, but for now remember that the angle of the voltage will be 90° more positive (more counterclockwise) than the angle of the current. This property is illustrated both in phasor form and in instantaneous form in Figure 4-9. Thus, the steady-state ac voltage across an inductance *leads* the current by 90°. An equivalent statement is that the steady-state ac current in an inductance *lags* the voltage by 90°.

Capacitance

The instantaneous voltage across a capacitance in terms of the current is

$$v = \frac{1}{C}\int_{-\infty}^{t} i(t) = \frac{1}{C}\int_{-\infty}^{t}\mathbf{I}e^{j\omega t} = \frac{1}{j\omega C}\mathbf{I}e^{j\omega t} = \mathbf{V}e^{jwt} \tag{4-49}$$

where $\mathbf{V}$ is the phasor representing the voltage.

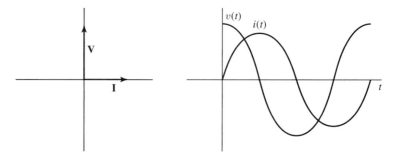

FIGURE 4-9
Phasor diagram for inductance and the instantaneous sinusoidal forms.

FIGURE 4–10
Phasor diagram for capacitance and the instantaneous sinusoidal forms.

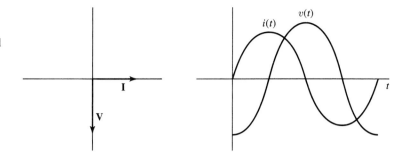

As explained earlier, it has been assumed that the lower limit value of the integration is zero. After suppressing the time variation based on the last two terms, we have

$$\mathbf{V} = \frac{1}{j\omega C}\mathbf{I} \qquad (4\text{-}50)$$

In this case, the integral relationship has been reduced to an algebraic relationship. The factor $1/j$ can be expressed as

$$\frac{1}{j} = \frac{1\angle 0°}{1\angle 90°} = 1\angle -90° = -j \qquad (4\text{-}51)$$

The negative value of the angle means that the angle of the voltage will be the angle of the current minus 90°.

For the capacitance, remember that the angle of the voltage will be 90° more negative (more clockwise) than the angle of the current. This property is illustrated both in phasor form and in instantaneous form in Figure 4-10. Thus, the steady-state ac voltage across a capacitance *lags* the current by 90°. An equivalent statement is that the steady-state ac current in an capacitance *leads* the voltage by 90°.

4-4 AC Impedance

We have seen in the preceding section how the voltage-current relationships for the three passive parameters become algebraic forms in the ac steady-state phasor representations. We are now ready to determine complete solutions for ac circuits using models developed from these relationships. To that end, we need to consider the concept of **impedance**.

Impedance

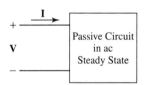

FIGURE 4–11
Block used to define impedance.

A simple way to describe impedance is that it is a parameter for steady-state ac that serves the same purpose as resistance for pure dc. Consider the block diagram shown in Figure 4-11 with terminal voltage and current represented in phasor forms as **V** and **I**, respectively. Assume that the circuit contains only the passive components of R, L, and C and that steady-state conditions exist. The impedance Z is defined as

$$\mathbf{Z} = \frac{\mathbf{V}}{\mathbf{I}} \qquad (4\text{-}52)$$

Stated in words, the impedance is the ratio of phasor voltage to phasor current.

Resistance and Reactance

Because the impedance Z is the ratio of two complex numbers, it can also be a complex number. In general, Z can be expressed in the form

$$\mathbf{Z} = R + jX \qquad (4\text{-}53)$$

The real part R is the **resistance**. The imaginary part X is called the **reactance**. It may be positive or negative. The reactance X is a real number, but the presence of the multiplier j makes this portion of the impedance imaginary. One popular misconception is to call jX

the reactance, but the standard is that the reactance is the real number X without the j-factor. Both R and X are measured in ohms.

Admittance

The impedance is denoted as Y. It is defined as

$$\mathbf{Y} = \frac{\mathbf{I}}{\mathbf{V}} = \frac{1}{\mathbf{Z}} \tag{4-54}$$

The use of admittance can often simplify ac computations, but we will defer consideration until Chapter 5 because its use is optional. In all subsequent work in this chapter, impedance will be used exclusively for ac circuit analysis.

We will now develop the impedance forms for each of the passive circuit parameters.

Resistance

The phasor voltage-current relationship for a resistor was developed in the previous section. It is

$$\mathbf{V} = R\mathbf{I} \tag{4-55}$$

The impedance Z_R is

$$\mathbf{Z_R} = \frac{\mathbf{V}}{\mathbf{I}} = R \tag{4-56}$$

This result is the simplest of all the passive parameters. It tells us that, as far as phasor forms are concerned, the ac impedance of a pure resistance is simply the value R. The equivalent circuit form is shown in Figure 4-12.

Inductive Impedance

The phasor voltage-current relationship for an inductor as developed in the previous section is

$$\mathbf{V} = j\omega L\mathbf{I} \tag{4-57}$$

The impedance $\mathbf{Z_L}$ is

$$\mathbf{Z_L} = \frac{\mathbf{V}}{\mathbf{I}} = j\omega L \tag{4-58}$$

The impedance is purely imaginary and has the j as a factor. The equivalent circuit model is shown in Figure 4-13. (Don't be misled by the term "purely imaginary." This term relates to the mathematical definition based on complex number theory. The impedance is definitely a *real quantity* that can be measured and used in practical applications.)

Capacitive Impedance

The phasor voltage-current relationship for a capacitor was shown to be

$$\mathbf{V} = \frac{1}{j\omega C}\mathbf{I} \tag{4-59}$$

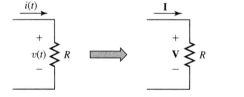

FIGURE 4–12
Resistance and its ac impedance form.

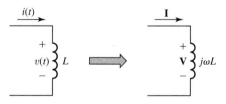

FIGURE 4–13
Inductance and its ac impedance form.

FIGURE 4–14
Capacitance and its ac impedance form.

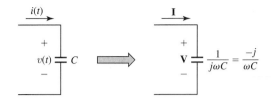

The impedance $\mathbf{Z_C}$ is

$$\mathbf{Z_C} = \frac{\mathbf{V}}{\mathbf{I}} = \frac{1}{j\omega C} = \frac{-j}{\omega C} \tag{4-60}$$

As in the case of the inductor, the capacitor impedance is purely imaginary and has the opposite sign. The equivalent circuit model is shown in Figure 4-14.

Steady-State AC Model

Assume that a circuit is excited by one or more sinusoidal sources with the *same frequency*. The following steps can be performed to permit a steady-state ac analysis:

1. Convert the sinusoidal sources (all at the same frequency) to phasors having magnitudes and angles.
2. Convert all elements in the circuit to their steady-state impedance forms as follows:
   (a) Resistance values remain the same.
   (b) Inductances are replaced by purely imaginary impedances of the form $j\omega L$.
   (c) Capacitances are replaced by purely imaginary impedances of the form $\frac{1}{j\omega C}$ or $\frac{-j}{\omega C}$.
3. Various circuit analysis methods may now be applied to the circuit to determine voltages or currents in phasor forms. The operations will all be algebraic in nature using complex numbers. For each voltage or current calculated, determine the magnitude and phase angle.
4. If it is necessary to determine the instantaneous forms, the various phasors may be converted to those forms. Depending on the application and the results desired, in many cases, all the information desired might be inferred directly from the phasor forms.

▮▮ EXAMPLE 4-6

The *RL* circuit shown in Figure 4-15(a) is assumed to be operating in the ac steady state. (a) Convert the circuit to the phasor form. (b) Determine the phasor values for the current, resistor voltage, and inductor voltage. (c) Determine the instantaneous forms of the three variables of (b).

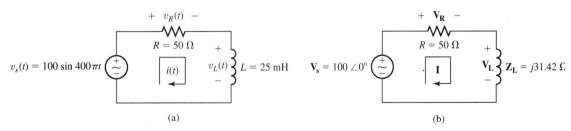

FIGURE 4–15
RL circuit of Example 4-6 and its phasor form.

SOLUTION

(a) Refer to both parts (a) and (b) of Figure 4-15 as the steady-state model is developed. Because the sinusoidal ac source given is a sine function, the angle of the source voltage will be 0° and its magnitude will be 100 V. Thus the phasor source voltage $\mathbf{V_s}$ is

$$\mathbf{V_s} = 100 \text{ V} \angle 0° \qquad (4\text{-}61)$$

The radian frequency is noted from the argument of the sine function as 400π rad/s, which corresponds to a cyclic frequency of $400\pi/2\pi = 200$ Hz. The inductive impedance is given by

$$\mathbf{Z_L} = j\omega L = j400\pi \times 25 \times 10^{-3} = j31.42 \text{ }\Omega \qquad (4\text{-}62)$$

The resistance retains the real value of 50 Ω. Finally, all the lower-case instantaneous time variables are replaced by their upper-case boldface phasor forms in Figure 4-15(b).

(b) A basic single-loop analysis could be used, but it is probably easier to determine first the impedance **Z**.

$$\mathbf{Z} = 50 + j31.42 = 59.05 \text{ }\Omega \angle 32.15° \qquad (4\text{-}63)$$

The current is then determined by applying a form of Ohm's law to the phasor model.

$$\mathbf{I} = \frac{\mathbf{V_s}}{\mathbf{Z}} = \frac{100}{59.05 \angle 32.15°} = 1.694° \text{ A} \angle -32.15° \qquad (4\text{-}64)$$

With the loop current now known, the voltages across the resistor and the inductor may be determined. The resistor voltage is

$$\mathbf{V_R} = \mathbf{Z_R}\mathbf{I} = (50) \times (1.694 \angle -32.15°) = 84.70 \text{ V} \angle -32.15° \qquad (4\text{-}65)$$

The inductor voltage is

$$\mathbf{V_L} = \mathbf{Z_L}\mathbf{I} = (j31.42) \times (1.694 \angle -32.15°)$$
$$= (31.42 \angle 90°)(1.694 \angle -32.15°) = 53.23 \text{ V} \angle 57.85° \qquad (4\text{-}66)$$

Several quick observations are noteworthy, the first two of which have been well predicted. First, the voltage across the resistor has the same phase angle as the current. Second, the voltage across the inductor has an angle that leads that of the current (and the resistive voltage) by 90°. Third, note carefully that the sum of the magnitudes of the two passive drops does *not* equal the magnitude of the source voltage. The voltages add in a vector fashion and a phasor diagram could illustrate this fact if desired.

(c) With the current expressed in amperes and the voltages expressed in volts, the instantaneous forms of the three variables determined in (b) are expressed as

$$i = 1.694 \sin(400\pi t - 32.15°) \qquad (4\text{-}67)$$
$$v_R = 84.70 \sin(400\pi t - 32.15°) \qquad (4\text{-}68)$$
$$v_L = 53.23 \sin(400\pi t + 57.85°) \qquad (4\text{-}69)$$

III EXAMPLE 4-7   The *RC* circuit shown in Figure 4-16(a) is assumed to be operating in the ac steady state. (a) Convert the circuit to the phasor form. (b) Determine the phasor values for the current, resistor voltage, and capacitor voltage. (c) Determine the instantaneous forms of the three variables of (b).

SOLUTION

(a) Refer to both parts of Figure 4-16. The source phasor in this case is

$$\mathbf{V_s} = 20 \text{ V} \angle 0° \qquad (4\text{-}70)$$

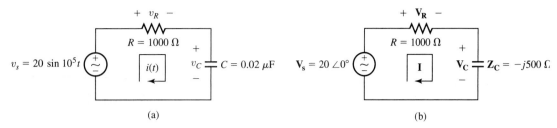

FIGURE 4–16
RC circuit of Example 4-7 and its phasor form.

The radian frequency is noted from the argument of the sine function as 10^5 rad/s, which corresponds to a cyclic frequency of $10^5/2\pi = 15.92$ kHz. The capacitive impedance is determined as

$$\mathbf{Z_C} = \frac{-j}{\omega C} = \frac{-j}{10^5 \times 0.02 \times 10^{-6}} = -j500 \; \Omega \tag{4-71}$$

The resistance retains the real value of 1000 Ω. The lower-case instantaneous time variables are replaced by their upper-case boldface phasor forms.

(b) Just to be a little different in this example as compared to the previous example, we will walk around the loop and apply KVL.

$$-20\angle 0° + 1000\mathbf{I} + (-j500)\mathbf{I} = 0 \tag{4-72}$$

Note the paradox for the last term on the left. We are adding a voltage drop, but the drop is the product of a negative imaginary impedance and a current. As long as you keep the signs straight, things should work out.

The current phasor can be determined from the preceding equation and is

$$\mathbf{I} = \frac{20\angle 0°}{1000 - j500} = \frac{20\angle 0°}{1118 \angle -26.57°} = 17.89 \times 10^{-3} \; \text{A} \angle 26.57° \tag{4-73}$$

The voltages across the resistor and the capacitor may now be determined. The resistor voltage is

$$\mathbf{V_R} = \mathbf{Z_R I} = (1000) \times (17.89 \times 10^{-3} \angle 26.57°) = 17.89 \; \text{V} \angle 26.57° \tag{4-74}$$

The capacitor voltage is

$$\begin{aligned}\mathbf{V_C} = \mathbf{Z_C I} &= (-j500) \times (17.89 \times 10^{-3} \angle 26.57°) \\ &= (500 \angle -90°)(17.89 \times 10^{-3} \angle 26.57°) = 8.945 \; \text{V} \angle -63.43° \end{aligned} \tag{4-75}$$

(c) The instantaneous forms of the three variables determined in (b) and expressed in their basic units are

$$i = 17.89 \times 10^{-3} \sin(10^5 t + 26.57°) \tag{4-76}$$
$$v_R = 17.89 \sin(10^5 t + 26.57°) \tag{4-77}$$
$$v_C = 8.945 \sin(10^5 t - 63.43°) \tag{4-78}$$

EXAMPLE 4-8 The phasor form of a series *RLC* circuit is shown in Figure 4-17. (a) Determine the phasor values of the current and voltages across the three elements. (b) Construct a phasor diagram to show that the source voltage is equal to the sum of the three voltage drops across the passive components.

SOLUTION This circuit model differs from those of the preceding two examples in that the circuit is already established as a phasor model. In many areas of electrical engineering, it is customary to work directly with the phasor forms and only occasionally revert to the

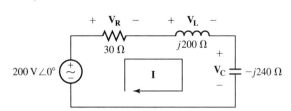

FIGURE 4–17
RLC circuit of Example 4-8 and its phasor form.

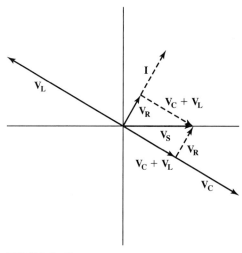

FIGURE 4–18
Phasor diagram for Example 4-8.

instantaneous time domain forms. We are not given the frequency, nor do we need to know it for the purpose of the task indicated. Of course, if it were desired to go back to the time domain form, we would need to know the frequency.

(a) We will use the impedance seen by the source as the first step in the analysis. It is

$$\mathbf{Z} = 30 + j200 - j240 = 30 - j40 \ \Omega = 50 \ \Omega \angle -53.13° \quad (4\text{-}79)$$

The current is

$$\mathbf{I} = \frac{200\angle 0°}{50\angle -53.13°} = 4 \ \text{A} \angle 53.13° \quad (4\text{-}80)$$

The three voltages are then determined as

$$\mathbf{V_R} = \mathbf{Z_R I} = (30) \times (4\angle 53.13°) = 120 \ \text{V} \angle 53.13° \quad (4\text{-}81)$$
$$\mathbf{V_L} = \mathbf{Z_L I} = (j200)(4\angle 53.13°) = 800 \ \text{V} \angle 143° \quad (4\text{-}82)$$
$$\mathbf{V_C} = \mathbf{Z_C I} = (-j240)(4\angle 53.13°) = 960 \ \text{V} \angle -36.87° \quad (4\text{-}83)$$

Note that both the inductor and capacitor voltage magnitudes are much larger than the source voltage magnitude. Indeed, in high-Q resonant circuits (to be studied in the next chapter), it is possible to have voltages across reactive components that are hundreds of times greater than the source voltage. Don't let the complex number definition of imaginary give you the wrong impression: These are *real* voltages.

(b) The phasor diagram for this circuit is shown in Figure 4-18. There is only one current, so it is chosen as a dashed line with a scale that is different than the voltages. Note that the inductor and capacitor voltages differ by 180°, so the sum $\mathbf{V_L} + \mathbf{V_C}$ will be in the direction of the larger of the two, which in this case is the latter. This net phasor plus the resistive voltage $\mathbf{V_R}$ must equal the source voltage $\mathbf{V_S}$ as shown.

PROBLEMS

4-1 A voltage $v(t)$ in volts is given by

$$v(t) = 170 \sin(2\pi \times 60t + 30°)$$

Determine the (a) peak value, (b) angular frequency, (c) cyclic frequency, (d) period, and (e) value of the voltage when $t = 2$ ms.

4-2 A current $i(t)$ in amperes in the *radio frequency* (*RF*) range is given by

$$i(t) = 0.02 \sin(10^7 t + 120°) \quad (4\text{-}84)$$

Determine the (a) peak value, (b) angular frequency, (c) cyclic frequency, (d) period, and (e) value of the voltage when $t = 0.5$ μs.

4-3 A sinusoidal voltage in volts is given by
$$v_1(t) = 20\sin(10^7 t + 45°)$$
Convert the voltage to a phasor.

4-4 A sinusoidal current in amperes is given by
$$i_1(t) = 2\cos(2\pi \times 1000t + 60°)$$
Convert the current to a phasor.

4-5 A voltage phasor is given by
$$\mathbf{V} = 2\,\text{V}\angle -30°$$
Determine the instantaneous voltage if the frequency is 1 kHz. Express as both a positive sine function and a positive cosine function.

4-6 A voltage phasor is given by
$$\mathbf{V} = 5\,\text{V}\angle \pi/3$$
Note that the angle in this case is expressed in radians. Determine the instantaneous voltage if the radian frequency is 2 Mrad/s. Express as both a positive sine function and a positive cosine function.

4-7 Use phasors to determine a single sinusoid equal to the sum
$$v_0(t) = v_1(t) + v_2(t)$$
where $v_1(t) = 12\sin(1000t + 30°)$ and $v_2(t) = 5\sin(1000t + 120°)$. Construct a phasor diagram showing each of the phasors.

4-8 Use phasors to determine a single sinusoid equal to the difference
$$v_0(t) = v_1(t) - v_2(t)$$
where $v_1(t) = 12\sin(1000t + 30°)$ and $v_2(t) = 5\sin(1000t + 120°)$. Construct a phasor diagram showing each of the phasors.

4-9 Use phasors to determine a single sinusoid equal to the sum
$$v_0(t) = v_1(t) + v_2(t)$$
where $v_1(t) = 12\sin(200t + 30°)$ and $v_2(t) = 8\cos(1000t - 30°)$. Construct a phasor diagram showing each of the phasors.

4-10 Use phasors to determine a single sinusoid equal to the sum
$$v_0(t) = v_1(t) + v_2(t) + v_3(t)$$
where $v_1 = 400\sin(377t)$, $v_2 = 400\sin(377t + 120°)$, and $v_3 = 400\sin(377t + 240°)$. Construct a phasor diagram showing each of the phasors.

4-11 The RL circuit shown in Figure P4-11 is assumed to be operating in the ac steady state. (a) Convert the circuit to the phasor form. (b) Determine the phasor values for the current, resistor voltage, and inductor voltage. (c) Determine the instantaneous forms of the three variables of (b).

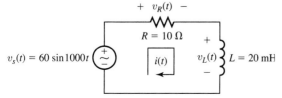

FIGURE P4–11

4-12 The RL circuit shown in Figure P4-12 is assumed to be operating in the ac steady state. (a) Convert the circuit to the phasor form. (b) Determine the phasor values for the three currents. (c) Determine the instantaneous forms of the three variables of (b).

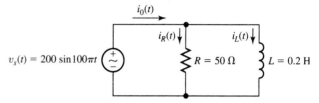

FIGURE P4–12

4-13 The RC circuit shown in Figure P4-13 is assumed to be operating in the ac steady state. (a) Convert the circuit to the phasor form. (b) Determine the phasor values for the current, resistor voltage, and capacitor voltage. (c) Determine the instantaneous forms of the three variables of (b).

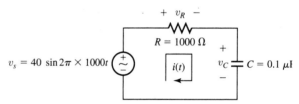

FIGURE P4–13

4-14 The RC circuit shown in Figure P4-14 is assumed to be operating in the ac steady state. (a) Convert the circuit to the phasor form. (b) Determine the phasor values for the three currents. (c) Determine the instantaneous forms of the three variables of (b).

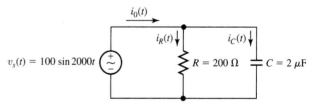

FIGURE P4–14

4-15 The series RLC circuit shown in Figure P4-15 is assumed to be operating in the ac steady state. (a) Convert the circuit to the phasor form. (b) Determine the phasor values

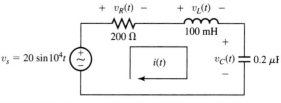

FIGURE P4–15

for the current, resistor voltage, inductor voltage, and capacitor voltage. (c) Sketch a phasor diagram to show that the source voltage is the sum of the three voltages across the passive elements.

4-16 The parallel *RLC* circuit shown in Figure P4-16 is assumed to be operating in the ac steady state. (a) Convert the circuit to the phasor form. (b) Determine the phasor values for the four currents. (c) Sketch a phasor diagram to show that the source current is the sum of the three currents through the passive elements.

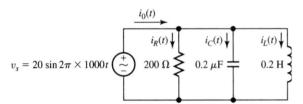

FIGURE P4–16

4-17 The phasor form of a series *RLC* circuit is shown in Figure P4-17. (a) Determine the phasor values of the current and voltages across the three elements. (b) Sketch a phasor diagram to show that the source voltage is the sum of the three voltages across the passive elements.

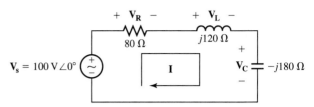

FIGURE P4–17

4-18 The phasor form of a parallel *RLC* circuit is shown in Figure P4-18. (a) Determine the phasor values of the currents indicated. (c) Sketch a phasor diagram to show that the source current is the sum of the three currents through the passive elements.

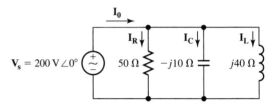

FIGURE P4–18

AC Circuits II   5

OVERVIEW AND OBJECTIVES

The basic concepts for analyzing ac circuits were developed in the previous chapter. In this chapter, these concepts will be extended to include other topics pertinent to alternating current, including some of the major applications. Specific topics include the power generated by ac, power factor correction, series and parallel equivalent circuits, and resonance.

Objectives

After completing this chapter, the reader should be able to

1. Define **dc** and **rms** values for a periodic waveform and discuss their significance.
2. Determine the dc and rms values for a **periodic waveform**.
3. State the dc and rms values for a sinusoidal waveform of either voltage or current.
4. Determine the **complex power** for ac voltage and current.
5. Construct a **power triangle** and discuss its components.
6. Apply **power factor correction** for an ac single-phase system.
7. Determine an ac parallel circuit from a given series circuit at a specific frequency.
8. Determine an ac series circuit from a given parallel circuit at a specific frequency.
9. Discuss and determine the various properties and parameters for a **series resonant circuit**.
10. Discuss and determine the various properties and parameters for a **parallel resonant circuit**.

5-1 DC and RMS Values

Before focusing on the sinusoidal steady-state ac case, we will generalize somewhat in this section and consider an arbitrary time-varying **periodic waveform**. Consider the voltage waveform $v(t)$ shown in Figure 5-1, which satisfies the condition that $v(t + T) = v(t)$. This is the requirement to be periodic, and the period is T. We will consider two important properties of an arbitrary periodic waveform: (a) dc value and (b) root-mean-square (rms) or effective value.

FIGURE 5–1
Example of a periodic voltage.

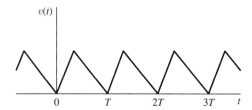

DC Value

The **dc value** V_{dc} is the *average value* of the waveform over a cycle. It is determined as follows:

$$V_{dc} = \frac{\text{area under curve in one cycle}}{\text{period}} = \frac{1}{T}\int_0^T v(t)\,dt \tag{5-1}$$

As a practical matter, the dc value for voltage is the value that would be read by an ideal dc voltmeter, and the dc value for current is the value that would be read by an ideal dc ammeter. This statement assumes that the fluctuations are fast enough that the meter provides the required smoothing for the reading to be stable. For an extremely slowly varying waveform, the meter would tend to follow the fluctuations.

RMS Value

The **root-mean-square** (rms) value V_{rms} is slightly more involved. It is determined as follows:

$$V_{rms} = \sqrt{\frac{\text{area under squared curve in one cycle}}{\text{period}}} = \sqrt{\frac{1}{T}\int_0^T v^2(t)\,dt} \tag{5-2}$$

The subscript *rms* provides an easy way to remember the formula. The rms value is the *root* (meaning square root) of the *mean* (referring to average) of the *square* of the function. The significance will be discussed in the next few paragraphs.

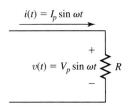

FIGURE 5–2
Circuit used to develop average ac power relationship for a resistor.

Power in a Resistance

When a voltage is varying with time, the resulting power generated is also varying with time. Assume that a periodic voltage $v(t)$ is applied across a resistance R as shown in Figure 5-2. Let $p(t)$ represent the instantaneous power and it is

$$p(t) = \frac{v^2(t)}{R} \tag{5-3}$$

Although the instantaneous power is an important quantity, more often the **average power** P is of major interest. The average power is determined by integrating the instantaneous power over a cycle and dividing by the period.

$$P = \frac{1}{T}\int_0^T p(t) = \frac{1}{T}\int_0^T \frac{v^2(t)}{R}\,dt = \frac{1}{R}\left[\frac{1}{T}\int_0^T v^2(t)\,dt\right] \tag{5-4}$$

Looking back at Equation 5-2, it can be deduced that Equation 5-4 can be expressed as

$$P = \frac{V_{rms}^2}{R} \tag{5-5}$$

If we had chosen to use an instantaneous periodic current as the basis for the development, the result could have been expressed as

$$P = R I_{rms}^2 \tag{5-6}$$

where I_{rms} is the rms current defined in the same manner as for the rms voltage. Finally, the result could also have been expressed as

$$P = V_{rms} I_{rms} \tag{5-7}$$

The significance of the rms value of a voltage or current can now be deduced. This value can be treated in the same way as a dc voltage or current in terms of determining the *average power* dissipated in a resistance.

Note that it is *not* the *dc value* of the waveform that is used to determine the *average power*, even though the dc value is the average value of the voltage or current. Rather the *rms* value of the voltage or current is used to determine the *average power*. These similar sounding terms can often lead to some confusion.

The rms value is also called the **effective value** in many applications. That term was briefly introduced in Chapter 1. However, we will now use the rms term because it provides a simpler way of remembering the formula.

DC Value of Sinusoid

We will now determine the dc value for a sine wave. Consider a voltage defined as

$$v(t) = V_p \sin \omega t \tag{5-8}$$

The dc value is

$$V_{dc} = \frac{1}{T} \int_0^T V_p \sin \omega t = \left. \frac{-V_p}{\omega T} \cos \omega t \right]_0^T = \frac{V_p}{\omega T} [1 - \cos \omega T] \tag{5-9}$$

The product $\omega T = 2\pi T/T = 2\pi$ and $\cos 2\pi = 1$. Hence,

$$V_{dc} = \frac{V_p}{2\pi} [1 - 1] = 0 \tag{5-10}$$

Although the basic sine function without an additive phase angle was used in the development for simplicity, the results apply to any sinusoid based on a full cycle. In fact, it will be stated that in general, *integration of a sinusoidal function over an* **integer number of cycles** *is zero*. This property will be used in subsequent developments to simplify the results.

RMS or Effective Value of Sinusoid

In working out the rms value, it is convenient from a notational point of view to work with the squared value until the last step. To that end, we have

$$V_{rms}^2 = \frac{1}{T} \int_0^T v^2(t)\, dt = \frac{1}{T} \int_0^T V_p^2 \sin^2 \omega t\, dt \tag{5-11}$$

A standard trigonometric identity is

$$\sin^2 \omega t = \tfrac{1}{2}(1 - \cos 2\omega t) \tag{5-12}$$

Substituting this result in Equation 5-11 and rearranging result in

$$V_{rms}^2 = \frac{V_p^2}{2T} \int_0^T dt - \frac{V_p^2}{2T} \int_0^T \cos 2\omega t\, dt \tag{5-13}$$

The second integral in Equation 5-13 represents the integral over two cycles of a sinusoid, and it is zero. The result of the complete integration is

$$V_{rms}^2 = \frac{V_p^2}{2T} \int_0^T dt - 0 = \left. \frac{V_p^2}{2T} t \right]_0^T = \frac{V_p^2 T}{2T} = \frac{V_p^2}{2} \tag{5-14}$$

The rms value is

$$V_{\text{rms}} = \sqrt{V_{\text{rms}}^2} = \frac{V_p}{\sqrt{2}} = 0.7071 V_p \qquad (5\text{-}15)$$

The same ratio holds for a current. Thus, a sinusoidal current of peak value I_p has an rms value given by

$$I_{\text{rms}} = \frac{I_p}{\sqrt{2}} = 0.7071 I_p \qquad (5\text{-}16)$$

The dc value of a sinusoid is zero, so an ideal dc meter would read zero provided that the frequency is high enough that the fluctuations are smoothed out. However, the rms value is about 0.7071 times the peak value. Thus, the *effectiveness* can be said to be about 0.7071 times a dc voltage or current whose value is the same as the peak value of the sinusoid. Because power is proportional to voltage or current squared, the actual average ac power level in a resistance would be $(0.7071)^2 = 0.5$ times that produced by the dc voltage or current referred to in this discussion.

The approximate 120-V value of the common ac household voltage supply in the United States is the rms value of the voltage. Thus, the peak value is about $\sqrt{2} \times 120 \approx 170$ V.

RMS Values as Phasor Magnitudes

Throughout Chapter 4, the peak values of sinusoids were used as the magnitudes of the voltage and current phasors. This is arguably the most basic process from a mathematical point of view. On the other hand, there is a widespread practice of using the rms value of a voltage or a current as the magnitude of a phasor representation. This practice is particularly common in the electrical power industry. It simplifies the computation of ac average power because the conversion from peak to rms values is not required in power computations. We believe that the reader should be familiar with both approaches, but it must be understood which value is being specified if a phasor is directly given in a problem and power is to be determined.

In this chapter, we will follow the practice of listing the term *rms* after the units of either V or A whenever the phasor is given in its rms value representation. However, in some later sections of the text, the term *ac* might follow the units of V or A, in which case it is assumed that these are rms values based on popular conventions. Moreover, whenever you encounter any reference outside of this book and a phasor voltage or current is given, be sure that you check to see whether it is a peak value or an rms value.

Just to illuminate the preceding discussion, suppose that a voltage is given as

$$v(t) = 100 \sin(\omega t + 30°) \qquad (5\text{-}17)$$

The value of 100 V is the peak value and it will always be the multiplier of the instantaneous sinusoidal form. If this voltage is converted to a phasor based on the peak value, as was the practice throughout Chapter 4, we would express it as

$$\mathbf{V} = 100 \text{ V} \angle 30° \qquad (5\text{-}18)$$

However, if we wish to work in rms units, we would express it as

$$\mathbf{V} = 70.71 \text{ V rms} \angle 30° \quad \text{or} \quad 70.71 \text{ V ac} \angle 30° \qquad (5\text{-}19)$$

Remember that the average power relationships developed in this section apply only for the average power dissipated in a pure resistance. As we will see in the next section, when there is reactance involved, the situation is more complex. Remember also that the ratio of rms to peak of $1/\sqrt{2} = 0.7071$ applies to a sinusoid, but each time-varying waveform will generally have a different ratio of rms value to peak value.

EXAMPLE 5-1

A sinusoidal voltage having a peak value of 30 V is applied across a 100-Ω resistor. (a) Determine the rms value of the voltage. (b) Determine the average power dissipated.

SOLUTION

(a) The rms value is

$$V_{\text{rms}} = \frac{30}{\sqrt{2}} = 21.21 \text{ V rms} \tag{5-20}$$

(b) The average power is

$$P = \frac{V_{\text{rms}}^2}{R} = \frac{(21.21)^2}{100} = 4.50 \text{ W} \tag{5-21}$$

It should also be noted that the power could be expressed directly in terms of the peak value squared by placing a 2 in the denominator. Thus, an alternate expression is

$$P = \frac{V_p^2}{2R} = \frac{(30)^2}{2 \times 100} = 4.50 \text{ W}$$

EXAMPLE 5-2

The ac voltage applied across a resistive heating element is 240 V rms. The element has a resistance of 40 Ω. (a) Determine the average power. (b) Determine the peak value of the voltage.

SOLUTION

(a) The rms voltage is given; therefore the average power is

$$P = \frac{V_{\text{rms}}^2}{R} = \frac{(240)^2}{40} = 1440 \text{ W} \tag{5-22}$$

(b) The peak voltage is

$$V_p = \sqrt{2} V_{\text{rms}} = \sqrt{2} \times 240 = 339.4 \text{ V} \tag{5-23}$$

5-2 Complex Power in AC Circuits

In the previous section, we considered the means for determining the average ac power dissipated in a resistance. When a circuit contains both resistance and reactance, the situation is more complex. Indeed, we can define the concept of **complex power** in much the same way that complex numbers have been used to define voltages and currents in steady-state ac.

Apparent, Real, and Reactive Power

In an ac circuit containing both resistance and reactance, resistors *dissipate power*. However, reactive elements *store energy* in one portion of a cycle and release it back to the circuit in another part of the cycle. It is useful to define three different forms of average power. They are (1) **apparent power**, (2) **real power**, and (3) **reactive power**. All three are dimensionally equivalent to watts, but it has become standard practice to refer to them with different unit forms. The units employed and definitions follow.

- *Apparent power*: The apparent power is the product of an rms voltage and an rms current measured at a set of terminals without regard to the phase difference between the voltage and the current. The basic unit of apparent power is the **volt-ampere (VA)**. In large power system applications, kilo-volt-ampere (**kVA**) and mega-volt-ampere (**MVA**) values are quite common.

- *Real power*: The real power is the actual power dissipated in the circuit and the unit is the watt (**W**). Often it is indicated simply as power without the adjective, but when the discussion is based on complex power, it is best to include the adjective *real* to minimize confusion.
- *Reactive power*: The reactive power is the power that is alternately stored and released in reactive components. Sometimes, it is referred to as "imaginary power." The unit of reactive power is the **volt-ampere reactive (VAR)**.

Power Development

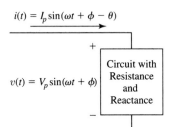

FIGURE 5–3
Circuit used to develop average ac power relationship for a circuit with resistance and reactance.

Consider the block diagram shown in Figure 5-3 representing some circuit containing both resistance and reactance. Assume that the instantaneous voltage is

$$v(t) = V_p \sin(\omega t + \phi) \tag{5-24}$$

Assume that the instantaneous current is given by

$$i(t) = I_p \sin(\omega t + \phi - \theta) \tag{5-25}$$

Note that there is a difference θ in phase between the voltage and current based on the circuit containing reactance. Note also that the current is assumed to *lag* the voltage. This choice is arbitrary, but it will assist in the interpretation of the results that follow. The instantaneous power $p(t)$ is given by

$$p(t) = v(t)i(t) = V_p I_p \sin(\omega t + \phi)\sin(\omega t + \phi - \theta) \tag{5-26}$$

By means of a useful trigonometric identity for the product of two sine functions, the instantaneous power may be expanded into the following two terms:

$$p(t) = \frac{V_p I_p}{2} \cos\theta - \frac{V_p I_p}{2} \cos(2\omega t + 2\phi - \theta) \tag{5-27}$$

As in the case of resistive ac power, the average power P is of primary interest. It is given by

$$P = \frac{1}{T}\int_0^T p(t)\,dt = \frac{1}{T}\int_0^T \left[\frac{V_p I_p}{2}\cos\theta - \frac{V_p I_p}{2}\cos(2\omega t + \theta)\right]dt \tag{5-28}$$

The first term within the expanded integral is the average value over the interval of a constant. The average of the second term is zero. Therefore, the average power is

$$P = \frac{V_p I_p}{2}\cos\theta \tag{5-29}$$

By substituting the relationships for the peak values of voltage and current in terms of the rms values, this result may be stated as

$$P = V_{\text{rms}} I_{\text{rms}} \cos\theta \tag{5-30}$$

This value is the *real power*. It represents power dissipated. It has its maximum value when $\theta = 0°$. This corresponds to the voltage and current having the same phase. This would either be a *purely resistive circuit* or one in which the inductive and capacitive effects cancel. Conversely, if $\theta = \pm 90°$, $\cos(\pm 90°) = 0$, and $P = 0$. This situation corresponds to a *purely reactive circuit*.

Power Factor

The factor $\cos\theta$ in Equation 5-30 is called the **power factor**. It is bounded by the range $0 \leq \cos\theta \leq 1$.

Reactive Power

Let Q represent the reactive power in VAR. It can be shown that this value is given by

$$Q = V_{rms} I_{rms} \sin \theta \qquad (5\text{-}31)$$

One standard is to define θ as positive if the current lags the voltage in accordance with the definition of voltage and current at the beginning of the section, and we will employ that standard. This would correspond to an inductive circuit, and $0 < \theta \leq 90°$. For this condition, $Q > 0$, and the power factor is said to be a **lagging power factor**. On the other hand, if θ is negative and in the range $-90° \leq \theta < 0$, $Q < 0$, and the power factor is said to be a **leading power factor**. The opposite convention also appears in the literature. As long as the reference direction is understood in applying the concept, it really doesn't make any difference which convention is used.

Power Triangle

A fairly simple geometric process can be used to analyze the complex power in an ac circuit. Refer to the triangle shown in Figure 5-4, which will be called a **power triangle**. The orientation shown corresponds to the convention just discussed, meaning that the power factor is lagging and the net reactance is inductive. For the opposite sign of reactive power, the triangle would be flipped downward.

The elements of the triangle and their definitions are as follows:

- The base of the triangle is the real power in watts and is denoted as P.
- The opposite side of the triangle is the reactive power in VAR and is denoted as Q.
- The hypotenuse of the triangle is the apparent power in VA and is denoted as S.
- The angle between the base and the hypotenuse is the power factor angle θ. It is also the angle between the voltage and current at the terminals of the circuit. For this particular reference, the current is assumed to lag the voltage, meaning that it is a lagging power factor.

It can be easily verified that Equations 5-30 and 5-31 can be validated from the triangle. The only variable not considered thus far is the **apparent power** S, which is

$$S = V_{rms} I_{rms} \qquad (5\text{-}32)$$

The logic behind this definition is this: If the voltage across a set of terminals and the current flowing into the circuit are known, it could be argued that the product of these two quantities *appears* to be the power. However, the real power is

$$P = S \cos \theta \qquad (5\text{-}33)$$

and the reactive power is

$$Q = S \sin \theta \qquad (5\text{-}34)$$

The preceding two relationships are alternate forms of Equations 5-30 and 5-31.

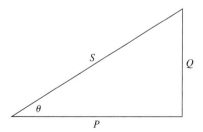

FIGURE 5–4
Power triangle.

EXAMPLE 5-3

The circuit of Figure 5-5(a) represents the ac phasor form of an *RL* circuit. Note that the phasor voltage is given in rms voltage units. Determine (a) the current phasor, (b) the apparent power, (c) the power factor, (d) the real power, and (e) the reactive power.

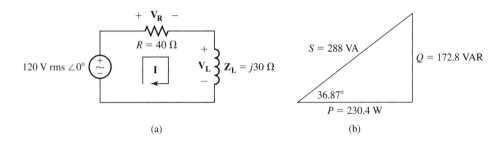

FIGURE 5-5
Circuit and power triangle of Example 5-3.

SOLUTION

(a) The current phasor is

$$\mathbf{I} = \frac{120\angle 0°}{40 + j30} = \frac{120\angle 0°}{50\angle 36.87°} = 2.4 \text{ A rms} \angle -36.87° \quad (5\text{-}35)$$

Note that because the source voltage is given in rms units, the current also will be in rms units.

(b) The apparent power is

$$S = V_{rms} I_{rms} = 120 \times 2.4 = 288 \text{ VA} \quad (5\text{-}36)$$

(c) Because the load is inductive, the power factor is lagging. It is

$$\cos\theta = \cos 36.87° = 0.8 \text{ lagging} \quad (5\text{-}37)$$

For computing the power factor, it makes no difference which sign is used for the angle because $\cos(-\theta) = \cos\theta$.

(d) The real power is

$$P = S\cos\theta = 288 \times \cos\theta = 288 \times 0.8 = 230.4 \text{ W} \quad (5\text{-}38)$$

An alternate method for determining the real power is to recognize that all the power is being dissipated in the resistance. Using this approach, we can say that

$$P = I^2 R = (2.4)^2 \times 40 = 230.4 \text{ W} \quad (5\text{-}39)$$

(e) The reactive power is

$$Q = S\sin\theta = 288 \sin 36.87° = 172.8 \text{ VAR} \quad (5\text{-}40)$$

A power triangle for the circuit is shown in Figure 5-5(b).

EXAMPLE 5-4

A wattmeter measures the real power being absorbed by a circuit. Consider the circuit shown in Figure 5-6. Assume that the effective values of voltage and current are measured as 120 V rms and 4 A rms, respectively. Assume that the wattmeter reads 200 W. Determine (a) the apparent power, (b) the power factor, and (c) the magnitude of the reactive power.

SOLUTION

(a) The apparent power is the product of the rms voltage and the rms current.

$$S = V_{rms} I_{rms} = 120 \times 4 = 480 \text{ VA} \quad (5\text{-}41)$$

FIGURE 5–6
Wattmeter connected to load.

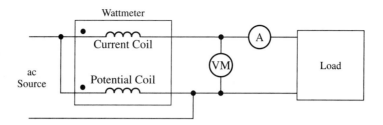

(b) The power factor is

$$\cos\theta = \frac{P}{S} = \frac{200}{480} = 0.4167 \tag{5-42}$$

(c) We do not know whether the load is inductive or capacitive, so we can only deduce the magnitude of the angle and not its sign. Thus,

$$|\theta| = \cos^{-1} 0.4167 = 65.37° \tag{5-43}$$

The magnitude of the reactive power is then

$$|Q| = S \sin|\theta| = 480 \sin 65.37° = 436.3 \text{ VAR} \tag{5-44}$$

5-3 Power Factor Correction

Imagine that an industrial plant has no resistive load connected, but that a large ideal inductance is placed across the power line. As we know, current is drawn by the inductance. This means that losses will be incurred by power dissipated in the transmission line connecting to the plant. Yet as far as the plant is concerned, no real power is being dissipated there and a watt-hour meter connected at the plant input would show no real energy dissipated.

The implication of the preceding scenario is that public utility companies do not like to feed loads that operate with zero or very small power factors. Therefore, the charge per kilowatt-hour is greater for lower power factors. Because many industrial loads are induction motors, which operate with lagging power factors, it is prudent for industrial customers to achieve high power factors.

Fortunately, it is possible to correct low power factors by adding reactive components of the opposite type to the load. This is a situation where "more is less," so to speak. Although in theory the situation could be in either direction, we will assume the common situation where the load is inductive; that is, a lagging power factor. Moreover, we will consider only the **single-phase** situation here, which is the case of a single ac line. Systems employing more than one phase (e.g., three-phase systems), will be considered in Chapter 13.

Power Triangle Revisited

We begin with the power triangle again as shown in Figure 5-7(a) based on an inductive load. Assume that the rms voltage V_{rms} across the load is given and that, for all practical purposes, it will not change with load. The strategy then is to connect a capacitance across the load that will generate reactive power of the opposite sign so that the net angle will be reduced as shown in Figure 5-7(b). Ideally, we should reduce the angle to 0°, but that might not be practical because the load could change with time. To provide a more general solution, we will assume a lagging phase angle θ_2, where $\theta_2 < \theta_1$.

FIGURE 5–7
Power triangles used in developing power factor correction.

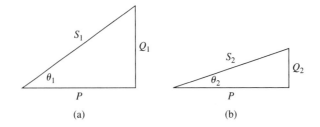

Assume that the initial value of the reactive power is Q_1. It can be expressed as

$$Q_1 = P \tan \theta_1 \tag{5-45}$$

The desired angle is θ_2 and the corresponding reactive power is

$$Q_2 = P \tan \theta_2 \tag{5-46}$$

We must then add a value Q_{add} so that

$$Q_2 = Q_1 + Q_{\text{add}} \tag{5-47}$$

or

$$Q_{\text{add}} = Q_2 - Q_1 \tag{5-48}$$

If proper signs are used for the respective values of the reactive power, the result will have the correct sign.

To determine the reactance necessary to achieve the added reactive power, we treat it in somewhat the same manner as resistance, except that we are working with reactance and reactive power. An appropriate equation is

$$Q_{\text{add}} = \frac{V_{\text{rms}}^2}{X} \tag{5-49}$$

and

$$X = \frac{V_{\text{rms}}^2}{Q_{\text{add}}} \tag{5-50}$$

In the case where capacitance is used to correct an inductive load, the value of Q_{add} will be negative and

$$X_C = -\frac{1}{\omega C} \tag{5-51}$$

Next, Equation 5-51 is substituted in Equation 5-50. However, the negative sign is eliminated by the use of magnitude bars and the capacitance is

$$C = \frac{|Q_{\text{add}}|}{\omega V_{\text{rms}}^2} = \frac{|Q_{\text{add}}|}{2\pi f V_{\text{rms}}^2} \tag{5-52}$$

■ **EXAMPLE 5-5**

An industrial process requires a power of 20 kW with a lagging power factor of 0.5 in a single-phase system. It is desired to add a parallel capacitance to raise the power factor to 0.95 lagging. For a line voltage of 240 V rms, determine the value of capacitance required.

SOLUTION The initial triangle is shown in Figure 5-8(a). The base is $P = 20$ kW and the angle θ_1 is

$$\theta_1 = \cos^{-1} 0.5 = 60° \tag{5-53}$$

The reactive power is

$$Q_1 = P \tan \theta_1 = 20 \tan 60° = 34.64 \text{ kVAR} \tag{5-54}$$

FIGURE 5–8
Power triangles of Examples 5-5 and 5-6.

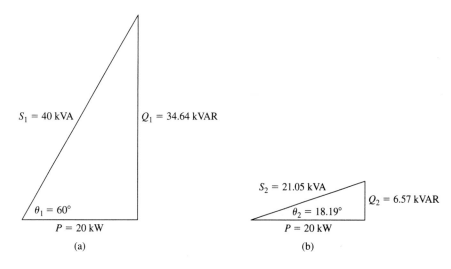

The final triangle is shown in Figure 5-8(b) and the angle θ_2 is given by

$$\theta_2 = \cos^{-1} 0.95 = 18.19° \tag{5-55}$$

The corresponding value of reactive power is

$$Q_2 = P \tan \theta_2 = 20 \tan 18.19° = 6.574 \text{ kVAR} \tag{5-56}$$

Thus, the capacitance must add a sufficient number of negative reactive power to reduce 34.64 kVAR down to 6.574 kVAR. The balance equation expressed in kVAR is

$$Q_1 + Q_{\text{add}} = 34.64 + Q_{\text{add}} = Q_2 = 6.57 \tag{5-57}$$

This leads to

$$Q_{\text{add}} = -28.07 \text{ kVAR} \tag{5-58}$$

The capacitance required is

$$C = \frac{|Q_{\text{add}}|}{\omega V_{\text{rms}}^2} = \frac{|Q_{\text{add}}|}{2\pi f V_{\text{rms}}^2} = \frac{28.07 \times 10^3}{2\pi \times 60 \times (240)^2} = 1.29 \times 10^{-3} \text{ F} = 1290 \text{ μF} \tag{5-59}$$

Note that it was appropriate to work in kW and kVAR until the last step, but we reverted to basic units to determine the capacitance.

The final triangle is shown in Figure 5-8(b). Some of the values on the triangles will be explored in the next example.

■ **EXAMPLE 5-6** For the system of Example 5-5, determine the line current (a) before power factor correction and (b) after power factor correction.

SOLUTION There is one assumption that must be made to determine a solution, and that could be challenged in some cases. We will assume that the line voltage both before and after correction is 240 V. Actually, if there were significant line losses, the line voltage drops would be different and that would cause some variation in the line voltage. However, for the type of application involved, it is reasonable to ignore any variation in the line voltage.

(a) The initial value of apparent power S_1 is the hypotenuse of the triangle of Figure 5-8(a). This could be determined in several ways. We will use the cosine of the included angle, and the value of S_1 is

$$S_1 = \frac{P}{\cos \theta_1} = \frac{20}{\cos 60°} = 40 \text{ kVA} \tag{5-60}$$

The current rms magnitude I_1 is

$$I_1 = \frac{S_1}{V_{\text{line}}} = \frac{40 \times 10^3}{240} = 166.7 \text{ A rms} \qquad (5\text{-}61)$$

(b) The final value of apparent power is determined with the triangle of Figure 5-8(b).

$$S_2 = \frac{P}{\cos\theta_1} = \frac{20}{\cos 18.19°} = 21.05 \text{ kVA} \qquad (5\text{-}62)$$

The current rms magnitude I_2 is

$$I_2 = \frac{S_2}{V_{\text{line}}} = \frac{21.05 \times 10^3}{240} = 87.7 \text{ A rms} \qquad (5\text{-}63)$$

The line current is reduced by nearly 50 percent in this particular case!

5-4 Admittance

The concept of **admittance** was briefly introduced in Chapter 4, but its application has been delayed to this chapter. That is because most engineers and technologists *think* in terms of impedance in solving most practical problems. However, the use of admittance simplifies certain types of problems, so the treatment of this section will be devoted to the various admittance forms and their interpretation.

Admittance Quantities

Let **Y** represent admittance, which, like impedance, is a complex quantity. It is defined in terms of the impedance **Z** as

$$\mathbf{Y} = \frac{1}{\mathbf{Z}} \qquad (5\text{-}64)$$

As a complex value, it can be defined in terms of real and imaginary parts as

$$\mathbf{Y} = G + jB \qquad (5\text{-}65)$$

The quantity G is the **conductance** and B is defined as the **susceptance**. The admittance, the conductance, and the susceptance are all measured in siemens (S).

Although **Y** is the reciprocal of **Z** and vice versa, the conductance is *not* the reciprocal of the resistance and the susceptance is *not* the reciprocal of the reactance, *except* for a purely resistance circuit and for the magnitude in the case of a purely reactive circuit. In general, it is necessary to form the reciprocal of the impedance and then determine the real and imaginary parts of the new quantity.

Current in Terms of Voltage and Admittance

A form of Ohm's law for ac phasors using admittance can be expressed as

$$\mathbf{I} = \mathbf{Y}\mathbf{V} \qquad (5\text{-}66)$$

This form is particularly useful when writing node voltage equations. Instead of expressing currents leaving nodes as voltages divided by complex impedances, it is often more convenient to express the currents as voltages times complex admittances. This will be demonstrated later.

Admittances of Circuit Parameters

We will now consider the admittance forms for each of three basic circuit parameters. We considered the reciprocal form of resistance back in Chapter 1, but for completeness, we begin with it. For a resistance of value $R = 1/G$, the admittance is

$$\mathbf{Y} = G \tag{5-67}$$

The result, of course, is a real number representing the conductance in siemens (S).

For an inductance whose impedance is $j\omega L$, the admittance is

$$\mathbf{Y} = \frac{1}{j\omega L} = \frac{-j}{\omega L} = jB_L \tag{5-68}$$

where the **inductive susceptance** is

$$B_L = \frac{-1}{\omega L} \tag{5-69}$$

For a capacitance whose impedance is $1/j\omega C$, the admittance is

$$\mathbf{Y} = \frac{1}{1/j\omega C} = j\omega C = jB_C \tag{5-70}$$

where the **capacitive susceptance** is

$$B_C = \omega C \tag{5-71}$$

Comparing the signs of susceptance with those of reactance, it is deduced that the signs are reversed. Thus, an *inductive reactance is positive* and an *inductive susceptance is negative*. In contrast, a *capacitive reactance is negative* and a *capacitive susceptance is positive*. The sign patterns are summarized in the table that follows.

Quantity	Inductive	Capacitive
Reactance	+	−
Susceptance	−	+

Even though these conventions are the most rigorous and are generally followed on most measurement systems such as ac bridges, some references use positive values for both reactance and susceptance of both types. However, we strongly recommend the conventions defined here because they follow naturally from the mathematical forms. When the reciprocal of a complex number with a positive real part is formed, there is an automatic sign reversal of the imaginary part.

Conversion from a Series to a Parallel Circuit

At any given frequency, a passive circuit containing both resistance and reactance may be represented either as a *series ac circuit* or a *parallel ac circuit*. Certain ac measuring instruments provide results in one form, and others provide the alternate form. Thus, a useful exercise in some areas of application is to convert from a series circuit to an equivalent parallel circuit, or vice versa.

Before getting into the details, it should be emphasized that *the type of reactance does not change when performing the conversion*. The authors have observed students switching a capacitive reactance to an inductive reactance in the conversion process, but this error was likely a result of either a misuse or a misunderstanding of the sign pattern previously discussed. Thus, if the reactance of one form is inductive, the reactance of the other form also will be inductive.

Let us begin with the series form. The circuit of Figure 5-9(a) will illustrate the process. We will somewhat arbitrarily assume an inductive reactance, but the development could be made with either form. The impedance $\mathbf{Z}$ is given by

$$\mathbf{Z} = R_s + jX_s \tag{5-72}$$

FIGURE 5–9
Conversion of a series circuit to an equivalent parallel circuit at a specific frequency.

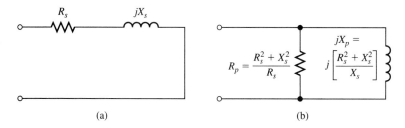

Two impedances in series fits very naturally with a series equivalent form because the impedance is immediately expressed as a sum. As one might expect, a parallel form fits more naturally with admittance. Therefore, as a starting point, the admittance **Y** corresponding to this impedance is expressed as

$$\mathbf{Y} = \frac{1}{\mathbf{Z}} = \frac{1}{R_s + jX_s} \tag{5-73}$$

The next step is to rationalize the denominator by multiplying both the denominator and the numerator by the complex conjugate of the denominator. This step yields

$$\mathbf{Y} = \frac{1}{(R_s + jX_s)}\frac{(R_s - jX_s)}{(R_s - jX_s)} = \frac{R_s - jX_s}{R_s^2 + X_s^2} = \frac{R_s}{R_s^2 + X_s^2} - \frac{jX_s}{R_s^2 + X_s^2} \tag{5-74}$$

In general, a parallel combination of a real and an imaginary admittance can be expressed as

$$\mathbf{Y} = G + jB \tag{5-75}$$

Equating the real and imaginary parts of the previous two expressions yields

$$G = \frac{R_s}{R_s^2 + X_s^2} \tag{5-76}$$

$$B = \frac{-X_s}{R_s^2 + X_s^2} \tag{5-77}$$

Some quick observations are in order. First, we see that the conductance is not the simple reciprocal of the conductance, but rather it is a more involved function of both the resistance and the reactance. Likewise, the equivalent susceptance is a function of both the resistance and the reactance.

Note how the sign of the susceptance is opposite to that of the reactance. This makes for a good argument on using the rigorous sign pattern defined earlier. Recall that an inductive reactance is positive and an inductive susceptance is negative and that this sign change occurs automatically in the reciprocation process.

One final important point before labeling the circuit: Although admittances in parallel add, we will follow the convention of *labeling* the equivalent parallel circuit in terms of the equivalent *impedance values* to minimize confusion on the circuit diagram. For this purpose, we will designate the impedance values as R_p and X_p. The values are

$$R_p = \frac{R_s^2 + X_s^2}{R_s} \tag{5-78}$$

$$X_p = \frac{R_s^2 + X_s^2}{X_s} \tag{5-79}$$

Note that the negative sign has been eliminated in the process of converting the susceptance back to a reactance. The two parallel impedances along with the *j*-factor for the reactance are shown in Figure 5-9(b).

Conversion from a Parallel to a Series Circuit

A process similar to that of the preceding section may be used to convert from a given parallel circuit to an equivalent series circuit at a specific frequency. The details will be

FIGURE 5–10
Conversion of a parallel circuit to an equivalent series circuit at a specific frequency.

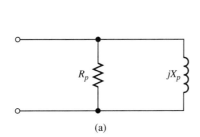
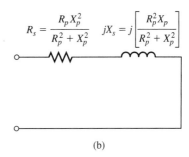

left as a guided exercise for the reader (Problem 5-31), but the results will be summarized here.

Consider the parallel ac circuit shown in Figure 5-10(a). As in the previous development, we will arbitrarily select an inductive reactance, but the process applies to either form. The corresponding parallel circuit at the specific frequency is shown in Figure 5-10. Based on a series resistance of R_s and a series reactance of X_s, the corresponding parallel values, denoted as R_p and X_p, are given by

$$R_s = \frac{R_p X_p^2}{R_p^2 + X_p^2} \tag{5-80}$$

$$X_s = \frac{R_p^2 X_p}{R_p^2 + X_p^2} \tag{5-81}$$

■ **EXAMPLE 5-7**

Use the admittance form of Ohm's law to determine the phasor current flowing into a 0.02-μF capacitor at a frequency of 2 kHz if the voltage across the capacitor is 50 V rms.

SOLUTION   No angle is given for the voltage, so the logical choice is to assume 0°. Thus, the phasor voltage is

$$\mathbf{V} = 50 \text{ V rms } \angle 0° = 50 \text{ V rms} \tag{5-82}$$

The admittance of the capacitor at 2 kHz is

$$\mathbf{Y} = j\omega C = j(2\pi \times 2000 \times 0.02 \times 10^{-6}) = j251.3 \times 10^{-6} \text{ S} \tag{5-83}$$

The current is

$$\mathbf{I} = \mathbf{YV} = j251.3 \times 10^{-6} \times 50 = 12.57 \times 10^{-3} \text{A} = 12.57 \text{ mA} \tag{5-84}$$

Of course, we could have calculated the impedance and divided the voltage by the impedance to obtain the same result, but the use of admittance, especially in the case of ac capacitive current, is somewhat simpler.

■ **EXAMPLE 5-8**

At a particular frequency, the series model of Figure 5-11(a) can represent the impedance of a certain circuit. Determine an equivalent parallel circuit.

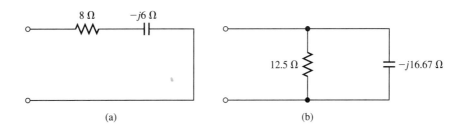

FIGURE 5–11
Circuits of Example 5-8.

SOLUTION The formulas were derived in this section and one could "plug in" the values. However, there is greater educational value in going through the steps with the numbers, so we will take that approach.

First, the impedance is

$$\mathbf{Z} = 8 - j6 \; \Omega \tag{5-85}$$

The admittance is

$$\mathbf{Y} = \frac{1}{\mathbf{Z}} = \frac{1}{8 - j6} \tag{5-86}$$

Next, the denominator is rationalized by multiplying it and the numerator by $8 + j6$. We have

$$\mathbf{Y} = \frac{1}{(8-j6)} \frac{(8+j6)}{(8+j6)} = \frac{8+j6}{64+36} = \frac{8+j6}{100} = 0.08 + j0.06 \tag{5-87}$$

In general, a complex admittance can be expressed as

$$\mathbf{Y} = G + jB \tag{5-88}$$

Equating real and imaginary parts, we obtain

$$G = 0.08 \; \text{S} \tag{5-89}$$
$$B = 0.06 \; \Omega \tag{5-90}$$

Note that the sign of the susceptance is positive, which is to be expected for a capacitive susceptance.

Before showing the circuit, we will convert the conductance and susceptance to resistance and reactance. The corresponding parallel resistance will be denoted as R_p and the parallel reactance will be denoted as X_p. The results are

$$R_p = \frac{1}{0.08} = 12.5 \; \Omega \tag{5-91}$$

$$jX_p = \frac{1}{j0.06} = -j16.67 \; \Omega \tag{5-92}$$

The parallel circuit with appropriate labeling is shown in Figure 5-11(b). It should be stressed that this equivalent circuit is valid only at one frequency.

5-5 Resonance

Resonance is a condition for an *RLC* circuit in which the net effects of the inductive and the capacitive reactances (or the corresponding susceptances) cancel at a given frequency, thus forcing the circuit to act purely as a resistive circuit at that particular frequency. Such circuits are used as components of **filters**, for which frequency selective properties may be exploited. For example, the tuning of a radio or a television typically involves changing the resonant frequency of one or more resonant circuits.

Whereas we could consider several variations under the heading of resonance, by far the two most common cases are **series resonance** and **parallel resonance**. Each case will be studied in its basic form.

So far in the analysis, the frequency of any ac source has been assumed as a constant value. However, to study resonance effectively, it is desirable to consider frequency as a variable. We are still using the single-frequency ac phasor concept, but we will leave the actual frequency unidentified so that we can study the behavior of the circuit for different frequencies. In some cases, the cyclic frequency f will be used and in other cases, the radian frequency $\omega = 2\pi f$ will be assumed.

FIGURE 5–12
Series resonant circuit and the steady-state phasor form.

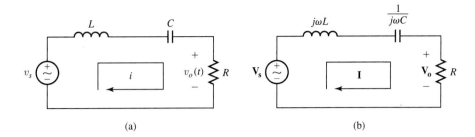

Series Resonance

The circuit of Figure 5-12(a) has the basic form of a series resonant circuit. It is assumed to be driven by a sinusoidal voltage source, and the output variable can be either the current in the loop or the voltage across the resistance. We will choose the loop current as the variable for consideration. The steady-state ac model for the circuit is shown in Figure 5-12(b).

The impedance seen by the source may be expressed as

$$\mathbf{Z} = R + j\omega L + \frac{1}{j\omega C} = R + j\omega L - \frac{j}{\omega C} = R + j\left(\omega L - \frac{1}{\omega C}\right) \quad (5\text{-}93)$$

The phasor current $\mathbf{I}$ may be expressed as

$$\mathbf{I} = \frac{\mathbf{V_s}}{\mathbf{Z}} = \frac{V_s \angle 0°}{R + j\left(\omega L - \frac{1}{\omega C}\right)} \quad (5\text{-}94)$$

Because V_s and R are assumed to be fixed, the maximum value of the current phasor magnitude will occur when the reactive part of the denominator is zero. Let ω_0 represent this particular radian frequency and it satisfies

$$\omega_0 L - \frac{1}{\omega_0 C} = 0 \quad (5\text{-}95)$$

Solving for ω_0, we obtain

$$\omega_0 = \frac{1}{\sqrt{LC}} \quad (5\text{-}96)$$

This frequency, called the **resonant frequency**, is expressed in rad/s in this form. The corresponding cyclic resonant frequency f_0 in Hz is

$$f_0 = \frac{1}{2\pi\sqrt{LC}} \quad (5\text{-}97)$$

At the resonant frequency, the impedance seen by the source is simply

$$\mathbf{Z} = R \quad (5\text{-}98)$$

The current at this frequency has the same phase angle as the voltage. The magnitude of this current is the maximum value for any frequency. It will be denoted as $I_{\max}$. It is given by

$$I_{\max} = \frac{V_s}{R} \quad (5\text{-}99)$$

By some rather involved algebraic manipulations, the magnitude of the current of Equation 5-94 can be expressed as

$$\frac{I}{I_{\max}} = \frac{1}{1 + jQ\left(\dfrac{f}{f_0} - \dfrac{f_0}{f}\right)} \quad (5\text{-}100)$$

where Q is a factor having significant implications in resonance and in filter design and for which an explanation will be provided later. Before we explain, it should be stressed that

FIGURE 5–13
Parallel resonant circuit and the steady-state phasor form.

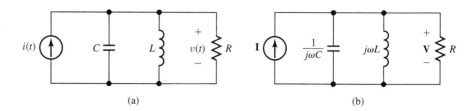

the use here is totally different from that employed in an earlier section for reactive power. For unknown reasons, Q is a somewhat "overworked" symbol in electrical theory and is widely used as a standard for several totally different applications. Fortunately, there is very little chance of getting confused between these different applications.

We will definitely revisit Q and its significance, but let us first look at the corresponding parallel resonant circuit.

Parallel Resonance

The circuit of Figure 5-13(a) has the basic form of a parallel resonant circuit and the steady-state phasor form is shown in (b). In the purest form as shown here, it is driven by a current source, and the output variable is usually the voltage across the circuit. If the circuit is driven by a voltage source in series with a resistance, a source transformation can be made to create an equivalent current source in parallel with a resistance, which will result in the form shown. We will choose the voltage across the circuit as the output variable of interest.

For the parallel circuit, it is more convenient to work in terms of admittance. The admittance of the parallel combination may be expressed as

$$\mathbf{Y} = G + j\omega C + \frac{1}{j\omega L} = G + j\omega C - \frac{j}{\omega L} = G + j\left(\omega C - \frac{1}{\omega L}\right) \tag{5-101}$$

where $G = 1/R$ is the parallel conductance.

The phasor voltage $\mathbf{V}$ may be expressed as

$$\mathbf{V} = \mathbf{Z}\mathbf{I} = \frac{\mathbf{I}}{\mathbf{Y}} = \frac{I_s\angle 0°}{G + j\left(\omega C - \frac{1}{\omega L}\right)} \tag{5-102}$$

Because I_s and G are assumed to be fixed, the maximum value of the voltage phasor magnitude will occur when the susceptive part of the denominator is zero. Let ω_0 represent this particular radian frequency, and it satisfies

$$\omega_0 C - \frac{1}{\omega_0 L} = 0 \tag{5-103}$$

The radian and cyclic resonant frequencies are

$$\omega_0 = \frac{1}{\sqrt{LC}} \tag{5-104}$$

and

$$f_0 = \frac{1}{2\pi\sqrt{LC}} \tag{5-105}$$

We pause to note that the resonant frequency for a parallel resonant circuit is the same as for a series resonant circuit for given values of inductance and capacitance. Some of the other properties, however, will be different, as will be seen shortly.

At the resonant frequency, the impedance seen by the source is simply

$$\mathbf{Z} = R \tag{5-106}$$

The voltage at this frequency has the same phase angle as the current. The magnitude of the voltage is maximum value for any frequency. It will be denoted as $V_{\max}$. It is

$$V_{\max} = RI_s \tag{5-107}$$

As in the case of current for the series resonant circuit, the voltage magnitude for the parallel resonant circuit can be manipulated into a similar form as given by

$$\frac{V}{V_{max}} = \frac{1}{1 + jQ\left(\dfrac{f}{f_0} - \dfrac{f_0}{f}\right)} \tag{5-108}$$

Opposite Properties for Series and Parallel Resonance

We will pause briefly from the development to note certain important properties of the two resonant cases that may be deduced by a careful study of the impedance and admittance properties:

- The *impedance* magnitude of a series resonant circuit is *minimum* at resonance. Below resonance, the impedance exhibits a capacitive reactance that increases as the frequency decreases. Above resonance, the impedance exhibits an inductive reactance that increases as the frequency increases.
- The *impedance* magnitude of a parallel resonant circuit is *maximum* at resonance. Below resonance, the impedance exhibits an inductive reactance that decreases as the frequency decreases. Above resonance, the impedance exhibits a capacitive reactance that decreases as the frequency increases.

Q-Factor for Resonant Circuits

Let us now discuss the Q-factor and its significance. The basic definition of Q in the context of resonance is proportional to the ratio of energy stored to power dissipated per cycle. Whereas that definition is very fundamental in a theoretical sense, the practical use of Q relates to its interpretation as a relative bandwidth factor.

In view of the similarity between Equations 5-100 and 5-108, one set of curves can be used to display both forms. Several curves for different values of Q are shown in Figure 5-14. The independent variable is indicated as "Normalized Frequency." It is defined as

$$\text{Normalized Frequency} = \frac{f}{f_0} \tag{5-109}$$

All of the curves have a level of unity for a normalized frequency of unity. However as Q increases, the relative amplitude decreases at frequencies below and above the resonant frequency. Thus, Q may be interpreted as a **selectivity factor**, meaning that the curves tend to have a narrower frequency range as Q increases. This property may be exploited in circuits known as **band-pass filters**, which will be considered in Chapter 8.

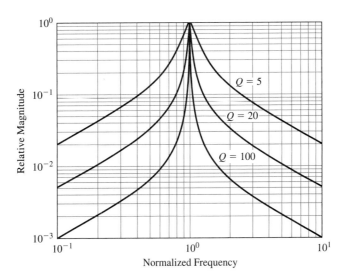

FIGURE 5–14
Resonant circuit reponse curves.

Relationship of Q to Parameter Values

It turns out that the value of Q is opposite for series and parallel resonance. A summary of the relationships follows:

Series Resonance

$$Q = \frac{\omega_0 L}{R} = \frac{1}{\omega_0 RC} = \frac{1}{R}\sqrt{\frac{L}{C}} \qquad (5\text{-}110)$$

Parallel Resonance

$$Q = \frac{R}{\omega_0 L} = \omega_0 RC = R\sqrt{\frac{C}{L}} \qquad (5\text{-}111)$$

Comparing the relationships, it can be deduced that a combination of components that would create a relatively large value of Q for a series circuit would create a relatively small value of Q for a parallel circuit and vice versa. This factor can make a significant difference in the practical application of resonant conditions.

Bandwidth

An important parameter in dealing with resonant circuits is the so-called bandwidth, which will be denoted as B. The bandwidth for resonant circuits is defined as the frequency increment between points at which the relative amplitude response is $1/\sqrt{2} = 0.7071$ times the peak value. There is a simple equation relating bandwidth, center frequency, and Q. It is

$$Q = \frac{f_0}{B} \qquad (5\text{-}112)$$

From this relationship, it is clear that a larger value of Q results in a smaller bandwidth relative to the center frequency.

This concept will be revisited in Chapter 8 in conjunction with active filters. It will be shown that the level of 0.7071 corresponds to a drop of about **3 decibels**, for which the meaning will be discussed there.

EXAMPLE 5-9

A series resonant circuit of the form of Figure 5-12(a) has $L = 20$ mH, $C = 0.03$ μF, and $R = 120$ Ω. Determine the (a) resonant frequency, (b) Q, and (c) bandwidth between points at which the response is 0.7071 times the peak response.

SOLUTION

(a) The resonant frequency expressed in Hz is

$$f_0 = \frac{1}{2\pi\sqrt{LC}} = \frac{1}{2\pi\sqrt{20 \times 10^{-3} \times 0.03 \times 10^{-6}}} = 6.497 \text{ kHz} \qquad (5\text{-}113)$$

(b) The Q is

$$Q = \frac{1}{R}\sqrt{\frac{L}{C}} = \frac{1}{120}\sqrt{\frac{20 \times 10^{-3}}{0.03 \times 10^{-6}}} = 6.804 \qquad (5\text{-}114)$$

(c) The bandwidth is determined from Equation 5-112 as

$$B = \frac{f_0}{Q} = \frac{6497}{6.804} = 954.9 \text{ Hz} \qquad (5\text{-}115)$$

PROBLEMS

5-1 A sinusoidal voltage having a peak value of 170 V is applied across a 150-Ω resistor. (a) Determine the rms value of the voltage. (b) Determine the average power dissipated.

5-2 A sinusoidal voltage having a value of 100 V rms is applied across a 200-Ω resistor. (a) Determine the peak value of the voltage. (b) Determine the average power dissipated.

5-3 A sinusoidal current having a peak value of 5 A is flowing through an 8-Ω resistor. (a) Determine the rms value of the current. (b) Determine the average power dissipated.

5-4 A sinusoidal voltage having an value of 10 A rms is flowing through a 4-Ω resistor. (a) Determine the peak value of the current. (b) Determine the average power dissipated.

5-5 The impedance of a steady-state ac series circuit is $Z = 6 + j8$ Ω and it is excited by an ac voltage source of value $\mathbf{V_s} = 40$ V rms $\angle 0°$. Determine (a) the current phasor, (b) the apparent power, (c) the power factor, (d) the real power, and (e) the reactive power.

5-6 The impedance of a steady-state ac series circuit is $Z = 30 - j40$ Ω and is excited by an ac voltage source of value $\mathbf{V_s} = 240$ V rms $\angle 0°$. Determine (a) the current phasor, (b) the apparent power, (c) the power factor, (d) the real power, and (e) the reactive power.

5-7 The voltage across an ac circuit is 120 V rms and the current is 10 A rms. The voltage *leads* the current by 30°. Determine the (a) apparent power, (b) real power, and (c) reactive power.

5-8 The voltage across an ac circuit is 240 V rms and the current 12 A rms. The voltage *lags* the current by 30°. Determine the (a) apparent power, (b) real power, and (c) reactive power.

5-9 The voltage across an ac circuit is 120 V rms and the current is 10 A rms. The power factor is 0.6 lagging. Determine the (a) angle between the voltage and the current and which one leads, (b) apparent power, (c) real power, and (d) reactive power.

5-10 The voltage across an ac circuit is 240 V rms and the current is 20 A rms. The power factor is 0.8 leading. Determine the (a) angle between the voltage and the current and which one leads, (b) apparent power, (c) real power, and (d) reactive power.

5-11 The variables for an ac circuit are measured by a voltmeter, an ammeter, and a wattmeter. The effective values of voltage and current are measured as 240 V rms and 4 A rms, respectively. Assume that the wattmeter reads 500 W. Determine (a) the apparent power, (b) the power factor, and (c) the magnitude of the reactive power.

5-12 The variables for an ac circuit are measured by a voltmeter, an ammeter, and a wattmeter. The effective values of voltage and current are measured as 480 V rms and 8 A rms, respectively. Assume that the wattmeter reads 2 kW. Determine (a) the apparent power, (b) the power factor, and (c) the magnitude of the reactive power.

5-13 A motor requires a power of 5 kW with a lagging power factor of 0.7 in a single-phase system. It is desired to add a parallel capacitance to raise the power factor to 0.9 lagging. For a line voltage of 240 V rms, determine the value of capacitance required.

5-14 An industrial process requires a power of 10 kW with a lagging power factor of 0.6 in a single-phase system. It is desired to add a parallel capacitance to raise the power factor to unity. For a line voltage of 240 V rms, determine the value of capacitance required.

5-15 For the system of Problem 5-13, determine the line current (a) before power factor correction and (b) after power factor correction.

5-16 For the system of Problem 5-14, determine the line current (a) before power factor correction and (b) after power factor correction.

5-17 Use the admittance form of Ohm's law to determine the phasor current in rms units flowing into a 2-μF capacitor at a frequency of 5 kHz if the voltage across the capacitor is 20 V rms.

5-18 Use the admittance form of Ohm's law to determine the phasor current in rms units flowing into a 2-mH inductor at a frequency of 1 kHz if the voltage across the inductor is 30 V rms.

5-19 At a particular frequency, the series impedance of an ac circuit is $Z = 6 + j8$ Ω. Determine the two values of impedances for an equivalent parallel circuit and draw the circuit.

5-20 At a particular frequency, the series impedance of an ac circuit is $Z = 80 - j60$ Ω. Determine the two values of impedances for an equivalent parallel circuit and draw the circuit.

5-21 At a particular frequency, the parallel admittance of an ac circuit is $Y = 0.4 - j0.3$ S. Determine the two values of impedances for an equivalent series circuit and draw the circuit.

5-22 At a particular frequency, the parallel admittance of an ac circuit is $Y = 0.08 + j0.06$ S. Determine the two values of impedances for an equivalent series circuit and draw the circuit.

5-23 A series resonant circuit has $L = 5$ mH, $C = 0.02$ μF, and $R = 50$ Ω. Determine the (a) resonant frequency, (b) Q, and (c) bandwidth between points at which the response is 0.707 times the peak response.

5-24 A series resonant circuit has $L = 82$ μH, $R = 40$ Ω and a resonant frequency of 1 MHz. Determine the (a) value of C, (b) Q, and (c) bandwidth between points at which the response is 0.707 times the peak response.

5-25 A parallel resonant circuit has $L = 5$ mH, $C = 0.02$ μF, and $R = 10$ kΩ. Determine the (a) resonant frequency, (b) Q, and (c) bandwidth between points at which the response is 0.707 times the peak response.

5-26 A parallel resonant circuit has $C = 100$ pF, $R = 10$ kΩ, and a resonant frequency of 2 MHz. Determine the (a) value of L, (b) Q, and (c) bandwidth between points at which the response is 0.707 times the peak response.

5-27 Several cycles of a *half-wave rectified* periodic voltage is shown in Figure P5-27. The waveform may be described over a cycle as

$$v(t) = V_p \sin \omega t = V_p \sin \frac{2\pi t}{T} \quad \text{for} \quad 0 \le t \le T/2$$

$$= 0 \quad \text{for} \quad T/2 \le t \le T$$

Prove that the dc and rms values are

$$V_{\text{dc}} = \frac{V_p}{\pi}$$

$$V_{\text{rms}} = \frac{V_p}{2}$$

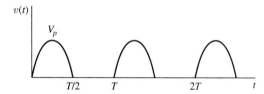

FIGURE P5–27

5-28 Several cycles of a *full-wave rectified* periodic voltage are shown in Figure P5-28. For convenience, the same time scale employed in Problem 5-27 is employed, which effectively results in a period of $T/2$. During this interval, the function is

$$v(t) = V_p \sin \omega t = V_p \sin \frac{2\pi t}{T} \quad \text{for} \quad 0 \le t \le T/2$$

Prove that the dc and rms values are

$$V_{\text{dc}} = \frac{2V_p}{\pi}$$

$$V_{\text{rms}} = \frac{V_p}{\sqrt{2}}$$

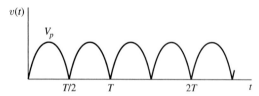

FIGURE P5–28

5-29 Several cycles of a *symmetrical square wave* are shown in Figure P5-29. Prove that the dc and rms values are

$$V_{\text{dc}} = 0$$

$$V_{\text{rms}} = V_p$$

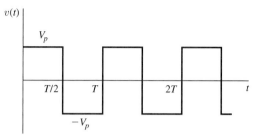

FIGURE P5–29

5-30 Several cycles of a *pulse train* are shown in Figure P5-30. A parameter called the *duty cycle* is defined as

$$d = \frac{\tau}{T}$$

Prove that the dc and rms values are

$$V_{\text{dc}} = dV_p$$

$$V_{\text{rms}} = \sqrt{d}\,V_p$$

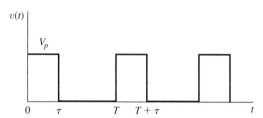

FIGURE P5–30

5-31 The objective of this exercise is to derive the equations for conversion from a parallel circuit at a specific frequency to an equivalent series circuit as summarized in Figure 5-10. Start with the parallel form of (a) and write an expression for the net admittance as the sum of the two parallel admittances. Form the reciprocal to determine the impedance and arrange over a common denominator. Then rationalize the denominator and simplify. The real part will represent the equivalent series resistance, and the imaginary part (the real number multiplied by j) will represent the equivalent series reactance.

ELECTRONIC DEVICES AND LINEAR ELECTRONICS

6 Diodes and Their Applications

7 Transistors

8 Operational Amplifiers

Diodes and Their Applications — 6

OVERVIEW AND OBJECTIVES

This chapter will begin the study of semiconductor devices by introducing the diode and some of its most important applications. The emphasis will be directed toward their external terminal characteristics and how they behave in circuits. A major portion of the chapter will be devoted to rectifier and power supply applications.

Objectives

After completing this chapter, the reader should be able to

1. Describe how the properties of a **semiconductor** differ from those of a **conductor** and an **insulator**.
2. Define **donors** and **acceptors** and discuss their functions in creating **P-type** material and **N-type** material.
3. Discuss the basic properties of a **PN-junction diode** and sketch the form of the terminal characteristic curve.
4. Describe the **ideal model** and the **constant voltage model** for a junction diode and draw the equivalent circuit forms.
5. Analyze simple circuit forms containing diodes and determine when a diode is **forward biased** or **reverse biased**.
6. State the input-output properties of an **ideal transformer** in terms of the turns ratio.
7. Draw the schematic diagram of a **half-wave rectifier** and analyze its operation.
8. Draw the schematic diagram of a **full-wave rectifier** with a center-tapped secondary and analyze its operation.
9. Draw the schematic diagram of a full-wave **bridge rectifier** and analyze its operation.
10. Describe the function of a **rectifier filter** and analyze the ripple level.
11. Define **peak inverse voltage** and determine its value for different rectifier circuits.
12. Discuss the properties of different types of diodes.
13. Draw the schematic diagram of a **zener regulator circuit** and analyze its operation.

6-1 Semiconductor Concepts

Before considering specific solid-state semiconductor devices, some terminology must be introduced as a prelude to the operation and classification of the components to be studied. Based on relative conductivity, there are three classes of materials: (1) **conductors**, (2) **insulators** (also called nonconductors), and (3) **semiconductors**.

Conductors are those materials that have very low resistance or high conductivity. Two examples are copper and aluminum. *Insulators* are those materials that have very high resistance or very low conductivity. Two examples are dry wood and plastic. Thus, conductors and insulators represent the two opposite ends of the conduction spectrum.

Semiconductors are materials that have properties that fall somewhere between conductors and insulators. Conduction can occur in semiconductors but not as easily as in conductors. This property can be exploited to create a wide variety of electronic devices, including complete systems having the equivalent of thousands of transistors.

Conduction in semiconductors can occur from the movement of either (1) **electrons** or (2) **holes**. *Electrons* have a negative charge and move in the direction opposite to that of conventional current flow. In contrast, *holes* have a positive charge and move in the direction of conventional current flow. Holes are peculiar to semiconductor devices and, in a sense, represent the absence of an electron in the particular band of the atom.

The processing of semiconductor materials for device construction is called **doping**. This is accomplished by adding **impurities**. Impurities are called either **donors** or **acceptors**. *Donor* materials result in an addition of *negative* charge carriers (electrons), and *acceptor* materials result in an addition of *positive* charge carriers (holes).

Any semiconductor material having an excess of positive charge carriers is called a **P-type** material, and a material having an excess of negative charge carriers is called an **N-type** material. Throughout the treatment of semiconductor devices in this chapter and later, frequent reference will be made to P-type and N-type materials.

6-2 Junction Diode

A semiconductor **junction diode** is created by combining a section of P-type material with a section of N-type material as illustrated in Figure 6-1(a). Based on interactions of the charge carriers, current flows easily in one direction but not in the other. The "easy" direction of conventional current flow is into the side composed of P-type material and out of the side with the N-type material as shown by the solid arrow. The arrow with dashed lines represents the direction offering large resistance to conventional current flow. A mechanical analogy to the diode is a **check valve**.

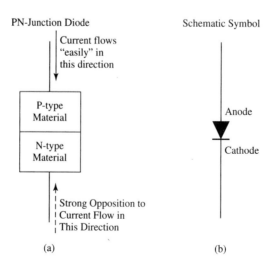

FIGURE 6–1
PN-junction diode and its schematic symbol.

Schematic Symbol

The schematic symbol of the junction diode is shown in Figure 6-1(b). It is aligned with the layout of (a). The best way to remember the schematic is that the "arrow" shown here at the top of the symbol corresponds to the easy direction of conventional current flow. The top terminal as shown here is called the **anode** and the bottom terminal is called the **cathode**. These terms date back to terminology associated with vacuum tubes.

Applications

There are hundreds of different diode types with a wide range of operating current values, some of which will appear in various applications throughout the remainder of the text. Some of the more basic applications will be considered in this chapter. Excluding for the moment some of the more specialized types, the major application of diodes is to allow current to flow in one direction while simultaneously blocking current flow in the opposite direction. This theme will appear in various circuits to be considered in subsequent sections.

Terminal Characteristic

As the first step in studying the behavior of a two-terminal electronic device, the terminal characteristic provides useful overall information. As a prelude, let us review the behavior of a simple resistance and its terminal characteristic. Consider the resistance of Figure 6-2(a) in which a voltage is assumed to exist across the resistance and a current is assumed flowing into it as shown. We will consider the voltage as the independent variable and the current as the dependent variable. We will also assume that the resistance is a constant value and that its maximum power rating is not reached. The current is related to the voltage by Ohm's law; that is,

$$i = \frac{v}{R} \tag{6-1}$$

In terms of the voltage and current variables, this is a linear equation with a slope of $1/R$. A plot is shown in Figure 6-2(b). The plot will be referred to as the **terminal characteristic** of the device. It is a plot of the current versus the voltage.

Diode Terminal Characteristics

Now consider the circuit shown in Figure 6-3, consisting of an adjustable dc power supply denoted as V_s, a current-limiting resistance R, and a diode. The resistance is desirable in an

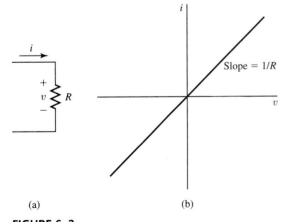

FIGURE 6–2
Ideal resistance and its terminal characteristic.

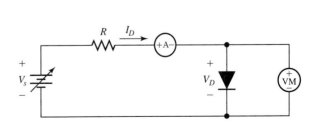

FIGURE 6–3
Experiment used to determine semiconductor diode terminal characteristic.

FIGURE 6–4
Form of the terminal characteristic of a semiconductor diode.

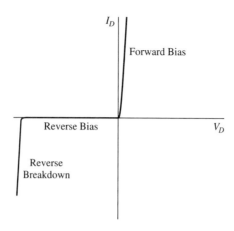

actual measurement setup such as this because the diode could be easily destroyed if the voltage or current exceeded the maximum ratings of the diode. Its presence allows an easier adjustment of the voltage across the diode in small increments. The ammeter measures the diode current I_D and the voltmeter measures the diode voltage V_D.

As the voltage is increased in the positive reference direction shown, the current increases rapidly as shown in the first quadrant of Figure 6-4. This region is considered as the **forward-bias region**. It represents the "easy" direction of current flow referred to in an earlier discussion.

Assume next that the voltage is changed to negative values. This could be achieved by reversing the terminals of the power supply along with those of the ammeter and voltmeter. However, an easier way is to simply reverse the terminals of the diode!

The behavior of the diode now moves to the third quadrant in Figure 6-4. This is the **reverse-bias region**. The scale is very small and not visible on Figure 6-4, but there is a small temperature-dependent reverse current at smaller values of the reverse voltage; that is, the diode is not a perfect open circuit when it is reverse biased.

If the negative voltage level is made to be sufficiently large, a phenomenon called **reverse breakdown** occurs. At this voltage level, the current increases rapidly in the negative region as shown and an ordinary diode may be easily destroyed. However, certain diodes known as **zener diodes** are made to exploit the nearly constant voltage in this region to regulate the voltage level. We will discuss their behavior later in the chapter. Of course, there is a maximum current level that must be controlled with this phenomenon.

6-3 Diode Circuit Models

To use diodes in actual circuit applications, it is necessary to employ models of their behavior that may be used in circuit analysis and design. From observation of the actual shape of the terminal characteristic in Figure 6-4, it is apparent that the behavior is nonlinear.

Diode models range from very simple to very sophisticated, depending on the application and the accuracy required. For routine applications where high accuracy is not required, two simple models are widely employed as the first step in the analysis, even when the more sophisticated techniques are to be used.

Ideal Diode Model

Consider the diode shown in Figure 6-5(a) with assumed voltage and current. The **ideal diode model** assumes that the diode is either a perfect short circuit or a perfect open circuit, depending on the directions of voltage or current.

The ideal diode model assumes that it acts as a short circuit when current is flowing in the direction of the arrow. This corresponds to forward bias as shown in the left-hand model of Figure 6-5(b). Conversely, the diode acts as an open circuit when the voltage at the cathode (bottom of figure) is more positive than the anode (top of figure), and this

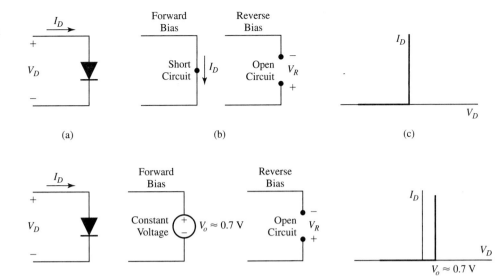

FIGURE 6–5
Ideal diode model and terminal characteristic.

FIGURE 6–6
Constant voltage diode model and terminal characteristic.

corresponds to reverse bias as shown on the right-hand side of Figure 6-5(b). The voltage V_R shown in the figure represents a reverse bias value.

The ideal diode terminal characteristic is shown in Figure 6-5(c). It reduces to the combination of the positive ordinate and the negative abscissa. It can be concluded that in the simplest diode model, *the voltage is zero when the current is positive, and the current is zero when the voltage is negative.*

Constant Voltage Model

A somewhat better model for many applications is that of the constant voltage model. Consider the diode shown in Figure 6-6(a) with voltage and current labeled. The reverse bias portion of this model is the same as that of the ideal model; that is, the current is assumed to be zero when the voltage is negative as shown in Figure 6-6(b).

It turns out that when a practical diode is forward biased, there is a small voltage across it that does not vary much over a wide range of current. The constant voltage model assumes that this voltage is a constant value, which is indicated as V_o in Figure 6-6(b).

Over the range of operation for small diodes with current ratings in the milliampere range or so, this voltage typically ranges from about 0.6 V to about 0.7 V for silicon diodes, although it may be larger for diodes having much larger currents. Various references use values such as 0.6 V, 0.65 V, and 0.7 V. We will choose in this text to use $V_o = 0.7$ V as a "standard value," but remember that this is only an approximation.

The terminal characteristic of the constant voltage model is shown in Figure 6-6(c). This characteristic indicates that reverse bias corresponds to the negative abscissa, but forward bias corresponds to a vertical line at an abscissa of V_o.

Note that the diode is a passive device and is not capable of supplying energy, as a voltage source would be. Rather, when it is forward biased, its behavior is similar to a voltage source with current flowing into its positive terminal, and thus it can be modeled by that element. However, the equivalent voltage source does not come "alive" until the diode is forward biased, and the fictitious voltage source is always absorbing power from some other source.

Other Models

A short description of some other models will be made here. One other linear model is to add a resistance in series with the constant voltage source to provide an increasing voltage as the current increases, but it is still approximate.

124 CHAPTER 6 • Diodes and Their Applications

One of the most accurate models in the forward biased region is based on the Shockley equation, which reads as follows:

$$I_D = I_S \left(e^{\frac{qV_D}{\eta kT}} - 1 \right) \tag{6-2}$$

The parameters in this equation are defined as follows:

V_D = diode forward voltage

I_D = diode forward current

I_S = reverse saturation current

q = magnitude of electron charge = 1.6×10^{-19} C

k = Boltzmann's constant = 1.38×10^{-23} joules/kelvin (J/K)

T = absolute temperature in kelvins (K)

η = nonideality factor, typically in the range from 1 to 2

We won't discuss this equation in any detail since it is not used in the text, but it is presented primarily for general information. In effect, its complexity and nonlinear nature suggest that the simpler models be employed whenever possible. However, computer models employ variations of this equation to predict results when high accuracy is required.

Which Model Is Appropriate?

A fundamental question is, which model should be used? The answer depends on the accuracy required for the particular application. As a quick initial check, the ideal model is frequently employed, especially where little or no quantitative results are needed. However, the constant voltage source model is probably the most widely employed model for a variety of routine applications. When higher accuracy is required, the more accurate nonlinear models often suggest using computer analysis. Most of the work in this text will employ the constant voltage model. We will see in the next chapter that this model can also be used with transistors.

Is a Diode Forward Biased or Reverse Biased?

Another question arises in analysis: Is a given diode forward biased or reverse biased? For fairly simple circuits, one can usually tell by intuition, and the examples at the end of this section will help you develop this intuition. For more complex circuits, it may be necessary to make assumptions and then justify the assumptions.

If a given diode is assumed to be forward biased, make a calculation to determine the current through the diode, which normally would be represented by the constant voltage source model. If the calculated current turns out to be in the direction of the diode arrow (into the positive terminal of the assumed constant voltage), the assumption was correct. However, if the current turns out to be in the opposite direction, the assumption was incorrect and an open circuit should be used to model the diode.

If a given diode is assumed to be reverse biased, make a calculation to determine the voltage across the assumed open circuit. If the voltage is positive at the cathode with respect to the anode, the assumption was correct, but if the more positive terminal is the anode, the assumption was incorrect and the diode should then be modeled by the constant voltage source.

■ **EXAMPLE 6-1**

For the circuit of Figure 6-7(a), determine the loop current, the voltage across the diode, and the voltage across the resistance.

SOLUTION The first question is whether the diode is forward biased or reverse biased. If we mentally visualize the diode as a constant voltage source of 0.7 V, the much larger

FIGURE 6–7
Circuit of Example 6-1.

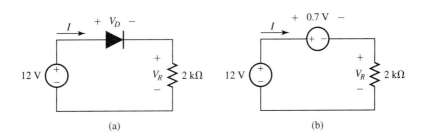

source voltage of 12 V clearly would dominate and would cause current to flow in a clockwise direction. Hence, the diode is forward biased and the equivalent circuit using the constant voltage source model is shown in Figure 6-7(b). Before continuing with the analysis, note that if the source voltage were less than about 0.7 V based on the assumed model, the net voltage would be insufficient to cause the current to flow in the direction of forward bias. Said differently, the voltage required to cause the diode to become forward biased must be on the order of about 0.7 V.

The circuit is readily analyzed using KVL, which reads

$$-12 + 0.7 + 2000I = 0 \tag{6-3}$$

Solving for the current, we have

$$I = \frac{12 - 0.7}{2000} = \frac{11.3}{2000} = 5.65 \times 10^{-3}\,\text{A} = 5.65\,\text{mA} \tag{6-4}$$

Based on the assumed model, the diode voltage is quite obvious. It is

$$V_D = 0.7\,\text{V} \tag{6-5}$$

The resistive voltage could be determined by Ohm's law, but a simpler approach is to recognize from KVL that

$$V_R = 12 - 0.7 = 11.3\,\text{V} \tag{6-6}$$

We will leave as an exercise for the reader to verify that if the ideal diode model had been employed, the current would be 6 mA, the diode voltage would be 0, and the resistive voltage would be 12 V. So which solution is correct? Actually, probably neither one, because the diode voltage drop will rarely be *exactly* 0.7 V. However, the solution obtained with the constant voltage drop should be quite close. If greater accuracy is required, a computer analysis using a nonlinear diode model could be used.

▌▌ EXAMPLE 6-2

For the circuit of Figure 6-8(a), determine the loop current, the voltage across the resistance, and the voltage across the diode.

SOLUTION This circuit has the same structure as in Example 6-1, but there is one important difference. The diode in this case has been turned around. For the diode to have

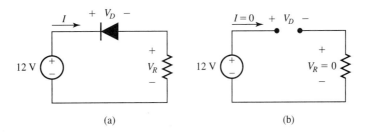

FIGURE 6–8
Circuit of Example 6-2.

forward bias, the voltage source would have to be reversed. Therefore, the diode in this case is reverse biased and the circuit model is shown in Figure 6-8(b).

The current in this case is immediately obvious. It is

$$I = 0 \tag{6-7}$$

Because there is no current flowing, the voltage across the resistance is

$$V_R = 0 \tag{6-8}$$

However, the entire source voltage will now appear across the reverse-biased diode. To help justify this conclusion, a KVL equation can be expressed as follows:

$$-12 + V_D + 0 = 0 \tag{6-9}$$

where the last 0 on the left-hand side of the equation represents the resistive voltage. This equation leads to

$$V_D = 12 \text{ V} \tag{6-10}$$

Thus, the entire source voltage appears across the reverse-biased diode. It is assumed here that this voltage magnitude is smaller than the reverse breakdown voltage or else the entire analysis would be invalid.

From this example and the previous one, some intuition should develop for predicting forward or reverse bias in simple circuits.

6-4 Diode Rectifier Circuits

One of the most widely employed applications of diodes is in **rectifier** circuits. A rectifier circuit is one in which alternating current (ac) is converted into some form of direct current (dc). Rectifier operation constitutes an integral part of **dc power supply** circuits.

Why Convert AC to DC?

Commercial power generation and transmission are usually performed with ac because of the ease of transforming voltages to different levels with transformers and the dramatic increase in efficiency resulting from such transformations. However, much electrical and electronic equipment requires dc voltages and currents for proper operation. Therefore, it is necessary to convert ac at the operational level to dc, and this is achieved with dc power supplies. Many power supplies such as in radio and television units and computers are integrated into the systems and are not visible from the exterior, but a typical modern household could have dozens or more ac-to-dc power supplies in operation. Many industrial operations may employ power supplies to run dc motors and other devices. A typical power supply will contain rectifier circuits, transformers, smoothing filters, a voltage regulator, one or more fuses or circuit breakers, and a switch.

Half-Wave Rectifier Circuit (Unfiltered)

We will first consider an unfiltered rectifier circuit for which the output is not constant but is a form of dc. Consider the circuit of Figure 6-9(a) consisting of a transformer, a diode, a resistive load, a switch, and a fuse. Transformers will be studied in great detail later in the text, but for the purpose here we will consider it as an ideal component. The ideal transformer either steps up an ac voltage to a higher value or to a lower value depending on the turns ratio. Let N_1 represent a number proportional to the number of turns on the left-hand side or input ac power source side of the transformer (often referred to as the *primary* side). Let N_2 represent a number proportional to the number of turns on the right-hand side to be converted to dc (often referred to as the *secondary* side). We say "proportional to" because

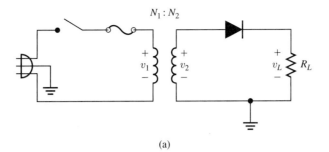

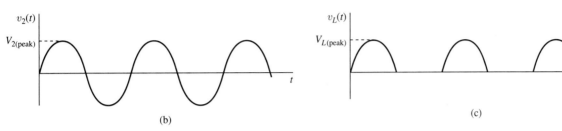

FIGURE 6–9
A half-wave rectifier without filtering and the waveforms.

it is rarely practical to know the actual number of turns on each side, but it is the ratio that is important. This will be illustrated on schematic diagrams by the notation $N_1 : N_2$.[1]

The primary and secondary voltages will be denoted as v_1 and v_2 and the primary and secondary currents will be denoted as i_1 and i_2, respectively. The ratio of the two voltages is given by the relationship

$$\frac{v_2}{v_1} = \frac{N_2}{N_1} \tag{6-11}$$

The ratio of the two currents is

$$\frac{i_2}{i_1} = \frac{N_1}{N_2} \tag{6-12}$$

Thus, if $N_2 > N_1$, $v_2 > v_1$, and vice versa. However, if $N_2 > N_1$, $i_2 < i_1$, and vice versa. Thus, if the voltage is increased, the current will be decreased and vice versa. These properties constitute some of the great advantages of ac. More detailed properties of transformers will be developed later. For the moment, accept these relationships and assume that the transformer is ideal.

The turns ratio equation in Equations 6-11 and 6-12 can be applied to instantaneous, peak, or rms values provided that *both* primary and secondary are expressed in the same form. For example, if the input variables are expressed in rms units, the output variables determined from the turns ratio will also be in rms units.

Waveforms on the output side of a transformer for some arbitrary input ac voltage are shown in Figure 6-9(b) and (c). The voltage across the secondary is shown in (b) and is given by

$$v_2(t) = \frac{N_2}{N_1} v_1(t) \tag{6-13}$$

If $N_2 > N_1$, the transformer is acting as a *step-up* transformer, and if the opposite inequality is true, the transformer is acting as a *step-down* transformer. There are some transformers in which $N_1 = N_2$, in which case they serve as *isolation transformers* to provide a form of electrical isolation between primary and secondary.

---

[1] In the power chapters later in the book, the turns ratio will be denoted by the simpler notation $a : 1$. Because both forms appear in the literature, we have chosen to introduce both forms in the text.

During the portion of the cycle that v_2 is less than about 0.7 V, the diode will be an open circuit and the load voltage will be zero. During the portion of the cycle that v_2 exceeds about 0.7 V, the load voltage can be expressed approximately as

$$v_L(t) = v_2(t) - 0.7 = \frac{N_2}{N_1}v_1(t) - 0.7 \quad \text{for} \quad v_2 > 0.7 \text{ V} \tag{6-14}$$

In many higher power circuits, the drop across the diode can be greater than 0.7 V. Moreover, there are also losses in the transformer, but to maintain a consistent approach without using nonlinear models, we will continue to assume the constant voltage drop of 0.7 V for the diode.

The resulting waveform is shown in Figure 6-9(c). There is a slight drop in the voltage because of the diode drop, but the result is nearly a half-wave replica of the transformer secondary sinusoidal voltage. It is either positive or zero at all points.

Is this waveform a dc voltage? Certainly it is not in the strictest sense of dc. However, it is nonnegative everywhere, which means that it has a non-zero average value and thus it has a dc component. This type of waveform has traditionally been called **pulsating dc**. It can be used for certain types of industrial processes requiring a voltage and current in one direction, but where the time variation is acceptable. This waveform is *not* suitable for applications requiring a constant dc voltage and current. That situation will be covered in the next section.

Let $V_{L(\text{peak})}$ represent the peak value of the load voltage. The dc value of this voltage will be denoted as $V_{L(\text{dc})}$. It is the time average of the waveform over a cycle. Problem 5-27 was given as an exercise for the reader to derive the value of the dc and rms values for the half-wave sinusoid. The dc value is given by

$$V_{L(\text{dc})} = \frac{V_{L(\text{peak})}}{\pi} = 0.318 V_{L(\text{peak})} \tag{6-15}$$

The dc value of the load current is then given by

$$I_{L(\text{dc})} = \frac{V_{L(\text{dc})}}{R_L} \tag{6-16}$$

Full-Wave Rectifier (Unfiltered)

The dc value of the unfiltered output voltage can be doubled by using a full-wave rectifier, whose circuit diagram is shown in Figure 6-10(a). The transformer must have a center tap as shown. Note that the number of secondary turns is proportional to N_2, which means that the number across each half is proportional to $0.5N_2$.

The resulting waveforms are shown in Figures 6-10(b) and (c). Refer to both of these figures and the circuit models shown in Figure 6-11 for the discussion that follows.

When the upper terminal of the transformer secondary is positive with respect to the ground center tap as shown in Figure 6-11(a) and larger than about 0.7 V, the upper diode will be forward biased, and the voltage across the load can be expressed as

$$v_L(t) = \frac{0.5N_2}{N_1}v_1(t) - 0.7 \tag{6-17}$$

During this portion of the cycle, the lower diode will be reverse biased and the lower half of the transformer and lower diode will act as an open circuit.

During the second half of the input cycle, the lower terminal of the secondary will be positive with respect to ground, and for a voltage of about 0.7 V and greater, this voltage is coupled to the output terminal through the lower diode as shown in Figure 6-11(b). However, the upper diode will be reverse biased for this part of the cycle and that portion of the circuit will be open.

The resulting waveform is shown in Figure 6-10(c). The negative half-cycles of the secondary ac voltage are in effect "flipped over" and become positive. The result is a full-wave rectified voltage.

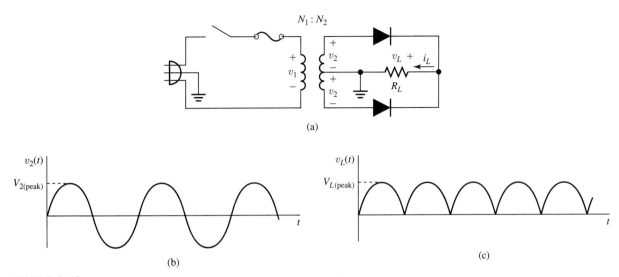

FIGURE 6–10
Schematic diagram of an unfiltered full-wave rectifier with a center-tapped transformer and the waveforms.

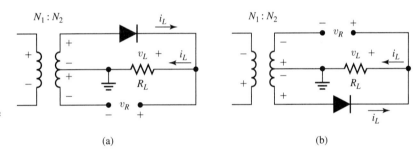

FIGURE 6–11
Conduction models for full-wave rectifier with center tap for alternate half-cycles.

Problem 5-28 was given as an exercise for the reader to derive the value of the dc and rms values for the full-wave sinusoid. The dc value is twice that of the half-wave rectifier and is given by

$$V_{L(\text{dc})} = \frac{2V_{L(\text{peak})}}{\pi} = 0.637 V_{L(\text{peak})} \tag{6-18}$$

Full-Wave Bridge Rectifier

A widely employed full-wave rectifier circuit is the **bridge rectifier**, whose basic unfiltered form is shown in Figure 6-12. It offers several advantages over the previous full-wave

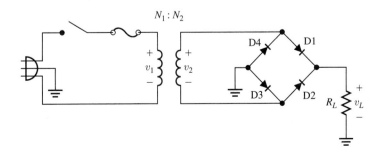

FIGURE 6–12
Schematic diagram of an unfiltered full-wave bridge rectifier circuit.

FIGURE 6-13
Conduction models for bridge rectifier for alternate half-cycles.

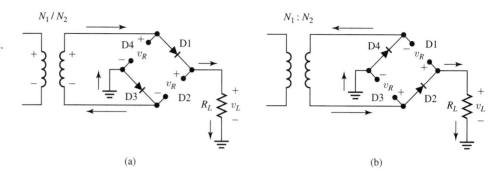

rectifier circuit, some of which follow: (1) The transformer does not require a center tap. (2) The entire secondary voltage of the transformer serves as the basis for the output voltage. (3) The peak reverse rating of the diodes need only be half as great as for the previous full-wave circuit. This last property is probably not obvious at this point but will be considered later in the chapter.

One disadvantage is that four diodes are required, but a matched bridge of diodes can be purchased as a complete unit. One other disadvantage is that there are two diode drops in the conducting path for each half-cycle, as will be seen shortly.

Refer to the circuit models shown in Figure 6-13 for the discussion that follows. When the secondary voltage is greater than about 1.4 V or so, diodes D1 and D3 are forward biased and the circuit model is shown in (a). Current flow is through D1 to the load to ground, up from the ground through D3 and into the lower side of the secondary of the transformer. Reference back to the original circuit will verify that D2 and D4 are reverse biased on that portion of the cycle. The two reserve voltages are essentially equal and will be considered later. Note, however, that there are two forward diode drops, or about 1.4 V or so in the current path.

On the opposite half-cycle, the action is depicted in Figure 6-13(b). Current flows out of the lower terminal of the secondary and now D2 and D4 are forward biased. However, the output voltage and the current through the load are both in the same direction as for the first half cycle. The diodes D1 and D3 are now reverse biased.

The output voltage of the bridge rectifier is about

$$v_L(t) = \frac{N_2}{N_1} v_1(t) - 1.4 \quad (6\text{-}19)$$

Aside from the increased diode drop, the unfiltered bridge rectifier circuit has similar waveforms to those of those of the rectifier with the center-tapped secondary. They are of the form shown back in Figure 6-10(c).

To provide calculator consistency for the reader and to minimize roundoff, we will continue to express values to three or four significant digits in the examples that follow. However, the reader should recognize that some of the values calculated are not that accurate because of the variation of the diode voltage drop, coupled with the transformer losses not considered at this point.

III EXAMPLE 6-3

An unfiltered half-wave rectifier circuit of the type shown in Figure 6-9(a) has the following parameters:

Input ac voltage = 120 V rms

Turns ratio $N_1 : N_2 = 5 : 1$

Load resistance = 80 Ω

Determine (a) peak value of primary voltage, (b) peak value of secondary voltage, (c) peak value of load voltage, (d) dc load voltage, and (e) dc load current.

SOLUTION Because there are more turns on the primary than on the secondary, the transformer is arranged as a *step-down* transformer. This means that the secondary voltage will be smaller than the primary voltage. Note that the input ac voltage is given in rms form, which is the usual way that an ac voltage is specified in most power applications.

(a) The first step will be to determine the peak value of the input voltage $V_{1(\text{peak})}$. From earlier work in the text, this voltage is related to the corresponding rms voltage $V_{1(\text{rms})}$ by

$$V_{1(\text{peak})} = \sqrt{2} V_{1(\text{rms})} = \sqrt{2} \times 120 = 169.7 \text{ V} \tag{6-20}$$

(b) The peak secondary voltage $V_{2(\text{peak})}$ is obtained by multiplying the peak primary voltage by the turns ratio arranged in the form that gives a smaller voltage. Therefore

$$V_{2(\text{peak})} = \frac{N_2}{N_1} V_{1(\text{peak})} = \frac{1}{5} \times 169.7 = 33.94 \text{ V} \tag{6-21}$$

(c) The peak load voltage is approximately

$$V_{L(\text{peak})} = V_{2(\text{peak})} - 0.7 = 33.94 - 0.7 = 33.24 \text{ V} \tag{6-22}$$

(d) The dc load voltage is

$$V_{L(\text{dc})} = \frac{V_{L(\text{peak})}}{\pi} = \frac{33.24}{\pi} = 10.58 \text{ V} \tag{6-23}$$

(e) The dc load current is then determined from Ohm's law as

$$I_{L(\text{dc})} = \frac{V_{L(\text{dc})}}{R} = \frac{10.58}{80} = 0.132 \text{ A} \tag{6-24}$$

■ **EXAMPLE 6-4**   Assume that the half-wave unfiltered rectifier circuit of Example 6-3 is converted to a full-wave unfiltered circuit of the form of Figure 6-10(a). The transformer with turns ratio $N_1 : N_2 = 5 : 1$ is replaced by a center-tapped transformer with turns ratio $N_1 : N_2 = 5 : 2$, meaning that the number of secondary turns is doubled relative to the primary and another diode is added to the circuit. Indicate any results that are the same as in Example 6-3 and perform any new required computations.

SOLUTION The number of secondary turns is doubled relative to the primary, so the value of the voltage across half the secondary is the same as the total secondary voltage in Example 6-3. Therefore, the results of parts (a), (b), and (c) remain the same and are

$$V_{1(\text{peak})} = \sqrt{2} V_{1(\text{rms})} = \sqrt{2} \times 120 = 169.7 \text{ V} \tag{6-25}$$

$$V_{2(\text{peak})} = 0.5 \frac{N_2}{N_1} V_{1(\text{peak})} = 0.5 \times \frac{2}{5} \times 169.7 = \frac{1}{5} \times 169.7 = 33.94 \text{ V} \tag{6-26}$$

$$V_{L(\text{peak})} = V_{2(\text{peak})} - 0.7 = 33.94 - 0.7 = 33.24 \text{ V} \tag{6-27}$$

Note that the actual formula for computing the peak secondary voltage is modified to fit the center-tapped transformer, but the result is the same as in the previous example.

The dc voltage in this case is

$$V_{L(\text{dc})} = \frac{2 V_{L(\text{peak})}}{\pi} = \frac{2 \times 33.24}{\pi} = 21.16 \text{ V} \tag{6-28}$$

The dc current is

$$I_{L(\text{dc})} = \frac{V_{L(\text{dc})}}{R} = \frac{21.16}{80} = 0.265 \text{ A} \tag{6-29}$$

EXAMPLE 6-5

Using the transformer of Example 6-3, which had a turns ratio of 5:1, assume that the circuit is converted to a full-wave bridge rectifier as shown in Figure 6-12. Indicate any results that are the same as in previous examples and perform any new computations.

SOLUTIONS The primary and secondary peak voltages are the same as in the preceding two examples. They are repeated here for convenience.

$$V_{1(peak)} = \sqrt{2} V_{1(rms)} = \sqrt{2} \times 120 = 169.7 \text{ V} \tag{6-30}$$

$$V_{2(peak)} = \frac{N_2}{N_1} V_{1(peak)} = \frac{1}{5} \times 169.7 = 33.94 \text{ V} \tag{6-31}$$

The peak voltage across the load in this case is about

$$V_{L(peak)} = V_{2(peak)} - 1.4 = 33.94 - 1.4 = 32.54 \text{ V} \tag{6-32}$$

The dc voltage is

$$V_{L(dc)} = \frac{2 V_{L(peak)}}{\pi} = \frac{2 \times 32.54}{\pi} = 20.72 \text{ V} \tag{6-33}$$

The dc current is

$$I_{L(dc)} = \frac{V_{L(dc)}}{R} = \frac{20.72}{80} = 0.259 \text{ A} \tag{6-34}$$

Comparing the results of this example with those of Example 6-4, it is noted that the load voltage and current are reduced slightly because two diode drops are in the path. Nevertheless, the advantages of the bridge rectifier circuit usually outweigh any disadvantages, so it is used widely.

6-5 Rectifier Circuits with Filtering

To obtain a constant or near-constant value of voltage, filtering or regulation are required. It could be said that it is theoretically impossible to obtain absolute dc from an alternating voltage, but the variation or **ripple** can be made to be sufficiently small that the voltage can be considered as pure dc for all practical purposes.

Half-Wave Rectifier with Capacitor Filter

For illustrative purposes, we will begin with a half-wave circuit. Refer to the circuit diagram of Figure 6-14. It is nearly the same as that of Figure 6-9(a) except that a capacitor has been placed in parallel. The value of the capacitor required will typically be of the order of hundreds of microfarads, which represents a very large capacitance. In some cases, a small resistance is placed between the diode and the capacitor to reduce the peak charging current, which can be very large.

The action of the filter is illustrated in Figure 6-15. The voltage that would appear if the capacitor were not present is shown in Figure 6-15(a). We will assume that the circuit

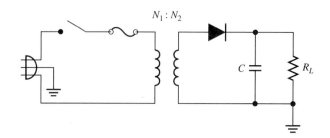

FIGURE 6–14
Schematic diagram of a basic half-wave rectifier circuit with transformer and capacitor filter.

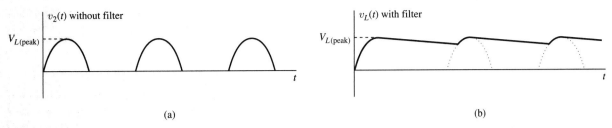

FIGURE 6–15
Waveforms for half-wave rectifier with transformer and filter.

is actually turned on at the point when $t = 0$. The capacitor quickly charges to the peak load voltage in a time equal to a quarter cycle as shown in Figure 6-15(b).

Assuming that the capacitance is sufficiently large, the secondary voltage of the transformer drops below the voltage level to which the capacitor is charged and the diode now becomes reverse biased. If there were no load, the capacitor would theoretically hold the voltage to which it has been charged indefinitely. However, the effect of the load is to draw charge from the capacitor so there is a small decrease, which is exaggerated here for visualization. Current flow from the transformer occurs when the capacitor load voltage drops below the value of the rectified secondary voltage and the capacitor is quickly charged back to the peak voltage. This process is repeated for each cycle. The resulting output voltage is therefore nearly constant, but with a ripple produced by a slight discharge during each cycle. Current flow occurs in short bursts during the charging portion of each cycle and this current can be very large. Diodes selected for a rectifier circuit must be capable of handling the large peak current.

Full-Wave Rectifiers with Capacitor Filter

Although the half-wave circuit has been useful for illustrative purposes, most practical power supply circuits use either the full-wave rectifier with center-tapped secondary or the bridge rectifier. A circuit diagram of a full-wave rectifier with a center-tapped transformer and a capacitor filter is shown in Figure 6-16. The corresponding bridge rectifier with a capacitor filter is shown in Figure 6-17.

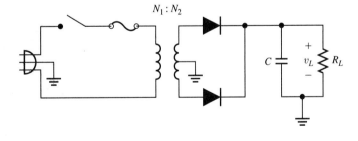

FIGURE 6–16
Schematic diagram of a full-wave rectifier circuit with transformer and capacitor filter.

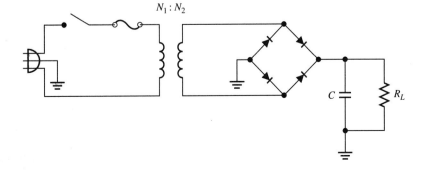

FIGURE 6–17
Schematic diagram of a full-wave bridge rectifier circuit with capacitor filter.

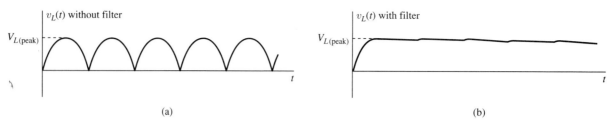

FIGURE 6–18
Waveforms for full-wave rectifier without and with filter.

Ripple Analysis

We will limit the analysis of ripple to that of a full-wave rectifier even though the analysis could be easily adapted to the half-wave case. The ripple process is essentially the same for both types of full-wave rectifiers, so one explanation should be adequate. If the input frequency to a full-wave rectifier is denoted as f, the frequency of the full-wave rectified voltage is $2f$. The doubling of the frequency means that for a given ripple level, the capacitor needs to be only half as large. Moreover, the peak value of the charging current need only be half as large for a given load current, because there are twice as many charging intervals per unit time.

Selected waveforms for a full-wave rectifier are shown in Figure 6-18. The voltage of Figure 6-18(a) represents the voltage that would appear across the load if there were no filter. However, the capacitor charges to the peak value during the first quarter-cycle of the original frequency, which is now the first half-cycle of the full-wave rectified voltage as illustrated in Figure 6-18(b). The capacitor is recharged to the peak value during each successive cycle of the rectified voltage.

Several approaches are used in the literature to estimate the ripple with slightly different answers depending on the assumptions. It is simplest to assume that the capacitance is sufficiently large that a linear decrease in the voltage can be used over the time interval of a half-cycle of the ac frequency. Moreover, as a first-order approximation, the current I_L will be assumed to be constant. With these assumptions, the change Δv in voltage can be approximated as

$$\Delta v = \frac{1}{C} I_L \Delta t \qquad (6\text{-}35)$$

The time interval Δt is half of one cycle of the ac frequency, and if the input frequency is denoted as f, we have

$$\Delta t = \frac{1}{2f} \qquad (6\text{-}36)$$

Substituting this value and redefining Δv as the peak-to-peak ripple level V_{rpp}, the result is

$$V_{\text{rpp}} = \frac{I_L}{2fC} \qquad (6\text{-}37)$$

These results indicate that the ripple is directly proportional to the load current and inversely proportional to the frequency and the capacitance.

For 60-Hz ac voltage as used in the United States, the full-wave ripple equation can be expressed as

$$V_{\text{rpp}} = \frac{I_L}{120C} \qquad (6\text{-}38)$$

The dc value of the output load voltage can be approximated by subtracting half of the peak-to-peak ripple voltage from the peak load voltage

$$V_{L(\text{dc})} = V_{L(\text{peak})} - 0.5 V_{\text{rpp}} = V_{2(\text{peak})} - \frac{I_L}{4fC} \qquad (6\text{-}39)$$

where $V_{L(\text{peak})}$ is determined by subtracting either one or two diode drops from the peak secondary voltage, depending on the type of rectifier. We remind the reader that there is some estimation involved in this formula so it should be considered as a reasonable approximation rather than an exact result.

Reverse Diode Voltage

Diode rectifier circuits present one problem: the peak reverse voltage that appears across a diode when it is reverse biased. As noted early in the chapter, semiconductor diodes have a reverse breakdown voltage rating. A diode can be easily destroyed if the voltage reaches that level.

By referring to the circuit models used in the analysis, the peak diode reverse voltage V_R can be calculated for each of the circuits. The values are tabulated in Table 6-1.

Table 6–1 Peak reverse voltage values for several rectifier circuits.

Circuit Type	Diode Peak Reverse Voltage
Half-wave rectifier (unfiltered)	$V_{2(\text{peak})}$
Half-wave rectifier (capacitor filter)	$2V_{2(\text{peak})}$
Full-wave rectifier with center tap (with or without capacitor filter)	$2V_{2(\text{peak})}$
Full-wave bridge rectifier (with or without capacitor filter)	$V_{2(\text{peak})}$

Note: $V_{2(\text{peak})}$ is the peak voltage across the transformer secondary for the half-wave rectifier and the full-wave bridge rectifier. It is the voltage across half of the transformer secondary for the full-wave rectifier with center tap.

More Complex Filter Circuits

It is possible to reduce the ripple level to a smaller level without using excessively large capacitors by using an inductance in series followed by a capacitor in parallel with the load or by more than one filter section. However, such power supply filters are not very common in newer equipment designs.

Power Supplies with a Voltage Regulator

The most common type of general purpose power supply is one that employs a solid-state voltage regulator. A typical layout is shown in Figure 6-19. From the input to the capacitor, the circuit form is that of a bridge rectifier circuit. As we have seen, a very large capacitance is required to reduce the ripple to an extremely small level. In a practical case using a regulator, the capacitor reduces the ripple to a moderate level, but the voltage regulator provides significant additional regulation of the voltage level and produces a nearly constant output dc voltage across the load. Voltage regulators are available as integrated-circuit (IC) chips. Internally, they employ operational amplifier circuit forms, which will be considered in Chapter 8.

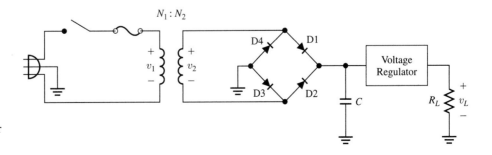

FIGURE 6–19
Schematic diagram of a rectifier circuit with voltage regulator.

EXAMPLE 6-6

Consider a full-wave rectifier circuit of the form shown in Figure 6-10 with $f = 60$ Hz, $N_1 : N_2 = 5 : 2$, $C = 1000$ μF, and a load current of 0.2 A. Determine the (a) peak-to-peak ripple voltage and (b) the dc load voltage.

SOLUTION The peak secondary and load voltages are the same as in Examples 6-3 and 6-4; that is,

$$V_{2(\text{peak})} = 33.94 \text{ V} \tag{6-40}$$

$$V_{L(\text{peak})} = 33.24 \text{ V} \tag{6-41}$$

(a) The peak-to-peak ripple voltage is given by

$$V_{\text{rpp}} = \frac{I_L}{120C} = \frac{0.2}{120 \times 1000 \times 10^{-6}} = 1.667 \text{ V} \tag{6-42}$$

(b) A reasonable estimate of the dc load voltage is given by

$$V_{L(\text{dc})} = V_{2(\text{peak})} - 0.5V_{\text{rpp}} = 33.24 - 0.5 \times 1.667 = 32.41 \text{ V} \tag{6-43}$$

The perceptive reader might have noticed a slight amount of "hand-waving" in this analysis. We assumed a constant value of load current for the purposes of computing the ripple voltage. Yet, because the voltage is varying, the load current is also varying in the same proportion for a resistive load. This is still another approximation in the analysis. However, as long as the ripple level is reasonably small, this assumption leads to workable results.

EXAMPLE 6-7

Assume in the system of Example 6-6 that it is desired to decrease the peak-to-peak ripple voltage to 1 V while maintaining a dc load current of 0.2 A. Determine the value of the capacitance required.

SOLUTION Setting the peak-to-peak ripple voltage to 1 V results in the equation that follows.

$$V_{\text{rpp}} = \frac{I_L}{120C} = \frac{0.2}{120C} = 1 \tag{6-44}$$

Solving for C, we obtain

$$C = \frac{I_L}{120V_{\text{rpp}}} = \frac{0.2}{120 \times 1} = 1.667 \times 10^{-3} \text{ F} = 1667 \text{ μF} \tag{6-45}$$

As can be seen, the net capacitance increases as the desired ripple level decreases.

6-6 Survey of Other Diode Types

Several other types of diodes will be surveyed in this section. Some are designed for very special applications; others are more common. Refer to Figure 6-20 for the schematic representations.

Zener Diode

The **zener diode** was mentioned earlier in the chapter. It exploits the reverse breakdown voltage. By setting specific doping levels in the manufacturing process, the value of the reverse breakdown can be controlled to provide a specific voltage level. This value is nearly constant over a range of current. It can be used to provide voltage reference levels to a reasonable degree of accuracy.

The schematic symbol for a zener diode is shown in Figure 6-20(a). Because the need frequently arises for a voltage reference, the next section will be devoted to an analysis of some of the properties of a zener voltage reference circuit.

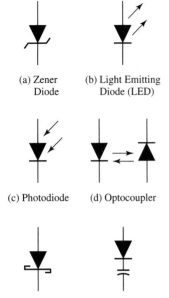

(a) Zener Diode
(b) Light Emitting Diode (LED)
(c) Photodiode
(d) Optocoupler
(e) Schottky Diode
(f) Varactor Diode

FIGURE 6–20
Schematic symbols for various types of diodes.

Light Emitting Diode

The **light emitting diode (LED)** forms the basis for many different types of displays ranging from calculators to television and computer screens. The schematic for a single LED is shown in Figure 6-20(b), and the arrows leaving the diode represent light. Light is emitted when the diode is forward biased. Many LED applications involve arrays of LEDs combined as an integrated circuit module for numerous types of display purposes.

Photodiode

A **photodiode** is roughly opposite to the LED. Its conductivity is a function of the light impinging on it. The schematic diagram is shown in Figure 6-20(c) and the arrows are directed toward it. Photodiodes find applications as light-sensitive transducers.

Optocoupler

An **optocoupler** implemented with diodes consists of a light emitting diode coupled with a photodiode. Its schematic symbol is shown in Figure 6-20(d). It transfers signals optically from one point to another without any direct electrical coupling. Optocouplers are also implemented with transistors.

Schottky Diode

The **Schottky diode** is a special-purpose diode that reduces the forward diffusion capacitance to a negligible value, which permits its use well into the microwave frequency range. The schematic symbol is shown in Figure 6-20(e).

Varactor Diodes

Junction diodes display a capacitance effect when the diode is reverse biased. The capacitance is a function of the reverse voltage across the junction. Diodes manufactured to exploit this effect are called **varactor diodes** or simply **varactors**. The schematic symbol is shown in Figure 6-20(f). The effect of causing the capacitance to change with a change in voltage can be used in tuning circuits at high frequencies. Many television sets use varactors in the tuning circuits.

6-7 Zener Regulator Circuits

In this section, we consider the application of a zener diode to produce a reasonably constant voltage reference. We note at the outset that the voltage will vary slightly with different operating conditions. However, zener circuits provide an important step in the process of voltage regulation. The integrated circuit voltage regulators that provide the highest degree of regulation often use a zener diode as part of the overall control strategy. The development in this section will ignore slight voltage variations and will focus primarily on how the circuit operates.

The form of a typical **zener regulator circuit** is shown in Figure 6-21. Note that the zener is arranged in the direction for reverse bias, which is required for proper operation.

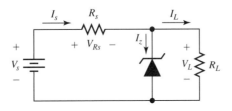

FIGURE 6–21
Typical zener regulator circuit.

138 CHAPTER 6 • Diodes and Their Applications

Zeners are rated at different voltage levels, so one is selected for the given circuit to provide the desired voltage.

The objective in this case is to provide a nearly constant voltage V_L across the load R_L. In the most general case, the load resistance may vary and the source voltage V_s, from which the power is obtained, may also vary. However, if the zener effect were perfect, the voltage V_L would remain constant over a range of operating conditions. The zener voltage rating of the diode must be V_L. It is also necessary that $V_s > V_L$ or the circuit won't work as intended.

Assume that the voltage across the load is V_L and thus the current through the load resistance R_L is

$$I_L = \frac{V_L}{R_L} \tag{6-46}$$

There must be a current I_z flowing into the reverse-biased zener to establish the necessary voltage level. In fact there is a range of current $I_{z(min)} \leq I_z \leq I_{z(max)}$ over which the zener will operate properly. Worst-case conditions must be considered in the design process to ensure that the current will remain within these bounds.

By KCL, the current flowing from the source to the node junction connecting the diode and the load is the sum of the zener current plus the load current.

$$I_s = I_z + I_L \tag{6-47}$$

The voltage drop across R_s is $R_s I_s$, and by KVL, the source voltage must satisfy

$$V_s = R_s I_s + V_L \tag{6-48}$$

Qualitative Explanation

Assume initially that the circuit is in equilibrium according to the equations previously indicated with a source voltage V_s and an output voltage V_L. Two possible separate changes in circuit conditions will be qualitatively analyzed: (1) a change in the source voltage V_s and (2) a change in the load current requirement I_L (meaning that the load resistance has changed). In discussing these changes, remember that the zener diode will adjust its current as necessary within limits to maintain the constant voltage V_L.

Consider first that the voltage V_s increases while the load current is assumed to remain constant. For V_L to remain constant, the voltage drop across R_s must increase. The load current is fixed by V_L so the only way to increase the voltage across R_s is for the zener current to increase to whatever value is required to cause the voltage across R_s to reach the difference between the source voltage and the load voltage. If the source voltage decreases, the opposite process occurs. Thus, an increase in source voltage causes an increase in zener current and a decrease in source voltage causes a decrease in zener current. Of course, there are limits in both directions. On the high side, the diode could be destroyed and on the low side, the regulating action could be lost.

Consider next that the voltage V_s is fixed, but there is a need for an increased load current at the fixed voltage V_L, corresponding to a decrease in the load resistance. Both the source voltage and the load voltage are fixed; therefore the voltage across the resistance R_s must remain fixed, meaning that the current I_s cannot vary. The only way that this can happen is for the current I_z to decrease so that the sum $I_L + I_z$ remains constant. If the load current decreases, the opposite process occurs. Thus, an increase in load current causes a decrease in zener current and a decrease in load current causes an increase in zener current. Again, the limits of operation must be considered for the circuit to work.

III EXAMPLE 6-8 Consider the regulator circuit of Figure 6-22. The zener diode is rated at 10 V. Assume that the initial value of the source voltage is 14 V as shown. Determine the values of V_{R_s}, I_L, I_s, and I_z.

FIGURE 6–22
Circuit of Example 6-8 with voltage and current values labeled.

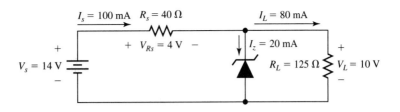

SOLUTION From KVL, the voltage V_{Rs} is determined as

$$V_{Rs} = V_s - V_z = 14 - 10 = 4 \text{ V} \tag{6-49}$$

The current I_s is determined as

$$I_s = \frac{V_{Rs}}{R_s} = \frac{4}{40} = 0.1 \text{ A} = 100 \text{ mA} \tag{6-50}$$

The load current is

$$I_L = \frac{10}{125} = 0.08 \text{ A} = 80 \text{ mA} \tag{6-51}$$

By KCL, the zener current is

$$I_z = I_s - I_L = 100 - 80 = 20 \text{ mA} \tag{6-52}$$

The preceding values are all labeled on the circuit diagram.

EXAMPLE 6-9 Assume in the circuit of Example 6-8 that the source voltage increases to 16 V as shown in Figure 6-23. Assuming that the zener voltage remains at 10 V, determine V_{Rs}, I_L, I_s, and I_z.

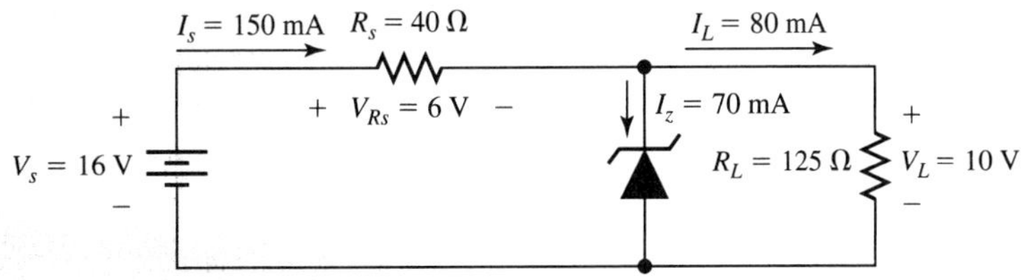

FIGURE 6–23
Circuit of Example 6-9 with voltage and current values labeled.

SOLUTION Note that the assumption of an absolutely constant zener voltage is an approximation. The actual voltage will vary slightly as the current through it changes. Specification sheets provide a value of **zener impedance** for estimating changes in that voltage. However, that is a second-order effect, and as long as other changes are not too great, it is reasonable as a first-order approximation to assume a constant zener voltage. The analysis here will be limited to that assumption.

The calculations for V_{Rs} and I_s follow.

$$V_{Rs} = V_s - V_z = 16 - 10 = 6 \text{ V} \tag{6-53}$$

$$I_s = \frac{V_{Rs}}{R_s} = \frac{6}{40} = 0.15 \text{ A} = 150 \text{ mA} \tag{6-54}$$

If the load voltage is assumed to remain at 10 V, the load current will remain at

$$I_L = 80 \text{ mA} \tag{6-55}$$

The zener current changes to

$$I_z = I_s - I_L = 150 - 80 = 70 \text{ mA} \tag{6-56}$$

Thus, when the source voltage increases, more current is required in the series resistance to maintain the load voltage at a constant value. The diode absorbs the additional current to maintain the assumed constant zener voltage, and the load current remains constant.

EXAMPLE 6-10

Consider the circuit of Example 6-8 with the original source voltage of 14 V. Assume now that the load changes from 125 Ω to 150 Ω as shown in Figure 6-24. Assuming again that the zener voltage remains at 10 V, determine V_{Rs}, I_L, I_s, and I_z.

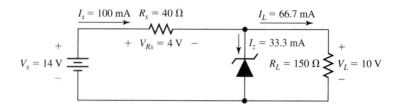

FIGURE 6–24
Circuit of Example 6-10 with voltage and current values labeled.

SOLUTION The voltage V_{Rs} and current I_s are the same as in Example 6-8. They are

$$V_{Rs} = V_s - V_z = 14 - 10 = 4 \text{ V} \tag{6-57}$$

$$I_s = \frac{V_{Rs}}{R_s} = \frac{4}{40} = 0.1 \text{ A} = 100 \text{ mA} \tag{6-58}$$

In this case, the load current will change. It is

$$I_L = \tfrac{10}{150} = 0.0667 \text{ A} = 66.7 \text{ mA} \tag{6-59}$$

The zener current is now

$$I_z = I_s - I_L = 100 - 66.7 = 33.3 \text{ mA} \tag{6-60}$$

When the load current decreases, the zener must absorb more current to maintain the constant current required in the series resistance, corresponding to a constant voltage drop.

To summarize the various cases considered here, it is helpful to think of the zener voltage as the "rigid" constraint. Neglecting the slight changes that have been referred to earlier, the circuit variables change in whatever manner is required to keep the zener voltage constant, but simultaneously they satisfy the basic requirements of KVL, KCL, and Ohm's law. However, all zener diodes are characterized by a range of minimum current to maximum current, and operation must be constrained to that range.

PROBLEMS

Note: Assume that the forward-biased voltage drop for each diode is 0.7 V. In Problems 6-1 through 6-4, note carefully the signs of the calculated voltages and currents based on the assumed reference directions given.

6-1 For the circuit of Figure P6-1, determine the loop current, the voltage across the diode, and the voltage across the resistance.

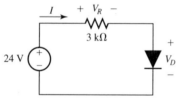

FIGURE P6–1

6-2 For the circuit of Figure P6-2, determine the loop current, the voltage across the diode, and the voltage across the resistance.

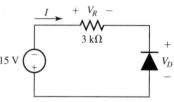

FIGURE P6–2

6-3 For the circuit of Figure P6-3, determine the loop current, the voltage across the diode, and the voltage across the resistance.

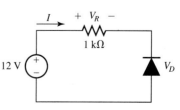

FIGURE P6–3

6-4 For the circuit of Figure P6-4, determine the loop current, the voltage across the diode, and the voltage across the resistance.

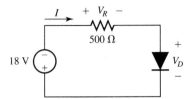

FIGURE P6-4

6-5 An unfiltered half-wave rectifier circuit of the type shown in Figure 6-9(a) has the following parameters:

Input ac voltage = 115 V rms

Turns ratio $N_1 : N_2 = 4 : 1$

Load resistance = 50 Ω

Determine (a) peak value of primary voltage, (b) peak value of secondary voltage, (c) peak value of load voltage, (d) dc load voltage, and (e) dc load current.

6-6 An unfiltered half-wave rectifier circuit of the type shown in Figure 6-9(a) has the following parameters:

Input ac voltage = 110 V rms

Turns ratio $N_1 : N_2 = 1 : 2$

Load resistance = 150 Ω

Determine (a) peak value of primary voltage, (b) peak value of secondary voltage, (c) peak value of load voltage, (d) dc load voltage, and (e) dc load current.

6-7 Assume that the half-wave unfiltered rectifier circuit of Problem 6-5 is converted to a full-wave unfiltered circuit of the form of Figure 6-10(a). The transformer with turns ratio 4 : 1 is replaced by a center-tapped transformer with turns ratio 4 : 2 and another diode is added to the circuit. Indicate any results that are the same as in Problem 6-5 and perform any new required computations.

6-8 Assume that the half-wave unfiltered rectifier circuit of Problem 6-6 is converted to a full-wave unfiltered circuit of the form of Figure 6-10(a). The transformer with turns ratio 1 : 2 is replaced by a center-tapped transformer with turns ratio 1 : 4 and another diode is added to the circuit. Indicate any results that are the same as in Problem 6-6 and perform any new required computations.

6-9 Using the transformer of Problem 6-5, which has a turns ratio of 4 : 1, assume that the circuit of that problem is converted to a full-wave bridge rectifier as shown in Figure 6-12. Indicate any results that are the same as in previous examples and perform any new computations.

6-10 Using the transformer of Problem 6-6, which has a turns ratio of 1 : 2, assume that the circuit of that problem is converted to a full-wave bridge rectifier as shown in Figure 6-12. Indicate any results that are the same as in previous examples and perform any new computations.

6-11 Consider a full-wave rectifier circuit of the form shown in Figure 6-10 with an input voltage of 120 V rms, $f = 60$ Hz, $N_1 : N_2 = 4 : 2$, $C = 1200$ μF, and a load current of 0.2 A. Determine the (a) peak-to-peak ripple voltage and (b) dc load voltage.

6-12 Consider a full-wave rectifier circuit of the form shown in Figure 6-10 with an input voltage of 120 V rms, $f = 60$ Hz, $N_1 : N_2 = 1 : 4$, $C = 820$ μF, and a load current of 0.25 A. Determine the (a) peak-to-peak ripple voltage and (b) dc load voltage.

6-13 Consider the regulator circuit of Figure P6-13. The zener diode is rated at 9 V. Assume that the initial value of the source voltage is 12 V as shown. Determine the values of V_{Rs}, I_L, I_s, and I_z.

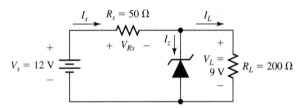

FIGURE P6-13

6-14 For the circuit of Problem 6-13 (Figure P6-13), assume that the source voltage increases to 14 V. Assuming that the zener voltage remains at 9 V, determine V_{Rs}, I_L, I_s, and I_z.

6-15 For the circuit of Problem 6-13 (Figure P6-13) with the original source voltage of 12 V, assume that the load changes from 200 Ω to 250 Ω. Assuming again that the zener voltage remains at 9 V, determine V_{Rs}, I_L, I_s, and I_z.

6-16 For the circuit of Problem 6-13 (Figure P6-13), assume that the source voltage changes to 14 V *and* the load resistance changes to 250 Ω. Assuming again that the zener voltage remains at 9 V, determine V_{Rs}, I_L, I_s, and I_z.

6-17 Assume that a certain process can be achieved with a full-wave rectified voltage without filtering. The average or dc value of the load voltage is to be 12 V. Based on the full-wave rectifier circuit of Figure 6-10(a) and an assumed input ac voltage of 120 V rms, determine the required turns ratio $N_1 : N_2$.

6-18 Assume that the design of Problem 6-17 is to be achieved with a bridge rectifier circuit of the form shown in Figure 6-12. Assuming again an input ac voltage of 120 V rms, determine the required turns ratio $N_1 : N_2$.

6-19 A full-wave bridge rectifier circuit with a capacitor filter of the form shown in Figure 6-17 is to be designed to meet the following specifications:

approximate dc output voltage: 12 V

load current: 0.5 A

maximum ripple: 0.2 V peak-to-peak

The transformer input ac voltage is 120 V rms. Determine (a) value of capacitance required, (b) peak rectifier output voltage, (c) peak secondary transformer voltage, and (d) turns ratio $N_1 : N_2$ required for the transformer.

6-20 Assume that the design of Problem 6-18 is to be achieved with the full-wave rectifier circuit of the form shown in Figure 6-16. Repeat the analysis of Problem 6-19 based on this configuration.

6-21 A zener diode regulator circuit of the form shown in Figure 6-21 is to be designed to establish a 15-V dc reference with a diode having that assumed voltage level. Some design assumptions are as follows:

> source voltage: 20 V
>
> load current requirement: 100 mA
>
> zener nominal bias current 40 mA

Determine the required value of the resistance R_s.

6-22 A zener diode regulator circuit of the form shown in Figure 6-21 is to be designed to establish a 5-V dc reference with a diode having that assumed voltage level. Some design assumptions are as follows:

> source voltage: 12 V
>
> load current requirement: 100 mA
>
> zener nominal bias current 30 mA

Determine the required value of the resistance R_s.

6-23 The circuit of Figure P6-23 is one form of a **split power supply** in which both a positive and a negative voltage can be simultaneously obtained at the output. The two voltages are denoted here as v_L^+ and v_L^-. Of course, filtering and regulation can be added to both sides to reduce the ripple. Provide an analysis of the circuit similar to that performed in the text to determine expressions for v_L^+ and v_L^- in terms of the input voltage v_1.

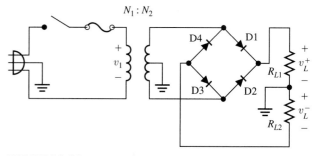

FIGURE P6–23

6-24 The circuit of Figure P6-24 is a **voltage doubler** circuit. (a) Construct equivalent circuits for both halves of the input cycle and explain how the circuit works. (b) Ignoring any ripple on the output, develop an expression for v_L in terms of the peak value $V_{1(\text{peak})}$ of the input voltage.

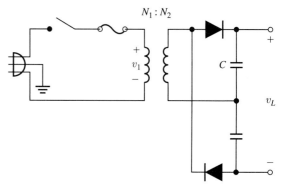

FIGURE P6–24

Transistors

OVERVIEW AND OBJECTIVES

This chapter will be devoted to an introductory treatment of transistors and their characteristics. Both the **bipolar junction transistor (BJT)** and the **field effect transistor (FET)** will be considered. As in the previous chapter, the terminal characteristics will be emphasized, and some of the common properties will be considered.

Objectives

After completing this chapter, the reader should be able to

1. Describe the basic properties of a **bipolar junction transistor (BJT)**, including the names of the terminals and the classifications as **NPN** or **PNP**.
2. Construct equivalent circuits of the **BJT** for both **cutoff** and **saturation**.
3. Describe the nature of both the **base** and **collector** characteristic curves for a BJT.
4. Describe the three regions of operation for a BJT.
5. Define **dc current gain** for a BJT and discuss its significance.
6. Design a BJT switch using worst-case conditions.
7. Discuss the family tree for the **field effect transistor (FET)** and show the various classifications.
8. Describe the general properties of the **junction field effect transistor (JFET)**.
9. Describe the three regions of operation for a JFET.
10. Describe the general properties of the **metal-oxide semiconductor field effect transistor (MOSFET)**.
11. Explain the differences between the **depletion-mode MOSFET** and the **enhancement-mode MOSFET**.
12. Discuss how FETs can be used as switches.
13. Provide a qualitative explanation of how BJTs and FETs can function as **amplifiers**.
14. Provide a general comparison between BJTs and FETs.

7-1 Introduction

The **bipolar junction transistor (BJT)** was developed in the late 1940s. It began a major revolution in the electronics industry. Previously all electronic amplification and control functions were performed with **vacuum tubes**. Transistors were smaller, more reliable, and used less power.

The transistor was only one step in the evolution of modern electronics. In the next several decades, the **integrated circuit (IC)** was introduced and expanded to become a major core of the electronics industry. The first integrated circuits represented the equivalent of several individual transistors. The evolution continued through the processes referred to as **large-scale integration (LSI)** and **very large-scale integration (VLSI)**. At the time this book is being developed, integrated circuits are available that contain the equivalent of hundreds of thousands of transistors, and they are found in virtually all areas of modern life. This includes personal computers, television sets, automobiles, cooking appliances, and many other areas.

Why Study Transistors?

This book is certainly not aimed at producing individuals who can design integrated circuits. Most nonspecialists primarily will be using complete integrated circuits for most electronic circuit applications, so a natural question is why transistors should be studied at the component level. Of course, knowing something about transistors is part of a broad general education in the electrical and electronics field. However, a more pragmatic answer is that discrete transistors still are used for many applications. Many of the very complex ICs must be interfaced with various electrical devices, and it is often necessary to use a discrete transistor to accomplish the necessary connections between the IC chip and the application device.

We will deal very little with classical transistor amplifier circuits because most applications of the type emphasized in this text can be performed much more easily with operational amplifier integrated circuits, which will be studied in some detail in the next chapter. Rather, the applications considered in this chapter are primarily aimed at the types of considerations inherent in interfacing, including switching and level shifting.

7-2 Bipolar Junction Transistor

The most basic form of the *bipolar junction transistor (BJT)* is composed of three sections of semiconductor material. The three sections are called the **emitter**, the **base**, and the **collector**. On circuit diagrams, they will usually be identified in this text with the symbols E, B, and C, respectively.

NPN and PNP

There are two basic configurations of the BJT. They are referred to as **NPN** and **PNP**. These terms will be evident after the brief discussion that follows.

NPN Layout

The basic construction of an NPN transistor is suggested by the layout shown in Figure 7-1(a) The emitter and collector portions are composed of N-type material. They constitute the outer sections. The base is composed of a thin layer of P-type material that is lightly doped. The emitter is heavily doped and the collector has a moderate level of doping. The collector region is larger than the emitter region and can dissipate more heat.

The schematic symbol for an NPN transistor is shown in Figure 7-1(b). Note that the arrow in the emitter is pointing in the direction of conventional current flow through the device.

PNP Layout

The basic construction of a PNP transistor is suggested by the layout shown in Figure 7-2(a). The doping properties discussed for the NPN type are applicable here except that the P and N sections are opposite to those for the NPN type. The schematic symbol is shown in Figure 7-2(b). Again, the direction of conventional current flow is in the direction of the arrow.

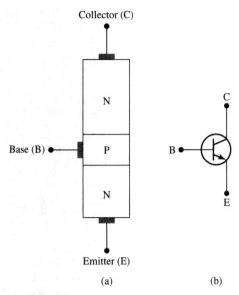

FIGURE 7–1
Layout of an NPN transistor and its schematic symbol.

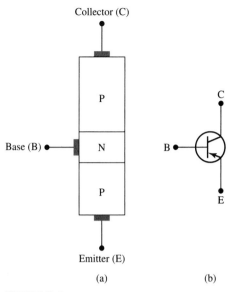

FIGURE 7–2
Layout of a PNP transistor and its schematic symbol.

Analogy to Two Diodes

A thought process that proves to be very useful will now be given. The junction between the base and emitter is called the **base-emitter junction** or **base-emitter diode**. Likewise, the junction between the base and collector is called the **base-collector junction** or **base-collector diode**.

From the standpoint of bias conditions and the previous definitions, the NPN and PNP transistors can be visualized in some sense as shown in Figure 7-3. However, it should be stressed that *a transistor is much more complex than the simple connection of two diodes*. However, the two models shown are very useful in assisting one to consider how to connect and bias transistors. Note the dashed line connecting to the base. As will be seen later, there is a degree of electrical isolation between the base and the external connection to it, and the dashed line serves as a reminder of that fact.

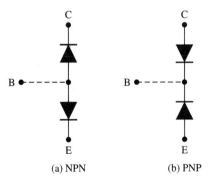

FIGURE 7–3
Models to assist in visualizing BJT as two diodes.

Basic Operation

The actual internal operation of the transistor requires some background in solid-state theory to explain fully. However, on a simplified level, think of the base current as a *controlling variable* and the collector and emitter currents as the *controlled variables*. Said differently, a small base current can control much larger collector and emitter currents.

Relationship for Currents

Note that in most applications, the collector current and the emitter current are nearly the same. Specifically, the emitter current I_E is given by the sum of the collector current I_C and the base current I_B; that is,

$$I_E = I_C + I_B \qquad (7\text{-}1)$$

However, the base current is typically much smaller than either of the other currents, so in some design and analysis steps in which in the base current is not known, it is often reasonable to assume that collector and emitter currents are approximately equal.

■ EXAMPLE 7-1

In parts (a) through (c), two of the three transistor currents are given. Determine the third in each case if possible: (a) $I_C = 40$ mA, $I_B = 200$ μA; (b) $I_E = 8$ mA, $I_B = 50$ μA, (c) $I_E = 5.02$ mA, $I_C = 5$ mA.

SOLUTION Each part of this solution will use Equation 7-1 arranged in whichever form is required. Unit conversions will be made in each case utilizing the most convenient forms.

(a) $I_C = 40$ mA, $I_B = 200$ μA

The emitter current is

$$I_E = I_C + I_B = 40 \text{ mA} + 0.2 \text{ mA} = 40.2 \text{ mA} \qquad (7\text{-}2)$$

(b) $I_E = 8$ mA, $I_B = 50$ μA

The collector current is

$$I_C = I_E - I_B = 8 \text{ mA} - 0.05 \text{ mA} = 7.95 \text{ mA} \qquad (7\text{-}3)$$

(c) $I_E = 5.02$ mA, $I_C = 5$ mA

The base current is

$$I_B = I_E - I_C = 5.02 \text{ mA} - 5 \text{ mA} = 0.02 \text{ mA} = 20 \text{ μA} \qquad (7\text{-}4)$$

These are typical values for small transistors. The actual characteristics of a given transistor type will often vary considerably from one unit to another, so the reader might see why the approximation that collector and emitter currents are nearly the same would be reasonable in many cases.

7-3 BJT Operating Regions

As discussed in the preceding section, the BJT can be represented from the standpoint of bias conditions as the equivalent of two junction diodes. Therefore, each can be forward biased or reverse biased, so there are four combinations of bias conditions for the two diodes: (1) Both junctions reverse biased, (2) both junctions forward biased, (3) base-emitter junction forward biased and base-collector junction reverse biased, and (4) base-emitter junction reverse biased and base-collector junction forward biased. Condition (4) is not commonly encountered, and when it is, it can be viewed somewhat in the same manner as condition (3), but with the emitter and collector reversed. The characteristics are not very desirable in that condition, because the collector and emitter have different properties. Hence, that condition will not be considered further. The first three conditions and the names of the corresponding operating regions are summarized in the table that follows. Each will be discussed.

Base-Emitter Junction	Base-Collector Junction	Region Name
Reverse bias (RB)	Reverse bias (RB)	Cutoff
Forward bias (FB)	Forward bias (FB)	Saturation
Forward bias (FB)	Reverse bias (RB)	Active

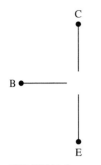

FIGURE 7–4
Ideal model of BJT in cutoff region.

Cutoff Region

The **cutoff region** corresponds to reverse bias for both the base-emitter and the base-collector junctions. This means that both diodes essentially act as open circuits. An ideal equivalent circuit is shown in Figure 7-4. This ideal circuit means that for many practical purposes, the three transistor terminals may be considered as open circuits. Actually, there may be a small temperature-dependent reverse leakage current for each junction, but these currents may be ignored in a large number of applications.

Saturation Region

The **saturation region** corresponds to forward bias for both the base-emitter and the base-collector junctions. In the same sense as for the simplest model for a forward-biased diode, a first approximation for a quick analysis is to assume short circuits for both of the junctions. This equivalent circuit model is shown in Figure 7-5 and, in effect, it brings all the terminals together.

More accurate models for saturation are shown in Figure 7-6 for both the NPN and PNP transistors. With these models, the base-emitter voltage is assumed to be a constant voltage drop of about 0.7 V, with the directions opposite for NPN and PNP transistors as shown.

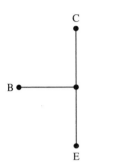

FIGURE 7–5
Ideal model of BJT in saturation region.

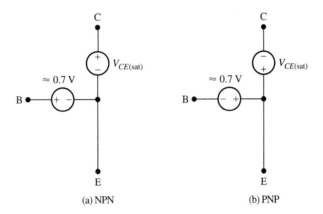

FIGURE 7–6
Better models for BJTs in saturation region.

While one could assume about 0.7 V for base-collector junction under some conditions, it is more useful to deal instead with the collector-emitter voltage drop in applications involving saturation. If base-emitter and base-collector drops were equal with saturation, the voltage between collector and emitter would be zero. This assumption will be made in various applications that follow. However, the collector current in saturation is typically much larger than the base current. The net result is to form a small difference voltage between collector and emitter. This voltage is denoted as $V_{CE(\text{sat})}$. It has opposite polarities for NPN and PNP transistors as shown in Figure 7-6. It varies from less than 0.1 V at small saturation currents to 1 V or greater at very large saturation currents for power transistors that deal with very large currents.

Active Region

The **active region** corresponds to forward bias for the base-emitter junction and reverse bias for the base-collector junction. It is the region where **linear operation** is assumed. This region corresponds essentially to all amplifier-type applications. Although this mode of operation is very important, it will not be pursued in depth in this chapter because virtually all applications involving linear operation (except at microwave frequencies) use integrated-circuit forms. Some consideration of the active region and linear operation will be considered in conjunction with operational amplifiers in the next chapter. Moreover, the analysis that follows in this chapter will often use some of the properties of the active region, so they need to be understood.

EXAMPLE 7-2 Classify each of the following conditions as *cutoff, saturation*, or *active region*. (a) NPN: $V_{BE} = 0.7$ V, $V_{CE} = 15$ V, (b) NPN: $V_{BE} = 0.7$ V, $V_{CE} = 0$ V, (c) NPN: $V_{BE} = -0.5$ V, $V_{CE} = 10$ V, (d) PNP: $V_{EB} = 0.7$ V, $V_{CE} = 0$ V.

SOLUTION

(a) For an NPN transistor, $V_{BE} = 0.7$ V and $V_{CE} = 15$ V correspond to the *active region*.

(b) For an NPN transistor, $V_{BE} = 0.7$ V and $V_{CE} = 0$ V correspond to *saturation*.

(c) For an NPN transistor, $V_{BE} = -0.5$ V and $V_{CE} = 10$ V correspond to *cutoff*.

(d) For a PNP transistor, $V_{EB} = 0.7$ V (or $V_{BE} = -0.7$ V) and $V_{CE} = 0$ V corresponds to *saturation*.

7-4 BJT Characteristic Curves

The external characteristics of a BJT are best explained by describing an experiment that can be conducted in any basic electronics laboratory and is often assigned to students studying electronics. For this analysis and many other applications, we will assume an NPN transistor because the polarities seem to be a little more natural in the learning process. For a PNP transistor, simply reverse all the voltage polarities and current directions and voltmeter and ammeter connections, and similar results can be obtained.

Common-Emitter Form

In most BJT applications, one of the three terminals is considered as common between input and output. Thus, the terms **common emitter**, **common base**, and **common collector** all appear as descriptive circuit terms throughout the applications of transistors. Of the three, the *common-emitter* configuration is by far the most widely employed. Most transistor characteristics are measured in that form. That will be the case in the development that follows.

Measurement Circuit

The circuit to be used in the experiment is shown in Figure 7-7. Note how the emitter terminal is common between the left-hand and right-hand sides of the circuit and is connected to the common ground. Therefore, the circuit is arranged in a *common-emitter* form and measurements of voltages on both sides of the circuit will be made with respect to the emitter.

The adjustable voltage source V_{BB} is used to forward bias the base-emitter junction. The ammeter on the left measures the base current I_B, which is often in the range of milliamperes or microamperes. The voltmeter connected between base and emitter measures the base-to-emitter voltage V_{BE}. The purpose of the resistance R_B is the same as in the circuit used to measure a diode characteristic, namely to allow fine adjustment of the base current without burning out the transistor by exceeding the base current.

FIGURE 7–7
Circuit that can be used to measure transistor characteristics.

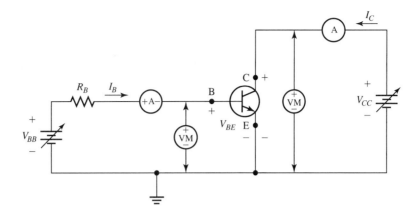

On the collector side, an adjustable dc source V_{CC} sets the voltage V_{CE}. Note that this voltage has a positive direction as shown, but as long as it exceeds about 0.7 V and a little more as the collector current increases, it will reverse bias the base-collector junction. The ammeter on the right measures the collector current I_C and the voltmeter on the right measures the collector-to-emitter voltage.

Base Characteristics

Assume first that the collector voltage supply is set to a value of 0 V. Then assume that the base current is measured as a function of the base-to-emitter voltage by varying the dc supply V_{CC} in small steps. A curve of the form shown in Figure 7-8 will be obtained.

Does this curve look familiar? It should, because it is essentially the forward-bias curve of a PN junction such as was encountered in the preceding chapter. Like the diode of that chapter, the voltage does vary and various references assume 0.6 V, 0.65 V, or 0.7 V. To be consistent with earlier assumptions, we will assume 0.7 V. Although the curve has the same form as for a diode, the base current will be *considerably smaller* than the current that would be flowing in a forward-biased PN junction diode. This is one of the properties that sets the transistor apart from a simple series connection of two diodes.

One question that will be addressed is whether or not the base characteristic curve varies with changes in collector-to-emitter voltage. Actually it does change slightly. It tilts a little toward the right as the collector-to-emitter voltage is increased. This phenomenon is referred to as the **Early effect**. For our purposes, we will neglect this phenomenon and assume that the base characteristic remains essentially fixed as the collector-to-emitter voltage is varied. This means that we will need to consider only one characteristic curve on the base side of the circuit.

FIGURE 7–8
Base characteristic with collector-to-emitter voltage constant.

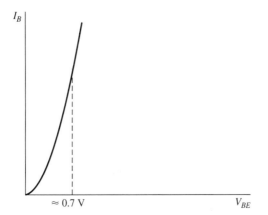

Collector Characteristics

The most interesting characteristics, and the ones that produce the important action of the transistor, are the collector characteristics. The process will be idealized somewhat in the discussion that follows. The values that will be assumed should be considered as typical because they will vary from one transistor to another. However, some possible values should help the reader visualize the process. While keeping a reference to the circuit of Figure 7-7, refer to Figure 7-9 for the discussion that follows.

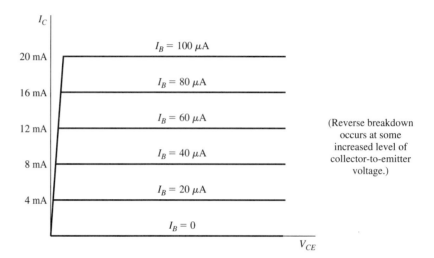

FIGURE 7–9
Collector characteristics of an ideal representative BJT.

Assume first that the base current is set to zero by adjusting V_{BB} to zero. While this value is set, assume that the collector-to-emitter voltage V_{CE} is adjusted from zero to some positive value. The result is essentially a value of zero collector current, and the curve can be considered as coinciding with the abscissa as shown. (Actually, there may be a small temperature-dependent reverse current, but it is usually negligible for most applications.)

For this curve and subsequent curves, it is assumed that the collector-to-emitter voltage is always kept below the breakdown voltage, at which point the transistor could be destroyed.

Next, assume that the left-hand dc voltage is adjusted to a level that establishes a base current $I_B = 20$ μA. As before, the voltage V_{CE} is adjusted from zero to some maximum positive level. Now, note that the collector current changes to 4 mA and remains at that level over a wide range.

This process is continued in steps of 20 μA for the base current. For each increase in base current, the collector current also increases. Note, however, that the increase in collector current is always considerably greater than the increase in base current. The control function of the transistor is coming into play here.

As noted earlier, these characteristics have been idealized. Actual transistor characteristics differ from these in at least two ways: (1) The increases in collector current for constant changes in base current are not exactly the same, as operation is varied over the set of operating curves. (2) Actual transistor characteristic curves have a slight positive slope, and the collector current on a given curve increases somewhat with increasing voltage. Nevertheless, the behavior here is a reasonable first approximation, that has a great deal of utility in practice.

Operating Regions

A general overview of the collector characteristics displaying the three operating regions is shown in Figure 7-10. The **cutoff region** corresponds to the abscissa and occurs when the base current is zero. The **saturation region** corresponds to the region near the ordinate in which the collector-to-emitter voltage is zero or a small positive value $V_{CE(\text{sat})}$. All of the

FIGURE 7-10
Three regions of operation for a BJT.

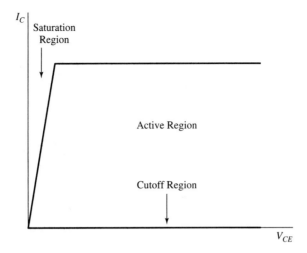

region above the abscissa and to the right of the saturation line corresponds to the **active region**. This is the area for linear operation. It corresponds to amplification. However, working with cutoff or saturation necessitates that some analysis be performed with respect to this region. That process will be illustrated as we progress.

DC Current Gain

One of the most important parameters used in characterizing the behavior of a BJT is the **dc current gain**, which will be denoted as β_{dc}. It is defined as

$$\beta_{dc} = \frac{I_C}{I_B} \tag{7-5}$$

Stated in words, the value of the dc current gain is the ratio of the collector current to the base current. Typically, the collector current is considerably larger than the base current, so β_{dc} can be viewed as a type of control or amplification factor. Typical values range from about 50 to 200 or more.

Polarities and Current Directions

The reference directions of voltage and current that will be used in the text for both NPN and PNP transistors are shown in Figure 7-11. Note that for NPN transistors, V_{BE} and V_{CE} are more convenient. For PNP transistors, V_{EB} and V_{EC} are more convenient.

Variations in the Current Gain

For a particular transistor, there will be some variation in the value of β_{dc} as the operating point on the curve is changed. However, an even more pronounced variation is that of

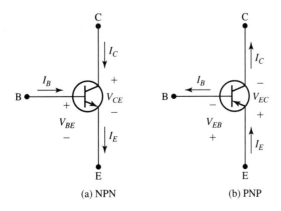

FIGURE 7-11
Reference positive directions for both NPN and PNP transistors.

(a) NPN   (b) PNP

different transistors of the same model number. It is not uncommon for the value of β_{dc} to vary by a factor of 5 to 1 or greater within a large stock of a particular type.

If one particular circuit is being designed, it would be possible to choose the components around the measured value of β_{dc} or even choose the values experimentally for a given design. A more difficult question, however, is how to design for production, in which transistor characteristics can vary considerably for a given model number within a large stock. The answer is that the concept of **worst-case design** is employed. This approach is based on picking the value of the variable that makes the design most vulnerable and selecting components that would make it work under that particular condition. This should ensure that the system will work under all conditions. This process will be illustrated for BJT switches in Section 7-6.

Another general approach for designing around parameter uncertainty is that of **negative feedback**. This approach will be illustrated in Chapter 8 in conjunction with operational amplifiers.

EXAMPLE 7-3

Determine the value of β_{dc} for the idealized transistor of Figure 7-9.

SOLUTION In an actual transistor, the value of β_{dc} will vary somewhat at different points, but for this idealized case, it is the same throughout the active region, as will be demonstrated. First, consider the curve corresponding to $I_B = 20$ μA. The collector current is $I_C = 4$ mA and the value of β_{dc} is

$$\beta_{dc} = \frac{4 \times 10^{-3}}{20 \times 10^{-6}} = 200 \tag{7-6}$$

Next, we will jump all the way to the curve $I_B = 100$ μA, in which case $I_C = 20$ mA. The value of β_{dc} is

$$\beta_{dc} = \frac{20 \times 10^{-3}}{100 \times 10^{-6}} = 200 \tag{7-7}$$

It can be readily verified that β_{dc} is the same everywhere in the active region. That was a deliberate move to illustrate the idealized form. The next example will illustrate a different situation.

EXAMPLE 7-4

Some measurements made with a nonideal BJT yield the data provided in the following table. Calculate the value of β_{dc} at each of the three points denoted as (a), (b), and (c).

	I_B(μA)	I_C(mA)	V_{CE}(V)
(a)	25	2.0	12
(b)	50	4.8	12
(c)	50	5.2	24

SOLUTION The value of β_{dc} at each point is

$$\beta_{dc} = \frac{I_C}{I_B} \tag{7-8}$$

(a) At the first point, the value of β_{dc} is

$$\beta_{dc} = \frac{2 \times 10^{-3}}{25 \times 10^{-6}} = 80 \tag{7-9}$$

(b) The value at the second point is

$$\beta_{dc} = \frac{4.8 \times 10^{-3}}{50 \times 10^{-6}} = 96 \qquad (7\text{-}10)$$

(c) Finally, the value at the third point is

$$\beta_{dc} = \frac{5.2 \times 10^{-3}}{50 \times 10^{-6}} = 104 \qquad (7\text{-}11)$$

These values are reasonably representative. They show that β_{dc} does vary with the operating point. Note that the variation from (a) to (b) is somewhat greater than the variation from (b) to (c). In the latter case, the increase illustrates that the collector characteristic does have a positive slope as the collector-to-emitter voltage is increased.

7-5 Brief Look at a Classical BJT Amplifier Circuit

For an educational and historical perspective, and to satisfy any "nit-pickers" who might be critical of its absence, we will take a brief look at a classical single-stage BJT amplifier circuit. Refer to Figure 7-12 for the discussion that follows. This circuit is a common-emitter BJT ac-coupled amplifier circuit.

This particular circuit has a coupling capacitor C_1 on the input and a coupling capacitor C_2 on the output. These capacitors block any dc voltage levels on either side so that the dc portion of the transistor circuit is independent of dc levels elsewhere. Of course, this means that the circuit cannot amplify very low frequency, or "dc-type" signals, and is usable only for time-varying signals.

The resistors R_1, R_2, and $R_{E1} + R_{E2}$ constitute a bias circuit that establishes a stable operating point for the transistor. The reason that $R_{E1} + R_{E2}$ is separated into two resistances is that a relatively large resistance is needed for bias stability, but the gain would be drastically reduced if one resistor were used. By splitting into two parts, the largest one R_{E2} can be bypassed with a relatively large capacitor C_E, which acts as a very low impedance path for time-varying signals. The resistance R_C serves as the collector load resistance, which is required to produce the desired output voltage.

Qualitatively, the circuit functions as follows: A small voltage at the input causes a small base current directly proportional to that voltage to flow into the base. The small base current is multiplied by the current gain of the transistor, resulting in a much larger collector

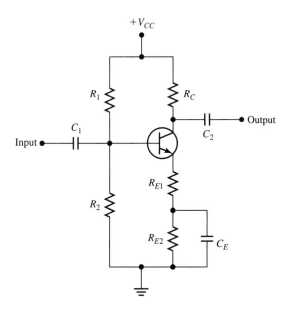

FIGURE 7–12
Classical BJT amplifier circuit that operates in active region.

current. That current flowing through the collector resistance results in an amplified voltage across the load resistance, and the capacitor on the right couples the time-varying portion to the load.

This type of amplifier circuit and its variations were widely employed in linear applications for many years and can still be found in older equipment. The detailed study of such circuits might be a worthwhile educational process for electronic specialists, but the authors believe that the coverage is not justified in a text such as this. Most amplifier circuits in the same frequency range can be much more easily achieved with operational amplifiers, and that will be one of the goals for the next chapter. In contrast, many amplifier circuits in the microwave frequency range use special-purpose transistors, but that technology is quite different and beyond the scope of this text.

7-6 BJT Switches

One of the primary applications of transistors to be considered at several places in the text is their function as **solid-state switches**. The idea behind their use as switches is to permit a small voltage or current to switch a much larger voltage or current without any moving parts in the switch. This is one major application that still arises in various interfacing functions. In particular, the later work with **programmable logic controllers** in Chapter 17 will require the use of transistors as switches.

Saturation and Cutoff in Switches

To use a BJT as a switch, it is usually necessary to change the operating region abruptly from cutoff to saturation or vice versa. This means that the active region serves simply as a transition region as the BJT moves from one extreme condition to the other.

NPN Switch

There are different variations in the manner in which the switching is arranged, but a common form using an NPN transistor is shown in Figure 7-13(a). The voltage v_i is the control voltage. Typically it assumes only two levels. Thus, it could be a digital signal, whose forms will be studied later in the text. For this particular situation, we will assume that the two levels are

$$v_i = 0 \quad \text{or} \quad v_i = V_i \tag{7-12}$$

where V_i is some positive voltage for this particular circuit. Later, when we study digital circuits, we will see that typical values for so-called **transistor-transistor logic (TTL)** circuits are in the range from about 3 to 5 V.

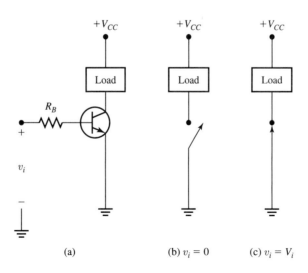

FIGURE 7–13
BJT switch using an NPN transistor.

(a)   (b) $v_i = 0$ (c) $v_i = V_i$

When $v_i = 0$, the base-emitter junction is reverse biased, and no base current flows. Therefore, the collector current is zero and there is no current through the load. An equivalent circuit model is shown in Figure 7-13(b) and the BJT is equivalent to an open switch.

When $v_i = V_i$, and assuming that $V_i > 0.7$ V, the base-emitter junction is forward biased and base current flows. Therefore, collector current will also flow due to the current gain factor β_{dc}. Strictly speaking, depending on the value of the current gain or the value of the base current, the transistor could be in the active region. However, the strategy for a good switch design is to ensure that circuit conditions move the operating region all the way to saturation. That process will be considered shortly.

Under ideal conditions of saturation, the model of Figure 7-13(c) could represent the approximate situation. In practice, there might be a small value of voltage $V_{CE(sat)}$, which has been neglected in this representation.

PNP Switch

A PNP transistor can also be used as a switch. A representative circuit is shown in Figure 7-14. This particular schematic utilizes an "upside-down" form, which is often confusing to beginners (and to many "old-timers"). The emitter appears at the top of the transistor and the collector is shown at the bottom. The power supply connection in this case is to a negative dc voltage $-V_{CC}$.

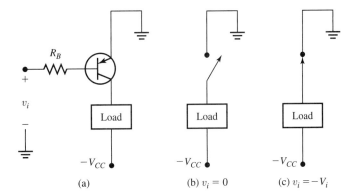

FIGURE 7–14
BJT switch using a PNP transistor.

For a control voltage $v_i = 0$, the transistor is reverse biased and the switch is open as shown in Figure 7-14(b). The signal voltage in this case must assume a negative value to saturate the transistor. Thus for some negative voltage $v_i = -V_i$, the transistor is moved into saturation and the switch is effectively closed. Assuming an ideal situation, the circuit of Figure 7-14(c) applies.

We have now seen how either a positive voltage with an NPN transistor or a negative voltage with a PNP transistor can be used to close an electronic switch. There are other forms that also can be used. The key strategy with any BJT switch is that one level of the control voltage must create cutoff for the transistor and the other level must result in saturation. The example that follows this section will illustrate a typical approach for designing a BJT switch.

Increase of Base Current after Saturation

An important point to remember in the design process for BJT switches is this: *Once saturation is reached, further increases in base current have little effect or no effect on collector current.* Said differently, the basic relationship of Equation 7-5 relating collector current to base current is applicable only in the active region.

The preceding point means that the value of base current that causes saturation should be determined under worst-case conditions. Strictly speaking, increasing the base current beyond the value that causes saturation will cause the collector-to-emitter saturation voltage to change slightly, and that change in conjunction with an external circuit could cause slight variations in collector current. However, most applications in this text will assume that the collector-to-emitter voltage with saturation is essentially zero, in which case the collector current may be assumed to remain constant.

■ **EXAMPLE 7-5**

The circuit of Figure 7-15 is to be used to switch on the 24-V dc source across the 200-Ω load and is to be designed for production, where many different transistors of the same type will be used. The controlling voltage assumes the two levels of 0 V and 5 V. Specifications on the BJT indicate that the value of β_{dc} can vary from 50 to 200 within the stock. Determine the minimum value of R_B that will ensure that the BJT will saturate. Assume that the value of the collector-to-emitter saturation voltage can be neglected.

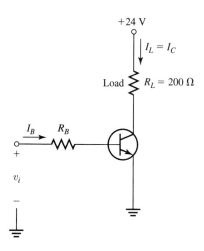

FIGURE 7–15
Circuit of Example 7-5.

SOLUTION Assuming that the saturation voltage is zero, the entire 24-V dc source will appear across the load when the switch is closed and the load current will be

$$I_L = I_C = \frac{24 \text{ V}}{200 \text{ }\Omega} = 0.12 \text{ A} = 120 \text{ mA} \tag{7-13}$$

The base current I_B is related to the collector current by

$$I_B = \frac{I_C}{\beta_{dc}} \tag{7-14}$$

The strategy here is to choose a base current that will "guarantee" that the collector current will be 120 mA. For a fixed collector current, the maximum value of base current required corresponds to the minimum value of β_{dc}. Therefore,

$$I_{B(\max)} = \frac{I_C}{\beta_{dc(\min)}} = \frac{0.12}{50} = 2.4 \text{ mA} \tag{7-15}$$

The loop equation for the base circuit is

$$-v_i + R_B I_B + 0.7 = 0 \tag{7-16}$$

Substituting $v_i = 5$ V and $I_B = 2.4 \times 10^{-3}$ A, we solve for R_B.

$$R_B = \frac{5 - 0.7}{2.4 \times 10^{-3}} = \frac{4.3}{2.4 \times 10^{-3}} = 1792 \text{ }\Omega \tag{7-17}$$

As a review of the process, the smallest possible value of β_{dc} within the stock of transistors has been used to determine the value of base current that will cause saturation. When transistors having larger values of β_{dc} are inserted in the circuit, it could be said that the collector current has the potential to be larger. However, it cannot increase because the collector-to-emitter voltage is saturated at an assumed value of zero and the entire collector dc supply is across the load resistance, thus fixing the load current.

7-7 Field Effect Transistor Family

A second category of transistor is collectively called the **field effect transistor (FET)**. However, the FET has a much wider variation in the types of characteristics than the BJT, and there are quite a few different acronyms that describe these variations. The treatment here will represent an exposure to some of the major categories and some of the types of characteristics.

Family Tree

We begin by displaying a somewhat simplified "family tree" that displays most of the general categories of FETs. Refer to Figure 7-16 for the discussion that follows.

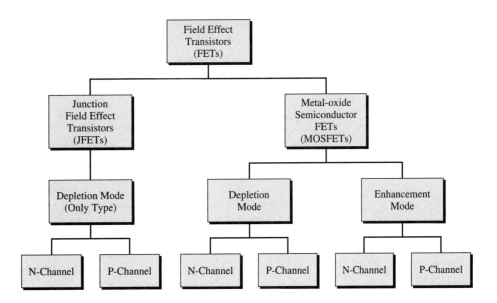

FIGURE 7–16
Classification scheme for field effect transistors.

At the broadest level, FETs can be classified as **junction field effect transistors (JFETs)** and **metal-oxide semiconductor field effect transistors (MOSFETs)**. The latter type is also referred to in some parts of the literature as **insulated gate field effect transistors (IGFETs)**. However, we will refer to them as MOSFETs.

JFETs

JFETs are considered as **depletion mode** devices for reasons dealing with their internal behavior. There are two types: (1) **N-channel** and (2) **P-channel**.

MOSFETs

MOSFETs have two subfamilies: (1) **depletion mode** and (2) **enhancement mode**. Each one of these subfamilies has two types: (1) **N-channel** and (2) **P-channel**.

It is obvious that all of these different types can be confusing, so any user of FETs needs to study the data sheets very carefully to understand the various voltage polarities and current directions associated with the different types.

7-8 Junction Field Effect Transistors

N-Channel Layout

The basic construction of an N-channel JFET is suggested by the layout shown in Figure 7-17(a). The **channel** consists of a lightly doped bar of N-type material such as silicon. Both ends have contacts. The lower end on the figure is called the **source** (S) and the upper end is called the **drain** (D). Two sections of heavily doped P-type material are embedded in the sides and are connected internally. This portion of the FET is called the **gate** (G). Current can flow along the channel, but the voltage at the gate heavily influences it. This voltage creates a so-called **depletion layer** that affects the current flow.

The connection between the P-type gate and the N-type channel forms a PN junction diode. However, the gate of the N-channel FET is normally biased with a negative voltage or zero with respect to the source terminal, and this diode then remains reverse-biased. Therefore, the small reverse current of the gate-channel diode will be the only current that flows in the gate.

A common schematic symbol for an N-channel JFET is shown in Figure 7-17(b). Note that the direction of the arrow at the gate is directed *inward*. As will be seen later, the direction of the arrow for the corresponding P-channel FET is directed *outward*. An alternate schematic for both types is to display the gate terminal at a lower position; that is, closer to the source. However, we will choose in this text to put the terminal midway between the source and drain terminals, which is probably the most common schematic form.

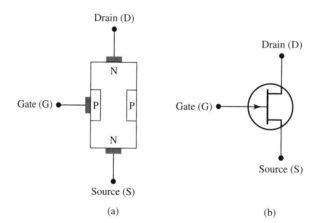

FIGURE 7–17
Construction of an N-channel JFET and the schematic symbol.

Three JFET Operating Regions

In a parallel but slightly different sense as compared with a BJT, a JFET may be considered to have three different regions of operation. They are called the (1) **cutoff region**, (2) **ohmic region**, and (3) **beyond pinchoff region**. These terms will be used in the measurement process that follows for determining the characteristic curves, and their meanings will then be discussed.

Although there are some minor differences, some analogies between the three BJT regions and the three FET regions are useful. They are illustrated as follows:

BJT Region	FET Region
Cutoff	Cutoff
Saturation	Ohmic
Active	Beyond pinchoff

FIGURE 7–18
Circuit that can be used to measure N-channel JFET characteristics.

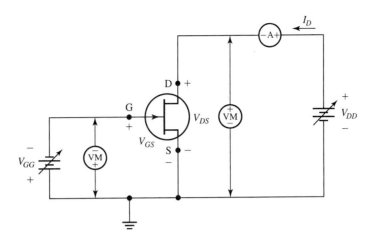

Obtaining JFET Characteristics

In the same spirit that we considered for the BJT, we will describe an experiment that can be performed in a basic electronics laboratory to obtain characteristic curves. A circuit configuration for an N-channel JFET is shown in Figure 7-18. Compared to the circuit used for an NPN BJT in Figure 7-7, there are both similarities and differences. A similarity is that the drain-source portion of the circuit for an N-channel JFET has the same voltage polarity and current direction as for the collector-emitter portion of an NPN BJT. However, the gate-source voltage polarity is reversed. Moreover, there is no ammeter shown in the gate. Although there may be a small reverse current flowing in the gate circuit, the gate input to a JFET is a very high resistance, and the current will be neglected in this experiment because it is a secondary effect.

The circuit is arranged in the form called the **common-source** configuration. Thus, the source terminal appears in common with both the input and output and establishes the reference ground. The *common-source* configuration for an FET is analogous to the *common-emitter* configuration for a BJT.

The adjustable voltage source V_{GG} is used to bias the gate. This corresponds to reverse bias for the gate-source junction, as previously noted. The voltmeter connected between gate and source measures the gate-to-source voltage V_{GS}. To maintain a negative sign on the curves that follow, however, the reference positive terminal of the gate-to-source voltage is considered to be at the gate, meaning that all voltage values will be expressed with a negative sign.

On the drain side, an adjustable dc source V_{DD} sets the drain-to-source voltage V_{DS}. The ammeter on the right measures the drain current I_D and the voltmeter on the right measures the drain-to-source voltage. Note that if the ammeter is considered to be ideal, $V_{DS} = V_{DD}$.

Gate-to-Source Cutoff Voltage and the Cutoff Region

For any N-channel JFET, there is some particular negative voltage that effectively reduces the current flow along the channel to zero; that is, $I_D = 0$. This voltage is called the **gate-to-source cutoff voltage**. It will be denoted as $V_{GS(off)}$. A typical value is of the order of $V_{GS(off)} = -4$ V. Active operation of the FET will then assume a voltage range of $V_{GS(off)} < V_{GS} \leq 0$. For $V_{GS} \leq V_{GS(off)}$ (meaning a more negative voltage of gate-to-source voltage), the JFET will be in the **cutoff region**. An ideal model of a JFET in the cutoff region is shown in Figure 7-19. The model is the same as for the cutoff region of a BJT.

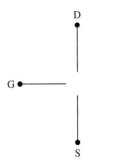

FIGURE 7–19
Ideal model of JFET in the cutoff region.

Drain Characteristics

The characteristic curves for the JFET that correspond to the collector characteristics for a BJT are the **drain characteristics**. They portray the most important properties of the JFET.

FIGURE 7–20
Idealized drain characteristics of N-channel JFET in linear portion of the ohmic region.

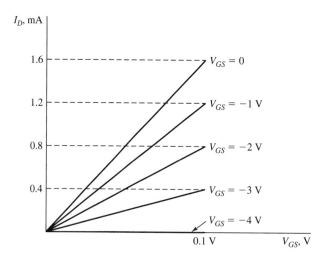

FIGURE 7–21
Drain characteristics for an ideal representative N-channel JFET.

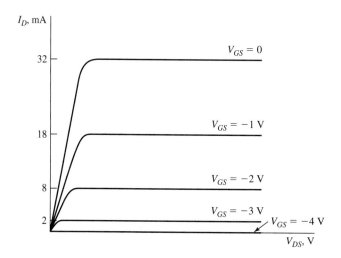

As was the case for the BJT, the process will be idealized somewhat in the discussion that follows. The values that will be assumed should be considered as typical because they will vary from one FET to another. However, some possible values should help the reader visualize the process. While keeping a reference to the circuit of Figure 7-18, refer to Figures 7-20 and 7-21 for the discussion that follows.

Ohmic Region

The **ohmic region** for the JFET corresponds to a very small drain-to-source voltage, which typically will be less than about 0.1 V or so. Assume first that the gate-to-source voltage is set to zero by adjusting V_{GG} to the gate-to-source cutoff voltage. While this value is set, assume that the drain-to-source voltage is adjusted from zero to a relatively small value The result is essentially a value of zero drain current, and the curve can be considered as coinciding with the abscissa as shown in Figure 7-20 based on $V_{GS(off)} = -4$ V.

Next, assume that $V_{GS} = -3$ V. As the drain-to-source voltage is increased over a very small interval, the drain current increases approximately linearly. Note the idealized curve for $V_{GS} = -3$ V on Figure 7-20. As changes in the gate-to-source voltage are continued in a positive direction toward zero, additional curves in the ohmic region are generated as shown on Figure 7-20. For normal operation, the gate-to-source voltage for an N-channel unit is not increased above 0 V.

In the ohmic region, the JFET is acting as a **voltage-controlled resistance**. The slope of each curve shown on Figure 7-20 represents the conductance of the JFET for that

Beyond Pinchoff Region

Changing the horizontal scale to observe larger values of drain-to-source voltage, the resulting curves for the same hypothetical FET will have the idealized curves shown in Figure 7-21. For these curves, it is assumed that the drain-to-source voltage is always kept below the breakdown voltage, at which point the transistor could be burned out.

The ohmic region on this set of scales is the narrow portion near the ordinate. Note that the linear assumption made earlier is valid only at very small values of drain-to-source voltage. Eventually, the characteristic curves start to become nonlinear in nature.

For a certain level of the drain-to-source voltage, the curves start to flatten out as shown. For reasons that relate to the physical phenomena internal to the JFET, this region is called the **beyond pinchoff region**. It should be noted that the vertical increments between successive curves are not equal.

Formula for JFET Drain Current

There is a widely employed formula that predicts the approximate drain current in the beyond pinchoff region. It reads

$$I_D = I_{DSS}\left(1 - \frac{V_{GS}}{V_{GS(\text{off})}}\right)^2 \tag{7-18}$$

The quantity I_{DSS} is called the **zero-bias drain current**. For the hypothetical JFET used in the experiment, $I_{DSS} = 32$ mA and the curves in the beyond pinchoff region were generated from this equation with $V_{GS(\text{off})} = -4$ V as previously discussed. Note that for an N-channel JFET, $V_{GS} \leq 0$ and $V_{GS(\text{off})} < 0$, and the ratio in the second term within the parentheses in Equation 7-18 is either positive or zero. The signs of both of these parameters are opposite for a P-channel JFET, so the ratio is again positive or zero. In fact, a different way to write the equation that eliminates confusion about the signs is

$$I_D = I_{DSS}\left(1 - \frac{|V_{GS}|}{|V_{GS(\text{off})}|}\right)^2 \tag{7-19}$$

■ EXAMPLE 7-6

For the hypothetical JFET whose idealized characteristics in the ohmic region were shown in Figure 7-20, determine the conductance and resistance for each one of the curves shown.

SOLUTION The conductance of a given curve will be denoted as g_{DS}. It is the slope of the curve. The corresponding resistance will be denoted as r_{DS}. It is $r_{DS} = 1/g_{DS}$. The values for $V_{GS} = 0$ are calculated as

$$g_{DS} = \frac{1.6 \times 10^{-3} \text{ A}}{0.1 \text{ V}} = 16 \times 10^{-3} \text{ S} = 16 \text{ mS} \tag{7-20}$$

$$r_{DS} = \frac{1}{g_{DS}} = \frac{1}{16 \times 10^{-3} \text{ S}} = 62.5 \text{ }\Omega \tag{7-21}$$

The values for $V_{GS} = -1$ V are calculated as

$$g_{DS} = \frac{1.2 \times 10^{-3} \text{ A}}{0.1 \text{ V}} = 12 \times 10^{-3} \text{ S} = 12 \text{ mS} \tag{7-22}$$

$$r_{DS} = \frac{1}{g_{DS}} = \frac{1}{12 \times 10^{-3} \text{ S}} = 83.3 \text{ }\Omega \tag{7-23}$$

The process is repeated at the other values of gate-to-source voltage and the results are summarized in the short table that follows.

V_{GS}, V	g_{DS}, mS	r_{DS}, Ω
0	16	62.5
−1	12	83.3
−2	8	125
−3	4	250
−4	0	∞

Note that the resistance increases with an increasing negative voltage. Eventually at cutoff, the JFET becomes an open circuit.

EXAMPLE 7-7

Write the equation for the drain current in the beyond pinchoff region for the hypothetical JFET whose characteristics were shown in Figure 7-21.

SOLUTION From the work of this section, we know that $V_{GS(\text{off})} = -4$ V and $I_{DSS} = 32$ mA. Working in basic units, we have

$$I_D = 0.032\left(1 - \frac{V_{GS}}{-4}\right)^2 = 0.032\left(1 + \frac{V_{GS}}{4}\right)^2 \qquad (7\text{-}24)$$

Don't get confused by the apparent positive sign that appears between the two terms within parentheses in Equation 7-24. When the negative values of V_{GS} are substituted in the equation, a negative sign will reappear. To avoid the sign confusion, we can also express the current in the alternate form

$$I_D = 0.032\left(1 - \frac{|V_{GS}|}{4}\right)^2 \qquad (7\text{-}25)$$

EXAMPLE 7-8

For the JFET of Examples 7-6 and 7-7, determine the drain current for $V_{GS} = -2.5$ V.

SOLUTION We substitute $|V_{GS}| = 2.5$ V in Equation 7-25 and obtain

$$I_D = 0.032\left(1 - \frac{2.5}{4}\right)^2 = 0.032(0.141) = 4.5 \times 10^{-3}\text{ A} = 4.5\text{ mA} \qquad (7\text{-}26)$$

EXAMPLE 7-9

For the hypothetical JFET used in the preceding development, determine the gate-to-source voltage that would be required to create a drain current of 10 mA in the beyond pinchoff region.

SOLUTION If there were a sufficient number of curves in Figure 7-21, we might be able to determine the value graphically. Alternately, we could try to extrapolate between curves, although the nonlinear nature would suggest otherwise. Instead, we will attempt an analytic solution by setting $I_D = 10$ mA in Equation 7-25 and solve for V_{GS}. We have

$$10 \times 10^{-3} = 0.032\left(1 - \frac{|V_{GS}|}{4}\right)^2 \qquad (7\text{-}27)$$

Dividing both sides by 0.032 and taking the square roots of both sides, we obtain

$$1 - \frac{|V_{GS}|}{4} = \pm 0.559 \qquad (7\text{-}28)$$

There are two solutions to this equation, one corresponding to the positive square root and one corresponding to the negative square root. The solution corresponding to the negative square root would be $|V_{GS}| = 6.24$ V. However, this is a superfluous solution resulting from the mathematics, because a negative voltage of this value would correspond to the cutoff region. Therefore, the correct solution would be based on the positive square root and would be $|V_{GS}| = 1.76$ V. The proper voltage is then

$$V_{GS} = -1.76 \text{ V} \tag{7-29}$$

7-9 Brief Look at Classical JFET Amplifier Circuit

Earlier in the chapter, we took a brief look at a BJT amplifier. To give equal treatment to the FET family, we will take a similar look at a typical N-channel **JFET amplifier circuit**. Refer to Figure 7-22 for the discussion that follows.

Like the corresponding BJT circuit, this particular FET circuit has a coupling capacitor C_1 on the input and a coupling capacitor C_2 on the output. These capacitors block any dc voltage levels on either side so that the dc portion of the transistor circuit is independent of dc levels elsewhere. As before, this means that the circuit cannot amplify very low frequency or "dc-type" signals and is usable only for time-varying signals.

For use as an amplifier, the FET is biased in the beyond pinchoff region. The resistances R_1, R_2, and $R_{S1} + R_{S2}$ constitute a bias circuit that establishes a stable operating point for the transistor. The reason for two resistances is so that a part of the net resistance can be left unbypassed. This provides some negative feedback for the signal. Although this process results in lower gain, it will reduce distortion and provide more gain stability. The gain would be reduced drastically if the entire source resistance were unbypassed. The capacitor C_S acts as a very low impedance path around R_{S2} for time-varying signals so that only R_{S1} is unbypassed. The resistance R_D serves as the collector load resistance, which is required to produce the desired output voltage.

Qualitatively, the circuit functions as follows: A small voltage at the input causes a change in the drain current. The change in the drain current multiplied by the drain resistance can result in an output voltage much greater than the input voltage and gain can occur. The capacitor on the right couples the time-varying signal over to the next stage and it blocks the dc.

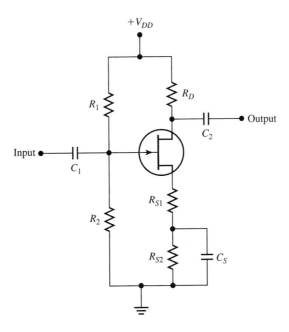

FIGURE 7–22

Classical JFET amplifier circuit that operates in beyond pinchoff region.

7-10 MOSFETs

The second general type of FET is the **metal-oxide semiconductor field effect transistor (MOSFET)**. The length of the name certainly suggests that the acronym should be used exclusively from this point forward! You might wish to review Figure 7-16, because the family structure for the MOSFET is more complex than for the JFET. There are two general classifications of MOSFETs: (1) **depletion mode** and (2) **enhancement mode**. Recall that the MOSFET is also referred to in the literature as an **insulated gate field effect transistor (IGFET)** for reasons that will be clearer shortly.

Depletion-Mode MOSFET

The physical layout of an N-channel depletion-mode MOSFET is suggested by the diagram in Figure 7-23(a). As is true for the JFET, a channel is formed between the source and the drain with contacts at both ends. A P-type region called the substrate reduces the channel width. Some MOSFETs have the substrate internally connected, but in others it can be connected externally as shown here.

A layer of silicon dioxide (SiO_2) provides insulation between the gate and the channel. This means that the PN diode that exists in the JFET is not present in the MOSFET. While the small gate current for the JFET is quite small, the gate current for the MOSFET is even smaller.

A schematic symbol for an N-channel depletion-mode MOSFET is shown in Figure 7-23(b). The substrate terminal is shown as a separate external connection. For the corresponding P-channel MOSFET, the arrow on the substrate connection is reversed.

As was true for JFETs, and as suggested by the preceding discussion, there are two types of depletion-mode MOSFETs based on polarities and current directions: (1) N-channel and (2) P-channel. The drain characteristics of depletion-mode MOSFETs are very similar to the corresponding JFET characteristics within the same region of gate-to-source voltages for the given type.

There is one difference between a JFET and a depletion-mode MOSFET. Refer to an N-channel type for reference. With a JFET, normally the gate-to-source voltage is maintained at a negative value or zero. A positive value would forward bias the gate-channel junction, and this would counteract the desired high input resistance property for an FET. However, the insulation layer within the MOSFET prevents any significant current from flowing, and it is possible to operate the N-channel MOSFET with positive gate-to-source voltages. Of course, the opposite polarity conditions hold for the P-channel MOSFET.

Enhancement-Mode MOSFET

The physical layout of an N-channel enhancement-mode MOSFET is illustrated in Figure 7-24(a). The P-type substrate provides a complete block between the two N-type sections. Because of this situation, an enhancement mode MOSFET is a *normally off*

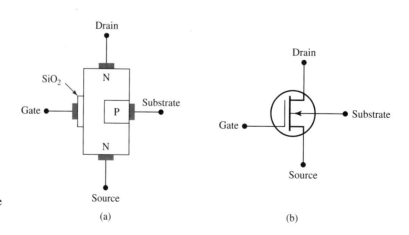

FIGURE 7-23
Construction of an N-channel *depletion-mode* MOSFET with external connection for substrate and the schematic symbol.

FIGURE 7–24
Construction of an N-channel *enhancement-mode* MOSFET with external connection for substrate and the schematic symbol.

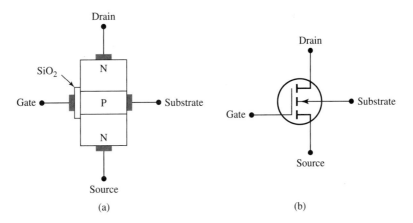

device. As shown for the depletion-mode MOSFET, a separate terminal is provided here for the substrate.

For positive values of the drain-to-source voltage, no conduction occurs in the channel for negative values of the gate-to-source voltage. However, a sufficiently positive value of the gate-to-source voltage will attract free electrons from the N-type material into the P-region. The result is a sort of solid-state bridge across the P-type substrate, and conduction occurs. The positive voltage has enhanced the conduction, and this suggests the description of *enhancement mode* for the device.

The schematic symbol for the N-channel enhancement-mode MOSFET with an external substrate connection is shown in Figure 7-24(b). A P-channel type would have the arrow reversed in direction.

At first glance, this symbol might appear to be the same as that of the N-channel depletion-mode type of Figure 7-23(b). However, a comparison will show that the relatively small difference is that the vertical line on the right is solid for the depletion-mode unit but is broken into several parts for the enhancement-mode unit.

Threshold Voltage

For the N-channel enhancement-mode MOSFET, there is a minimum value of positive voltage that results in drain current flow. This voltage is called the **threshold voltage**, and it will be denoted as V_T. This means that $I_D = 0$ for $V_{GS} < V_T$ in an N-channel unit. Thus, the threshold voltage for an enhancement mode MOSFET corresponds to the gate-to-source cutoff voltage for a JFET. Note, however, that the voltage polarity is opposite.

The drain characteristics for an idealized representative N-channel enhancement-mode MOSFET are shown in Figure 7-25. This particular MOSFET is assumed to have a threshold voltage of 2 V.

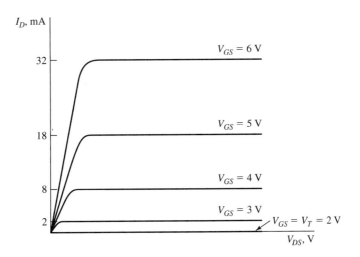

FIGURE 7–25
Drain characteristics for an ideal representative N-channel enhancement-mode MOSFET.

Drain Current Equation

The drain current in the beyond pinchoff region for an N-channel enhancement-mode MOSFET can be closely approximated by the equation that follows.

$$I_D = K \left(\frac{V_{GS}}{V_T} - 1 \right)^2 \tag{7-30}$$

The constant K represents a curve-fitting constant that applies to a given MOSFET. The signs of V_{GS} and V_T are both positive for an N-channel type and both are negative for a P-channel type, so the ratio is positive.

EXAMPLE 7-10

For the idealized drain characteristics of the N-channel enhancement-mode MOSFET of Figure 7-25, determine an equation for predicting the drain current in the beyond pinchoff region.

SOLUTION The expression will be of the form of Equation 7-30 and the value of the threshold voltage is known to be $V_T = 2$ V. The expression will then be of the form

$$I_D = K \left(\frac{V_{GS}}{2} - 1 \right)^2 \tag{7-31}$$

The constant K can then be determined based on a known value from the curves. A real MOSFET deviates from the ideal form, so the constant should be determined in the desired region of operation. However, this problem has been "rigged" to be ideal, so any one of the curves can be used. We will pick the curve corresponding to $I_D = 8$ mA and $V_{GS} = 4$ V.

$$8 \times 10^{-3} = K \left(\frac{4}{2} - 1 \right)^2 = K(1)^2 \tag{7-32}$$

This yields

$$K = 8 \times 10^{-3} \tag{7-33}$$

The current in amperes is then given as

$$I_D = 8 \times 10^{-3} \left(\frac{V_{GS}}{2} - 1 \right)^2 \tag{7-34}$$

The reader is invited to show that this equation predicts the values of drain current for all of the curves shown. But remember that this is an idealized situation.

7-11 FET Switches

As a result of the nearly linear current versus voltage behavior in the ohmic region, FETs can be used either as a switch or as a resistance that can be controlled by a voltage; that is, a **voltage-controlled resistance**. (We could use VCR as an acronym, but that particular one would be a confusing choice!) Some FETs are manufactured specifically for these functions and many have voltage levels that work easily with digital circuit voltage values.

FETs have certain advantages over BJTs for some switching applications. Recall that with BJTs, there is a small collector-to-emitter saturation voltage, which varies somewhat with base current. However, a properly designed FET switch acts more like a constant value of resistance.

JFET Resistance in Ohmic Region

We have seen earlier that for very small values of drain-to-source voltage, there is close to a linear relationship between the current and voltage in the ohmic region. The value of the resistance between drain and source for a JFET will be denoted as r_{DS}. It can be approximated as

$$r_{DS} = \frac{r_{DSO}}{1 - |V_{GS}|/|V_{GS(\text{off})}|} \tag{7-35}$$

where r_{DSO} is the resistance corresponding to zero bias voltage and is given by

$$r_{DSO} = \frac{|V_{GS(\text{off})}|}{2I_{DSS}} \tag{7-36}$$

N-Channel JFET Switch

A possible circuit form using an N-channel JFET is shown in Figure 7-26(a). The input voltage is assumed to have two levels. When $v_i < V_{GS(\text{off})}$ (more negative), the JFET is in the cutoff region and the switch is assumed to be open as shown in (b). However, when $v_i = 0$, the minimum resistance for the ohmic region is realized and the equivalent circuit is shown in (c). The resistance shown is r_{DSO} corresponding to $v_i = 0$.

N-Channel MOSFET Switch

A possible circuit form using an N-channel enhancement-mode MOSFET is shown in Figure 7-27(a). This form is easier to interface with many digital circuits, because the two voltage levels are 0 and a positive value. The FET switch will effectively close, and the equivalent circuit is shown in Figure 7-27(c). The particular MOSFET has the substrate internally connected, so there are three external terminals.

Analog Multiplexer

In many communication system applications, it is desired to take samples of several different signals in a sequential fashion. This process is called **time-division multiplexing (TDM)**, and a circuit that performs this function is called a **multiplexer**. There are both

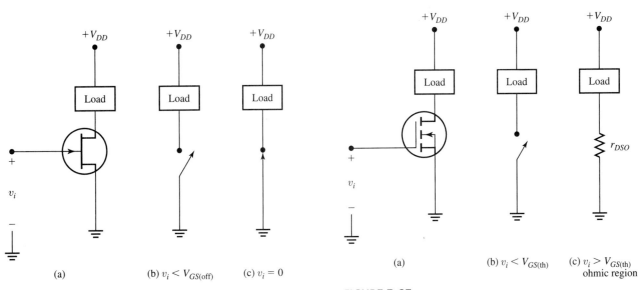

FIGURE 7–26
N-channel JFET switch.

FIGURE 7–27
Switch using an N-channel enhancement-mode MOSFET.

FIGURE 7–28
Implementation of a four-channel multiplexer using P-channel JFET switches and an operational amplifier.

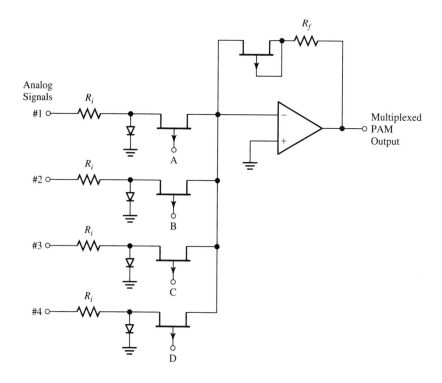

analog and digital multiplexers, and FET switches may be used to implement an analog multiplexer.

Refer to the circuit shown in Figure 7-28 using four input channels, each of which contains a P-channel JFET switch along with an operational amplifier combination circuit. Operational amplifiers will be studied extensively in the next chapter, but for the moment, accept the idea that any signals that pass through the input JFETs will be summed and routed to the output. The four input analog switches are controlled by digital logic signals, with a given switch open for a typical digital positive level and closed for a level of zero. The diodes are used to protect the switches and do not perform a signal processing function. The JFET on the top right is used to compensate for the resistance displayed by a given JFET when it is turned on.

The digital signals that control the JFETs are arranged to turn each one on in a periodic sequential fashion so that each one of the inputs is sampled once for each sampling cycle. The output is a multiplexed **pulse amplitude modulated (PAM)** signal. The concept of TDM will be discussed in more detail in Chapter 18.

7-12 FET Schematic Summary

It is very likely that the reader will be somewhat dazzled at this point with all the different types of FETs, the schematic symbols, and operating voltage polarities and current directions. For reassurance, it will be stated that electronic engineers and technologists have the same difficulty. Indeed, it would take many chapters to fully explore all of the minute details, so the treatment here was intended primarily as an overview and basic familiarization. Actually, there are other types that build upon the foundations discussed here; for example, DMOS, CMOS, and other categories. In later chapters concerning digital circuits, some of these concepts will reappear.

To end this chapter, a summary of some of the common schematic diagram symbols is provided in Figure 7-29. All of the MOSFETs shown have the substrates internally connected.

FIGURE 7–29
Schematic diagrams for different types of field effect transistors.

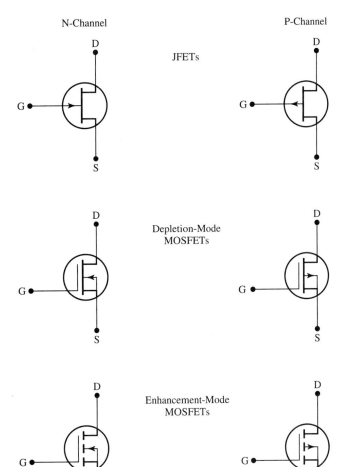

PROBLEMS

7-1 In parts (a) through (c), two of three BJT currents are given. Determine the third in each case if possible:
(a) $I_B = 1$ mA, $I_E = 120$ mA, (b) $I_C = 2$ mA, $I_E = 2.04$ mA, (c) $I_C = 60$ mA, $I_B = 250$ μA.

7-2 In parts (a) through (c), two of three BJT currents are given. Determine the third in each case if possible:
(a) $I_C \approx 1$ mA, $I_E \approx 1$ mA, (b) $I_C = 2$ A, $I_B = 50$ mA, (c) $I_B = 10$ μA, $I_E = 4$ mA.

7-3 Classify each of the following conditions as *cutoff, saturation,* or *active region*. (a) NPN: $V_{BE} = -0.7$ V, $V_{CE} = 15$ V, (b) NPN: $V_{BE} = 0.7$, $V_{CE} = 0$ V, (c) NPN: $V_{BE} = 0.7$ V, $V_{CE} = 10$ V, (d) PNP: $V_{EB} = 0.7$ V, $V_{CE} = 0$ V.

7-4 Classify each of the following conditions as *cutoff, saturation,* or, *active region*. (a) PNP: $V_{BE} = 0.7$ V, $V_{EC} = 15$ V, (b) PNP: $V_{BE} = -0.7$ V, $V_{EC} = 20$ V, (c) PNP: $V_{BE} = -0.7$ V, $V_{CE} = 0$, (d) NPN: $V_{EB} = 1$ V, $V_{EC} = 15$ V.

7-5 The collector characteristics for an idealized NPN transistor are shown in Figure P7-5. Determine the value of β_{dc}.

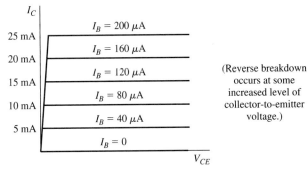

(Reverse breakdown occurs at some increased level of collector-to-emitter voltage.)

FIGURE P7–5

7-6 The collector characteristics for an idealized NPN transistor are shown in Figure P7-6. Determine the value of β_{dc}.

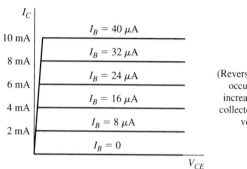

(Reverse breakdown occurs at some increased level of collector-to-emitter voltage.)

FIGURE P7-6

7-7 For the idealized BJT of Problem 7-5 (Figure P7-5), determine the values of (a) collector current and (b) emitter current when $I_B = 100\ \mu A$.

7-8 For the idealized BJT of Problem 7-6 (Figure P7-6), determine the values of (a) collector current and (b) emitter current when $I_B = 20\ \mu A$.

7-9 Some measurements made on a nonideal BJT yield the data provided in the following table. Calculate the value of β_{dc} at each of the three points denoted as (a), (b), and (c).

	$I_B (\mu A)$	$I_C (mA)$	$V_{CE}(V)$
(a)	20	2.0	15
(b)	40	3.8	15
(c)	40	4.2	24

7-10 Some measurements made with a nonideal BJT yield the data provided in the following table. Calculate the value of β_{dc} at each of the three points denoted as (a), (b), and (c).

	$I_B (\mu A)$	$I_C (mA)$	$V_{CE}(V)$
(a)	100	4.0	10
(b)	100	4.8	20
(c)	200	8.2	20

7-11 Refer back to the BJT switch design of Example 7-5. Assume that the load is 300 Ω and that the transistor used has a β_{dc} range of 60 to 200. In addition, assume that the value of base current is to be set to a value 50 percent above the minimum value. Determine the value of R_B required to ensure saturation.

7-12 Refer back to the BJT switch design of Example 7-5. Assume that the positive value of the input voltage corresponding to the on-condition of the switch can vary from 4 to 5 V. Assume all other conditions to be the same as in Example 7-5. Determine the value of R_B to ensure saturation.

7-13 The ohmic region characteristics for an idealized N-channel JFET are shown in Figure P7-13. Prepare a table providing the value of the conductance and resistance associated with each value of gate-to-source voltage.

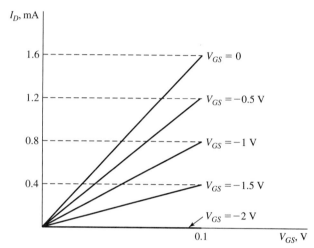

FIGURE P7-13

7-14 The ohmic region characteristics for an idealized N-channel JFET are shown in Figure P7-14. Prepare a table providing the value of the conductance and resistance associated with each value of gate-to-source voltage.

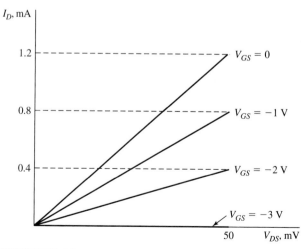

FIGURE P7-14

7-15 The drain characteristics for an idealized N-channel JFET are shown in Figure P7-15. Determine an equation for predicting the beyond pinchoff drain current.

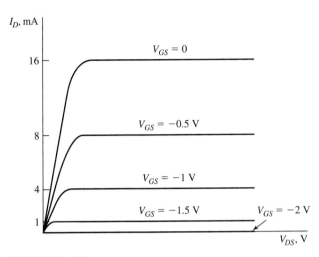

FIGURE P7–15

7-16 The drain characteristics for an idealized N-channel JFET are shown in Figure P7-16. Determine an equation for predicting the beyond pinchoff drain current.

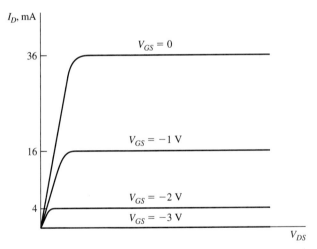

FIGURE P7–16

7-17 For the JFET of Problem 7-15 (Figure P7-15), determine the drain current in the beyond pinchoff region for $V_{GS} = -0.8$ V.

7-18 For the JFET of Problem 7-16 (Figure P7-16), determine the drain current in the beyond pinchoff region for $V_{GS} = -0.5$ V.

7-19 For the JFET of Problem 7-15 (Figure P7-15), determine analytically the gate-to-source voltage that results in a drain current of 10 mA.

7-20 For the JFET of Problem 7-16 (Figure P7-16), determine analytically the gate-to-source voltage that results in a drain current of 20 mA.

7-21 Using the results of Equations 7-35 and 7-36, determine an equation for the resistance in the ohmic region for the FET of Problems 7-13 and 7-15.

7-22 Using the results of Equations 7-35 and 7-36, determine an equation for the resistance in the ohmic region for the FET of Problems 7-14 and 7-16.

7-23 The circuit shown in Figure P7-23 is a *voltage-controlled voltage divider*, in which the negative dc voltage applied to the gate controls the *attenuation* or reduction in the signal level between v_i and v_o. Assume that the JFET has the ohmic region characteristics of Problem 7-13 (Figure P7-13) and that the signal level will be sufficiently small that ohmic region operation can be assumed. Let $A = v_o/v_i$ represent the voltage divider ratio and assume that $R = 3$ kΩ. Determine the range of A as the gate-to-source voltage is varied from $V_{GS(\text{off})}$ to 0.

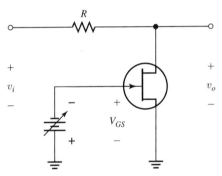

FIGURE P7–23

7-24 The circuit form of Problem 7-23 (Figure P7-23) is to be used in a design to provide a given range of attenuation at small signal levels and the JFET is assumed to have the ohmic region characteristics of Problem 7-14 (Figure P7-14). Assume that the desired range of A is $0.05 \leq A \leq 1$. Determine the required value of R.

Operational Amplifiers

OVERVIEW AND OBJECTIVES

This chapter will be devoted to an introductory treatment of **operational amplifiers** and their characteristics. Operational amplifiers (abbreviated as **op-amps**) are integrated-circuit modules that scan be used to amplify small signals to levels that can be used in many functions, including those associated with music and audio, instrumentation, sensing devices, and other applications. They are easy to use, and nonelectronic specialists can learn quickly to design and implement various basic circuits for many engineering disciplines.

Objectives

After completing this chapter, the reader should be able to

1. Describe the properties of a **voltage amplifier** and use a circuit model.
2. Define **open circuit gain, input resistance**, and **output resistance**.
3. Determine the **loaded gain** of an amplifier when connected between source and load.
4. Describe the properties of an **operational amplifier**.
5. State and apply the ideal conditions used in analyzing and designing op-amp circuits.
6. Analyze and describe the functions of various op-amp circuits such as the **inverting amplifier, noninverting amplifier**, and various **controlled sources**.
7. Design op-amp circuits to perform amplification and to function as controlled sources.
8. Analyze and design an op-amp linear combination circuit and a difference amplifier.
9. Describe various **integrator** and **differentiator** circuits using op-amps.
10. Discuss the advantages of **active filters** using op-amps.
11. Design basic **low-pass** and **band-pass** active filter circuits.
12. Describe how an op-amp can be used to create a **voltage regulator**.

8-1 Amplifier Properties

Before considering operational amplifiers, we should discuss some of the general properties that apply to many types of amplifiers. The treatment will focus on the external properties and the types of specifications that will be significant in most applications. Although amplifiers may be used to amplify voltage, current, power, or a combination of these variables, most discussions begin with the concept of a **voltage amplifier**, because it is the most common type.

174 CHAPTER 8 • Operational Amplifiers

FIGURE 8–1
Block diagrams of voltage amplifiers showing input and output voltages.

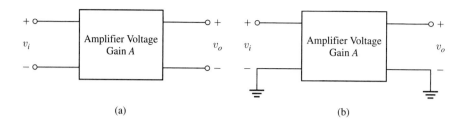

Voltage Amplifier Input-Output Characteristic

Consider the block diagram shown in Figure 8-1(a). Assume that there is some type of signal, usually timevarying, across the input terminals on the left, indicated here as v_i. The signal across the output terminals in denoted as v_o. A **linear amplifier** is characterized by the property that the output voltage is a constant times the input voltage; that is,

$$v_o = Av_i \tag{8-1}$$

The constant A is the **voltage gain**. It is related to the input and output voltages as

$$A = \frac{v_o}{v_i} \tag{8-2}$$

In a wide range of applications, one of the terminals on both the input and output will be connected to a common ground. A block diagram illustrating this form is shown in Figure 8-1(b).

General Properties

The voltage gain may be either positive or negative. If the gain is positive, the amplifier is said to be a **noninverting amplifier**. If the gain is negative, the amplifier is said to be an **inverting amplifier**.

The input-output characteristic of a representative noninverting amplifier over the linear range is shown in Figure 8-2(a). A representative characteristic of an inverting amplifier is shown in (b). The numbers shown relate to examples that will be considered at the end of this section. They illustrate typical values. The slope of either curve represents the corresponding gain. It should be obvious from the values shown that a small input voltage produces a much larger output voltage.

No amplifier can maintain these linear properties over an infinite range, so at some maximum input level (not shown), the curves tend to flatten out, and eventually **saturation**

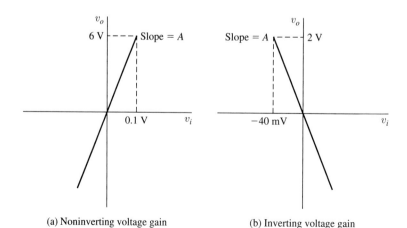

FIGURE 8–2
Typical input-output characteristics of noninverting and inverting amplifiers in linear region.

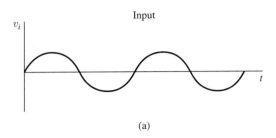

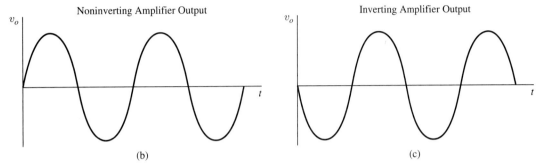

FIGURE 8–3
Amplifier input and output waveforms based on a sinusoidal input.

will occur. Depending on the unit, reaching this level could cause the device to become defective, or **distortion** of the signal could result. However, some special-purpose amplifier circuits exploit saturation conditions, and the linear relationships discussed earlier do not apply in such cases.

The nature of the input and output voltages using sinusoidal waveforms are shown in Figure 8-3 based on either a moderate gain level or on different scales. The input waveform is shown in (a). Output waveforms for a noninverting amplifier and an inverting amplifier are illustrated in (b) and (c), respectively. Some applications dictate that one or the other types must be used, while in other applications, it might not make any difference.

Most amplifiers are characterized by the fact that $|A| > 1$, meaning that there is a true voltage amplification of the signal. However, there are some circuits in which $|A| \leq 1$, which are used primarily for isolation. Such circuits typically have a current gain or a power gain, so it is still proper to call them "amplifiers."

How can there be more power coming out of the amplifier than was provided at the input? It is because we are showing here only the signal input and output terminals. All amplifiers must have a dc source or **bias supply** that provides power to the unit. The signal processing capability of the unit converts the dc power to time-varying power, so that ultimately the output power can never exceed the total input power. However, the signal output power, voltage, current, or a combination of these variables might be much greater than the corresponding properties of the input signal.

Equivalent Circuit Model

In general, all of the properties of an amplifier will vary with frequency. However, many amplifiers are designed to operate over a frequency range in which the reactive effects of capacitance or inductance can be ignored, and for now we will restrict our consideration to that case. In the frequency range where reactive effects are negligible, and over the amplitude range where linear operation can be assumed, the model shown in Figure 8-4 can represent many amplifiers.

In this model, there are three parameters that are significant: (1) **input resistance**, (2) **open-circuit voltage gain**, and (3) **output resistance**. The two resistances are often

FIGURE 8–4
Block diagram of amplifier showing input resistance, open-circuit gain, and output resistance.

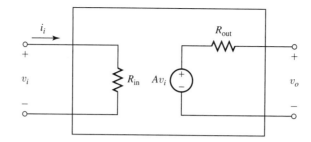

called **input impedance** and **output impedance**, because they could have reactive properties, but we will use the resistive terms in this discussion.

Input Resistance

The input resistance is denoted as R_{in}. It represents the load "seen" by a source at the input. Specifically, it can be defined in terms of the input voltage and current on the schematic as

$$R_{in} = \frac{v_i}{i_i} \qquad (8\text{-}3)$$

Open-Circuit Voltage Gain

The open-circuit voltage gain has been defined previously, although no adjective was attached to it at that point. As we will see later, loading on the output can affect the net voltage gain.

Note that the voltage source in this model has the value Av_i. This is an example of a **dependent** or **controlled source**, in that its value is a function of some other variable in the circuit. In this case it is called a **voltage-controlled voltage source (VCVS)**, and the value is the gain factor A multiplied by the input voltage. Under open-circuit conditions, the voltage at the output is simply Av_i as expected. Some references use a diamond-shaped source to represent a dependent source, but we will use the same symbol as other sources.

Output Resistance

The resistance R_{out} represents the Thevenin equivalent resistance appearing in series with the dependent source. Under open-circuit conditions, it has no effect on the output voltage, but with a load connected, it will have an effect, as will be discussed shortly.

Connecting an Amplifier to a Source and a Load

To use an amplifier for its intended function, it will normally be connected to a source and a load. For example, the source could represent the output of some measurement transducer having a very small signal level and the load could represent the monitoring instrumentation. The purpose of the amplifier is to magnify the small signal to a level sufficient for observation and other possible applications.

Refer to Figure 8-5 for the discussion that follows. The Thevenin model of the source is an open-circuit voltage v_s in series with a source resistance R_s. The output of the amplifier is connected to a load resistance R_L.

The effect of the source resistance coupled with the input resistance of the amplifier is to form a voltage divider at the input, which means that the voltage v_i across the amplifier input is smaller than the open-circuit source voltage. The same effect occurs at the output because of the voltage divider effect between the amplifier output resistance and the load resistance.

FIGURE 8-5
Model of amplifier with source and load connected.

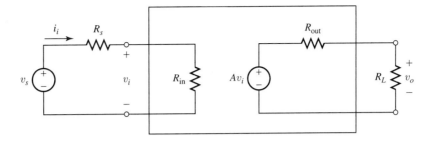

The voltage across the amplifier input terminals can be calculated by the voltage divider rule. It is

$$v_i = \frac{R_{in} v_s}{R_{in} + R_s} \tag{8-4}$$

The voltage across the load can be calculated in terms of the dependent voltage source by another application of the voltage divider rule. It is

$$v_o = \frac{R_L(Av_i)}{R_L + R_{out}} \tag{8-5}$$

Substitution of Equation 8-4 in 8-5 results in

$$v_o = \frac{R_L}{R_L + R_{out}} \times \frac{AR_{in} v_s}{R_{in} + R_s} \tag{8-6}$$

Let us define the voltage gain relating the loaded output voltage to the open-circuit source voltage as A_{so}. This quantity is calculated from Equation 8-6, and the result, after some rearrangement for descriptive purposes, is

$$A_{so} = \frac{v_o}{v_s} = \frac{R_{in}}{R_{in} + R_s} \times A \times \frac{R_L}{R_L + R_{out}} \tag{8-7}$$

This result can be expressed as

$$A_{so} = (\text{input loading factor}) \times A \times (\text{output loading factor}) \tag{8-8}$$

Therefore, the loaded voltage gain relating the output voltage to the open-circuit source voltage is less than the open-circuit voltage gain of the amplifier by the multiplication of two factors. The first is the voltage divider effect occurring at the input to the amplifier, and the other is the voltage divider effect at the output.

Ideal Voltage Amplifier

How do we obtain maximum voltage gain under loaded conditions? The answer is to make the two loading factors as close to unity as possible. The input loading factor increases as the input resistance increases relative to the source output resistance. The output loading factor increases as the load resistance increases relative to the output resistance of the amplifier.

The two loading factors would each be unity if the input resistance to the amplifier were infinite and the output resistance were zero. Therefore, an ideal voltage amplifier is characterized by (1) infinite input resistance (open circuit) and (2) zero output resistance. In this case, the voltage gain between source and load is simply

$$A_{so} = A \quad \text{for ideal amplifier} \tag{8-9}$$

All amplifiers have a finite input resistance and a nonzero output resistance, but many can be approximated as ideal amplifiers. We will see shortly how the operational amplifier can be approximated by that condition in many circuits.

EXAMPLE 8-1

Refer back to the input-output characteristic curve of Figure 8-2(a) and calculate the voltage gain.

SOLUTION The output voltage changes by 6 V when the input voltage changes by 0.1 V, so the voltage gain is

$$A = \frac{6 \text{ V}}{0.1 \text{ V}} = 60 \tag{8-10}$$

We will assume that this is an open-circuit voltage gain for reference in a later example.

EXAMPLE 8-2

Refer back to the input-output characteristic curve of Figure 8-2(b) and calculate the voltage gain.

SOLUTION The output voltage changes by 2 V when the input voltage changes by -40 mV $= -0.04$ V and the voltage gain is

$$A = \frac{2 \text{ V}}{-0.04 \text{ V}} = -50 \tag{8-11}$$

EXAMPLE 8-3

Assume that the amplifier of Example 8-1 has an input resistance of 10 kΩ and an output resistance of 500 Ω. Draw an equivalent circuit model.

SOLUTION Combining the result of Example 8-1 and the data provided in this example, the equivalent circuit model is shown in Figure 8-6.

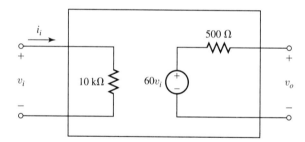

FIGURE 8-6
Amplifier model of Example 8-3.

EXAMPLE 8-4

The amplifier of Example 8-3 is connected between a transducer having an output resistance of 2 kΩ and an instrumentation system having an equivalent input resistance of 4.5 kΩ, as shown in Figure 8-7. Calculate the loaded voltage gain between the source open-circuit voltage and the output voltage across the load.

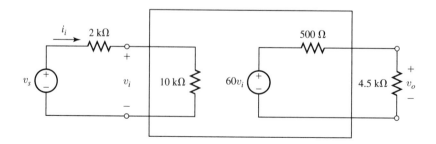

FIGURE 8-7
Circuit model of Example 8-4.

SOLUTION We will apply the concept of Equations 8-7 and 8-8 directly and express all resistance values in ohms, although we could just as easily work in kilohms as long as the amplifier output resistance were expressed as 0.5 kΩ. The loaded voltage gain is expressed as

$$A_{so} = \left(\frac{10{,}000}{10{,}000 + 2000}\right) \times 60 \times \left(\frac{4500}{4500 + 500}\right) = \left(\frac{5}{6}\right) \times 60 \times \left(\frac{9}{10}\right) = 45 \quad (8\text{-}12)$$

Note that there is a reduction factor of 5/6 at the input and a reduction factor of 9/10 at the output. As a net result, the open-circuit gain for the amplifier of 60 is reduced to 45 under loaded conditions.

8-2 Operational Amplifiers

We are now ready to begin the study of **operational amplifiers**, for which the abbreviation **op-amp** will be used in many subsequent discussions. Operational amplifiers are integrated circuits that contain the equivalent of several dozen transistors, along with resistors and one or more capacitors. We could show a schematic diagram to dazzle the reader, but it wouldn't make much sense unless one devoted at least a chapter to the internal details. Even most electrical engineers and technologists who are not integrated-circuit specialists only have a cursory knowledge of the internal workings.

Thankfully, you don't have to understand what goes on inside the chip to use an op-amp properly. Most common applications of the op-amp require a basic understanding of the terminal characteristics, and that will be our focus. If specifications become too rigid for this treatment to meet, it is best to call in an expert anyway.

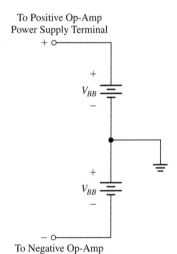

FIGURE 8–8
Dual power supply configuration to provide power for an op-amp.

Power Supply Connections

Before introducing the characteristics of the op-amp, we will examine the connection scheme to the dc bias power supplies. Figure 8-8 shows the most common manner for providing power to an op-amp. It involves two separate power supplies, which are represented here as batteries. The midpoint between the series connection of the power supplies is connected to the common ground reference as shown. The upper power supply provides a positive voltage $+V_{BB}$ with respect to ground, and the lower power supply provides a negative voltage $-V_{BB}$, also with respect to ground. By using the symbol V_{BB} for both supplies, we are assuming that the voltage magnitudes are the same, and that is not always the case. Indeed, it is even possible to power an op-amp with only one power supply, and some are designed with that goal in mind. However, the most common case is that dual power supplies have equal bias voltages, and we make that assumption throughout the chapter unless otherwise indicated. Typical values are ±15 V for a variety of common op-amps.

Block Diagram of Op-Amp with Power Supply Connections

A block diagram of an op-amp showing the power supply connections is shown in Figure 8-9(a). The positive power supply terminal is on the top and the negative power supply terminal is shown on the bottom. These terminals are connected to the two power supply connections with the same designations as explained earlier. The other three terminals will be discussed shortly. One or more additional terminals can be used in some demanding applications to minimize certain offset effects, but they will not be considered here.

Note that there is no ground connection on the op-amp itself. The ground reference is established at the midpoint connection between the two power supply voltages. Most voltages are referred to that common reference point.

FIGURE 8-9
Operational amplifier schematic diagrams.

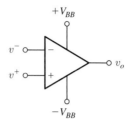
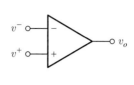
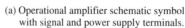

(a) Operational amplifier schematic symbol with signal and power supply terminals.

(b) Operational amplifier schematic symbol showing only signal terminals.

Unless otherwise indicated, the power supply connections will be understood to exist in subsequent circuits, but they will be omitted from the schematics to simplify the diagrams. Therefore, the developments will focus on the signal properties.

Block Diagram of Op-Amp Signal Terminals

A simplified block diagram of an op-amp with the power supply connections understood is shown in Figure 8-9(b). There are two input terminals and one output terminal. The two input terminals are shown on the left and the output terminal is shown on the right.

First, in sharp contrast to the power supply terminals previously discussed, the negative and positive terminals for the input do *not* refer to polarities of signal voltages. Rather they refer to the concepts of **inverting** and **noninverting**.

Inverting Input

The negative sign on the top left identifies that input as an **inverting input**. This means that any voltage applied to that input (either positive or negative) will cause the output to change in the opposite direction. Thus, an increase in the input voltage at that terminal will cause the output voltage to decrease.

Noninverting Input

The positive sign on the bottom left identifies that input as a **noninverting input**. This means that any voltage applied to that input will cause the output to change in the same direction.

Symbol Direction

Some references reverse the schematic order on the left and show the noninverting input on top and the inverting input at the bottom. Indeed, that may seem more natural and it may occur at some point within this text to simplify a drawing. However, most references tend to show it with the inverting input on the top as shown here, because that form is a little more convenient in most circuit diagrams. Just be alert to the possibility of finding it in the opposite form.

Defining Relationship

Let v^+ represent the voltage at the noninverting input with respect to ground and let v^- represent the voltage at the inverting input with respect to ground. The output signal voltage v_o with respect to ground is given by

$$v_o = A(v^+ - v^-) \tag{8-13}$$

Stated in words, the output voltage of the op-amp is a gain factor A (which is defined as positive for this purpose) times the difference between the two voltages at the input terminals, with the voltage at the noninverting input assuming the positive reference.

Differential Input Voltage

The difference between the two input voltages is referred to as the **differential input voltage** v_d. It is

$$v_d = v^+ - v^- \tag{8-14}$$

In terms of the differential input voltage, the output voltage is readily expressed as

$$v_o = A v_d \tag{8-15}$$

A reminder: The signal voltage at either terminal may either be positive or negative and the superscripts of $+$ and $-$ indicate noninverting and inverting, respectively.

DC Coupling and Offsets

One way to classify amplifiers is as **dc coupled** or **ac coupled**. A *dc coupled* amplifier can process signals at frequencies all the way down to dc, meaning that a constant change at the input causes a constant change at the output. However, an *ac coupled* amplifier will amplify only signal components above a lower frequency limit. Traditionally, ac coupled amplifiers were easier to design because they were less susceptible to problems caused by dc drift in the various voltage levels within the circuit. Moreover, many of the early applications of electronics in the 20th century, such as in audio, radio, and television applications, did not require dc response. Much of the design using vacuum tubes and transistors used ac coupling, usually by the use of capacitors between stages, to remove the dc levels involved. Indeed, the design of stable dc amplifiers for applications requiring dc in the early days of electronics was a real challenge.

Most integrated circuit op-amps are dc coupled, and therefore can process both dc changes at the input and time-varying signals. This condition is implied by the input-output defining relationships of Equations 8-13, 8-14, and 8-15. In fact, the reason to have two separate power supplies is to permit the output voltage to be zero when the input voltage is zero, without requiring a coupling capacitor.

With the advantage of dc coupling, there is the problem of dc changes at the output, which could be interpreted in some applications as signal levels. The phenomena that cause these problems are referred to as **voltage offsets** and **current offsets**. These phenomena are most troublesome when dc-type signals (i.e., slowly varying signals) are being processed at very small voltage and current levels. These effects will be assumed as negligible for the applications considered in this chapter.

Open-Loop Voltage

When discussing general amplifier properties, we referred to A as the open-circuit gain. For op-amps, we will define it as the **open-loop gain** for reasons that will be clear later. A typical value is of the order of 100,000 or greater. As we will see shortly, for linear applications, op-amps are rarely operated under open-loop conditions because of stability problems. Therefore, linear amplifier operation is almost always performed with **negative feedback** under so-called **closed-loop conditions**.

Op-Amp Input-Output Characteristic

A somewhat idealized representative input-output characteristic of an op-amp is shown in Figure 8-10. The values shown are typical and will be considered in the example at the end of this section. It is clear from the values shown that the voltage gain is very high.

FIGURE 8–10
Typical idealized input-output characteristic of a representative op-amp.

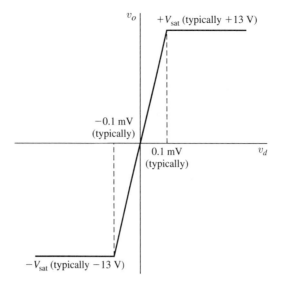

Saturation Voltages

Without the presence of any types of reactive components, the output voltage will be constrained to the limits of the power supply voltages. In fact, there is some "backoff" because of transistor or diode drops. For power supply voltages of ±15 V, conservative values of the maximum limits are about ±13 V. Those are the values assumed on the idealized curves of Figure 8-10. These values are the **saturation voltages**. We will assume that the magnitudes are equal. The magnitude will be denoted as V_{sat}, and for the idealized characteristic shown, the two saturation voltages are $\pm V_{sat} = \pm 13$ V.

Negative Feedback

Although the phenomenon is largely obscure to the general public, **negative feedback** has to rank as one of the greatest developments of the 20th century in the evolution of modern electronics. Many chapters could be devoted to the process, but the treatment here will be limited to its application in creating operational amplifier linear applications.

The general concept of negative feedback in an amplifier is that a sample of the output signal is fed back to the input and subtracted from the input signal. If the gain starts to increase due to drift, a larger signal is fed back and this will reduce the effective gain. The opposite occurs if the gain starts to decrease.

The result of this negative feedback is that the net gain can be stabilized to a high degree of accuracy. The resulting gain with feedback will be called the **closed-loop gain** (as contrasted with the **open-loop gain**). The advantages that occur with negative feedback include (1) greater stability of the circuit, (2) more precise control of the closed-loop gain, (3) reduction of distortion and noise, (4) more precise control of the input and output resistances, and (5) increased bandwidth. It is safe to say that without negative feedback, the electronics industry would never have advanced to the level that exists today.

■ EXAMPLE 8-5

Determine the open-loop voltage gain of the idealized op-amp input-output characteristic of Figure 8-10.

SOLUTION The open-loop voltage gain is the slope of the straight line. As the differential input voltage changes from 0 to 0.1 mV, the output voltage changes from 0 to 13 V. The open-loop voltage gain is then

$$A = \frac{13 \text{ V}}{0.1 \times 10^{-3}} = 130{,}000 \tag{8-16}$$

This is a typical value, and it illustrates how large the gain can be. An actual op-amp would exhibit some rounding in the curve as the level approached saturation.

■ EXAMPLE 8-6

Assume that there is a closed-loop application of the op-amp of Example 8-5 in which the output voltage is 2.6 V. Determine the differential input voltage.

SOLUTION This problem sounds like the "cart is pulling the horse," but in the next section we will explore how this happens. If the output is 2.6 V, the input that produces it is given by

$$v_d = \frac{v_o}{A} = \frac{2.6}{130{,}000} = 20 \text{ μV} \tag{8-17}$$

This is an extremely small voltage, and the analysis performed here might seem a little strange to the reader. The next section should shed some light on why it was performed. ■

8-3 Operational Amplifier Circuit Analysis

Many of the circuits that utilize op-amps can be analyzed by a simple approach that will be explained in this section. The analysis is based on certain idealized assumptions. In practice, the designs are established around strategies that satisfy these conditions to a very high degree of accuracy, and, therefore, the assumptions are generally valid in a wide variety of applications.

The assumptions are based on linear operation with negative feedback. For nonlinear applications, or for those that do not employ negative feedback, the assumptions are generally not valid.

Assumptions

Assuming linear operation with negative feedback, the three major assumptions are

- The input resistance of the op-amp as measured between the noninverting and the inverting input is infinite, meaning that it can be considered as an open circuit.
- The output resistance of the op-amp looking back from the output terminals is zero, meaning that the output dependent source may be considered as an ideal voltage source.
- The voltage gain of the op-amp may be considered as infinite.

Quantitative statements of the preceding three assumptions are as follows:

$$R_{\text{in}} = \infty \tag{8-18}$$

$$R_{\text{out}} = 0 \tag{8-19}$$

$$A = \infty \tag{8-20}$$

The first two assumptions are those desired in all amplifier circuits, but the last one might seem a bit puzzling. Let's study the implications of all three to see how this works.

Implications

- The implication of the infinite input resistance is that the current flowing into or out of either op-amp input terminal may be assumed to be zero.

- The implication of the zero output resistance is that any load connected to the output terminal with respect to ground will not alter the voltage.
- The assumption of infinite gain is a little harder to visualize, but let us assume that somehow, the presence of negative feedback causes the output voltage v_o to be finite. The differential input voltage v_d may then be expressed as

$$v_d = \frac{v_o}{A} \quad (8\text{-}21)$$

Next, assume that the open-loop gain A is permitted to approach infinity, in which case the differential input voltage v_d approaches

$$\lim_{A \to \infty} v_d = \lim_{A \to \infty} \frac{v_o}{A} = 0 \quad (8\text{-}22)$$

This means that the differential input voltage approaches zero in the limit as the gain increases without limit, assuming, of course, that the output voltage remains finite.

The assumption of zero differential voltage may be stated in either of two ways:

$$v_d \approx 0 \quad (8\text{-}23)$$

or

$$v^+ \approx v^- \quad (8\text{-}24)$$

Both points of view are useful and will be used in various developments that follow.

Caution

Although the assumption that the voltages at the two input terminals are equal is valid for analysis purposes, *the two inputs cannot be connected*. There will always be a small voltage across the two terminals which, when multiplied by the open-loop gain, will result in the actual output voltage. Refer back to Example 8-6 for reference to a typical value.

8-4 Inverting Amplifier Circuit

Probably the simplest circuit to analyze using the procedure of the previous section is the closed-loop **inverting amplifier** configuration shown in Figure 8-11. It has been labeled with various variables to be used in the analysis that follows.

First, we should point out that negative feedback has been applied to this circuit. The resistor R_f is a feedback resistor and R_i is an input resistor. Some portion of the output voltage is being sensed through the combination of the two resistors, but we need not elaborate on that point, because the analysis that follows will cover the action of the feedback circuit delineate indirectly. However, this is the first example of a *closed-loop* circuit with negative feedback, and many others will follow.

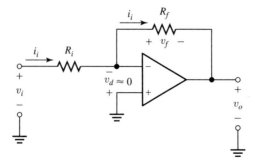

FIGURE 8–11
Inverting amplifier circuit with labeling of circuit variables.

Step-by-Step Analysis

In many closed-loop circuits, the first step is to assume that the differential voltage across the input is zero. The noninverting input is grounded, so we can assume that the inverting input is also at ground potential. The inverting input is not actually connected to ground (and shouldn't be), so it is referred to as a **virtual ground**.

For this circuit, the next step is to realize that because the inverting input can be considered at ground potential, the input voltage v_i appears across the input resistance R_i. By Ohm's law, the current i_i flowing from the source into this resistance is

$$i_i = \frac{v_i}{R_i} \tag{8-25}$$

This current cannot flow into the op-amp itself because it appears as an open circuit. Therefore, it must flow through the resistance R_f as shown. The voltage appearing across this resistance is determined by Ohm's law as

$$v_f = R_f i_i \tag{8-26}$$

Note that the positive polarity is on the left based on the assumed current direction.

Next, a loop will be formed as follows: Start at the grounded noninverting amplifier terminal, move up to the inverting terminal, then move over to the right through the feedback resistance, and finally move back to ground based on the assumed output voltage v_o. This equation reads

$$0 + v_f + v_o = 0 \tag{8-27}$$

where the first 0 corresponds to the assumed differential voltage value. This leads to

$$v_o = -v_f \tag{8-28}$$

Substitution of Equations 8-26 and 8-25 into Equation 8-28 leads to

$$v_o = -\frac{R_f}{R_i} v_i \tag{8-29}$$

Let A_{CL} represent the **closed-loop voltage gain**. This value is determined as

$$A_{CL} = \frac{v_o}{v_i} = -\frac{R_f}{R_i} \tag{8-30}$$

With a little practice, some of these steps can be shortened, but we have chosen to show each one in some detail for clarity.

Before further discussion, we remind the reader that although we assume that the differential input voltage is zero, the two input terminals should *not* be connected because the circuit will not work properly for that condition. There has to be a small voltage existing between the two input terminals with a value such that the product of that small voltage and the very large open-loop gain will equal the output voltage given by Equation 8-29.

Discussion of Closed-Loop Gain

Let us pause to investigate the nature of the closed-loop voltage gain given by Equation 8-30. First, it has a negative value, which was expected based on its definition as an inverting amplifier. However, the major virtue of the circuit is that the gain is a function of the ratio of two resistance values. This might not seem striking to someone without an electronics background, but the design of amplifiers to have a fixed gain is difficult for open-loop circuits. This is because the gain factors of transistors and other active devices vary significantly from one unit to another and they are heavily temperature dependent. However, resistors can be manufactured with very low error tolerances, and by choosing the resistance values appropriately, stable gain values can be achieved.

The simplicity of Equation 8-30 might suggest that any gain could be realized by choosing the correct values of resistance, but the equation is valid only if the magnitude of the closed-loop gain is much smaller than that of the open-loop gain. A good rule of thumb used in some cases is to design for closed-loop gain values that are no greater than about 1 percent of open-loop values although there also are other factors. For example, the bandwidth or range of frequencies that must be processed with the amplifier decreases as the closed-loop gain magnitude increases.

Input and Output Resistances

Two other properties of the closed-loop circuit are significant: The input resistance and the output resistance. The inverting input assumes a virtual ground level, so the input resistance R_{in} is the resistance between the source and the inverting input; that is,

$$R_{in} = R_i \tag{8-31}$$

Don't confuse the input resistance of the closed-loop circuit with that of the op-amp alone. The latter value is assumed to be infinite, but the net circuit input resistance has the finite value R_i.

The output resistance of the circuit is nearly zero, so for most practical purposes

$$R_{out} \approx 0 \tag{8-32}$$

Note that this property holds as long as the maximum output current rating of the op-amp is not exceeded. There is always a maximum rating, and if it is exceeded, the linear model is no longer valid.

Summary

A schematic with only input and output variables labeled, along with a summary of the properties, is shown in Figure 8-12. In some cases, a **bias compensation resistor** is connected between the noninverting input and ground. It does not change the basic operation of the circuit, but it minimizes one of the secondary effects resulting from the presence of very small bias currents that actually flow into or out of the input terminals, but we will neglect that effect in this analysis. It is important primarily in circuits in which extremely small dc-type signals need to be processed.

Choosing Resistance Values for Design

When choosing resistance values for a practical design, certain guidelines should be followed. Consider, for example, that we desire a gain of –10. From Equation 8-30, it would appear that two possible solutions are (a) $R_i = 10 \, \Omega$ and $R_f = 100 \, \Omega$ and (b) $R_i = 10 \, M\Omega$

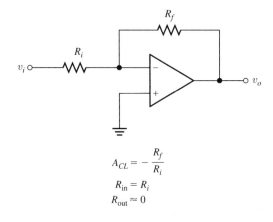

FIGURE 8–12
Inverting amplifier circuit and a summary of its properties.

and $R_f = 100$ MΩ. Yet both of these solutions are BAD, but for different reasons. (1) If resistance values are too small, excessive loading is placed on both the source and the output of the op-amp. (2) If resistance values are too large, problems with offset effects and thermal noise are greater. It turns out that the thermal noise power contained in a resistance is directly proportional to the resistance.

Some exceptions may appear in this text, but a good design strategy is to try to use resistances in the range from about 1 kΩ to several hundred kΩ. Moreover, the inverting amplifier has an input resistance of R_i, so that value should be selected in consort with the output resistance of any possible source. These are suggestions rather than rigid rules, and each design should be studied based on its particular constraints.

■ EXAMPLE 8-7

Consider the inverting amplifier circuit of Figure 8-13. Determine the (a) closed-loop voltage gain, (b) input resistance, and (c) output resistance.

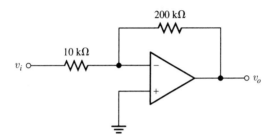

FIGURE 8–13
Circuit of Examples 8-7 and 8-8.

SOLUTION

(a) The closed-loop voltage gain is

$$A_{CL} = -\frac{R_f}{R_i} = -\frac{200 \text{ k}\Omega}{10 \text{ k}\Omega} = -20 \tag{8-33}$$

Note that both numerator and denominator were expressed in kilohms, so the ratio works out without changing to ohms.

(b) The input resistance is

$$R_{in} = 10 \text{ k}\Omega \tag{8-34}$$

(c) The output resistance is assumed to be

$$R_{out} \approx 0 \tag{8-35}$$

■ EXAMPLE 8-8

For the circuit of Example 8-7 (Figure 8-13), assume that the saturation voltages are ±13 V. (a) Determine the maximum magnitude of the input voltage before saturation occurs. (b) Determine the output voltage for input voltages of 0.2 V, −0.6, and 1 V.

SOLUTION

(a) The magnitude of the output voltage cannot exceed 13 V. Let $|v_{i,\text{max}}|$ represent the maximum input voltage magnitude, and the sign can be ignored for this purpose.

$$|v_{i,\text{max}}| = \frac{13}{|A_{CL}|} = \frac{13}{20} = 0.65 \text{ V} \tag{8-36}$$

This means that the input voltage must be bounded by the range $-0.65 \text{ V} \leq v_i \leq 0.65 \text{ V}$ for linear operation.

(b) The output voltage for any input voltage in the linear range is

$$v_o = -20 v_i \tag{8-37}$$

For $v_i = 0.2$ V,
$$v_o = -20 \times 0.2 = -4 \text{ V} \tag{8-38}$$
For $v_i = -0.6$ V,
$$v_o = -20 \times (-0.6) = 12 \text{ V} \tag{8-39}$$

Note that a positive input voltage produces a negative output voltage and vice versa.

For $v_i = 1$ V, substitution in Equation 8-37 would yield $v_o = -20 \times 1 = -20$ V, but this is impossible because the saturation voltages are ± 13 V. Of course, the input has exceeded the maximum level of 0.65 V for linear operation, so the output in this case is

$$v_o = -13 \text{ V} \tag{8-40}$$

This result means that an input voltage exceeding the maximum level forces the output to reach a saturation level. Because the amplifier is an inverting circuit, the positive voltage involved forces the output to assume the negative saturation voltage.

▌▌ EXAMPLE 8-9

Using a general-purpose op-amp, design an amplifier circuit with an inverting gain of -10 and an input resistance of 20 kΩ.

SOLUTION If only the gain were specified, there could be many solutions. However, the specification of the input resistance mandates that

$$R_i = 20 \text{ k}\Omega \tag{8-41}$$

Because

$$|A_{CL}| = \frac{R_f}{R_i} \tag{8-42}$$

Then

$$R_f = |A_{CL}|R_i = 10 \times 20 \text{ k}\Omega = 200 \text{ k}\Omega \tag{8-43}$$

Thus, a gain of -10 with an input resistance of 20 kΩ can be achieved with $R_i = 20$ kΩ and $R_f = 200$ kΩ.

8-5 Noninverting Amplifier Circuit

Next, we will consider the non-inverting amplifier circuit. As in the previous section, we will show each step for clarity. Refer to the circuit diagram of Figure 8-14, which has the different circuit variables labeled on the schematic.

Step-by-Step Analysis

In the noninverting amplifier circuit, the signal input v_i is applied to the noninverting amplifier input as shown. The differential input voltage across the op-amp input terminals is

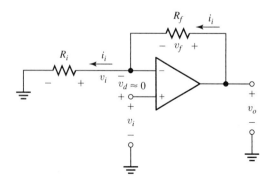

FIGURE 8–14
Noninverting amplifier circuit with labeling of circuit variables.

assumed to be zero, so the voltage at the inverting input is also v_i. This voltage appears across the resistance R_i, and by Ohm's law, a current i_i will flow from that point to ground. It is given by

$$i_i = \frac{v_i}{R_i} \tag{8-44}$$

This current cannot flow out of the op-amp itself because it appears as an open circuit. Therefore, it must flow from the output of the op-amp through the resistance R_f as shown. The voltage appearing across this resistance is determined by Ohm's law as

$$v_f = R_f i_i \tag{8-45}$$

Note that the positive polarity is on the right based on the assumed current direction.

Next, a loop will be formed as follows: Start at ground, move up to the noninverting terminal, then up to the inverting terminal, then over to the output, and finally back to ground based on the assumed output voltage v_o. This equation reads

$$-v_i + 0 - v_f + v_o = 0 \tag{8-46}$$

where the first 0 corresponds to the assumed differential voltage value. This leads to

$$v_o = v_i + v_f \tag{8-47}$$

Substitution of Equations 8-45 and 8-44 in Equation 8-47 leads to

$$v_o = \left(1 + \frac{R_f}{R_i}\right) v_i \tag{8-48}$$

Let A_{CL} represent the closed-loop voltage gain. This value is determined as

$$A_{CL} = \frac{v_o}{v_i} = 1 + \frac{R_f}{R_i} = \frac{R_i + R_f}{R_i} \tag{8-49}$$

As in the case of the inverting amplifier, the gain is a function of resistance values alone. However, there is a slight difference in the ratio for this case.

Input and Output Resistances

The signal input for the closed-loop noninverting amplifier is the op-amp noninverting input, so it is essentially an open circuit. Thus, for most practical purposes, we can say that

$$R_{\text{in}} \approx \infty \tag{8-50}$$

The output resistance of the noninverting amplifier can be approximated as

$$R_{\text{out}} \approx 0 \tag{8-51}$$

Summary

A schematic with only input and output variables labeled, along with a summary of the properties, is shown in Figure 8-15(a). In some cases, a **bias compensation resistor** is placed in series between the signal input and the op-amp input. It does not change the basic operation of the circuit, but it minimizes one of the secondary effects resulting from the presence of very small bias currents flowing into or out of the input terminals, but we will neglect that effect in this analysis. It is important primarily in circuits in which extremely small dc-type signals are processed.

Alternate Configuration

An alternate schematic diagram of the noninverting amplifier is shown in Figure 8-15(b). In this diagram, the two inputs of the op-amp are reversed in position, and the signal input is shown at the top. The diagram of Figure 8-15(a) will be the preferred one for this text, but the reader should be aware of the other form because it does appear in some places in the literature.

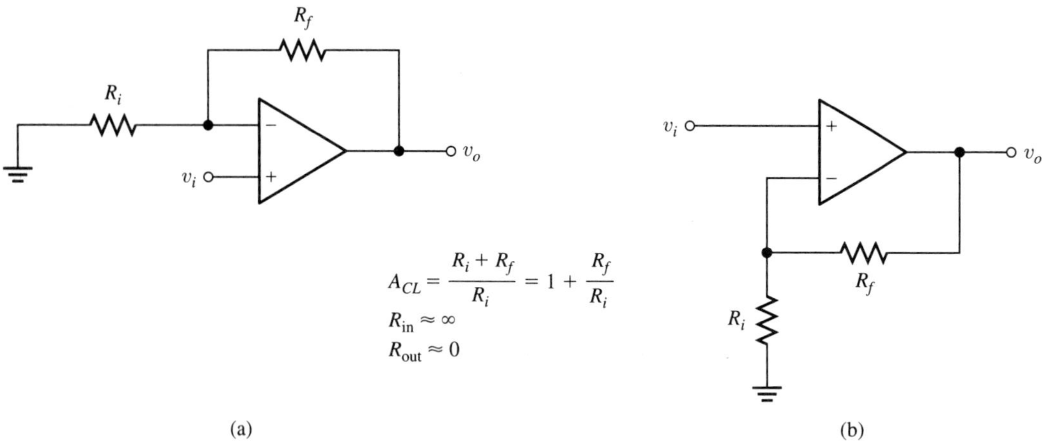

190 CHAPTER 8 • Operational Amplifiers

$$A_{CL} = \frac{R_i + R_f}{R_i} = 1 + \frac{R_f}{R_i}$$
$$R_{in} \approx \infty$$
$$R_{out} \approx 0$$

(a)   (b)

FIGURE 8–15
Noninverting amplifier circuit shown two ways and a summary of its properties.

Voltage Follower

A special case of the noninverting amplifier is the so-called *voltage follower* shown in Figure 8-16. Because $R_f = 0$ and $R_i = \infty$ (an open circuit to ground), the closed-loop voltage gain is

$$A_{CL} = 1 \qquad (8\text{-}52)$$

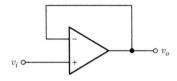

FIGURE 8–16
Voltage follower circuit.

This circuit is used primarily to isolate a signal source from a load while maintaining the same voltage level. Any source will "see" essentially an open circuit at the input, so virtually no loading will occur at that point. However, the output of the op-amp is essentially an ideal voltage source so output current and power can be delivered to a load.

■ EXAMPLE 8-10

Consider the noninverting amplifier circuit of Figure 8-17. Determine the (a) closed-loop voltage gain, (b) input resistance, and (c) output resistance.

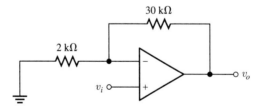

FIGURE 8–17
Circuit of Example 8-10.

SOLUTION

(a) The closed-loop voltage gain is

$$A_{CL} = 1 + \frac{R_f}{R_i} = 1 + \frac{30\,\text{k}\Omega}{2\,\text{k}\Omega} = 16 \qquad (8\text{-}53)$$

For the same resistance values, the gain of an inverting amplifier circuit would be -15.

(b) The input resistance is assumed to be

$$R_{in} = \infty \qquad (8\text{-}54)$$

(c) The output resistance is assumed to be

$$R_{out} \approx 0 \qquad (8\text{-}55)$$

EXAMPLE 8-11 For the circuit of Example 8-10 (Figure 8-17), assume that the saturation voltages are ± 13 V. (a) Determine the maximum magnitude of the input voltage before saturation occurs. (b) Determine the output voltage for input voltages of 0.2 V, -0.6, and 1 V.

SOLUTION

(a) The magnitude of the output voltage cannot exceed 13 V. Let $|v_{i,\max}|$ represent the maximum input voltage magnitude.

$$|v_{i,\max}| = \frac{13}{|A_{CL}|} = \frac{13}{16} = 0.813 \text{ V} \tag{8-56}$$

This means that the input voltage for linear operation must be bounded by the range $-0.813 \text{ V} \leq v_i \leq 0.813 \text{ V}$.

(b) The output voltage for any input voltage in the linear range is

$$v_o = 16 v_i \tag{8-57}$$

For $v_i = 0.2$ V,

$$v_o = 16 \times 0.2 = 3.2 \text{ V} \tag{8-58}$$

For $v_i = -0.6$ V,

$$v_o = 16 \times (-0.6) = -9.6 \text{ V} \tag{8-59}$$

Note that a positive input voltage produces a positive output voltage and vice versa.

For $v_i = 1$ V, substitution in Equation 8-57 would yield $v_o = 16 \times 1 = 16$ V, but this is impossible because the saturation voltages are ± 13 V. The input has exceeded the maximum level of 0.813 V for linear operation, so the output in this case is

$$v_o = 13 \text{ V} \tag{8-60}$$

In the case of a noninverting amplifier, a positive voltage exceeding the maximum linear limit results in the output assuming the positive saturation voltage.

EXAMPLE 8-12 Using a general-purpose op-amp, design an amplifier circuit with a noninverting gain of 25.

SOLUTION The relationship that must be satisfied is the following:

$$A_{CL} = 25 = 1 + \frac{R_f}{R_i} \tag{8-61}$$

This leads to

$$R_f = 24 R_i \tag{8-62}$$

Theoretically, there is an infinite number of solutions to this equation. The only constraint is that R_f must be 24 times as large as R_i. We will select

$$R_i = 1 \text{ k}\Omega \tag{8-63}$$

The value of the feedback resistance is then determined as

$$R_f = 24 \times 1 \text{ k}\Omega = 24 \text{ k}\Omega \tag{8-64}$$

8-6 Operational Amplifier Controlled Sources

There are many different op-amp closed-loop circuits, and entire books are devoted to the subject. In this section, we will explore some of the circuits called **controlled sources**.

For the most part, we will state the properties in this section and later sections without proof. However, some of the end-of-chapter problems are provided as exercises for the reader to verify some of the properties stated.

Voltage-Controlled Voltage Sources

The two amplifier circuits considered thus far; that is, the inverting and non-inverting amplifier configurations, may be considered as **voltage-controlled voltage sources**, meaning that an input signal voltage controls an output voltage through the amplification process. As explained earlier in the chapter, this is the most common amplifier situation; the abbreviation **VCVS** identifies it.

In addition to a *VCVS*, an op-amp can also be used to create a **current-controlled voltage source (ICVS)**, a **voltage-controlled current source (VCIS)**, and a **current-controlled current source (ICIS)**. All of these forms are useful in certain applications and will be considered in this section.

Current-Controlled Voltage Source

The simplest form of an ICVS is shown in Figure 8-18. At first glance, it appears to be similar to the inverting amplifier circuit, but there is no input resistance. Instead, the current flowing into the circuit is assumed to be the controlling variable, and the output voltage is simply

$$v_o = -R_f i_i \tag{8-65}$$

Thus, the property of significance here is that a current flowing toward a virtual ground controls an output voltage and the constant $-R_f$ is referred to in the application of this circuit as the **transresistance**. To avoid saturation, the following inequality must be satisfied:

$$R_f i_i \leq V_{\text{sat}} \tag{8-66}$$

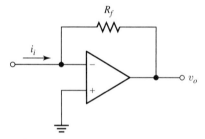
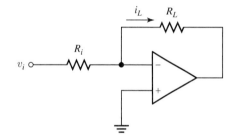

FIGURE 8–18
Current-controlled voltage source.

FIGURE 8–19
Voltage-controlled current source.

Voltage-Controlled Current Source

The simplest form of a VCIS is shown in Figure 8-19. This circuit has exactly the same form as the inverting amplifier, with one difference. Instead of the voltage at the output of the op-amp, the output variable is the current i_L through the load R_L that is of interest. The value of this current is

$$i_L = \frac{v_i}{R_i} \tag{8-67}$$

Note that the current is not a function of the load resistance R_L but is a function only of the input voltage and the input resistance R_i. The value $1/R_i$ is called the **transconductance**. It is measured in siemens (S).

Operation in the linear region requires that

$$R_L i_L \leq V_{\text{sat}} \tag{8-68}$$

Current-Controlled Current Source

The next circuit to be considered is an ICIS, for which a basic form is shown in Figure 8-20. It will be left as a guided exercise (Problem 8-37) to show that the current i_L through the

FIGURE 8–20
Current-controlled current source.

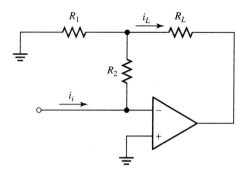

load resistance R_L is given by

$$i_L = \left(1 + \frac{R_2}{R_1}\right) i_i = \beta i_i \tag{8-69}$$

where β is the current gain and is given by

$$\beta = 1 + \frac{R_2}{R_1} \tag{8-70}$$

The condition for linear operation is developed in Problem 8-38 and the requirement is

$$(\beta R_L + R_2) i_i \leq V_{\text{sat}} \tag{8-71}$$

III EXAMPLE 8-13

The output of a certain analog-to-digital converter is a current source that should be connected to a virtual ground. It is desired to convert the current into a voltage for further processing. Design a circuit that will accomplish the task with a calibration scale of -1 V/mA.

SOLUTION The solution is almost shorter than the problem statement. The ICVS of Figure 8-18 is employed and the magnitude of the transresistance is equal to the conversion factor. Thus,

$$R_f = \frac{1 \text{ V}}{1 \times 10^{-3} \text{ A}} = 1000 \, \Omega = 1 \text{ k}\Omega \tag{8-72}$$

Thus, the output voltage is -1 V per mA of input current. If the inversion is a problem, an inverting amplifier with a gain of -1 could follow this circuit.

Based on saturation voltages of ± 13 V, the input current magnitude should not exceed 13 mA for linear operation.

III EXAMPLE 8-14

(a) Design a voltage-controlled current source with a conversion factor of 1 mA/V.
(b) Assuming saturation voltage of ± 13 V, determine the maximum value of the load resistance when the input voltage is 5 V.

SOLUTION

(a) The circuit will have the form shown in Figure 8-19. The conversion factor is the transconductance. It is $1/R_i$ so the value of that resistance is

$$R_i = \frac{1 \text{ V}}{1 \times 10^{-3} \text{ A}} = 1000 \, \Omega = 1 \text{ k}\Omega \tag{8-73}$$

(b) When the input voltage is 5 V, the value of the current is

$$i_L = \frac{5 \text{ V}}{1 \text{ k}\Omega} = 5 \text{ mA} \tag{8-74}$$

Saturation occurs when $R_L i_L = 13$ V or

$$R_L = \frac{13}{5 \times 10^{-3}} = 2600 \ \Omega \tag{8-75}$$

This means that the load will have a constant current of 5 mA as the load is varied provided that the load resistance does not exceed 2600 Ω.

8-7 Circuits That Combine Signals

In this section, we will consider some circuits that combine two or more signals in either an additive or a subtractive fashion.

Linear Combination Circuit

The circuit of Figure 8-21 will be called a **linear combination circuit**. It can be used to form a voltage that is proportional to the sum of several different voltages, while simultaneously adjusting the weighting or gain factors for the sum.

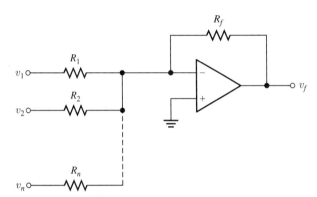

FIGURE 8–21
Linear combination circuit.

Let $v_1, v_2, \ldots, v_n$ represent n different voltages, each of which is referred to a common ground, and let v_o represent the output voltage. It will be left as a guided exercise (Problem 8-39) to show that the output voltage is given by

$$v_o = -\frac{R_f}{R_1}v_1 - \frac{R_f}{R_2}v_2 - \cdots - \frac{R_f}{R_n}v_n \tag{8-76}$$

This equation has the form

$$v_o = A_1 v_1 + A_2 v_2 \cdots + A_n v_n \tag{8-77}$$

where

$$A_1 = -\frac{R_f}{R_1} \tag{8-78}$$

$$A_2 = -\frac{R_f}{R_2} \tag{8-79}$$

$$\vdots$$

$$A_n = -\frac{R_f}{R_n} \tag{8-80}$$

Thus, the output voltage is a weighted sum of all the input voltages, for which the weighting factor or gain associated with each input can be adjusted by setting the value of the particular input resistance.

Note that each gain factor has a negative sign, which means that all the voltages are weighted with the same sign. Positive values for all the gain factors can be achieved by following the circuit with an inverting amplifier having a gain of -1.

FIGURE 8–22
Difference amplifier configuration.

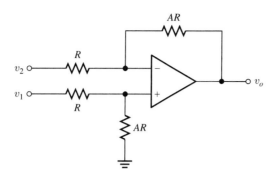

The negative sum of all the voltages can be achieved by setting all the resistances in the circuit to equal values; that is, $R_f = R_1 = R_2 = \cdots = R_n$. In this case, the output voltage is given by

$$v_o = -v_1 - v_2 - \cdots - v_n \tag{8-81}$$

Difference Amplifier

Some applications require that the difference between two voltages be formed. A circuit to achieve this condition while simultaneously multiplying the difference by a constant A is shown in Figure 8-22. Problem 8-40 provides a procedure to derive an expression for the output voltage and it is given by

$$v_o = A(v_1 - v_2) \tag{8-82}$$

If all the resistances are made equal, the gain constant is $A = 1$, and the output voltage is

$$v_o = v_1 - v_2 \tag{8-83}$$

■ EXAMPLE 8-15

The circuit of Figure 8-23 is used to combine two signals and provide a separate gain factor for each signal. Determine the relationship between the output voltage and the two input voltages.

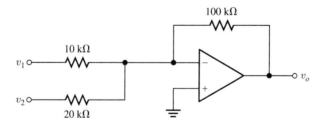

FIGURE 8–23
Circuit of Example 8-15.

SOLUTION The circuit is a linear combination circuit of the form shown in Figure 8-21 with two inputs. The output will be of the form

$$v_o = A_1 v_1 + A_2 v_2 \tag{8-84}$$

The two gain factors are determined as

$$A_1 = -\frac{R_f}{R_1} = -\frac{100 \text{ k}\Omega}{10 \text{ k}\Omega} = -10 \tag{8-85}$$

$$A_2 = -\frac{R_f}{R_2} = -\frac{100 \text{ k}\Omega}{20 \text{ k}\Omega} = -5 \tag{8-86}$$

The output voltage can then be expressed as

$$v_o = -10v_1 - 5v_2 \tag{8-87}$$

▌ EXAMPLE 8-16

For the circuit of Example 8-15 (Figure 8-23), determine the output voltage for each of the following combinations of input voltages: (a) $v_1 = 0.5$ V, $v_2 = 0.8$ V, (b) $v_1 = -1.2$ V, $v_2 = 2$ V, (c) $v_1 = 1$ V, $v_2 = 2$ V. Assume that the saturation voltages are ± 13 V.

SOLUTION From Example 8-15, the output voltage for linear operation is

$$v_o = -10v_1 - 5v_2 \tag{8-88}$$

(a) $v_1 = 0.5$ V, $v_2 = 0.8$ V

$$v_o = -10(0.5) - 5(0.8) = -5 - 4 = -9 \text{ V} \tag{8-89}$$

(b) $v_1 = -1.2$ V, $v_2 = 2$ V

$$v_o = -10(-1.2) - 5(2) = 12 - 10 = 2 \text{ V} \tag{8-90}$$

(c) $v_1 = 1$ V, $v_2 = 2$ V

If we substitute directly into the equation, we would obtain $v_o = -10(1) - 5(2) = -20$ V. Of course, this is impossible because the saturation voltages are ± 13 V. Based on the negative level obtained, the output will be driven into negative saturation and the output voltage is

$$v_o = -13 \text{ V} \tag{8-91}$$

▌ EXAMPLE 8-17

A closed-loop control system is required to form the difference between an input signal and an error signal and multiply the difference by 10. Using a general-purpose op-amp, design a circuit that will accomplish the task.

SOLUTION Because no further specifications are provided, the difference amplifier circuit of Figure 8-22 will be used. Although there are many possible solutions, we will choose $R = 10$ kΩ for the two input resistances and $AR = 10 \times 10$ kΩ $= 100$ kΩ for the other two resistances. The specified input signal is considered as v_1 and the error signal is considered as v_2.

8-8 Integration and Differentiation Circuits

Within certain limits and restrictions, op-amps can be configured to provide the basic calculus operations of integration and differentiation. In fact, the earliest forms of operational amplifiers used back in the 20th century were in so-called **analog computers**. They used op-amp circuits to model various industrial processes involving differential and integral equations. Some of the same concepts were adapted to circuit design as IC op-amps became readily available.

Ideal Op-Amp Integrator Circuit

The first op-amp integrator circuit is shown in Figure 8-24(a). Although some of the circuit variables are not labeled, the analysis follows a format similar to that of the inverting amplifier circuit. Assuming a positive reference for the input voltage v_i, the current i_i flowing from the input toward the virtual ground is

$$i_i = \frac{v_i}{R} \tag{8-92}$$

This current cannot flow into the op-amp so it flows into the capacitor. Assume that the capacitor is initially uncharged. The positive reference for the voltage across the capacitor is on the left and the output voltage is then

$$v_o = -\frac{1}{C} \int_0^t i_i \, dt \tag{8-93}$$

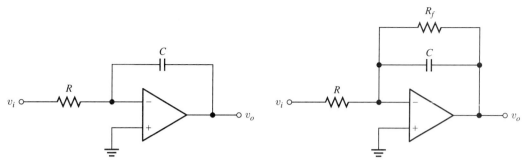

(a) Ideal op-amp integrator circuit.   (b) AC integrator for use with general purpose op-amps.

FIGURE 8–24
Op-amp integrator circuits.

Substitution of Equation 8-92 into 8-93 yields

$$v_o(t) = -\frac{1}{RC} \int_0^t v_i(t)\, dt \qquad (8\text{-}94)$$

where the functional forms have been used for clarity.

Possible Problems with Ideal Integrator

Throughout the chapter, we have used the term *general-purpose op-amp* to refer to common op-amps such as the ubiquitous 741, one of the least expensive and most widely available units. However, an ideal integrator circuit requires that an op-amp have extremely small input bias and offset parameters, some secondary effects not considered in this analysis. Therefore, if pure integration is required, a special purpose op-amp meeting rather stringent bias and offset specifications must be selected.

AC Integrator

It is possible to produce an integrator that will integrate the time-varying portion of an input signal using a more modest op-amp. The form of the circuit is shown in Figure 8-24(b). We will not cover the details of this circuit, but with appropriate design values, it can be used to integrate the "ac portion" of a signal. For dc, the circuit reverts to an amplifier form. More detailed information about this circuit is provided in a book by the first author.[1]

Ideal Op-Amp Differentiator Circuit

The first op-amp differentiator circuit is shown in Figure 8-25(a). Although some of the circuit variables are not labeled, the analysis follows a format similar to that of the inverting amplifier circuit. Assuming a positive reference for the input voltage v_i, the current i_i flowing from the input toward the virtual ground is

$$i_i = C\frac{dv_i}{dt} \qquad (8\text{-}95)$$

This current cannot flow into the op-amp so it flows into the resistor. The positive reference for the voltage across the capacitor is on the left and the output voltage is then

$$v_o = -Ri_i \qquad (8\text{-}96)$$

---

[1] W. D. Stanley, *Operational Amplifiers with Linear Integrated Circuits*, 4th edition, Prentice Hall, 2002.

FIGURE 8–25
Op-amp differentiator circuits.

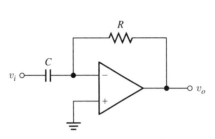
(a) Ideal op-amp differentiator circuit.

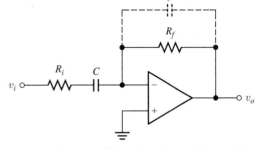
(b) Low-frequency differentiator for use with general purpose op-amps.

Substitution of Equation 8-95 into 8-96 yields

$$v_o(t) = -RC\frac{dv_i(t)}{dt} \tag{8-97}$$

where the functional forms have been used for clarity.

Possible Problems with Ideal Differentiator

As in the case of the ideal integrator, the ideal differentiator can be troublesome, but for different reasons. Differentiation tends to emphasize high-frequency components of a signal. This can result in a significant amount of additive noise on the output. Consequently, pure differentiator circuits are not used very often in practice.

Low-Frequency Differentiator

Differentiation over a limited frequency range can be achieved by the "low-frequency differentiator" shown in Figure 8-25(b). More details concerning this circuit are provided in the reference cited for the ac integrator circuit.

EXAMPLE 8-18

Assuming that an op-amp with sufficient specifications is available, design an ideal integrator whose output is related to the input by

$$v_o(t) = -100 \int_0^t v_i(t)\, dt \tag{8-98}$$

SOLUTION The circuit will be of the form shown in Figure 8-24(a). Comparing Equation 8-98 with the general result of Equation 8-94, the constant multiplier must satisfy

$$\frac{1}{RC} = 100 \tag{8-99}$$

or

$$RC = 0.01 \tag{8-100}$$

In theory, any values of the components that satisfy this relationship would work, but in practice, the range of available values and tolerances must be considered. We will somewhat arbitrarily select

$$C = 0.1\ \mu\text{F} \tag{8-101}$$

The resistance value is then determined as

$$R = \frac{0.01}{C} = \frac{0.01}{1 \times 10^{-7}} = 100\ \text{k}\Omega \tag{8-102}$$

8-9 Active Filters

One of the most useful applications of op-amps is in creating active filters. An **active filter** is one that uses an active device such as an op-amp or transistor as a major component of the filter. This is in contrast to a **passive filter**, which uses only passive components such as inductance, capacitance, and resistance.

Why Active Filters?

Active filters have the advantage that no inductance is required; that is, the filters use resistance, capacitance, and an active device with some feedback. Although inductors are widely used in high-frequency applications, they are very difficult to use at very low frequencies because of the large inductance values required, along with the associated physical size, weight, and significant resistive losses. Active filters can be used to create very good filter characteristics without the use of inductors.

Brief Filter Primer

Before discussing active filters, we will discuss some of the general properties of all filters. Filters are frequency selective circuits that pass certain frequencies while rejecting others. If you have turned on a television or radio today, it is likely that you have adjusted a filter in the process. When you change a channel on a TV or change a station on a radio, you are tuning a filter to pass the desired signal while rejecting others.

Filters are some of the most important components in the field of electronic communications and in other areas, and entire books have been devoted to the subject. Therefore, the treatment here is limited to a brief exposure, but it should be adequate to permit the reader to actually design certain types.

Filter Response Characteristic Forms

Based on the types of response functions obtained, most filters can be classified in four ways: (1) **low-pass**, (2) **band-pass**, (3) **high-pass**, and (4) **band-rejection**. Their behavior will be defined in terms of ideal models, but realize that no real filters meet these ideal conditions.

Low-Pass

An ideal low-pass filter would pass frequencies below a certain reference or **cutoff frequency** while rejecting frequencies above that value.

Band-Pass

An ideal band-pass filter would pass frequencies within a certain range while rejecting those rejecting those below and above that range.

High-Pass

An ideal high-pass filter would pass frequencies above a certain reference or **cutoff frequency** while rejecting frequencies below that value.

Band-Rejection

An ideal band-rejection filter would reject frequencies within a certain range while passing those below and above that range.

All of these types have varied uses in electronic circuits and all can be implemented with either passive circuits or with active circuits. However, the low-pass and band-pass types are the most common and the treatment here will be limited to certain active low-pass and band-pass types.

Decibel Response Forms

It is virtually impossible to deal with filter characteristics without some understanding of decibel forms. The use of decibel measurements evolved very early in the history of electrical and electronics engineering. It is widely employed in the communications industry. Without discussing all of the variations involved, we should understand some of the definitions.

Consider a relative voltage gain response $A(f)$ that is some function of the frequency f and let A_o represent a reference level. The relative decibel gain will be denoted as $A_{dB}(f)$, and it will be defined as

$$A_{dB}(f) = 20 \log_{10} \frac{A(f)}{A_o} \tag{8-103}$$

Strictly speaking, the original definition is based on a power ratio and the constant multiplier in that case would be 10 rather than 20. However, the limited exposure in this text will be based on a voltage ratio. The constant in that case is 20, as shown in Equation 8-103.

Butterworth Low-Pass Characteristics

Several different types of characteristics approximate ideal filter characteristics. One of the most common types is the **Butterworth characteristic**. A parameter used to specify the filter characteristic is the **order** or **number of poles**. The latter name refers to a mathematical property inherent in the development of the filter response. From the viewpoint of this limited treatment, it can be interpreted as the *number of capacitors* in the circuit.

Let us define the *normalized frequency u* as

$$u = \text{Normalized Frequency} = \frac{\text{Actual Frequency}}{\text{3-dB Cutoff Frequency}} = \frac{f}{f_c} \tag{8-104}$$

The Butterworth filter amplitude characteristics in decibels are defined by the function

$$A_{dB}(f) = -20 \log_{10}(1 + u^{2n}) \tag{8-105}$$

Normalized Butterworth filter amplitude response curves for three cases are shown in Figure 8-26. The left-hand half of the curve corresponds to the **passband**. At some arbitrary point on the right-hand half, the **stopband** is defined. Between these limits is a **transition band**.

The parameter n is the number of poles or order. Observe that as the order increases, (1) the response in the passband is "flatter." (2) The response in the stopband is smaller. This trend increases as the order or number of poles increases.

The response curves here are *normalized* with respect to the so-called **cutoff frequency**, which is defined for a Butterworth filter as the point at which the response is

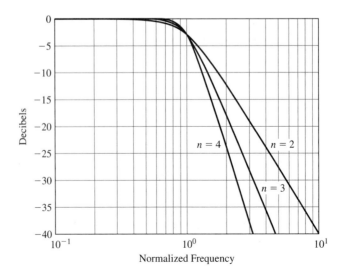

FIGURE 8–26
Several Butterworth low-pass amplitude response forms.

about 3 dB below the dc response. This corresponds to an absolute level of 0.707 times the dc response.

This normalization scheme makes the curves somewhat universal in nature. Thus, the midpoint $10^0 = 1$ corresponds to the cutoff frequency; $10^{-1} = 0.1$ corresponds to one tenth of the cutoff frequency and $10^1 = 10$ corresponds to 10 times the cutoff frequency.

Several Low-Pass Butterworth Filter Design Forms

Active filter low-pass Butterworth designs are shown in Figures 8-27, 8-28, and 8-29. These correspond to $n = 2, n = 3$, and $n = 4$, respectively. Obviously, this is a limited

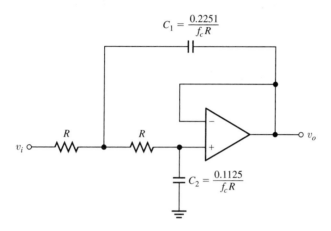

FIGURE 8–27
Second-order low-pass Butterworth filter.

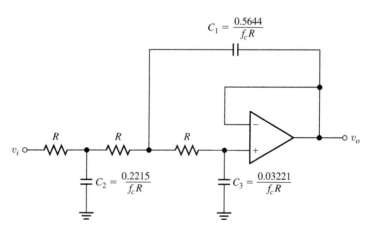

FIGURE 8–28
Third-order low-pass Butterworth filter.

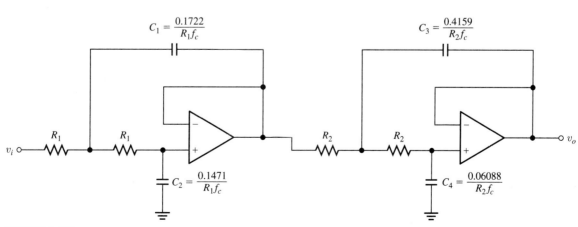

FIGURE 8–29
Fourth-order low-pass Butterworth filter.

catalogue of designs, but it should provide the reader with a representative group that can serve many practical purposes. Examples 8-19 and 8-20 at the end of this section will illustrate the design process.

For each design, there are more variables than design constraints. This situation permits the designer to make some arbitrary choices in the design, in which case the other variables will be constrained according to the equations. For example, a resistance value can be selected and the capacitance values can then be determined. Conversely, a capacitance value may be selected and the resistance values can then be determined. The latter procedure is often better because it is usually easier to adjust resistance values than capacitance values.

Two-Pole Band-Pass Responses Revisited

Back in Chapter 5, resonance circuits requiring inductance, capacitance, and resistance were considered. The same types of response functions may be obtained with active filters without the use of inductance. Repeats of some of the curves of Chapter 5, but now expressed in decibel form, are shown in Figure 8-30.

An active circuit that will achieve the same response as an RLC resonant circuit is shown in Figure 8-31. Note that the factor Q and the center frequency f_0 appear as design parameters.

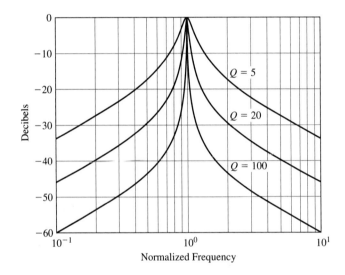

FIGURE 8-30
Several two-pole band-pass amplitude response forms.

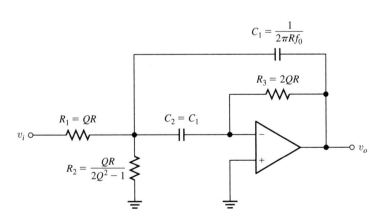

FIGURE 8-31
Second-order band-pass active filter that exhibits the properties of a resonant circuit.

EXAMPLE 8-19

An active low-pass filter is desired to satisfy the following specifications:

3-dB cutoff frequency = 1 kHz

Response at 2 kHz to be down by at least 18 dB relative to the dc response

Determine the minimum order for the filter.

SOLUTION This type of problem can be solved either by the mathematical form of the response or by the use of curves if sufficient data are available. Let's try the latter approach by referring to Figure 8-26. Because 1 kHz is the desired 3-dB cutoff frequency, 2 kHz corresponds to a normalized frequency of 2 kHz/1 kHz = 2. Therefore, see if the response is down by 18 dB for either of the curves when the normalized frequency is 2. Note that since the response is measured downward from the reference level of 0 dB, the value sought on the curves should be no higher than -18 dB.

For $n = 2$, the level is about -12 dB at a normalized frequency of 2, so a two-pole filter is insufficient. However, for $n = 3$, the response appears to be at about -18 dB. Of course, $n = 4$ would do even better, but the three-pole should be sufficient. Obviously, this problem was "rigged" to come out with a nice clean value.

The reader is invited to show that the exact value at the given frequency can be determined with Equation 8-105 with $u = 2$ and $n = 3$, and is about -18.13 dB. Thus, the specifications are met with a little to spare. For values not on curves such as these, Equation 8-105 can be used to assist in the design. However, some handbooks are available that display many response curves for different types of filters.

EXAMPLE 8-20

Design a low-pass Butterworth filter that will satisfy the specifications of Example 8-19.

SOLUTION It was determined that a three-pole filter was required and the cutoff frequency should be 1000 Hz. The form of the circuit is that of Figure 8-28. One parameter must be arbitrarily selected and some trial and error is involved if standard stock values must be used. Based on the objective of this book, we certainly don't expect readers to know what values can and cannot be found, so we will compute some exact values and then provide some guidance on the final design.

We will somewhat arbitrarily select the three equal resistance values to be

$$R = 100 \text{ k}\Omega \tag{8-106}$$

The exact values for the three capacitances are then determined as follows:

$$C_1 = \frac{0.5644}{Rf_c} = \frac{0.5644}{100 \times 10^3 \times 1000} = 5.644 \times 10^{-9} \text{ F} = 5644 \text{ pF} \tag{8-107}$$

$$C_2 = \frac{0.2215}{Rf_c} = \frac{0.2215}{100 \times 10^3 \times 1000} = 2.215 \times 10^{-9} \text{ F} = 2215 \text{ pF} \tag{8-108}$$

$$C_3 = \frac{0.03221}{Rf_c} = \frac{0.03221}{100 \times 10^3 \times 1000} = 322.1 \times 10^{-12} \text{ F} = 322.1 \text{ pF} \tag{8-109}$$

None of these three values are standard stock values. Of course, one could combine different capacitance values in series or parallel combinations to create almost any value desired, but that is seldom necessary. Low-pass Butterworth filters are fairly "forgiving" and will usually work reasonably well with some leeway on the values.

For this particular case, some standard 5 percent tolerance values for the capacitors should be tried, and these values follow.

$$C_1 = 5600 \text{ pF} \tag{8-110}$$

$$C_2 = 2200 \text{ pF} \tag{8-111}$$

$$C_3 = 330 \text{ pF} \tag{8-112}$$

It should be noted that narrow band-pass filters (those with high Q values) are less "forgiving," and this often mandates that some tuning of the final design be performed.

8-10 Voltage Regulation with an Op-Amp

In the consideration of rectifier circuits in Chapter 6, a reference was made at several points to the use of a solid-state regulator circuit to stabilize the dc output voltage and to reduce the ripple to a much smaller value than could be practically attained with a single capacitor filter. An integrated circuit designed specifically to accomplish this objective is called a **voltage regulator**. Voltage regulators are available in a wide variety of voltage levels and power ratings.

The internal operation of many common regulators is based on a closed-loop control mechanism employing an op-amp circuit form. We will study a typical form in which that operation is achieved in this section.

Regulator Circuit

Consider the circuit shown in Figure 8-32. Although the layout is different than we have considered in other applications in the chapter, the op-amp is operating as a noninverting amplifier. A zener diode is used to establish a voltage reference V_z, and the current that establishes the zener level is achieved with the resistance R_z. The op-amp feedback arrangement involving R_i and R_f forces the load voltage V_L to take the value

$$V_L = \left(1 + \frac{R_f}{R_i}\right) V_z \tag{8-113}$$

The output voltage of the op-amp will be slightly greater because of the base-emitter drop for the transistor. Note that there will be minimum variation of the zener voltage because its current supply is based on the regulated output voltage. Note also that the input voltage v_i is shown in lower-case because it is assumed to vary somewhat, but the output load voltage V_L should be a constant value over the range of operation.

Some regulators allow for the regulated output voltage to be adjustable. This can be achieved if either R_i or R_f is adjustable.

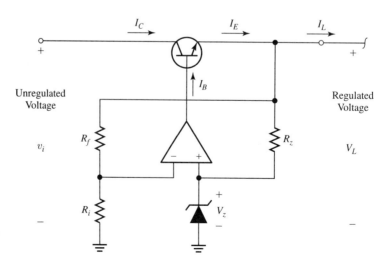

FIGURE 8-32
Voltage regulator circuit using an op-amp and a pass transistor.

PROBLEMS

8-1 An amplifier has an open-circuit gain of 80, an input resistance of 4 kΩ, and an output resistance of 1 kΩ. Draw an equivalent circuit model.

8-2 An amplifier has an open-circuit gain of −60, an input resistance of 8 kΩ, and an output resistance of 2 kΩ. Draw an equivalent circuit model.

8-3 The amplifier of Example 8-1 is connected between a source having an output resistance of 1 kΩ and a load of 3 kΩ. Calculate the loaded voltage gain between the source open-circuit voltage and the output.

8-4 The amplifier of Example 8-2 is connected between a source having an output resistance of 2 kΩ and a load of 10 kΩ. Calculate the loaded voltage gain between the source open-circuit voltage and the output.

8-5 An op-amp has a linear input-output characteristic over the input voltage range of ±0.04 mV and reaches output saturation voltages of ±13 V at the limits. Determine the open-loop voltage gain.

8-6 An op-has a linear input-output characteristic over the input voltage range of ±20 μV and reaches output saturation voltages of ±14 V at the limits. Determine the open-loop voltage gain.

8-7 Assume that there is a closed-loop application of the op-amp of Problem 8-5 in which the output voltage is −4 V. Determine the differential input voltage.

8-8 In a certain closed-loop application, the differential input of the amplifier of Problem 8-6 is 5 μV. Determine the output voltage.

8-9 An inverting amplifier circuit has $R_i = 10$ kΩ and $R_f = 220$ kΩ. Determine the (a) closed-loop voltage gain, (b) input resistance, and (c) output resistance.

8-10 An inverting amplifier circuit has $R_i = 15$ kΩ and $R_f = 120$ kΩ. Determine the (a) closed-loop voltage gain, (b) input resistance, and (c) output resistance.

8-11 For the circuit of Problem 8-9, assume that the saturation voltages are ±14 V. (a) Determine the maximum magnitude of the input voltage before saturation occurs. (b) Determine the output voltage for input voltages of −0.2 V, 0.6, and −1 V.

8-12 For the circuit of Problem 8-10, assume that the saturation voltages are ±14 V. (a) Determine the maximum magnitude of the input voltage before saturation occurs. (b) Determine the output voltage for input voltages of −0.2 V, 0.6, and −2 V.

8-13 Design an inverting amplifier circuit with a gain of −5 and an input resistance of 30 kΩ.

8-14 Design an amplifier circuit with an inverting gain of −25 and an input resistance of 3 kΩ.

8-15 A noninverting amplifier circuit has $R_i = 2$ kΩ and $R_f = 33$ kΩ. Determine the (a) closed-loop voltage gain, (b) input resistance, and (c) output resistance.

8-16 A noninverting amplifier circuit has $R_i = 10$ kΩ and $R_f = 20$ kΩ. Determine the (a) closed-loop voltage gain, (b) input resistance, and (c) output resistance.

8-17 For the circuit of Problem 8-15, assume that the saturation voltages are ±14 V. (a) Determine the maximum magnitude of the input voltage before saturation occurs. (b) Determine the output voltage for input voltages of −0.2 V, 0.6, and −1 V.

8-18 For the circuit of Problem 8-16, assume that the saturation voltages are ±14 V. (a) Determine the maximum magnitude of the input voltage before saturation occurs. (b) Determine the output voltage for input voltages of −2 V, 4 V, and −5 V.

8-19 Design a noninverting amplifier circuit with a gain of 40.

8-20 Design a noninverting amplifier circuit with a gain of 50.

8-21 Design a current-controlled voltage source with a transresistance magnitude of 10 kΩ. For saturation voltages of ±13 V, determine the maximum magnitude of the input current for linear operation.

8-22 Design a current-controlled voltage source with a conversion factor of 3 V/mA. For saturation voltages of ±14 V, determine the maximum magnitude of the input current for linear operation.

8-23 Design a voltage-controlled current source with a transconductance magnitude of 0.1 mS. For an input voltage of 12 V and saturation voltages of ±13 V, determine the maximum value of the load resistance for linear operation.

8-24 Design a voltage-controlled current source with a conversion factor of 0.5 mA/V. For a maximum load resistance of 8 kΩ and saturation voltages of ±14 V, determine the maximum value of the input voltage for linear operation.

8-25 Design a current-controlled current source with a current gain (β) of 4. Select $R_1 = 1$ kΩ. For an input current of 1 mA and saturation voltages of ±14 V, determine the maximum value of the load resistance for linear operation.

8-26 Design a current-controlled current source with a current gain (β) of 10. Select $R_1 = 3$ kΩ. For a load resistance of 1 kΩ and saturation voltages of ±13 V, determine the maximum value of the input current for linear operation.

8-27 Design a linear combination circuit that will combine two signals according to the relationship

$$v_o = -5v_1 - 2v_2$$

Select the feedback resistance to be 15 kΩ.

8-28 Design a linear combination circuit that will combine three signals according to the relationship

$$v_o = -5v_1 - 10v_2 - v_3$$

Select the input resistance for v_1 to be 20 kΩ.

8-29 Design a difference amplifier to form

$$v_o = v_1 - v_2$$

8-30 Design a circuit to form

$$v_o = v_1 + v_2$$

You may use more than one op-amp.

8-31 Assuming that you have an op-amp with sufficient specifications, design an ideal integrator circuit to satisfy the equation
$$v_o(t) = -200 \int_0^t v_i(t)\, dt$$
The input resistance to the circuit should be at least 100 kΩ.

8-32 Design an ideal op-amp differentiator circuit to satisfy the equation
$$v_o = -0.001 \frac{dv_i}{dt}$$

8-33 Design a low-pass active two-pole Butterworth filter with a 3-dB cutoff frequency of 1 kHz. Select the resistances as 100 kΩ.

8-34 Perform a new design for Problem 8-33 in which the largest capacitor is selected as 2000 pF.

8-35 Design a low-pass active four-pole Butterworth filter with a 3-dB cutoff frequency of 1 kHz. Select the resistances as 100 kΩ.

8-36 Perform a new design for Problem 8-35 in which the largest capacitor in *each section* is selected as 2000 pF.

8-37 Refer to the ICIS of Figure 8-20. Starting with the input current i_i, provide a step-by-step development to prove that the load current in the linear region is given by Equation 8-69. *Hint:* The voltage across R_2 is equal to the voltage across R_1, with the positive reference at ground and the current through R_L is the sum of the other two resistive currents.

8-38 From the development of Problem 8-37, derive Equation 8-71 for operation in the linear region.

8-39 Refer to the linear combination circuit of Figure 8-21. Derive the expression for the output voltage in terms of the various input voltages as given by Equation 8-76. *Hint:* The inverting input is at virtual ground so each input resistive current can be calculated independently and the current through the feedback resistance is the sum of all of the input currents.

8-40 Refer to the difference amplifier of Figure 8-22. Derive the expression for the output voltage in terms of the two input voltages as given by Equation 8-82. *Hint:* First calculate the voltage at the non-inverting input in terms of v_1 using the voltage divider rule. This voltage must also equal the voltage at the inverting input and the output voltage can then be determined in terms of both input voltages.

8-41 An active low-pass filter is desired to satisfy the following specifications:

3-dB cutoff frequency = 2 kHz

Response at 8 kHz to be down by at least 35 dB relative to the dc response

Determine the minimum order for the filter.

8-42 An active low-pass filter is desired to satisfy the following specifications:

3-dB cutoff frequency = 500

Response at 1 kHz to be down by at least 24 dB relative to the dc response

Determine the minimum order for the filter.

8-43 Design a low-pass active Butterworth filter that will satisfy the specifications of Problem 8-41 with the largest capacitance selected as 0.01 μF.

8-44 Design a low-pass active Butterworth filter that will satisfy the specifications of Problem 8-42 with all resistance selected as 100 kΩ.

8-45 Design a two-pole band-pass active filter with $f_0 = 1$ kHz, $Q = 10$, and $R = 10$ kΩ.

8-46 Repeat the design of Problem 8-45 with $C = 0.01\ \mu$F.

DIGITAL ELECTRONICS

9 Digital Circuits: Basic Combinational Forms

10 Digital Circuits: Advanced Combinational Forms

11 Digital Circuits: Sequential Forms

Digital Circuits: Basic Combinational Forms

OVERVIEW AND OBJECTIVES

In today's world of computers, personal wireless messaging and communications, music storage technology, digitally processed audio, and many other areas, one does not have to look very far to find digital circuits. Although in the past few decades we have seen an explosion of digital applications in every area of electronics, digital circuit technology is actually as old as electrical circuits. From the time when Samuel Morse sent messages over telegraph lines using a form of digital coding, to the present-day sophisticated digital systems, digital technology has taken on an increasing role in all areas of electronics. In this chapter we begin a study of digital circuits by investigating special number systems used in computing machines, and then move on to fundamental logic circuits, how they operate, and the special mathematics used in their design.

Objectives

After completing this chapter, you will be able to

- Convert numbers between various number systems including binary, octal, decimal, and hexadecimal.
- Perform addition and subtraction in numbers systems other than decimal.
- Convert a binary number to its twos-complement equivalent, and vice versa.
- Identify and write the truth tables for logical operations such as AND, OR, invert, NAND, NOR, exclusive OR, and exclusive NOR.
- Manipulate and reduce Boolean functions to their simplest form.
- Reduce switching functions using Karnaugh maps.
- Draw logic diagrams from switching function equations, and vice versa.

9-1 Binary Number System

Although most digital computing equipment has decimal inputs and outputs, a vast majority of such equipment uses the binary number system internally. In fact, for ease of use in digital computing equipment, designers must go to great lengths to have the equipment "talk" to us in decimal numbers. This is because the decimal number system is rather cumbersome for digital computers. To understand the operation of the logical devices contained in digital computing equipment, first we must understand how the binary number system is constructed and how mathematical operations are done in binary.

210 CHAPTER 9 • Digital Circuits: Basic Combinational Forms

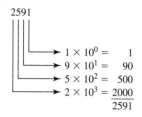

FIGURE 9–1
Analysis of a decimal number.

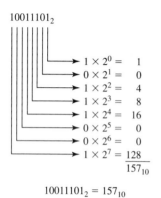

$10011101_2 = 157_{10}$

FIGURE 9–2
Binary-to-decimal conversion.

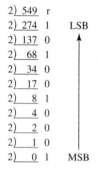

$549_{10} = 1000100101_2$

FIGURE 9–3
Decimal-to-binary conversion.

It is easier to understand the binary number system if we first analyze how the decimal number system is constructed. In any number system, numbers are constructed of digits that are arranged side by side. In the decimal number system (also called the **base-10** number system) each digit can take on any value of 0 through 9, and the weight (or multiplier) of each digit is determined by its placement in the number. The digit in the rightmost column has a multiplier (or weight) of 10^0, the one in the next column to the left has a multiplier of 10^1, and so forth in increasing powers of 10. Figure 9-1 shows how this construction works, using as an example the number 2591_{10}. Once we multiply each digit in the original number by its column weight, we then add the products to achieve the number.

Although this may seem very simple and obvious, it helps if we apply this same technique to binary numbers. Binary numbers are constructed in a similar manner, except that instead of the weight of each column in our decimal numbers having increasing powers of 10, in the binary number system the columns increase in weight by powers of 2. Additionally, although in the decimal number system the highest digit value is 9, in the binary number system the highest digit value is 1 (a *binary digit* is called a **bit**). Figure 9-2 shows how a binary number is constructed and how it can be converted to a decimal number.

Conversion of numbers between the binary and decimal number systems involves simple multiplication and division. As shown in Figure 9-2, binary numbers are converted to decimal by multiplying each digit by the appropriate power of 2. Converting from decimal to binary involves successive integer division by 2 as illustrated in Figure 9-3. In this method, we continue dividing by 2, writing the remainders (either 0 or 1) to the right of the quotients. The process continues until the quotient is zero. Then the binary equivalent is the remainders, read from bottom to top, with the bottom number being the **most significant bit (MSB)** and the top number being the **least significant bit (LSB)**.

Binary addition is performed much like decimal addition, except that one must be mindful that the largest binary digit is 1. Therefore, there are four rules for the addition of binary digits:

$$1: 0 + 0 = 0$$
$$2: 0 + 1 = 1$$
$$3: 1 + 0 = 1$$
$$4: 1 + 1 = 0 \text{ with a carry of } 1$$

Binary subtraction has similar rules, but with a borrow instead of a carry.

$$1: 0 - 0 = 0$$
$$2: 0 - 1 = 1 \text{ with a borrow of } 1$$
$$3: 1 - 0 = 1$$
$$4: 1 - 1 = 0$$

▐ EXAMPLE 9-1

Calculate the sum $11100101_2 + 10110100_2$ and the difference $11100101_2 - 10110100_2$.

SOLUTION Using the addition and subtraction rules above, the results are

```
  11100101      11100101
+ 10110100    - 10110100
──────────    ──────────
 110011001      00110001
```

Note that the above subtraction resulted in a 6-bit number. In this instance, the leading zeros are optional.

Like decimal numbers, it is also possible to have binary numbers with fractional parts. In this case, instead of a decimal point, we have a **binary point**, and as we move to the right of the binary point the column values have increasing negative powers of 2 (that is, 2^{-1}, 2^{-2}, 2^{-3}, and so on). Therefore, the first bit to the right of the binary point has a weight of ½, the next has a weight of ¼, the next ⅛, and so forth.

To convert a decimal fraction to a binary number, we perform the conversion in two parts. First the integer part is converted to its binary integer equivalent (in the usual manner). Next, we convert the fractional part. This is done by successively multiplying the fractional part by 2. After each multiplication, we discard the integer value and multiply again. We continue the process until the fractional part is zero. The fractional binary result is made up of the discarded integer values of the products, read from the top.

EXAMPLE 9-2

Convert the number 1001.1101_2 to decimal. Then check your answer by converting the decimal result to binary.

SOLUTION The binary-to-decimal conversion is done by multiplying each bit by its positional weight.

$$1001.1101_2 = 1 \times 2^3 + 0 \times 2^2 + 0 \times 2^1 + 1 \times 2^0 + 1 \times 2^{-1}$$
$$+ 1 \times 2^{-2} + 0 \times 2^{-3} + 1 \times 2^{-4}$$
$$= 1 \times 8 + 0 \times 4 + 0 \times 2 + 1 \times 1 + 1 \times 0.5 + 1 \times 0.25$$
$$+ 0 \times 0.125 + 1 \times 0.0625$$
$$= 9.8125_{10}$$

To convert 9.8125_{10} to a binary number, we perform the conversion in two parts. First the integer part, 9, is converted to 1001_2 (in the usual manner). Next, we convert the fractional part, 0.8125.

$$\downarrow$$
$$0.8125 \times 2 = 1.625$$
$$0.625\ \ \times 2 = 1.250$$
$$0.250\ \ \times 2 = 0.50$$
$$0.50\ \ \ \ \times 2 = 1.0$$
$$0.0$$

The fractional binary result is made up of the integer values of the products, read from the top, in this case 0.1101_2. Assembling the integer and fractional parts, our binary result is 1001.1101_2.

In decimal numbers, there are irrational numbers (numbers that cannot be exactly represented by decimal fractions). For example, ⅓ results in the repeating decimal fraction 0.333.... Likewise, there are irrational binary numbers. However, rational decimal numbers do not always convert to rational binary numbers. This is why digital computers often have difficulty calculating the exact result in a mathematical operation. When we input the number to the computer in decimal, the computer first converts it to an equivalent binary number. If the binary equivalent is an irrational number, the final result of the calculation will contain a small error.

■ EXAMPLE 9-3 Convert the rational decimal number 0.3_{10} to binary.

SOLUTION Using our standard method for conversion of a decimal fraction to a binary fraction, we have

$$\downarrow$$
$$0.3 \times 2 = 0.6$$
$$0.6 \times 2 = 1.2$$
$$0.2 \times 2 = 0.4$$
$$0.4 \times 2 = 0.8$$
$$0.8 \times 2 = 1.6$$
$$0.6 \times 2 = 1.2$$
$$0.2 \times 2 = 0.4$$
$$0.4 \times 2 = 0.8$$
$$0.8 \times 2 = 1.6$$
$$\ldots$$

Obviously, this is a repeating calculation that has no end. The result is the irrational binary number $0.010011001100\ldots$.

■

Binary multiplication and division are performed in the same manner as their decimal counterparts.

Some binary numbers consisting of a specific number of bits have special names. Four-bit binary numbers are called **nibbles** (sometimes spelled **nybble**), and 8-bit numbers are **bytes**. The term **word** is most commonly used to indicate 16-bit numbers; however, it is also used to mean the number of bits that are processed simultaneously by a digital system. For example, desktop computers generally process data in 32- or 64-bit words, whereas powerful supercomputers use 128- or 256-bit words.

9-2 Negative Binary Numbers

In the binary number system, there are several ways to represent negative numbers. The two most popular notation methods are **sign and magnitude** and **twos complement**. Of these, the twos-complement form is by far the most widely used and will be discussed here.

When numbers are converted to twos-complement notation, we first explicitly state the number of bits that will appear in the twos-complement number. We always choose a number of bits that is at least one bit larger than the unsigned number. For example, the twos complement of the 4-bit binary number 1010_2 will be at least a 5-bit number. The twos-complement notation for positive binary numbers requires no action other than adding extra zeros to the MSB end of the number to make it the appropriate length. Therefore, the 8-bit twos-complement representation of the number $+1010_2$ is 00001010_2. To convert a negative number to twos-complement form, we first add leading zeros to make the number the appropriate length, then complement all the bits in the number (that is, change the ones to zeros and change the zeros to ones), and then add 1.

As an example, to convert the number -1010_2 to 8-bit twos-complement form, we first add leading zeros to make it an 8-bit number (00001010_2). Then we complement the number (11110101_2), and add 1 ($11110101_2 + 1 = 11110110_2$). Although the resulting twos-complement number appears to be nothing like the original number, as we will see, it makes subtraction extremely easy for computers. Converting a twos-complement number back to a signed binary number requires simply taking the twos complement of the number again. The reader is invited to do this with the previous example to illustrate this principle. The MSB of all twos-complement numbers will indicate the sign of the number, with a zero MSB indicating a positive number and a one indicating a negative number.

Using twos-complement notation restricts the range of numbers that can be represented by a fixed number of binary bits. To illustrate this, let us compare the range of 8-bit numbers in both unsigned and twos-complement notation. The range of unsigned 8-bit binary numbers is 00000000_2 to 11111111_2, or 0_{10} to 255_{10}. For 8-bit twos-complement numbers, we will determine the range in two parts, first positive numbers and then negative numbers. Beginning with 00000000_2 (zero), we begin counting in the positive direction as with unsigned binary numbers. However, because the MSB is a sign indicator, when we reach the number 01111111_2 (127_{10}) we can go no farther because the next "higher" number is 10000000_2, which is a negative number in twos-complement notation. Therefore, the positive range of 8-bit twos-complement numbers is 127_{10}. For the negative numbers, we begin counting from zero in the negative direction. First is the number 11111111_2 (-1_{10}), then 11111110_2 (-2_{10}), and so forth, counting downward until we reach the number 10000000_2 (-128_{10}). We can go no farther negative because the next step will borrow one from the MSB, which will change the MSB to a zero, indicating a positive number. Therefore, we can conclude that the number range of an 8-bit twos-complement number is -128 to $+127$. We can further generalize this conclusion to say that *for any n-bit twos-complement number m, the range is* $-2^{n/2} < m < +2^{n/2} - 1$.

9-3 Subtraction Using Twos-Complement Numbers

Generally, when digital computing machines perform mathematical subtraction, it is not done in the same manner in which we manually do subtraction. Instead subtraction is performed by adding numbers in twos-complement form. In effect, when we perform twos-complement subtraction, we are leaving the minuend as a positive number while converting the subtrahend to a negative number so that the computer can perform the subtraction by adding. To see how this is done, we will perform the subtraction $11_{10} - 3_{10} = 8_{10}$ using 6-bit twos-complement subtraction. First, convert the numbers 11_{10} and -3_{10} to twos-complement form, resulting in the numbers 001011_2 and 111101_2. Now, simply add the numbers, discarding any carry bit that would make the result larger than 6 bits.

$$\begin{array}{r} 001011 \\ +111101 \\ \hline 1\!\!\!/001000 \end{array}$$

Notice that the result is 001000_2, or 8_{10}.

When performing twos-complement subtraction, the result can have two forms, with corresponding actions.

1. If there is a carry that makes the result larger than the required number of bits, discard the carry. The result is a positive number.

2. If there is no carry (that is, the result is the same number of bits as the minuend and subtrahend), the result is negative and is in twos-complement form. To convert it from twos-complement form, simply take the twos complement of it, and place a negative sign on the result.

It is important to recognize that when using twos-complement numbers, the number of bits represented by the numbers is large enough to contain the value of the number without overflow. Otherwise, there is a possibility of an error caused by an overflow.

EXAMPLE 9-4   Use 8-bit twos-complement subtraction to calculate (a) $57_{10} - 29_{10}$ and (b) $29_{10} - 57_{10}$. Convert the results to decimal notation.

SOLUTION

(a) Convert 57_{10} to 8-bit twos-complement form, which is 00111001_2. Next convert -29 to 8-bit twos-complement form, which is 11100011_2. Then add the two numbers.

$$\begin{array}{r} 00111001 \\ +11100011 \\ \hline 1\!\!\!/00011100 \end{array}$$

Because there is a 9th-bit carry, we discard it. The result is positive. Converting it to decimal, we have 28_{10}.

(b) Convert -57_{10} to 8-bit twos-complement form, which is 11000111_2. Next, convert 29 to 8-bit twos-complement form, which is 00011101_2. Then add the two numbers.

$$\begin{array}{r} 11000111 \\ +00011101 \\ \hline 11100100 \end{array}$$

There is no 9th-bit carry, and the 8th bit is a 1, so the result is negative, and in twos-complement form. We convert it from twos-complement form by taking the twos complement of it, resulting in -00011100_2, which is -28_{10}.

One special case can create an error when adding twos-complement numbers. To illustrate this potential problem, we will perform the subtraction $(-10_{10}) - (+11_{10})$ in 5-bit twos-complement notation. This will be accomplished by converting $+11_{10}$ to -11_{10}, and then adding it to -10_{10}. In twos-complement notation, -10_{10} is 10110 and -11_{10} 10101. The result of the addition is 101011. If we discard the carry, the result is 01011, which is $+11_{10}$. Obviously, the result of $(-10_{10}) + (-11_{10})$ should be -21_{10}, not $+11_{10}$. The error occurred because the number of bits used is insufficient to handle the twos-complement result; that is, the value -21_{10} cannot be represented as a 5-bit twos-complement number. This is called a **twos-complement overflow**. Had we performed this addition in 6-bit notation, the result would have been correct.

This illustration demonstrates that one must be vigilant when performing standard twos-complement addition to recognize when an overflow has occurred. Identifying an overflow is very simple. It involves inspecting the most significant bits (the signs) of the two addends, and the sum.

When performing twos-complement addition, the following sign rules apply:

- If the addends are both positive numbers (MSBs = 0), the result must be positive (MSB = 0).
- If the addends are both negative numbers (MSBs = 1), the result must be negative (MSB = 1).

If the addends have opposite signs, there will be no overflow and the result will always be correct. However, if one of the above rules is violated, a twos-complement overflow has occurred. All computing machines (calculators and computers) that use twos-complement mathematics have built-in circuitry to examine the MSBs of the addends and the sum to assure that an overflow has not occurred, and to generate an appropriate warning if needed.

9-4 Hexadecimal Number System

The binary number system is the fundamental number system used in all digital computing machines. However, writing large numbers using binary numbers can be tedious because they require many digits to represent relatively small decimal numbers. For example, to represent the 7-digit decimal number 1,000,000 in binary requires 20 binary bits (11110100001001000000). Additionally, working between binary numbers and decimal numbers can not only be cumbersome, it can also introduce numerical errors when working with fractions. Therefore, scientists and engineers resort to a simpler way to represent binary numbers that is error free, called the **hexadecimal** or **base-16** number system. Because the base of the hexadecimal (or "hex" for short) number system is 16, we need one-digit characters to represent all 16 values. Therefore, we use the digits 0 through 9 followed by the letters A through F. Table 9-1 shows the equivalent binary and decimal numbers for the hexadecimal numbers 0 through F.

Notice in Table 9-1 that we have represented every hex value with a 4-bit binary number. This makes conversion from binary to hex and vice versa relatively simple. To convert from binary to hex, we partition the binary number into groups of four bits (nibbles) starting from the LSB. Then we convert each nibble to its hex equivalent.

As with the decimal and binary number systems, in the hex number system, each digit of a number has a weight. In this case, the least significant (rightmost) digit has a multiplier of 16^0. The next digit to the left has a weight of 16^1. This pattern continues with increasing powers of 16. Conversion from hex to decimal is done by multiplying each digit by its power of 16 weight and then summing the results. Conversion from decimal to hex is generally not done directly (although it could be done by successively dividing by 16 and recording the remainders). Instead it is easier to use the binary number system as an intermediate step. Therefore, to convert from decimal to hex, first convert from decimal to binary, and then convert from binary to hex.

Table 9–1 Hex, binary, and decimal numbers.

Hexadecimal	Binary	Decimal
0	0000	0
1	0001	1
2	0010	2
3	0011	3
4	0100	4
5	0101	5
6	0110	6
7	0111	7
8	1000	8
9	1001	9
A	1010	10
B	1011	11
C	1100	12
D	1101	13
E	1110	14
F	1111	15

■ **EXAMPLE 9-5**

The decimal number 1,000,000 can be represented by the binary number 11110100001001000000. (a) Convert this binary number to its hexadecimal equivalent. (b) Check your answer by converting the result back to a decimal number.

SOLUTION

(a) First partition the binary number into nibbles starting with the LSB, and then write the hex digit for each nibble. If the leftmost nibble does not have four bits, add leading zeros to make it 4 bits.

$$11110100001001000000_2$$
$$\underbrace{1111}_{F}\ \underbrace{0100}_{4}\ \underbrace{0010}_{2}\ \underbrace{0100}_{4}\ \underbrace{0000}_{0}$$
$$F4240_{16}$$

(b) To convert from hex to decimal, we multiply each hex digit by its weight:

$$F4240_{16} = F_{16} \times 16_{10}^4 \quad + 4_{16} \times 16_{10}^3 \quad + 2_{16} \times 16_{10}^2 + 4_{16} \times 16_{10}^1 + 0_{16} \times 16_{10}^0$$
$$= 15_{10} \times 65{,}536_{10} + 4_{10} \times 4{,}096_{10} + 2_{10} \times 256_{10} + 4_{10} \times 16_{10} + 0_{10} \times 1_{10}$$
$$= 983{,}040_{10} \quad + 16{,}384_{10} \quad + 512_{10} \quad + 64_{10} \quad + 0_{10}$$
$$= 1{,}000{,}000_{10}$$

Conversion from hex to binary uses exactly the opposite method. Convert each hex digit into its binary nibble, and then put the nibbles together to form the binary number.

9-5 Octal Number System

In some digital computing machines, the designers choose to avoid the use of digits above 9 because inputting and outputting hexadecimal requires the use of special logic circuits and special numeric keypads and displays. However, decimal numbers are also a bad choice because it is cumbersome for the computer to convert between decimal and binary, and when fractional values are used, they are prone to irrational number errors. In this case, the **octal** or **base-8** number system is a good choice. In the octal number system, the weight of each digit is a power of 8 (8^0, 8^1, 8^2, 8^3, etc.). Conversion between octal and binary uses a similar method as that used to convert between hexadecimal and binary, except that instead of partitioning the binary number into groups of 4 bits, we partition them into groups of 3 bits. Because each octal digit can have a value from 0 to 7, each digit can also be represented by three binary bits.

▌ EXAMPLE 9-6

Again, the decimal number 1,000,000 can be represented by the binary number 11110100001001000000. (a) Convert this binary number to its octal equivalent. (b) Check your answer by converting the result back to a decimal number.

SOLUTION

(a) First partition the binary number into 3-bit groups starting with the LSB, and then write the octal digit for each group. The leftmost group does not have three bits, so we add leading zeros to make it 3 bits.

$$11110100001001000000_2$$
$$\underbrace{011}_{3} \; \underbrace{110}_{6} \; \underbrace{100}_{4} \; \underbrace{001}_{1} \; \underbrace{001}_{1} \; \underbrace{000}_{0} \; \underbrace{000}_{0}$$
$$3641100_8$$

(b) To convert from octal to decimal, we multiply each octal digit by its weight:

$$3641100_8 = 3_8 \times 8_{10}^6 \quad + 6_8 \times 8_{10}^5 \quad + 4_8 \times 8_{10}^4 \quad + 1_8 \times 8_{10}^3 \quad + 1_8 \times 8_{10}^2 \quad + 0_8 \times 8_{10}^1 + 0_8 \times 8_{10}^0$$
$$= 3_{10} \times 262{,}144_{10} + 6_{10} \times 65{,}768_{10} + 4_{10} \times 4{,}096_{10} + 1_{10} \times 512_{10} + 1_{10} \times 64_{10} + 0_{10} \times 8 \; + 0_{10} \times 1$$
$$= 786{,}432_{10} \quad + 196{,}608_{10} \quad + 16{,}384_{10} \quad + 512_{10} \quad + 64_{10} \quad + 0_{10} \quad + 0_{10}$$
$$= 1{,}000{,}000_{10}$$

Conversion between octal and hexadecimal does not require any mathematical operations. Instead, convert the number to binary, repartition the bits for the new number system, and convert each digit.

EXAMPLE 9-7

(a) Convert the hexadecimal $AC39D_{16}$ to octal. (b) Check your answer by converting both numbers to decimal.

SOLUTION

(a) Write the binary equivalent of each hexadecimal digit, put all the binary bits together, repartition them into groups of 3 bits, and convert each group to octal.

$$AC39D_{16}$$

$$\underbrace{A}\,\underbrace{C}\,\underbrace{3}\,\underbrace{9}\,\underbrace{D}$$
$$1010\;1100\;0011\;1001\;1101$$
$$10101100001110011101$$
$$\underbrace{010}_{2}\,\underbrace{101}_{5}\,\underbrace{100}_{4}\,\underbrace{001}_{1}\,\underbrace{110}_{6}\,\underbrace{011}_{3}\,\underbrace{101}_{5}$$
$$2541635_8$$

(b)
$$AC39D_{16} = A_{16} \times 16^4 + C_{16} \times 16^3 + 3_{16} \times 16^2 + 9_{16} \times 16^1 + D_{16} \times 16^0$$
$$= 705{,}437$$
$$2541635_8 = 2_8 \times 8^6 + 5_8 \times 8^5 + 4_8 \times 8^4 + 1_8 \times 8^3 + 6_8 \times 8^2 + 3_8 \times 8^1 + 5_8 \times 8^0$$
$$= 705{,}437$$

9-6 Binary Coded Decimal

In many cases we find ourselves in need of a method to represent decimal numbers using binary bits. In these cases, we represent decimal numbers using four binary bits, much as we do with hexadecimal numbers, except that our numbers only range from 0_{10} (0000_2) to 9_{10} (1001_2). Such a numbering system is called **binary coded decimal**, or **BCD**. Conversion from decimal to BCD is extremely simple. It is done by converting each decimal digit to its 4-bit binary equivalent. Then when we write the BCD number, we leave a space between every 4 bits to delineate each decimal digit, and we use a "BCD" subscript. Conversion from BCD to decimal is done by converting each of the 4-bit BCD number groups to its decimal equivalent.

EXAMPLE 9-8

Perform the following conversions:

(a) 249_{10} to BCD
(b) 4317_{10} to BCD
(c) 5_{10} to BCD
(d) $1001\;0110\;0111\;0001_{BCD}$ to decimal
(e) $1000\;0001\;0111_{BCD}$ to decimal
(f) 1001_{BCD} to decimal

SOLUTION

(a) Convert each digit in 249_{10} to binary. The result is $0010\;0100\;1001_{BCD}$.
(b) 4317_{10} is $0100\;0011\;0001\;0111_{BCD}$.
(c) 5_{10} is 0101_{BCD}.
(d) $1001\;0110\;0111\;0001_{BCD}$ is $9{,}671_{10}$.
(e) $1000\;0001\;0111_{BCD}$ is 817_{10}.
(f) 1001_{BCD} is 9_{10}.

9-7 Fundamental Logic Operations

As mentioned earlier, all digital computing machines internally use the binary number system. The binary system is the best choice because it is relatively easy to process the numbers 0 and 1 with electronic circuits that switch OFF and ON, called **switching circuits**. In most cases, we assign an ON condition to be a binary one and an OFF condition to be a binary 0.[1] Digital computing machines process binary bits using logical operations. These logical operations have inputs and outputs, much like amplifiers. However, the input and output values will be discrete voltages. In the most common type of digital logic circuits, called transistor-transistor logic or TTL, the voltage levels are 0 volts and 5 volts dc, with 0 volts corresponding to a binary 0 and 5 volts corresponding to a binary 1. However, for analysis purposes, it turns out that the voltage levels are unimportant. Instead we will concern ourselves with the binary values, the 1's and 0's.

There are three fundamental logical operations used in all digital computing machines, the AND, the OR, and the INVERT. All other logical operations are based upon these three operations. In fact, given an unlimited number of AND, OR, and INVERT circuits, one could build any computing machine, including a supercomputer.

AND

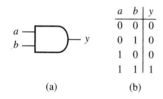

FIGURE 9–4
(a) Two-input AND gate logic symbol and (b) truth table.

The AND circuit (called an **AND gate**) has at least two inputs and one output, the most common variety being the two-input AND gate, which is shown in Figure 9-4(a). This gate has two inputs a and b, and one output y. The inputs and output a, b, and y are variable names that are assigned to digital signals. They can have binary values of 0 or 1. Stated in words, *the output of an AND operation will be a logical one when, and only when, all inputs are also a logical one.*

We can also illustrate the AND operation with a table, called a **truth table**. Generally speaking, a truth table has a column for each input and output on the logic gate. It also has row entries showing all possible combinations of 1's and 0's for the inputs, and the resulting output for each combination. The truth table for the two-input AND gate is shown in Figure 9-4(b).

A third and very useful way of representing an AND gate operation is with a **Boolean**[2] **expression**. A Boolean expression looks much like a conventional algebraic expression. However, the operational symbols used are somewhat different. There are two ways to express the AND operation in a Boolean expression, by using a dot (·) between the variables to be ANDed, or to omit the dot and simply write the two ANDed variables side by side. In this text, the latter method will be used. For example, the Boolean expression for out two-input AND gate in Figure 9-4 is

$$y = ab \tag{9-1}$$

Expression 9-1 is read, "y is equal to a AND b." Many times, for brevity, we leave out the word AND and simply say, "y is equal to $a\ b$."

It is permissible to have AND gates with as many inputs as desired. When drawing the logic symbol, if there are too many inputs to fit on the gate symbol, we simply extend the input side of the symbol above and below the gate to accommodate them, as illustrated with the six-input AND gate in Figure 9-5. The Boolean expression for the AND gate in Figure 9-5 is

$$y = abcdef \tag{9-2}$$

FIGURE 9–5
Six-input AND gate.

---

[1] Sometimes it improves performance and makes the digital circuits simpler if we assign an ON state to a binary zero and an OFF state to a binary one, an approach called **inverted logic** or **negative logic**.
[2] British mathematician George Boole (1815–1864) wrote *An investigation into the Laws of Thought, on Which are Founded the Mathematical Theories of Logic and Probabilities*, in which he described the postulates and theorems of what would later be named Boolean algebra.

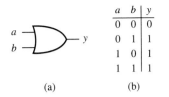

FIGURE 9–6
(a) Two-input OR gate logic symbol and (b) truth table.

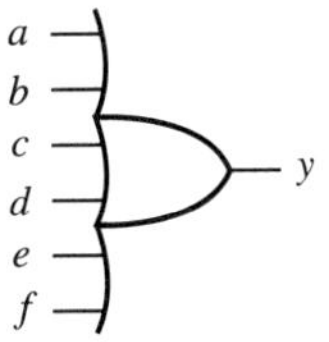

FIGURE 9–7
Six-input OR gate.

OR

Like the AND gate, the **OR gate** has two or more inputs and one output. However, with the OR operation we say that *the output of an OR operation will be a logical one when any one or more of the inputs is a logical one*. Therefore, we could also say that the output of an OR gate will be a zero only when all inputs are zero. The logic symbol for a two-input OR gate is shown in Figure 9-6(a), and Figure 9-6(b) shows its truth table. If the gate has many inputs, such that they will not fit on the input side of the logic symbol, the input side can be extended as shown in Figure 9-7.

INVERT (or NOT)

The **INVERT** operation has one input and one output. The gate used to perform the INVERT function is called an **inverter**. It is represented by the logic symbol and truth table shown in Figures 9-8(a) and (b), respectively. Notice that the logic symbol for the inverter has a small circle (called a **bubble**) on the output. In logic symbols and logic diagrams, all bubbles indicate logical inversion. *The inverter produces an output that is the logical opposite of the input*.

The Boolean expression for the inverter is

$$y = \overline{a} \qquad (9\text{-}3)$$

In Equation (9-3), there is a bar (called an **overbar**) over a indicating that a is inverted. The expression is read, "y equals not-a." In Boolean expressions, the word NOT is used interchangeably with INVERT. Figure 9-8(c) shows the alternate symbol for the inverter, with the inversion bubble moved to the input. With the inverter, it makes no difference whether the bubble appears on the input or output side of the gate; however, it is most commonly drawn on the output side.

As we will see next, the three fundamental logical operations AND, OR, and INVERT can be used to construct more complex and powerful logic gates.

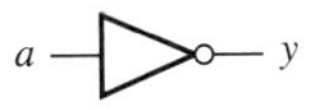

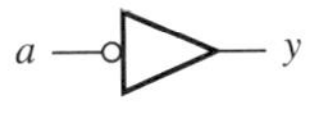

FIGURE 9–8
(a) INVERT logic symbol, (b) truth table, and (c) alternate symbol.

9-8 NAND and NOR Operations

In some cases, it is necessary to connect an AND gate in series with an inverter, to invert the logical output of the AND gate. To conserve on circuit complexity and reduce the number of gates in a logic circuit, the AND-INVERT operation, shown in Figure 9-9(a), is combined into one logical Not-AND gate called the **NAND**, shown in Figure 9-9(b). The NAND gate is available off-the-shelf as a logical circuit; it is not necessary to construct it as shown in Figure 9-9(a). Notice that the two-input NAND truth table in Figure 9-9(c) has an output y that is inverted from that of the AND truth table in Figure 9-4(b).

The Boolean expression for the two-input NAND is

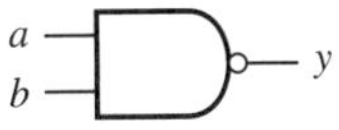

$$y = \overline{ab} \qquad (9\text{-}4)$$

FIGURE 9–9
(a) AND-INVERT, (b) NAND gate, and (c) truth table.

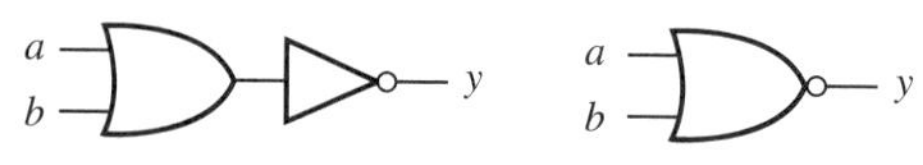

FIGURE 9–10
(a) OR-INVERT, (b) NOR gate, and (c) truth table.

a	b	y
0	0	1
0	1	0
1	0	0
1	1	0

The expression is read, "**y** equals *a* NAND *b*." Notice in Equation 9-4 that the overbar extends over both *a* and *b*. This indicates the order in which the Boolean operations are performed; that is, *a* and *b* are first ANDed, resulting in the term *ab*, and then the result of the AND operation is inverted. This order of operation is also indicated by the logic NAND symbol in Figure 9-9(b) by the inversion bubble appearing on the output of the gate. As we will see later, it is possible to have an overbar over each individual variable, which indicates a different order of Boolean operations with different results.

In a similar manner, in cases where it is necessary to invert the output of an OR gate as illustrated in Figure 9-10(a), we have available the Not-OR or **NOR gate** shown in Figure 9-10(b). Again, note that the *y* output in the truth table of Figure 9-10(c) is the logical inversion of the *y* output of the OR gate in Figure 9-6(b). The Boolean expression for the two-input NOR operation is

$$y = \overline{a + b} \tag{9-5}$$

In Equation 9-5 the overbar extends over the OR operation indicating that the OR is performed first, followed by the invert. The expression is read, "*y* equals *a* NOR *b*."

9-9 Boolean Algebra and Switching Functions

In digital circuits, we sometimes use the terms **Boolean function** and **switching function** interchangeably. In effect, they are the same; that is, they are algebraic functions in which the dependent and independent variables can take on only two values, with the two values being zero and one, or a false and true. All switching function operations adhere to Boolean postulates and theorems that were developed by British mathematician George Boole and published in 1854.

Postulate 1

Boolean algebra is a closed algebra. There are fundamentally two operators, the AND, indicated with a dot, (·), or with an implied dot, and the OR, indicated with a plus, (+). The variables can take on only two values, 1 and 0 (or true and false).

Each of the postulates 2 through 5 has a dual. Therefore, both are listed for each postulate.

Postulate 2

$$\begin{align} \text{(a) } x + 0 &= x \\ \text{(b) } x \cdot 1 &= x \end{align} \tag{9-6}$$

Postulate 3

$$\begin{align} \text{(a) } x + y &= y + x \\ \text{(b) } xy &= yx \end{align} \tag{9-7}$$

Postulate 4

$$\begin{align} \text{(a) } x(y + z) &= xy + xz \\ \text{(b) } x + (yz) &= (x + y)(x + z) \end{align} \tag{9-8}$$

Postulate 5

$$\text{(a) } x + \overline{x} = 1 \quad (9\text{-}9)$$
$$\text{(b) } x \cdot \overline{x} = 0$$

From the five Boolean postulates, six Boolean theorems have been developed. Generally, theorems require proofs. For brevity, those proofs will not be shown, but they can be found in any digital logic text. Again, each is listed with its dual (with the exception of Theorem 3, which does not have a dual).

Theorem 1

$$\text{(a) } x + x = x \quad (9\text{-}10)$$
$$\text{(b) } x \cdot x = x$$

Theorem 2

$$\text{(a) } x + 1 = 1 \quad (9\text{-}11)$$
$$\text{(b) } x \cdot 0 = 0$$

Theorem 3

$$\overline{\overline{x}} = x \quad (9\text{-}12)$$

Theorem 4

$$\text{(a) } x + (y + z) = (x + y) + z \quad (9\text{-}13)$$
$$\text{(b) } x(yz) = (xy)z$$

Theorem 5

$$\text{(a) } \overline{(x + y)} = \overline{x}\,\overline{y} \quad (9\text{-}14)$$
$$\text{(b) } \overline{x\,y} = \overline{x} + \overline{y}$$

Theorem 6

$$\text{(a) } x + xy = x \quad (9\text{-}15)$$
$$\text{(b) } x(x + y) = x$$

When evaluating a switching function, there is a hierarchy of evaluation. This order is (1) parentheses, (2) inversion, (3) AND, and (4) OR. If we wish to modify the order in which an expression is evaluated, we simply use parentheses. For example, consider the Boolean function

$$w = a\overline{b} + (x + y)z \quad (9\text{-}16)$$

When evaluating the function, the order of evaluation would be

1. Evaluate $(x + y)$.
2. Invert b.
3. AND the inverted b with a.
4. AND $(x + y)$ with z.
5. OR the result of 3 and 4 above, resulting in the solution w.

When a switching function is stated, the function name on the left side of the equation (which is also the dependent variable) is followed by a list of the independent variables (that is, the variables that determine the value of the function). These variables are listed in order by binary weight, with the LSB on the right. For example, we could define the Boolean function in Equation 9-16 as a switching function by writing

$$f(a, b, x, y, z) = a\overline{b} + (x + y)z \qquad (9\text{-}17)$$

Also, if we wish to define the value of the function, we can define the value of each of the variables. For our example in Equation 9-17, if the values of the independent variables are $a = 0, b = 1, x = 1, y = 0$, and $z = 1$, then the value of the function would be

$$\begin{aligned}
f(a, b, x, y, z) &= a\overline{b} + (x + y)z \\
f(0, 1, 1, 0, 1) &= 0 \cdot \overline{1} + (1 + 0) \cdot 1 \\
&= 0 \cdot 0 + (1) \cdot 1 \\
&= 0 + 1 \qquad (9\text{-}18) \\
&= 1
\end{aligned}$$

■ EXAMPLE 9-9

For the switching function

$$f(a, b, c, d) = ab + (\overline{a}\,c\,\overline{d}) + d \qquad (9\text{-}19)$$

find the value of (a) $f(1, 0, 0, 1)$, (b) $f(0, 0, 1, 0)$, (c) $f(0, 1, 0, 0)$.

SOLUTION

(a)

$$\begin{aligned}
f(a, b, c, d) &= ab + (\overline{a}\,c\,\overline{d}) + d \\
f(1, 0, 0, 1) &= 1 \cdot 0 + (\overline{1} \cdot 0 \cdot \overline{1}) + 1 \\
&= 0 + (0 + 0 + 0) + 1 \\
&= 0 + 0 + 1 \qquad (9\text{-}20) \\
&= 1
\end{aligned}$$

(b)

$$\begin{aligned}
f(a, b, c, d) &= ab + (\overline{a}\,c\,\overline{d}) + d \\
f(0, 0, 1, 0) &= 0 \cdot 0 + (\overline{0} \cdot 1 \cdot \overline{0}) + 0 \\
&= 0 + (1 \cdot 1 \cdot 1) + 0 \\
&= 0 + \overline{1} + 0 \qquad (9\text{-}21) \\
&= 1
\end{aligned}$$

(c)

$$\begin{aligned}
f(a, b, c, d) &= ab + (\overline{a}\,c\,\overline{d}) + d \\
f(0, 1, 0, 0) &= 0 \cdot 1 + (\overline{0} \cdot 0 \cdot \overline{0}) + 0 \\
&= 0 + (1 \cdot 0 \cdot 1) + 0 \\
&= 0 + 0 + 0 \qquad (9\text{-}22) \\
&= 0
\end{aligned}$$

■

9-10 Boolean Reduction and Karnaugh Maps

There are fundamentally two ways of representing a switching function, which are (1) in a nonreduced form called **canonical form**, and (2) in reduced form. Canonical form is a way of representing a switching function in which every independent variable appears once (either noninverted or inverted) in every term of the equation. If a switching function is not in canonical form, it is in reduced form. A reduced-form equation can appear as a partially reduced equation or a minimally reduced equation. Therefore, the switching function

$$f(a,b,c) = a\bar{b}\bar{c} + a\bar{b}c + ab\bar{c} + abc \qquad (9\text{-}23)$$

is in canonical form. Canonical form can be quickly recognized by comparing the independent variables listed in the function on the left side of the equation with the variables in each term on the right side. Notice that every term contains the variables a, b, and c.

The equation

$$f(a,b,c) = a\bar{c} + ac \qquad (9\text{-}24)$$

is in a partially reduced (not minimal) form, and

$$f(a,b,c) = a \qquad (9\text{-}25)$$

is reduced to minimal form. All three of the switching functions, Equations 9-23, 9-24, and 9-25, are equivalent, but each one is in a different stage of reduction.

■ EXAMPLE 9-10

Using the Boolean postulates and/or theorems, show the process to reduce (9-23) to the minimal form shown in Equation 9-25.

SOLUTION

$$
\begin{aligned}
f(a,b,c) &= a\bar{b}\bar{c} + a\bar{b}c + ab\bar{c} + abc \\
&= \bar{b}[a\bar{c} + ac] + b[a\bar{c} + ac] & \text{Postulate 4} \\
&= \bar{b}[a(\bar{c} + c)] + b[a(\bar{c} + c)] & \text{Postulate 4} \\
&= \bar{b}[a(1)] + b[(a(1)] & \text{Postulate 5} \\
&= \bar{b}(a) + b(a) & \text{Postulate 2} \\
&= a\bar{b} + ab & \text{Postulate 3} \qquad (9\text{-}26)\\
&= a(\bar{b} + b) & \text{Postulate 4} \\
&= a(1) & \text{Postulate 5} \\
&= a & \text{Postulate 2}
\end{aligned}
$$

When a switching function is written in canonical form, every term in the equation represents what is called a **minterm**. Each minterm has a decimal number that is the decimal equivalent of the binary number represented by the term. For example, let's say that we are working with the four-variable canonical-form switching function

$$f(a,b,c,d) = \bar{a}\bar{b}\bar{c}\bar{d} + \bar{a}\bar{b}cd + a\bar{b}c\bar{d} + abc\bar{d} \qquad (9\text{-}27)$$

This particular style of writing the switching function is called algebraic form. If we replace all inverted variables on the right side the function with a zero, and replace all non-inverted variables with a one, we have the minterm numbers in binary form. Then we simply convert the binary numbers to decimal values.

$$
\begin{aligned}
f(a,b,c,d) &= \bar{a}\bar{b}\bar{c}\bar{d} + \bar{a}\bar{b}cd + a\bar{b}c\bar{d} + abc\bar{d} \\
&= 0000 + 0011 + 1010 + 1110 \qquad (9\text{-}28)\\
&= 0 \ \ + 3 \ \ \ + 10 \ \ + 14
\end{aligned}
$$

Once we have the minterm numbers, we can additionally write the function in one of two different ways. First there is the m-form, which is

$$f(a,b,c,d) = m_0 + m_3 + m_{10} + m_{14} \qquad (9\text{-}29)$$

or we could use the Σm-form (said, "Sigma-m form"), which is

$$f(a,b,c,d) = \Sigma m(0, 3, 10, 14) \qquad (9\text{-}30)$$

Some texts abbreviate the sigma-m form by omitting the m, resulting in

$$f(a,b,c,d) = \Sigma(0, 3, 10, 14) \qquad (9\text{-}31)$$

Note that no matter which form we use, Equations 9-27, 9-29, 9-30, or 9-31, all are in canonical form because each one contains terms that defines every minterm in the switching function; none are in reduced form. Because these are all written as a sum (OR) of products (ANDs), these are called canonical **sum of products** form, or abbreviated as **SOP** form.

EXAMPLE 9-11

Write the switching function

$$f(a,b,c,d,e) = m_3 + m_6 + m_9 + m_{13} + m_{15} + m_{18} + m_{27} \qquad (9\text{-}32)$$

in both (a) the Σm-form and (b) the SOP algebraic form.

SOLUTION

(a) The Σm-form is simply the sum of minterms using the minterm (subscripted) numbers in Equation 9-32.

$$f(a,b,c,d,e) = \Sigma m(3, 6, 9, 13, 15, 18, 27) \qquad (9\text{-}33)$$

(b) To obtain the SOP algebraic form, we first convert the minterm numbers to their corresponding binary numbers (in this case they are 5-bit numbers because the left side of the equation lists five independent variables, a, b, c, d, and e). Then write each corresponding variable either inverted or noninverted, depending on whether it is represented by a binary zero or one.

$$\begin{aligned} f(a,b,c,d,e) &= 00011 + 00110 + 01001 + 01101 + 01111 + 10010 + 11011 \\ &= \bar{a}\bar{b}\bar{c}de + \bar{a}\bar{b}cd\bar{e} + \bar{a}b\bar{c}c\bar{d}e + \bar{a}bc\bar{d}e + \bar{a}bcde + a\bar{b}\bar{c}d\bar{e} + ab\bar{c}de \end{aligned}$$
$$(9\text{-}34)$$

The switching function is a very important stage in digital circuit design. As we will see later, digital logic diagrams can be drawn directly from complex switching functions. Generally, the complexity of the resulting digital circuitry will depend on the complexity of the switching function from which it was produced. Because we usually like our digital circuits to be as simple as possible, we will need for our switching functions to be reduced to minimal form. Using the minimal form of a switching function to draw a digital logic diagram results in a circuit that uses fewer gates (fewer gates means we will have fewer chips, which in turn translates to a less expensive and more reliable circuit). Although reducing switching expressions to minimal form is imperative when performing digital logic design, with larger equations the reduction process is often very prone to errors. Therefore, a graphical approach has been developed to reduce switching functions to minimal form. This method is called the **Karnaugh map**,[3] or simply **K-map**.

Two-Variable Karnaugh Maps

Consider the two-variable canonical-form switching function

$$f(a,b) = ab + a\bar{b} \qquad (9\text{-}35)$$

---

[3] The Karnaugh map method of reducing Boolean equations to minimal form was developed in 1950 by Maurice Karnaugh, a telecommunications engineer for Bell Telephone Laboratories.

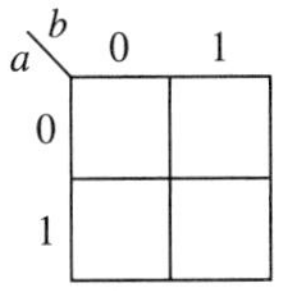

FIGURE 9–11
Two-variable Karnaugh map.

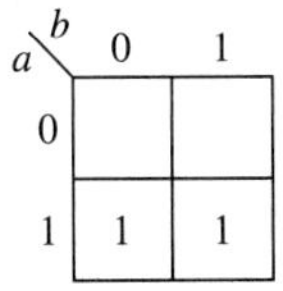

FIGURE 9–12
K-map for $f(a,b) = ab + a\bar{b}$.

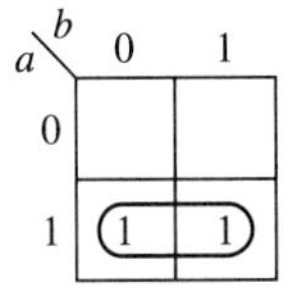

FIGURE 9–13
K-map solution for
$f(a,b) = ab + a\bar{b}$.

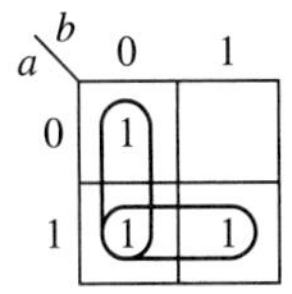

FIGURE 9–14
K-map solution for
$f(a,b) = ab + a\bar{b} + \bar{a}\bar{b}$.

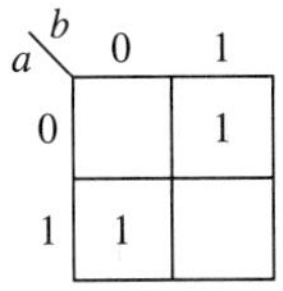

FIGURE 9–15
K-map solution for
$f(a,b) = a\bar{b} + \bar{a}b$.

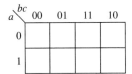

FIGURE 9–16
K-map for three-variable function $f(a, b, c)$.

We can graphically represent the minterms in this expression by drawing a map as shown in Figure 9-11. For our map, we will have two rows and two columns, resulting in four boxes. Outside the map we label the left side with the most significant (that is, leftmost) variable in the switching function, which is a. We label the top with the least significant variable, b. On the left side we label the rows with the two possible values of a, which are zero and one. We do likewise with the columns for variable b. Note that as labeled, each of the internal boxes in the map has a value for a (the row number) and a value for b (the column number). For example, the upper left box in our map represents $a = 0$ and $b = 0$, or the minterm $\bar{a}\bar{b}$ (also called minterm zero, or m_0). In a similar manner, the upper right box in our map represents $a = 0$ and $b = 1$, or the minterm $\bar{a}b$ (also called minterm one, or m_1). The minterm numbers of any box in the map can be determined using this method.

Now that we have the K-map constructed, we can enter the minterms of equation into the map as shown in Figure 9-12. Note that we do not enter zeros into the map for the unused minterms. Although it would not be incorrect to do so, we usually omit the zeros in an effort to keep the map as simple as possible, and to make it easy to spot the ones. After all the minterms have been entered, the simplification process is done by examining the map to attempt to find ones that are either horizontally or vertically adjacent (diagonally adjacent ones do not qualify). If we find a pair of adjacent ones in our K-map, we circle the pair as shown in Figure 9-13. Once we have identified and circled the pair, we write the reduced term indicated by the pair. For our example in Figure 9-13, notice that the circle is located entirely in row $a = 1$. Therefore, we know that the variable a will be present in our reduced solution. However, notice that the circle spans the columns $b = 0$ and $b = 1$. Because it covers both possible values of the variable b, this variable will be canceled from the reduced term. Therefore, our reduced term is simply a, and our reduced switching function is

$$f(a,b) = a \tag{9-36}$$

When circling pairs of minterms in a K-map, minterms can be used more than once. For example, consider the switching function

$$f(a,b) = ab + a\bar{b} + \bar{a}\bar{b} \tag{9-37}$$

In this case, our K-map will appear as illustrated in Figure 9-14. Notice that the minterm m_2 (lower left) has been used twice. Not only is this technique permissible, it is mandatory if we wish to arrive at the minimally reduced expression. Now that we have two circled pairs, we write the term for each and OR them together. For the vertical pair, note that the circle is entirely contained in the $b = 0$ column and that the variable a will cancel. In this case, our reduced (and minimal) switching function is

$$f(a,b) = a + \bar{b} \tag{9-38}$$

Occasionally, entering a switching function into a K-map will result in minterms that are not at all adjacent. An example of this is the switching function

$$f(a,b) = a\bar{b} + \bar{a}b \tag{9-39}$$

which results in the K-map illustrated in Figure 9-15. In this case we cannot circle any pairs of minterms because they are neither horizontally nor vertically adjacent. This is not an indication of an error; instead it means that our original switching function is irreducible, and therefore the canonical form and the minimal form are one and the same.

Three-Variable Karnaugh Maps

Now that we have explored the two-variable K-map, we will move on to the three-variable map. Because there are eight possible minterms in a three-variable function ($2^3 = 8$), a three-variable K-map will have eight minterm boxes. The most commonly used method of constructing the three-variable K-map is shown in Figure 9-16. This map is constructed for a three-variable function $f(a, b, c)$. The left side is again labeled with the most significant variable a. The top of the map is labeled with the remaining two variables b and c. Since we now have two variables represented by each column, the columns will be labeled with the four possible combinations of two variables, which are 00, 01, 11, and 10. Notice that the numbering scheme is not in numerical order (00, 01, 10, and 11). This is because *in a K-map, only one variable can change value between any two adjacent columns (or rows).* If we had labeled the columns in numerical order, we would have had 01 next to 10, which means that both *b* and *c* would have changed value.

Now that we have constructed our three-variable K-map, we can use it to reduce the switching function

$$f(a,b,c) = \overline{a}\,\overline{b}\,c + \overline{a}\,b\,c + \overline{a}\,b\,\overline{c} + a\,\overline{b}\,\overline{c} + a\,\overline{b}\,c + a\,b\,c \qquad (9\text{-}40)$$

to minimal canonical form.

Once we enter the minterms of our switching function into the map (in the same manner as with the two-variable function), we are ready to begin grouping minterms. However, with maps larger than two variables, a new rule must be introduced. *When grouping minterms in K-maps, the number of minterms in any group may be 2, 4, 8, 16, or any power of 2. Also, the groups must be arranged in a square or rectangle.* Therefore, the K-map for our function with the groups marked will appear as shown in Figure 9-17. Notice that for the four minterms in the center group, variables *a* and *b* will be canceled. When we OR the three terms together, the reduced result is

$$f(a,b,c) = \overline{a}\,b + a\,\overline{b} + c \qquad (9\text{-}41)$$

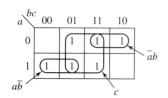

FIGURE 9–17
K-map solution for Equation 9-40.

■ EXAMPLE 9-12

Reduce the switching function

$$f(x,y,z) = \Sigma m(0, 1, 3, 5) \qquad (9\text{-}42)$$

to minimal form.

SOLUTION We could spend the time necessary to convert this function to algebraic form so that we could put each minterm into our map. However, because the original problem statement is given in minterm numbers, why not instead take a shortcut and number the minterm boxes in the K-map? When numbering the boxes, it is imperative that the variables be ordered properly. Because the original function is $f(x, y, z)$, the variable order is defined with *x* as the MSB and *z* the LSB. Therefore, the resulting minterm numbers are arranged as shown in Figure 9-18.

Once the minterm boxes in the K-map are numbered, we simply place a one in the boxes corresponding to the minterm numbers in the given equation, namely boxes 0, 1, 3, and 5, as shown in Figure 9-19. Next, in Figure 9-20, we group the appropriate minterms and then extract the reduced equation, which is

$$f(x,y,z) = \overline{x}\,\overline{y} + \overline{x}\,z + \overline{y}\,z \qquad (9\text{-}43)$$

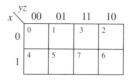

FIGURE 9–18
Three-variable K-map with minterm numbers.

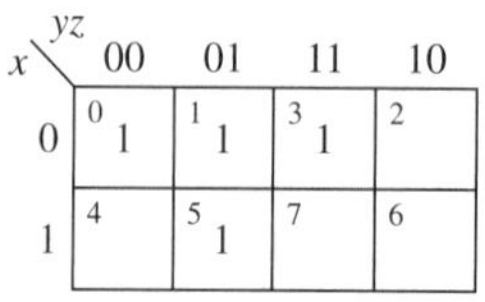

FIGURE 9–19
K-map for Equation 9-42.

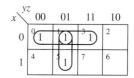

FIGURE 9–20
K-map solution for Equation 9-42.

K-Map Topology

Earlier, we learned that in a three-variable map, each column is numbered so that adjacent columns differ by only one variable change. However, in the standard three-variable map shown in Figure 9-16, notice the difference between the leftmost and rightmost columns. The leftmost column represents the variable combination $\overline{b}\,\overline{c}$ and the rightmost column represents $b\,\overline{c}$. Between these two columns, only the variable b changes value. Therefore, we can conclude that the leftmost and rightmost columns of the map are theoretically adjacent. It turns out that, although we draw the three-variable K-map as a two-dimensional graphic, it helps to visualize it as a cylinder with the left and right edges of the map connected.

EXAMPLE 9-13

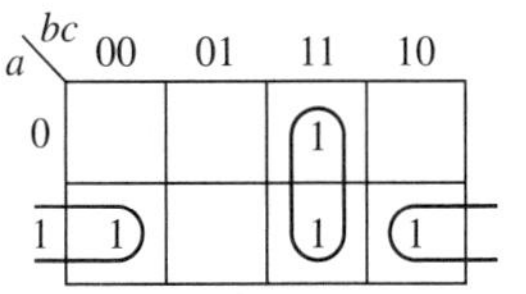

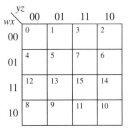

FIGURE 9–21
Solution of Equation 9-44.

Reduce the switching function

$$f(a, b, c) = \Sigma m(3, 4, 6, 7) \tag{9-44}$$

to minimal form.

SOLUTION The K-map for the function is shown in Figure 9-21. Note that we are grouping the adjacent minterms 4 (lower left) and 6 (lower right) together. The solution therefore is

$$f(a, b, c) = c\,d + a\,\overline{c} \tag{9-45}$$

Four-Variable Karnaugh Maps

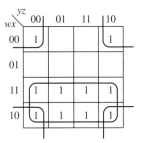

FIGURE 9–22
Karnaugh map for four variables.

The four-variable K-map is constructed much like the three-variable map, except that there are now two variables represented by the rows, requiring four rows. Figure 9-22 shows the four-variable K-map for a function $f(w, x, y, z)$. The minterm boxes have been numbered to illustrate the numbering system used in a four-variable map. Although the numbering system can be determined by combining the row and column numbers, after working a few problems using K-maps the reader will likely find it easier to simply memorize the numbering system and forgo writing the numbers into the map.

To illustrate the use of a four-variable K-map, we will reduce the switching function

$$f(w, x, y, z) = \Sigma m(0, 2, 8, 9, 10, 11, 12, 13, 14, 15) \tag{9-46}$$

to minimal form. Figure 9-23 shows the K-map with the minterms entered. Here we have a group of eight minterms in the lower half of the map. Also, notice the minterm numbers in the four corners of the map. *In a four-variable map, not only are the left and right edges of the map adjacent, but the top and bottom edges of the map are also adjacent.* Therefore, the four corners of the map form a four-minterm square. In this example, the four corners are occupied by ones, so they are considered to be an adjacent group of four. Therefore, the reduced form of the switching function is

$$f(w, x, y, z) = w + \overline{x}\,\overline{z} \tag{9-47}$$

FIGURE 9–23
Solution for Equation 9-46.

It is important to understand that if a Karnaugh map is used to reduce the function and it is done correctly, the resulting reduced function will be minimal; that is, it cannot be further reduced. However, a Karnaugh map will not factor the reduced result. If factoring is desired, it must be done by inspection. In some cases the simplest logic circuit (one with the minimum number of gates) is one that is also factored, if possible. For example, if our reduced function is

$$f(a, b, c) = ab + bc \tag{9-48}$$

we would need two two-input AND gates (one to AND ab, and the other to AND bc) and one two-input OR gate (to OR ab with bc) to construct the circuit. However, if we factor the function into the form

$$f(a, b, c) = b(a + c) \tag{9-49}$$

then we would need one two-input OR gate (to OR a with c), and one two-input AND gate (to AND the ORed result with b), thereby reducing the circuit from three gates to two.

Five- and Six-Variable Karnaugh Maps

Karnaugh maps for functions with five or six variables cannot be drawn as two-dimensional figures. Instead, the map is a three-dimensional graphic. Naturally, this makes the map much more difficult to solve because in addition to considering minterms that lie above, below, and to each side of a minterm of interest, we must also consider minterms behind and in front of the minterm of interest. The five-variable K-map is two four-variable maps stacked one on top of the other (4 × 4 × 2), and a five-variable map is a 4 × 4 × 4 cube. Because of the difficulty in solving three-dimensional K-maps, it generally is not done. Instead, other methods, such as computer programs, can reduce functions of more than four variables that are faster and less error prone.

9-11 K-Maps with Don't Care Conditions

Occasionally we encounter switching functions in which some of the possible input conditions will not occur in actual operation. For example, assume that we have a logic circuit that is connected to a six-cylinder engine and is controlling the firing of each of the six spark plugs. To determine internally which spark plug we are controlling, we will assign a binary number to each, 001_2 through 110_2. When we write any of the switching functions for the circuit, we will be using three-variable K-maps to reduce the functions, because our spark plugs have 3-bit numbers assigned to them. However, we are using 3-bit numbers, so there are eight possible binary values, 0 through 7, and we are only using six of these values, so two (spark plug numbers 0 and 7) will be unused. When we draw the K-maps for our functions, minterms 0 and 7 will become what are called **don't care conditions**, or simply **don't cares**. Because they will never occur, we can write either a 0 or a 1 into minterms 0 and 7 of our K-maps. In effect, the don't cares become "wild cards" in our K-map that we can use as either 0 or 1 as we wish.

In canonical form switching functions, we represent don't cares by separating them from the normal minterms of the function. For example, if we have a four-variable function $f(a, b, c, d)$ with minterms 3, 4, 7, 10, and 11, and don't care terms 5, 9, and 15, then we would write the function as

$$f(a, b, c, d) = \Sigma m(3, 4, 7, 10, 11) + \Sigma d(5, 9, 15) \qquad (9\text{-}50)$$

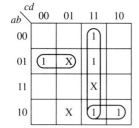

FIGURE 9-24
K-map with don't cares.

The K-map for this function is shown in Figure 9-24. Notice that the don't cares are entered into the map as X's. When we solve the map, we may choose to use or not to use the don't cares. In this example, we will be using don't cares 5 and 15 as ones, and don't care 9 as a zero. Therefore, the reduced result is

$$f(a, b, c, d) = \overline{a}\,\overline{b}\,\overline{c} + a\,\overline{b}\,c + c\,d \qquad (9\text{-}51)$$

9-12 DeMorgan's Theorem

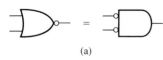

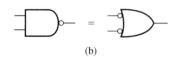

FIGURE 9-25
(a) NOR and (b) NAND gate equivalencies.

In an earlier section of this chapter we investigated the Boolean postulates and theorems. One theorem in particular, Theorem 5, is specifically named DeMorgan's theorem.[4] It is restated here for convenience.

$$\begin{aligned}&\text{(a) } \overline{(x + y)} = \overline{x}\,\overline{y} \\ &\text{(b) } \overline{\overline{x}\,\overline{y}} = \overline{x} + \overline{y}\end{aligned} \qquad (9\text{-}52)$$

Expression (a) states that a NOR gate is equivalent to an AND gate with its input signals inverted (called a **negative AND**), and the dual, expression (b), states that a NAND gate is equivalent to an OR gate with its input signals inverted (called a **negative OR**). These are graphically illustrated in Figures 9-25(a) and (b), respectively.

---

[4] DeMorgan's theorem is named after British mathematician Augustus DeMorgan, who did further development of George Boole's initial work on Boolean algebra.

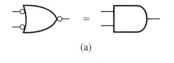

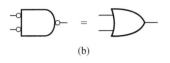

FIGURE 9–26
(a) Negative-NOR and (b) negative-NAND gate equivalencies.

As an extension of Equation 9-52, if we invert every variable in both (a) and (b), we will have

(a) $\overline{(\bar{x} + \bar{y})} = \bar{\bar{x}}\,\bar{\bar{y}} = x\,y$

(b) $\overline{\bar{x}\,\bar{y}} = \bar{\bar{x}} + \bar{\bar{y}} = x + y$

(9-53)

Expression (a) above states that if we invert the inputs of a NOR gate (called a **negative NOR**), it is equivalent to an AND gate, and (b) states that if we invert the inputs of a NAND gate (called a **negative NAND**), it is equivalent to an OR gate. The logic gates illustrating these are shown in Figure 9-26.

9-13 Boolean Expansion

Sometimes we wish to convert a reduced function to its canonical form. Such an operation is called **Boolean expansion**. In the early stages of learning this operation, the easiest way is to put the reduced expression into a K-map and then read the minterm numbers or algebraic terms directly from the map. However, with some experience and understanding of the technique, it will become easier to do the expansion mentally and simply write the expansion directly from the reduced expression, thereby skipping the K-map step. We will investigate both methods.

K-Map Expansion

To illustrate how to expand a switching function using a K-map, we will expand the reduced function

$$f(a, b, c) = \bar{a}\,\bar{b} + a\,b + c \quad (9\text{-}54)$$

Beginning with the first term, $\bar{a}\,\bar{b}$, find all of the minterm boxes in the K-map that satisfy the term and write a one in each of those boxes. For this term, it will be boxes 0 and 1. Now move on to the second term, $a\,b$. This will satisfy minterms 6 and 7 in the K-map. In a similar manner, the term c will satisfy the minterms 1, 3, 5, and 7. Next, we combine all the minterm numbers to get 0, 1, 6, 7, 1, 3, 5, and 7. Now remove the redundant ones and arrange them in ascending order to get 0, 1, 3, 5, 6, and 7. Therefore, the three canonical forms of the switching function are

$$\begin{aligned} f(a, b, c) &= \Sigma m(0, 1, 3, 5, 6, 7) \\ &= \bar{a}\,\bar{b}\,\bar{c} + \bar{a}\,\bar{b}\,c + \bar{a}\,b\,c + a\,\bar{b}\,c + a\,b\,\bar{c} + a\,b\,c \\ &= m_0 + m_1 + m_3 + m_5 + m_6 + m_7 \end{aligned} \quad (9\text{-}55)$$

Direct Expansion

Consider the reduced switching function

$$f(a, b, c, d) = \bar{a}\,\bar{b} + abcd + \bar{a}\,b + d \quad (9\text{-}56)$$

Starting with the first term, $\bar{a}\,\bar{b}$, notice how many variables are missing from the term. In this case the two variables c and d are missing. *When expanding a reduced term into canonical form, the term will expand into 2^n canonical terms, where n is the number of variables that have been reduced out of the original term.* Because our first term is missing two variables, c and d, it will expand into four canonical terms (which, of course are four minterms). To determine the four canonical terms, write the known variables $\bar{a}\,\bar{b}$ four times with each followed by the missing variables cd. Then add overbars to c and d so that all four possible combinations of c and d are formed. These are $\bar{a}\,\bar{b}\,\bar{c}\,\bar{d}, \bar{a}\,\bar{b}\,\bar{c}\,d, \bar{a}\,\bar{b}\,c\,\bar{d}$, and $\bar{a}\,\bar{b}\,c\,d$.

The second term $abcd$ is already a canonical-form term and cannot be expanded.

The third term $\bar{a}\,b$ expands into the four canonical terms $\bar{a}\,b\,\bar{c}\,\bar{d}, \bar{a}\,b\,\bar{c}\,d, \bar{a}\,b\,c\,\bar{d}$, and $\bar{a}\,b\,c\,d$.

The fourth term, d, is missing three variables, a, b, and c. Therefore, it will expand into 2^3, or eight canonical minterms, which are $\bar{a}\bar{b}\bar{c}d$, $\bar{a}\bar{b}cd$, $\bar{a}b\bar{c}d$, $\bar{a}bcd$, $a\bar{b}\bar{c}d$, $a\bar{b}cd$, $ab\bar{c}d$, and $abcd$.

The full expanded expression is now formed by combining all of the expanded minterms and deleting the redundant ones.

$$f(a,b,c,d) = \bar{a}\bar{b}\bar{c}\bar{d} + \bar{a}\bar{b}\bar{c}d + \bar{a}\bar{b}c\bar{d} + \bar{a}\bar{b}cd + \bar{a}b\bar{c}\bar{d}$$
$$+ \bar{a}b\bar{c}d + \bar{a}bc\bar{d} + \bar{a}bcd + a\bar{b}\bar{c}d + a\bar{b}cd + ab\bar{c}d + abcd$$
$$= \Sigma m(0,1,2,3,4,5,6,7,9,11,13,15)$$
$$= m_0 + m_1 + m_2 + m_3 + m_4 + m_5 + m_6 + m_7 + m_9 + m_{11} + m_{13} + m_{15}$$

(9-57)

9-14 Exclusive OR and Exclusive NOR

So far, we have investigated three fundamental logic gates, the AND, OR, and inverter, and two more complex gates, the NAND and NOR. There are two additional gates, the **exclusive OR** and **exclusive NOR**, which, although they are not specifically fundamental gate functions, are treated as such by designers and digital chip manufacturers.

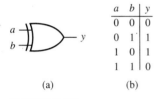

FIGURE 9–27
(a) Exclusive OR gate and (b) truth table.

Exclusive OR

The exclusive-OR gate is similar to the simple OR function except that it outputs a zero for the case when both inputs are ON. The exclusive OR (also called XOR and pronounced "ex-or") gate symbol and truth table are shown in Figure 9-27. Notice in the truth table that the output y is true when either input a OR input b are true, but not both. The XOR is also sometimes called a **disagreement gate** because the output is true when the input values do not agree. All exclusive OR gates have only two inputs.

The XOR symbol is the OR symbol enclosed in a circle, $\oplus$. The XOR can be replaced by its Boolean equivalent, which is

$$x \oplus y = x\bar{y} + \bar{x}y \tag{9-58}$$

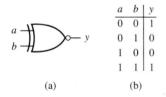

FIGURE 9–28
(a) Exclusive NOR gate and (b) truth table.

Exclusive NOR

As illustrated in Figure 9-28, the exclusive NOR gate (also called XNOR) is simply an XOR with an inverted output. It is sometimes called an **agreement gate** because the output is true when the input values agree. Like the XOR, all XNOR gates have only two inputs.

The XNOR symbol is a dot enclosed in a circle, $\odot$. The XNOR can be replaced by its Boolean equivalent, which is

$$x \odot y = \overline{x \oplus y} = xy + \bar{x}\bar{y} \tag{9-59}$$

9-15 Logic Diagrams and Combinational Logic

When performing a design of a digital circuit, there are several steps that must be performed, including formulating the problem, drawing a truth table, writing a canonical-form switching function, and reducing the function to minimal form. The final step in the design process is to draw the logic diagram.

Generally, logic diagrams are drawn directly from the reduced switching function. When doing so, it is important to remember the hierarchy of execution of Boolean algebra: (1) parentheses, (2) inversion, (3) AND, and (4) OR. With this ordering of execution in mind, let us begin by drawing the logic diagram for the function

$$f(a,b,c) = ab + ac \tag{9-60}$$

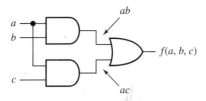

FIGURE 9–29
Circuit of Equation 9-60.

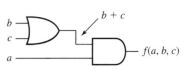

FIGURE 9–30
Circuit of Equation 9-61.

This particular equation is in SOP form and, as will become apparent, is sometimes called a two-level AND-OR circuit. Using the hierarchy of execution, we begin by looking for parentheses. There are none, so we move on to inversions. Finding none, next we look for AND operations. There are two, each involving two variables, so we begin our drawing with two two-input AND gates on the left side of the drawing with inputs ab and ac. Next, we OR these two products together to produce the function. The resulting logic diagram is shown in Figure 9-29. Notice that there are two levels of logic gates, with the AND level on the left and the OR level on the right; hence the term two-level AND-OR.

Earlier we mentioned that in some cases, a simpler logic circuit can be achieved by factoring the reduced switching function. If we factor the variable a on the right side of Equation 9-60, we have

$$f(a, b, c) = a(b + c) \tag{9-61}$$

The resulting logic circuit, shown in Figure 9-30, contains one less gate than the equivalent circuit in Figure 9-29.

■ **EXAMPLE 9-14**   Draw a minimal logic diagram for the switching function

$$f(x, y, z) = \Sigma m(0, 1, 3, 7) \tag{9-62}$$

SOLUTION   The K-map for the function is shown in Figure 9-31. The reduced switching function is

$$f(x, y, z) = \overline{x}\,\overline{y} + yz \tag{9-63}$$

and the logic diagram appears in Figure 9-32.

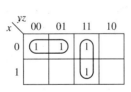

FIGURE 9–31
K-map for Example 9-14.

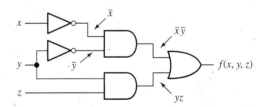

FIGURE 9–32
Logic circuit for Example 9-14.

PROBLEMS

9-1 Perform the following number base conversions.
  (a) 1011_2 to decimal   (b) 10001000_2 to decimal
  (c) 126_{10} to binary   (d) 1025_{10} to binary

9-2 Perform the following number base conversions.
  (a) 5_{10} to binary   (b) 14_{10} to binary
  (c) 11111111_2 to decimal   (d) 0010100101010001_2 to decimal

9-3 Perform the following binary additions and subtractions.
  (a) $1001 + 0011$ (b) $11001001 + 10000110$
  (c) $10110 - 00111$ (d) $00111011 - 10000001$

9-4 Perform the following binary additions and subtractions.
  (a) $00111011 + 10000001$ (b) $10110 + 00111$
  (c) $1001 - 0011$ (d) $11001001 + 10000110$

9-5 Convert the following signed decimal numbers to 8-bit twos-complement numbers.
(a) +5   (b) −23
(c) −159 (d) +123
(e) −1   (f) +1
(g) −15   (h) −123

9-6 Convert the following 8-bit twos-complement numbers to signed decimal numbers.
(a) 00000001   (b) 10010000
(c) 11111111   (d) 10000000
(e) 10000001   (f) 00001111
(g) 10101010   (h) 00000000

9-7 Perform the following decimal operations using 6-bit twos-complement numbers. Convert your answers to signed decimal numbers.
(a) 10 − 3   (b) 3 − 10
(c) 29 − 30   (d) 23 − 1
(e) 16 − 15   (f) −16 − 2
(g) 20 − 25   (h) 0 − 1
(i) −10 − (−5) (j) −16 − 17

9-8 Perform the following decimal operations using 12-bit twos-complement numbers. Convert your answers to signed decimal numbers.
(a) 2000 − 1   (b) 0 − 1
(c) 29 − 30   (d) 1 − 2000
(e) −1555 − 1   (f) 0 − 1024
(g) 100 − 100   (h) 100 − 101
(i) 943 − 2005 (j) −1030 − 1050

9-9 Perform the following number-base conversions.
(a) 10010100_2 to octal   (b) 754_8 to binary
(c) $3E7_{16}$ to binary   (d) 1011001_2 to hexadecimal
(e) 3460_8 to hexadecimal   (f) $3AB9_{16}$ to octal

9-10 Perform the following number-base conversions.
(a) 1100000100100_2 to octal (b) 777_8 to binary
(c) $A09_{16}$ to binary   (d) 1010111_2 to hexadecimal
(e) 1707_8 to hexadecimal   (f) $10DC_{16}$ to octal

9-11 Draw a single AND, NAND, OR, or NOR logic gate for each of these Boolean functions. Label the input and output variables.
(a) $f(a,b) = ab$ (b) $f(x,y) = \overline{xy}$
(c) $f(a,b,c) = a+b+c$ (d) $f(r,s,t) = \overline{r+s+t}$

9-12 Draw a single AND, NAND, OR, or NOR logic gate for each of these Boolean functions. Label the input and output variables.
(a) $f(a,b) = \overline{a}\,\overline{b}$ (b) $f(x,y) = \overline{\overline{x}\,\overline{y}}$
(c) $f(a,b,c) = \overline{a}+\overline{b}+\overline{c}$ (d) $f(r,s,t) = \overline{\overline{r}+\overline{s}+\overline{t}}$

9-13 Reduce each of the following canonical-form expressions to minimal algebraic form.
(a) $f(x,y) = xy + x\overline{y}$
(b) $f(a,b,c) = abc + a\overline{b}c + ab\overline{c}$
(c) $f(w,x,y,z) = \overline{w}\,x\,\overline{y}\,z + \overline{w}\,x\,y\,z$
$+ \overline{w}\,x\,y\,\overline{z} + w\,x\,y\,\overline{z} + w\,x\,y\,z$
(d) $f(a,b,c,d) = \overline{a}\,\overline{b}\,\overline{c}\,\overline{d} + \overline{a}\,\overline{b}\,c\,\overline{d} + \overline{a}\,\overline{b}\,c\,\overline{d}$
$+ \overline{a}\,\overline{b}\,c\,d + a\,\overline{b}\,\overline{c}\,d$

9-14 Reduce each of the following canonical-form expressions to minimal algebraic form.
(a) $f(x,y) = \Sigma m(0,1,3)$
(b) $f(r,s,t) = \Sigma m(0,1,2,7)$
(c) $f(w,x,y,z) = \Sigma m(0,2,8,10,12,13)$
(d) $f(a,b,c,d) = \Sigma m(0,2,4,5,6,7,9,11)$

9-15 Extract the reduced switching functions from the Karnaugh maps in Figure P9-15.

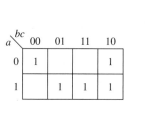
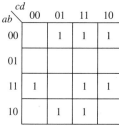

FIGURE P9–15

9-16 Extract the reduced switching functions from the Karnaugh maps in Figure P9-16.

FIGURE P9–16

9-17 Expand the following reduced switching functions into canonical Σm form.

(a) $f(a, b) = a$
(b) $f(x, y, z) = \bar{y} z$
(c) $f(a, b, c) = \bar{a} + \bar{b} + \bar{c}$
(d) $f(a, b, c, d) = a b + \bar{b} c + c \bar{d}$

9-18 Expand the following reduced switching functions into canonical Σm form.

(a) $f(a, b, c) = a$
(b) $f(w, x, y, z) = \bar{y} z$
(c) $f(a, b, c, d) = a \bar{d}$
(d) $f(a, b, c, d) = a b c + \bar{c}$

9-19 Draw a logic diagram of each of the following switching functions. Label all inputs and outputs.

(a) $Y = a \bar{b}$
(b) $Y = \overline{x \bar{y}}$
(c) $Y = a \bar{b} + \bar{c}$
(d) $Y = a b + b d + \bar{c}$

9-20 Draw a logic diagram of each of the following switching functions. Label all inputs and outputs.

(a) $Y = a(b + c)$
(b) $Y = w [x + \overline{(y + z)}]$
(c) $Y = \overline{a b + \overline{c d}}$
(d) $Y = a b \oplus b(\bar{c} + d)$

9-21 Write the switching function for the logic circuit in Figure P9-21.

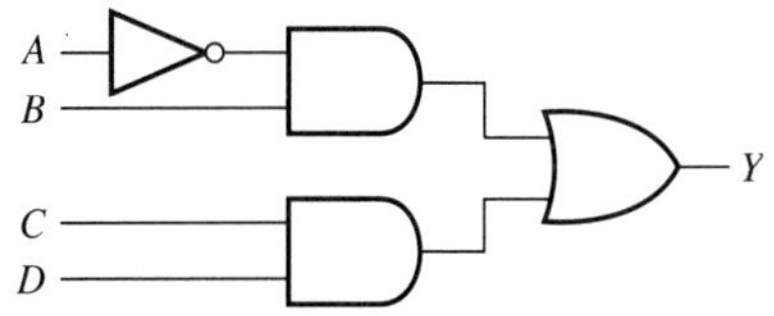

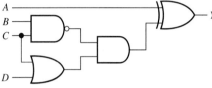

FIGURE P9–21

9-22 Write the switching function for the logic circuit in Figure P9-22.

FIGURE P9–22

Digital Circuits: Advanced Combinational Forms

OVERVIEW AND OBJECTIVES

In the previous chapter, we learned to use various logic design methods to design simple logic circuits. These methods included Boolean algebra, truth tables, and Karnaugh maps. In this chapter we will continue our study of combinational logic by investigating more complex digital circuitry that is capable of performing more powerful operations, both logical and numeric.

Objectives

After completing this chapter, you will be able to

- Design and draw logic circuits for half-adders and full-adders.
- State applications for decoders and encoders and design logic circuits using them.
- Analyze and use priority encoders.
- Design, draw, and apply various multiplexers.
- State the advantages and disadvantages of using the TTL and CMOS logic families.

10-1 Adders

Obviously, among the most important operations a digital circuit can perform are mathematical operations. It can be shown that all mathematical operations, no matter how complex, can be reduced to a series of additions. Therefore, in a digital computing machine, the addition operation is the most fundamental and most important. Fortunately, it is also relatively simple.

Consider the case where we wish to simply add two binary bits. As we learned earlier, binary addition has four possible combinations, which are

$$1: 0 + 0 = 0$$
$$2: 0 + 1 = 1$$
$$3: 1 + 0 = 1$$
$$4: 1 + 1 = 0 \text{ with a carry of } 1$$

x	y	s
0	0	0
0	1	1
1	0	1
1	1	0

(a)

x	y	c
0	0	0
0	1	0
1	0	0
1	1	1

(b)

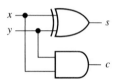

FIGURE 10–1
Truth tables for 2-bit adder, (a) sum and (b) carry.

FIGURE 10–2
Half-adder.

Note that in the preceding expressions, the plus symbol indicates arithmetic addition, not the logical OR. We can develop Boolean switching functions for the sum and the carry bits that result from the addition of 2 bits. The truth tables for these two operations are shown in Figure 10-1. Notice that the truth table for the sum operation is the same as that for the exclusive OR, and the carry operation is simply an AND operation. Therefore, we can create a simple binary adder using one XOR gate and one AND gate, as shown in Figure 10-2. This particular adder circuit is called the **half-adder**.

The half-adder is useful when we simply want to add 2 bits and produce a sum and a carry. However, consider the more complex case in which we will add two multiple-bit numbers. For example, if we add two 2-bit numbers, we can use a half-adder to add the least significant bits (called LSBs). However, when we add the most significant bits (MSBs), we also need to add the carry that might have been generated from the addition of the LSBs. Therefore, the addition of the MSBs becomes a 3-bit addition of the two MSBs and the carry. If the two 2-bit numbers are $x_1 x_0$ and $y_1 y_0$, the operation would be

$$\begin{array}{r} x_1\,x_0 \\ +y_1\,y_0 \\ \hline s_2\,s_1\,s_0 \end{array} \tag{10-1}$$

Note that there are three sum bits, s_2, s_1, and s_0, where s_2, is the possible carry from the second-column sum. Because the sum bit s_0 is the simple addition of x_0 and y_0, it can be generated by a half-adder. However, when adding the MSBs we are adding 3 bits (x_1, y_1, and the LSB carry c_0), so a half-adder will not suffice. Instead, we will need an adder capable of summing 3 bits, called a **full-adder**.

Consider the truth table and K-maps in Figure 10-3 that shows the addition of 3 bits, x, y, and z, which produces the resulting sum s and carry c. The Boolean expressions for the sum and carry are

$$s = \overline{x}\,\overline{y}\,z + \overline{x}\,y\,\overline{z} + x\,\overline{y}\,\overline{z} + x\,y\,z \tag{10-2}$$

and

$$c = x\,y + x\,z + y\,z \tag{10-3}$$

It can be shown using Boolean postulates and theorems that Equation 10-2 can be reduced to

$$s = (x \oplus y) \oplus z \tag{10-4}$$

The resulting logic circuit for the full-adder using Equations 10-4 and 10-3 is shown in Figure 10-4. It should be noted that the circuit shown here is only one of several variations on this circuit.

In many cases, it is easier to represent half-adders and full-adders by block symbols with appropriate inputs and outputs as shown in Figure 10-5. When using block symbols,

x	y	z	s	c
0	0	0	0	0
0	0	1	1	0
0	1	0	1	0
0	1	1	0	1
1	0	0	1	0
1	0	1	0	1
1	1	0	0	1
1	1	1	1	1

s

x \ yz	00	01	11	10
0		1		1
1	1		1	

c

x \ yz	00	01	11	10
0			1	
1		1	1	1

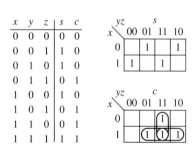

FIGURE 10–3
Full-adder truth table and K-maps.

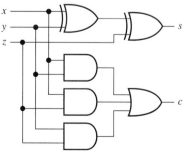

FIGURE 10–4
Full-adder.

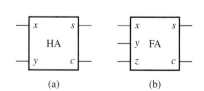

FIGURE 10–5
Block symbols, (a) half-adder, (b) full-adder.

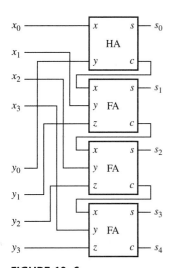

FIGURE 10–6
Nibble-adder.

the configuration of logic gates inside the blocks is unimportant, so long as the blocks perform their desired logical operations. Using our block symbols, we will now design a nibble-adder; that is, an adder that will add two 4-bit binary numbers, $x_3\,x_2\,x_1\,x_0$ and $y_3\,y_2\,y_1\,y_0$, and produce a sum $s_4\,s_3\,s_2\,s_1\,s_0$. First, from our previous example, we know that the two LSBs can be added using a half-adder, and that all other bit-pairs must be added using a full-adder because the column to the right of each bit pair might generate a carry bit. Also, when we finally add the two MSBs, any carry that is generated will simply become the fifth sum bit, s_4. Therefore, our nibble adder will appear as shown in Figure 10-6. Note that we did not necessarily connect the x and y signals to the corresponding x and y inputs on the adders. Because an adder simply adds the input values, the order in which they are added makes no difference. Therefore, we could have connected the inputs of the adders in any fashion without affecting the performance of the circuit. A digital circuit that has inputs that can be connected in any fashion without changing the circuit operation is called a **symmetrical circuit**, and the switching function that defines it is called a **symmetrical function**. Equations 10-3 and 10-4 are symmetrical functions.

10-2 Decoders

A decoder is a circuit with multiple outputs, one corresponding to each possible minterm that can be input to the circuit. For example, assume we have a 2:4 (pronounced "two-to-four") decoder as shown in the block symbol in Figure 10-7(a). This decoder has two inputs, A and B, and four outputs D_0 through D_3. One and only one of the outputs will be on at any given time as indicated in the truth table of Figure 10-7(b). Notice that the particular output selected is determined by the binary values applied to inputs A and B. The Boolean equations for the outputs are

$$D_0 = \overline{a}\,\overline{b} \tag{10-5}$$

$$D_1 = \overline{a}\,b \tag{10-6}$$

$$D_2 = a\,\overline{b} \tag{10-7}$$

and

$$D_3 = a\,b \tag{10-8}$$

Therefore, the 2:4 decoder block would internally consist of the logic circuit shown in Figure 10-8.

Decoders are generally used to numerically select logical systems. For example, assume a central processing unit (CPU) in a computer that is connected to memory. To make memory flexible, we will divide it into four banks, so that the user can add additional banks of memory as needed. We would like for the CPU to use the memory as if it were one seamless memory block, no matter how many memory banks we have installed. To keep our example relatively simple, let's assume that we have memory in banks of 64 bits each. Therefore, the memory addresses in each memory bank will be 6-bit binary numbers

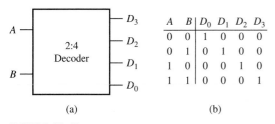

FIGURE 10–7
2:4 Decoder (a) symbol, (b) truth table.

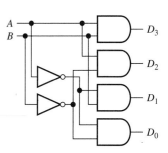

FIGURE 10–8
2:4 Decoder logic diagram.

FIGURE 10–9

Memory bank addressing using a 2:4 decoder.

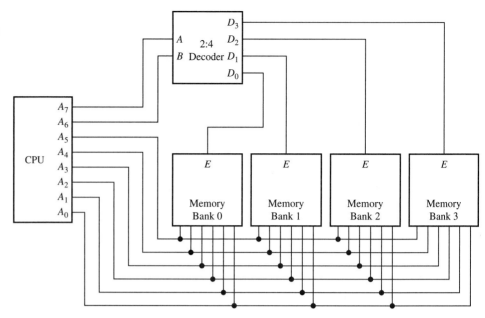

$A_5 A_4 A_3 A_2 A_1 A_0$ and will range from 000000_2 to 111111_2. We can have the CPU select the desired memory bank by using a 2:4 decoder. The two input bits of the decoder are connected to two additional memory address bits A_7 and A_6 on the CPU (making a total of eight memory address bits) and the four outputs of the decoder are connected to enable inputs on each of the memory banks (each memory bank is switched on when its enable input is on, and when the enable is off the bank is dormant). The memory system would be connected as shown in Figure 10-9. Note that the CPU has no special connections to select the memory banks because it is simply accomplished by the two MSBs of the memory address. Therefore memory addresses 00000000_2 through 00111111_2 reside in bank 0, addresses 01000000_2 through 01111111_2 reside in bank 1, addresses 10000000_2 through 10111111_2 reside in bank 2, and addresses 11000000_2 through 11111111_2 reside in bank 3. To the CPU, this memory system operates exactly like one contiguous 256-bit memory. Because we generally specify memory addresses in hexadecimal numbers, our memory ranges would be 00_{16} to $3F_{16}$ for bank 0, 40_{16} to $7F_{16}$ for bank 1, 80_{16} to BF_{16} for bank 2, and $C0_{16}$ to FF_{16} for bank 3.

Some families of logic chips inherently require more power supply current and will dissipate more power when a signal is at a low logic level. As we can see from the truth table of the 2:4 decoder, a majority of the outputs of the decoder (that is, all except one) will be low at any given time. Therefore, to reduce the power supply current and make the circuits operate cooler, most decoders have outputs that are normally high and go to a low logic level when an output is selected. We can also realize further current and power savings if we provide a way to totally disable all of the outputs. Therefore, most decoders are designed with an enable input (which is inverted) and inverted outputs. For the 2:4 decoder with enable input and inverted outputs, the circuit symbol and truth table are shown in Figure 10-10. Notice in the truth table of Figure 10-10(b) that when the enable input is 1, all outputs are 1 no matter what values are input for A and B. Figure 10-11 shows that internally this change is made simply by changing the AND gates to NAND gates, using NAND gates with three inputs, and adding an inverter for the enable input.

Decoders are also available with a larger number of outputs. For example, the 3:8 decoder shown in Figure 10-12 has three binary inputs, an enable input, and eight decoded outputs. In a similar manner, an off-the-shelf 4:16 decoder is also available, which has 4 binary inputs A, B, C, and D, an enable input, and 16 decoded outputs D_0 through D_{15}. If larger decoders are needed, or if the desired decoder size is not in stock, it is possible to build larger decoders by paralleling smaller ones. For example, we can construct a 3:8 decoder

(a)

E	A	B	D_0	D_1	D_2	D_3
0	0	0	0	1	1	1
0	0	1	1	0	1	1
0	1	0	1	1	0	1
0	1	1	1	1	1	0
1	0	0	1	1	1	1
1	0	1	1	1	1	1
1	1	0	1	1	1	1
1	1	1	1	1	1	1

(b)

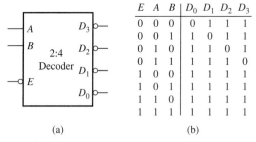

FIGURE 10–10
2:4 Decoder with enable and inverted outputs (a) symbol, (b) truth table.

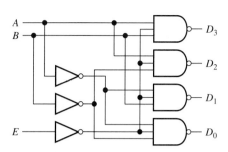

FIGURE 10–11
2:4 Decoder logic diagram with enable and inverted outputs.

(a)

E	A	B	C	D_0	D_1	D_2	D_3	D_4	D_5	D_6	D_7
0	0	0	0	0	1	1	1	1	1	1	1
0	0	0	1	1	0	1	1	1	1	1	1
0	0	1	0	1	1	0	1	1	1	1	1
0	0	1	1	1	1	1	0	1	1	1	1
0	1	0	0	1	1	1	1	0	1	1	1
0	1	0	1	1	1	1	1	1	0	1	1
0	1	1	0	1	1	1	1	1	1	0	1
0	1	1	1	1	1	1	1	1	1	1	0
1	X	X	X	1	1	1	1	1	1	1	1

(b)

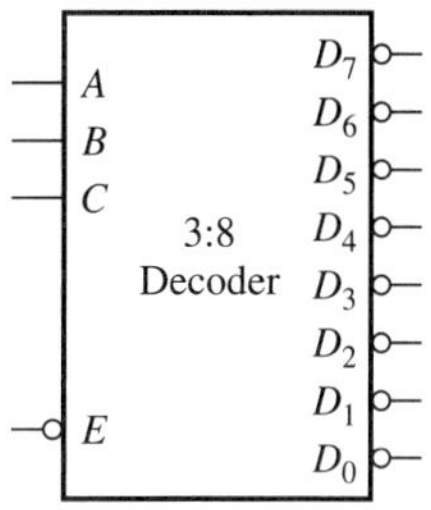

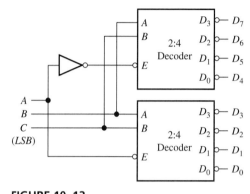

FIGURE 10–12
3:8 Decoder (a) symbol, (b) truth table.

FIGURE 10–13
3:8 Decoder constructed from two 2:4 decoders.

from two 2:4 decoders, as shown in Figure 10-13. In this circuit, notice that the *A* input operates the enable on each of the two 2:4 decoders, and the inverter allows the *A* input to enable only one of the decoders at any time.

10-3 Encoders

Functionally, an encoder is the reverse of a decoder. For example, the 4:2 encoder in Figure 10-14 has four inputs and two outputs. As shown in the truth table of Figure 10-14(b), if we apply a logical 1 to one of the input terminals, the encoder will output a binary number corresponding to that input. As simple as this seems, there are two fundamental problems that can occur that are not present in the decoder.

Again using the 4:2 encoder example, if we apply a logical 1 to input D_0, then according to the truth table we would expect the binary output to be 00_2 ($A = 0$ and $B = 0$). However, what is the output value if we do not apply a logical 1 to any of the inputs? To avoid this predicament, we will need an additional output from the encoder that will switch on when an input is on and will switch off when none of the inputs are on, in effect indicating that the binary output value is not valid. This is called a **data valid** output.

(a)

D_0	D_1	D_2	D_3	A	B
1	0	0	0	0	0
0	1	0	0	0	1
0	0	1	0	1	0
0	0	0	1	1	1

(b)

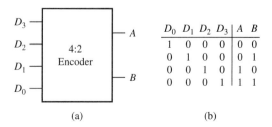

FIGURE 10–14
Simple 4:2 encoder (a) symbol, (b) truth table.

The other problem in the encoder occurs when more than one input is switched on simultaneously. In this case, the encoder finds itself in a situation where it needs simultaneously to output two different binary values, an impossible task. For this condition, encoders are designed so that no matter how many inputs are switched on, the binary output value will be that of the highest input. For our 4:2 encoder example, if we switch on inputs D_0 and D_2, the binary output will be 10_2, because the highest number input that is switched on is input 2. Encoders give priority to the higher-numbered inputs; therefore they are generally called **priority encoders**.

■ **EXAMPLE 10-1**

Design a 4:2 priority encoder. The encoder inputs are D_3, D_2, D_1, and D_0. The outputs are A (the MSB), B (the LSB), and V (valid, a logical 1 when at least one input line is selected). All inputs and outputs are to be noninverted. If more than one input is turned on simultaneously, the encoder should output the highest numbered input that is on.

SOLUTION The truth table and corresponding K-maps are shown in Figure 10-15. Notice that in the case where no inputs are on, the valid V output is 0 and the A and B outputs are don't cares. Otherwise, V is a logical 1. The equations for the three outputs that are extracted from the K-maps are

$$A = D_2 + D_3 \tag{10-9}$$

$$B = D_3 + D_1 \overline{D}_2 \tag{10-10}$$

$$V = D_0 + D_1 + D_2 + D_3 \tag{10-11}$$

$$= D_0 + D_1 + A$$

Notice in Equation 10-11, that $D_2 + D_3$ is equal to A, from Equation 10-9, and has therefore been replaced by A. As we will see, this will reduce the complexity of the gates in our circuit.

Figure 10-16 shows the logic circuitry to construct the 2:4 priority encoder using Equations 10-9, 10-10, and 10-11. Note that by using the simplified version of Equation 10-11, we avoided the use of a four-input OR gate, instead using a three-input OR gate.

D_3	D_2	D_1	D_0	A	B	V
0	0	0	0	X	X	0
0	0	0	1	0	0	1
0	0	1	0	0	1	1
0	0	1	1	0	1	1
0	1	0	0	1	0	1
0	1	0	1	1	0	1
0	1	1	0	1	0	1
0	1	1	1	1	0	1
1	0	0	0	1	1	1
1	0	0	1	1	1	1
1	0	1	0	1	1	1
1	0	1	1	1	1	1
1	1	0	0	1	1	1
1	1	0	1	1	1	1
1	1	1	0	1	1	1
1	1	1	1	1	1	1

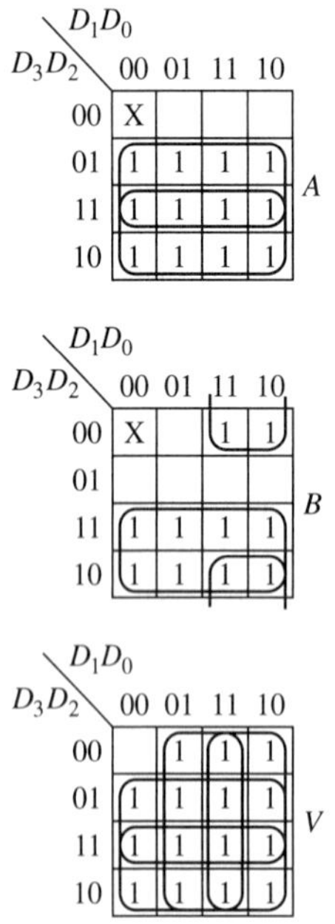

FIGURE 10–15
4:2 Priority encoder design.

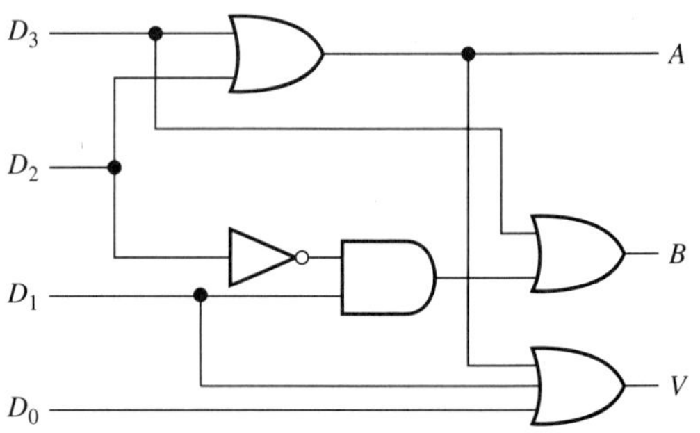

FIGURE 10–16
4:2 Priority encoder internal logic.

The most common type of priority encoder is the 10:4 priority encoder. This device has 10 inputs D_0 through D_9 and 5 outputs A, B, C, D, and V. The binary output is simply a BCD number representing the highest-numbered input that is on.

10-4 Multiplexers

A **multiplexer** (also called a **data selector** or **mux**) is a digitally controlled selector switch. Consider the functional diagram of a 4:1 multiplexer shown in Figure 10-17(a). Although this figure shows a selector switch inside the multiplexer, the switching is actually done using logic gates. The two select inputs S_1 and S_0 are the control lines that determine which of the inputs D_0 through D_3 will be connected to the output Y. For example, if $S_1 = 1$ and $S_0 = 0$, the input D_2 will be selected and any digital signal on input D_2 will appear on output Y.

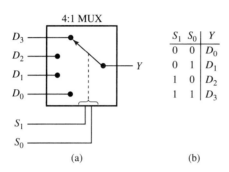

FIGURE 10–17
Multiplexer (a) functional diagram and (b) truth table.

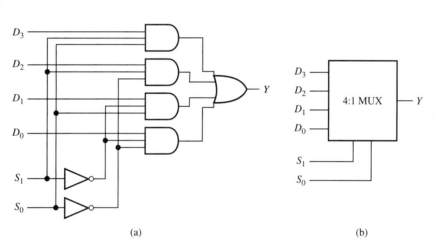

FIGURE 10–18
4:1 MUX (a) internal logic diagram, (b) logic symbol.

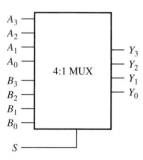

FIGURE 10–19
Quad-2:1 MUX.

The actual logic circuitry to construct a 4:1 mux is shown in Figure 10-18. Notice that the select inputs S_0 and S_1 are connected to inverters, thus creating both inverted and non-inverted polarities of the signals, which are then used uniquely to activate each of the AND gates.

Multiplexers are available in configurations of 4:1, 8:1, and 16:1. These work well when the requirement is to select only one input line. However, it is sometimes necessary to switch multiple bits at the same time and transfer them to a multiple-bit-wide output. For example, consider the quad-2:1 mux in Figure 10-19. For this device, when the select line S is zero, the nibble A (A_0 through A_3) will be connected to the outputs Y_0 through Y_3, and when the select line S is 1, the nibble B (B_0 through B_3) will be connected to the outputs Y_0 through Y_3. In effect, this is a digital four-pole double-throw switch. Combinations of multiplexers allow the switching of larger numbers of bits. For example, two quad-2:1 multiplexers would make it possible to select between 2 bytes.

10-5 Integrated-Circuit Digital Logic Families

When using digital logic, the designer must not only design the logic circuitry, but must also determine the logic type (called **logic family**) that will be used. An understanding of the characteristics of each of the more popular logic families is imperative because the logic family selection determines the speed, power consumption, available logic devices, and even the layout of the circuit. There are two fundamental logic families, TTL and CMOS, and many subfamilies within these major families. We will examine the TTL and CMOS families and investigate the characteristics of each.

Transistor-Transistor Logic (TTL)

Transistor-transistor logic (commonly called **TTL**) was developed in the 1960s and quickly became the backbone of integrated-circuit logic. From this beginning, many other TTL families have been developed, each with a different attractive characteristic such as increased speed or lower power consumption. As of this writing, there are more than 30 TTL families.

Most TTL families operate from a single +5.0 V dc power supply called V_{cc}. Within TTL, we use voltage levels to indicate logical ones and zeros. Ideally, these voltages are 0 volts and +5 volts. However, for most varieties of TTL, any voltage within the range of +2.4 volts to V_{cc} is considered to be a high logic level and any voltage within the range of +0.4 volts to ground is considered to be a low logic level. With the exception of a special TTL family called tri-state logic, a signal voltage within the range of +0.4 volts to +2.4 volts is not allowed and indicates either a defective chip or a wiring error. The TTL logic families are typically high-speed logic families, with maximum clock speeds ranging from 20 MHz to more than 200 MHz depending on the TTL family.

TTL chips consist entirely of silicon bipolar transistors, silicon diodes, and resistors, all packed onto a silicon wafer and encased in a plastic (commercial version) or ceramic (military version) integrated-circuit package, which we commonly call a **chip**. Most TTL chips have 14 or 16 connection pins, although TTL chips that contain some of the more powerful functions have 18, 20, 24, 28, or 40 pins. Two of the pins are reserved for V_{cc} (power) and ground connections, while the remaining pins are used for input and output (**I/O**) signals.

The output circuitry in a TTL chip is designed to provide a bias voltage to the input of all subsequent chips, so no biasing resistors or coupling capacitors are needed, and they are connected directly. However, there is a maximum limit to the number of subsequent chips to which a TTL output can be connected, called **fanout**, which is in the range of 10 to 50 depending on the TTL variety. If the fanout of a TTL output is exceeded, it is possible that the output will be overloaded. The result is that the output voltages will no longer fall within the normal ranges for high and low logic levels, thereby increasing the possibility of logical errors in the circuit.[1]

Chips within the standard TTL family have part numbers that begin with the digits 74 followed by a two- or three-digit number indicating the logical function performed by the chip. For example, Figure 10-20 shows (a) a 7400 TTL chip which contains four two-input NAND gates (called a **quad two-input NAND**), (b) a 7486 containing four two-input exclusive-OR gates (called a **quad two-input XOR**), and (c) a 7410 containing three three-input AND gates (called a **triple three-input AND**). Other TTL logic families have part numbers beginning with the digits 74 followed by letters designating the TTL family, and ending with a two- or three-digit number indicating the logical function performed by the chip. For example, a 74LS00 has the fundamental part number 7400, which means it is a quad two-input NAND. That is, it is constructed using the same chip package, has the same

---

[1] A logical error is defined as a logical one that is interpreted by the circuit as a zero, or a logical zero that is interpreted by the circuit as a 1. Errors are generally caused by overloading (exceeding the fanout), or by electrical noise on the V_{cc}, ground, or signal conductors.

FIGURE 10–20
TTL Chips, (a) 7400, (b) 7486, (c) 7410.

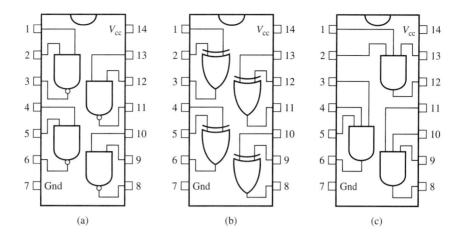

pin connections (called **pinouts**), and performs the same logical function as the 7400. However, the LS designation within the part number indicates that it is internally constructed using low-power Schottky design methods, resulting in lower power consumption and higher speed than the basic 7400 circuit. Some of the other popular TTL families are 74L (low power), 74S (Schottky—very high speed), and 74ALS (advanced low-power Schottky—low power consumption and very high speed). Because the TTL logic families exhibit higher speed than most other logic families, they are generally used in applications that require moderate to high speeds that are not constrained by high power requirements, such as desktop computers and mainframe computers.

Complementary Metal Oxide Semiconductors (CMOS)

CMOS is a radically different logic family in that the integrated circuits are constructed entirely of P-channel and N-channel MOSFET transistors. Because both P- and N-channel transistors are used, the term **complementary** is applied to this logic family, hence the name **complementary metal oxide semiconductors** or **CMOS**. Like TTL, CMOS operates from a single power supply, but the supply voltage can range from 3 to 18 V dc, making it an ideal logic family for battery-operated applications. CMOS chips are housed in standard integrated circuit packages with power (called V_{dd}), ground (called V_{ss}), and I/O pins.

When MOSFETs are utilized in switching applications, they exhibit extremely high output impedances in the off condition and moderately low output impedances in the on condition. The result is that the logic devices consume extremely low power, typically on the order of microwatts or nanowatts. For this reason, CMOS is used exclusively in all applications requiring low power consumption such as calculators, pagers, cellular phones, satellite instrumentation, medical pacemakers, and a long list of others. Because of its extremely low power consumption, CMOS generates very little heat and operates at near ambient temperature. Therefore it is not subjected to the constant fluctuations in temperature when power is switched on and off, which results in low mechanical stress on the devices. This, in turn, results in an extremely long life and high reliability for the devices. For example, the *Voyager I* and *Voyager II* spacecraft, which, as of this writing are in their 29th year of operation, contain computers and other logic circuitry made of CMOS.

PROBLEMS

10-1 Design a logic circuit that will take the twos complement of a 4-bit binary number.

10-2 Design a logic circuit that will add two 3-bit binary numbers.

10-3 Using half-adders or full-adders and any desired additional gates, design a circuit with five inputs and one output that will generate a logical 1 when a majority of the inputs are logical 1. Use block symbols for the adders.

10-4 Design a logic circuit that will perform the 3-bit binary subtraction $A_2 A_1 A_0 - B_2 B_1 B_0$. Because the result could be negative (if $B > A$), have the result shown in two's-complement form. (*Hint*: Have your circuit take the twos complement of *B*.)

10-5 Draw the internal logic diagram of a 3:8 decoder with inverted enable input and inverted outputs.

10-6 Draw the internal logic diagram of a 4:16 decoder with inverted enable input and inverted outputs.

10-7 Modify the circuit in Figure 10-13 to also provide an inverted enable input.

10-8 Using the 3:8 decoder block symbol of Figure 10-12, show that two 3:8 decoders can be interconnected (with additional logic gates) to create a 4:16 decoder that has an inverted enable and inverted outputs.

10-9 Design and draw the internal logic diagram of an 8:1 multiplexer.

10-10 If the multiplexer in Figure 10-18(b) is connected such that $D_0 = 1$, $D_1 = 1$, $D_2 = 1$, and $D_3 = 0$, what logical function does the circuit perform? For this circuit, S_1 and S_0 are the inputs and Y is the output. (*Hint*: Draw a truth table.)

10-11 If the multiplexer in Figure 10-18(b) is connected such that $D_0 = 0$, $D_1 = 1$, $D_2 = 1$, and $D_3 = 0$, what logical function does the circuit perform?

10-12 Give an advantage of using TTL logic when compared to CMOS.

10-13 Give an advantage of using CMOS logic when compared to TTL.

Digital Circuits: Sequential Forms

11

OVERVIEW AND OBJECTIVES

Our study of digital circuits so far has been limited to **combinational logic circuits**; that is, logic circuits that logically combine input signals based on Boolean rules and provide an output result. These combinational logic gates have no capability to store binary bits. In this chapter, we will investigate digital circuits that have the capability to store binary bits of data. Then, with an understanding of how these storage devices function, we will see how they can be interconnected with combinational logic circuits to construct more complex devices such as multibit storage registers, numerical counters, and memories.

Objectives

After completing this chapter, you will be able to

- Describe the difference between a latch and a flip flop, and state the difference between the two types of *RS* latches.
- Identify the three fundamental types of synchronous flip flops and describe how they operate.
- Describe the difference between an asynchronous counter and a synchronous counter.
- Analyze the operation of a simple counter circuit and determine how it operates.
- Describe the fundamental types of memory devices, their basic operating differences, and how they are best applied for storage applications.

11-1 Introduction

In addition to static logic gates, another important building block of digital systems is the storage element, in particular the **latch** and the **flip flop**. Each digital storage element is capable of storing one binary bit of data. As we will see, there are several different types of storage elements, each with different operating characteristics. However, no matter which type of element, all are capable of storing one and only one bit of binary data. Generally, the storage element is active as long as power remains applied to the devices. Such devices are said to have **volatile** memory because once power is removed, the stored data are lost.

11-2 Latch and Flip Flop Initialization

Because latches and flip flops are dynamic[1] devices with volatile memory, we encounter the dilemma of determining what information is stored in the device when power is applied to it. Because we have not yet input any data, and whereas we have lost the previous data that were stored in it the last time power was applied, what will be its state when power is applied? The answer is that it is indeterminate, and it depends on the device itself. Because of variations in the manufacturing processes, there will be slight variations in the internal electrical characteristics of a digital chip. When power is applied, the device might default to either the 0 or 1 state. We have no way of knowing the internal characteristics of the individual transistors on the chip, so the resulting state will be unpredictable. Therefore, all latches and flip flops are designed so they can be initialized to some desired state, so that their performance from that point is predictable.

11-3 Latches

There are fundamentally two types of latches, the RS latch and the $\overline{R}\,\overline{S}$ latch (pronounced "not-R, not-S latch"). Neither of these latches requires a clock or other synchronizing input. Instead, the latch responds to inputs at the instant they are applied. For this reason, RS and $\overline{R}\,\overline{S}$ latches are **asynchronous** devices.

RS Latch

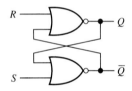

FIGURE 11–1
RS Latch.

The RS latch has two inputs, R and S, and two outputs, Q and $\overline{Q}$. The R input is the reset and the S input is the set. The Q output indicates the current binary value (0 or 1) stored in the latch, and the $\overline{Q}$ output is simply the inverse of the Q.

Referring to the circuit for the RS latch shown in Figure 11-1, note that if the R input is true and the S input is false, the R input will cause the output of the upper NOR gate, the Q output, to be false. The Q output applies a false input to one of the inputs of the lower NOR gate. With the S input also false, the $\overline{Q}$ output will be true. After applying this input condition, if the R input switches false, the Q and $\overline{Q}$ values of the latch do not change. The latch stores the zero that was loaded by resetting it because the Q output keeps the lower gate off while the $\overline{Q}$ output keeps the upper gate on. As long as power remains applied to the gates and the inputs remain false, the latch will remain in the zero state.

If a true signal is applied to the S input, it will cause the $\overline{Q}$ output to be false. This $\overline{Q}$ output causes the output of the upper NOR gate to switch true, thereby making the Q output true. After the S input is switched false, the latch retains the "1" that was loaded into it.

S	R	Q	$\overline{Q}$
0	0	Q	$\overline{Q}$
0	1	0	1
1	0	1	0
1	1	0	0*

FIGURE 11–2
RS Latch excitation table.

The excitation table for the RS latch is shown in Figure 11-2. Note that, as with all excitation tables, it addresses all the possible combinations of input conditions that could be applied to the device. It also shows how the device will respond to these input combinations. Notice in the first row that when a zero is applied to both the R and S inputs, the Q output is shown as Q and the $\overline{Q}$ output is shown as $\overline{Q}$. This simply means that the latch retains whatever value is stored in it. Also notice in the last line of the table there appears to be an error because the Q and $\overline{Q}$ are both zero. However, referring to the logic diagram of the device in Figure 11-1, note that if both the R and S inputs are true, then both the Q and $\overline{Q}$ outputs will be false. Because under these input conditions the latch does not respond as it should, we mark this row of the table with an asterisk, indicating that this input condition is not allowed.

$\overline{R}\,\overline{S}$ Latch

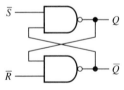

FIGURE 11–3
$\overline{R}\,\overline{S}$ Latch.

The $\overline{R}\,\overline{S}$ latch, shown in Figure 11-3, is similar to the RS latch except that, as the name implies, the inputs are inverted. Therefore, the input condition in which the latch remains

---

[1]A dynamic (or sequential) logic device is one in which the device stores information that is input to it, as opposed to static (or combinational) devices that have no storage capability. Latches and flip flops are dynamic devices. Simple gates such as AND, OR, and Invert are static logic devices.

$\overline{S}$	$\overline{R}$	Q	$\overline{Q}$
0	0	1	1*
0	1	1	0
1	0	0	1
1	1	Q	$\overline{Q}$

FIGURE 11-4
$\overline{R}\,\overline{S}$ Latch excitation table.

in a stable state is $\overline{R} = 1$ and $\overline{S} = 1$. The excitation table for the $\overline{R}\,\overline{S}$ latch is shown in Figure 11-4. Notice that the $\overline{R}$ and $\overline{S}$ inputs will respectively reset and set the latch when they are each zero. Also, the input condition that is not allowed, $\overline{R} = 0$ and $\overline{S} = 0$, is noted with an asterisk.

11-4 Flip Flops

Like latches, flip flops are also 1-bit storage devices. However, all flip flops have one additional input called the **clock**. The clock input is a signal that causes the flip flop to act upon other inputs only at a predetermined instant. As long as the clock input is false, we are allowed to change the other inputs in any way we wish without having any effect on the flip flop. However, when the clock occurs, the inputs are sampled and the flip flop instantly reacts to those inputs. Because the clock has a synchronizing effect on the flip flop operation, flip flops are synchronous devices. There are fundamentally three types of flip flops, the *D*, *T*, and *JK*. Additionally, there are two categories of *D* flip flop, the level triggered and the edge triggered.

Level-Triggered *D* Flip Flop

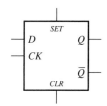

FIGURE 11-5
D Flip flop.

The **level-triggered *D* flip flop**, shown in Figure 11-5, has a single *D* input (in addition to the clock *CK* input) that is used to load data into the device. Because it has a single data input, there are no disallowed input conditions, as we encountered with the *RS* and $\overline{R}\,\overline{S}$ flip flops. Notice in the excitation table in Figure 11-6 that as long as the clock input is zero, the flip flop remains in its previous state and ignores the *D* input. However, when the clock input is 1, the flip flop stores the value of the *D* input, and the *Q* output of the flip flop takes on the value stored in the flip flop.

D	CK	Q	$\overline{Q}$
0	0	Q	$\overline{Q}$
0	1	0	1
1	0	Q	$\overline{Q}$
1	1	1	0

FIGURE 11-6
D Flip flop excitation table.

The *D* flip flop has two additional inputs on the top and bottom labeled *SET* and *CLR*. These are asynchronous inputs that set and clear the flip flop; that is, they will load a 1 or 0 into the flip flop even if the clock input is zero. Actually, if we connected the *D* and *CK* inputs of our flip flop to zero, we could use the *SET* and *CLR* inputs to act as *S* and *R* inputs, effectively making it into an *RS* latch. Similarly, we can have *D* flip flops with inverted *SET* and *CLR* inputs as shown in Figure 11-7. For this flip flop, the *SET* and *CLR* inputs are connected to a one when not used, and connected to zero when used, much like the $\overline{R}$ and $\overline{S}$ inputs of an $\overline{R}\,\overline{S}$ latch.

The *D* flip flops in Figure 11-5 and Figure 11-7 are called **level-triggered *D* flip flops** because each flip flop loads the *D* value whenever the clock input is at a 1 level. If we connect the clock input of a level-triggered *D* flip flop to a continuous one value, the *Q* output will be a buffered *D* and the $\overline{Q}$ will be an inverted *D*. When this occurs, the flip flop is said to be **transparent**, with the output continuously following the input value.

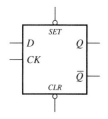

FIGURE 11-7
D Flip flop with inverted *SET* and *CLR*.

Flip Flop Timing Diagrams

When analyzing flip flop performance, it is often useful to use a graphical representation called a **timing diagram** that illustrates how the flip flop will perform when given various combinations of input signals. Timing diagrams are much the same as what one would see on a multichannel oscilloscope; that is, it is a plot showing time on the horizontal scale and signal voltage on the vertical scale. To show timing relationships between the inputs and outputs of the device, we usually stack several plots one above the other. The inputs and outputs of digital circuits are either a high- or low-voltage level, so we usually do not scale the vertical axis, nor do we show a vertical axis at all. Instead, we simply assume a high is a logical 1 and a low is a logical 0. Because digital circuits respond the same no matter what time scale we use (assuming we are operating the circuits within their specified operating frequency range), we generally omit the horizontal time scale also. Because timing

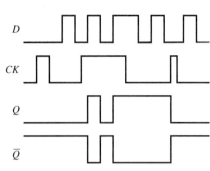

FIGURE 11–8
Timing diagram for a level-triggered D flip flop.

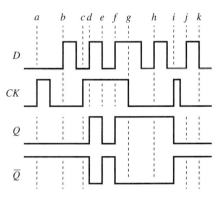

FIGURE 11–9
Timing diagram for a level-triggered D flip flop with callouts.

diagrams show the performance of storage devices (such as flip flops), we must assume that at the beginning of the timing diagram, the device is in some initial state, usually loaded with a zero.

Figure 11-8 is a timing diagram for a positive level-triggered D flip flop. It has four plots showing the synchronous inputs D and CK and the Q and $\overline{Q}$ outputs. In most timing diagrams (and most subsequent timing diagrams in this text), we do not show the $\overline{Q}$ output because it is considered trivial; it is simply an inverted Q output. Also, in many cases, the asynchronous inputs are shown if they are used, but for clarity they are not shown on this diagram. When we use timing diagrams to illustrate how a device functions, it is necessary to show all possible input conditions so that the readers can gain a full understanding of how the device will react to any and all input conditions in their applications.

To gain a full understanding of the level-triggered D flip flop timing diagram, we have modified Figure 11-8 by adding vertical grid lines to the diagram and labeling each with a callout letter as shown in Figure 11-9. The reader is invited to follow along on the timing diagram of Figure 11-9 for the following explanation of the flip flop performance.

The device is initialized to zero as can be seen from the Q output plot.

(a) At time *a*, the CK input is pulsed. Because the D input is zero, this loads a zero into the device. Whereas it was previously loaded with a zero, the outputs do not change.

(b) The D input is pulsed on, then off. Because the CK input is zero at this time, the flip flop ignores the input and remains in the zero state.

(c) The CK is switched on. Because D is zero, this loads another zero into the flip flop.

(d) The D is switched on. Because the CK is on, this loads a one into the device.

(e, f) Timing points *e* and *f* illustrate that while the CK is on, the flip flop is transparent; that is, the Q output follows the D input whenever CK is on.

(g) CK is switched off while D is a one, latching a one into the flip flop.

(h) While CK is off, the flip flop ignores any changes in the D input.

(i) CK is pulsed while D is zero, loading another zero into the device.

(j) The flip flop ignores changes in the D while CK is off.

Edge-Triggered D Flip Flop

D flip flops are also available with edge triggering and are called **edge-triggered D flip flops**. Figure 11-10 shows two edge-triggered D flip flops, one with noninverted SET and CLR, and the other with inverted SET and CLR. The edge-triggered D flip flop functions in much the same manner the level-triggered D flip flop, except that data are loaded into the flip flop only on the transition (edge) of the clock. This edge triggering is indicated on the

FIGURE 11–10
Edge-triggered *D* flip flops.

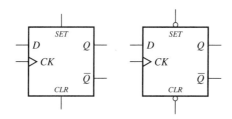

D	CK	Q	$\bar{Q}$
0	0	Q	$\bar{Q}$
0	1	Q	$\bar{Q}$
0	⌐	0	1
1	0	Q	$\bar{Q}$
1	1	Q	$\bar{Q}$
1	⌐	1	0

FIGURE 11–11
Edge-triggered *D* flip flop excitation table.

logic symbol by the ">" symbol on the *CK* input. It is also indicated in the flip flop's excitation table in Figure 11-11 by the rising edge symbol ⌐. Note that for the flip flop in Figure 11-10 and its excitation table in Figure 11-11, the loading of data occurs only on the zero-to-one, or positive edge of the clock input. The flip flop is not triggered by a one-to-zero, or negative edge of the clock. For this reason, this type of flip flop is more accurately called a **positive edge-triggered *D* flip flop**.

Figure 11-12 shows a timing diagram for a positive edge-triggered *D* flip flop. The times when the *CK* input signal is a positive-going transition are marked with arrows. These are the only instants when the flip flop will load the value present on the *D* input. During all other times, the flip flop ignores the signal values on the *D* input.

It is possible to purchase negative edge-triggered flip flops like those shown in Figure 11-13. As indicated in the excitation table, these flip flops load data when the clock changes from a one to a zero. If we subject the negative edge-triggered *D* flip flop to the same input signals as a positive edge-triggered flip flop as in Figure 11-12, the output waveforms would be different, as illustrated in Figure 11-14.

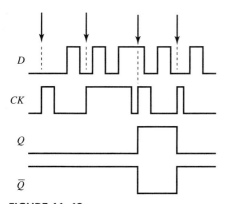

FIGURE 11–12
Timing diagram for a positive edge-triggered *D* flip flop.

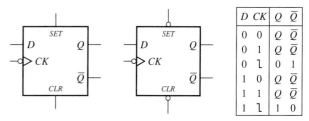

FIGURE 11–13
Negative edge-triggered *D* flip flops and excitation table.

FIGURE 11–14
Timing diagram for a negative edge-triggered *D* flip flop.

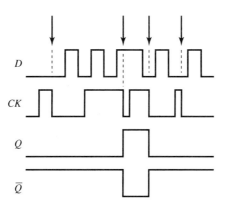

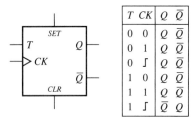

FIGURE 11-15
Positive edge-triggered *T*-flip flop logic symbol and excitation table.

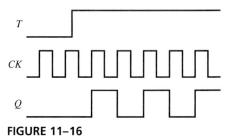

FIGURE 11-16
Positive edge-triggered *T* flip flop timing diagram.

T Flip Flops

The logic symbol and excitation table for the **T flip flop** are shown in Figure 11-15. This particular flip flop is positive edge-triggered; however, because all *T* flip flops are edge triggered, it would be sufficient to call it a positive-triggered *T* flip flop. The letter "*T*" is an abbreviation for "toggle." As indicated in the excitation table, the only input condition that causes the *T* flip flop to change states occurs when the *T* input is a one and the *CK* input changes from a zero to a one. When this occurs, the state of the flip flop toggles; that is, the *Q* takes on the previous value of the $\overline{Q}$ and vice versa. This is simply another way of saying that the flip flop changes to its opposite state. When the *T* input is a zero, the flip flop ignores all inputs and simply remains in its previous state.

The *T* flip flop makes an excellent digital frequency divider. Notice in the positive edge-triggered *T* flip flop timing diagram in Figure 11-16 that whenever the *T* input is a one, the *Q* output toggles at one-half of the *CK* frequency. For example, if we input a 440-Hz signal (the standard A musical note) on the *CK* input, the *Q* output frequency would be exactly one octave lower, or 220 Hz. This characteristic of the *T* flip flop is used extensively in electronic musical instruments, such as electronic pianos. In the piano, the designer only needs to build a frequency generator to produce the top octave (the rightmost 12 keys on the keyboard). Once these frequencies are available, it is a simple matter to use 12 *T* flip flops to divide these frequencies by 2 to produce the next lower octave. Then 12 more *T* flip flops can divide that octave by two, producing the next lower octave, and so forth until all the desired frequencies are produced. This technique has one other advantage: Because the frequencies of all the lower octaves are slaved to the top octave, adjusting only the top octave tunes the entire piano.

JK Flip Flops

As the name implies, the *JK* flip flop has two inputs (in addition to the clock), the *J* and the *K*. The logic symbol and excitation table for the *JK* flip flop are shown in Figure 11-17. Note in the first two rows of the excitation table for this device that we have don't care ("X") conditions for the *J* and *K* inputs. These indicate that if the *CK* is at a zero or one level, the values of the *J* and *K* inputs are ignored by the flip flop.

When the clock transition occurs, the new state of the flip flop is determined by the values of the *J* and *K* inputs and is as follows:

$J = 0, K = 0$ The flip flop does not change state. The previous data remain in the flip flop.

$J = 0, K = 1$ No matter what its previous state, the flip flop goes to the zero state.

$J = 1, K = 0$ No matter what its previous state, the flip flop goes to the one state.

$J = 1, K = 1$ The state of the flip flop toggles.

Note that whenever both $J = 1$ and $K = 1$, the flip flop toggles. Therefore, we can construct a *T* flip flop using a *JK* flip flop with the *J* and *K* inputs tied together to form the *T*

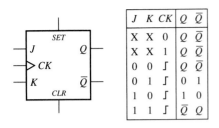

FIGURE 11–17
Positive edge-triggered JK flip flop logic symbol and excitation table.

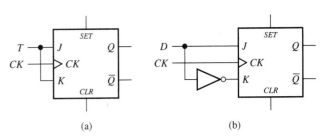

FIGURE 11–18
Using a JK flip flop as (a) a T flip flop, and (b) a D flip flop.

input. We can also construct a D flip flop from a JK flip flop by connecting the D input signal directly to the J input and the inverse of the D input signal to the K input. These adaptations are illustrated in Figure 11-18.

11-5 Parallel Registers

Registers are constructed of two or more flip flops. They are designed for the temporary storage of data that are being moved from one place to another. The data can be moved in either a serial or parallel fashion. Typical register sizes are nibbles (4 bits), bytes (8 bits), and words, although any size is possible, with all the flip flops in the register contained in one integrated-circuit chip. Parallel registers are generally constructed of D flip flops in which the clock inputs of all the flip flops are connected together to form one common clock. In this way, data are clocked into all the flip flops in the register simultaneously. Once data are clocked into the register, the data become available at the Q outputs of the flip flops in parallel form. Figure 11-19 shows a 4-bit parallel D-type register. Note that, as with many parallel registers, this 4-bit register does not have $\overline{Q}$ outputs. Because we are only interested in storing an n-bit number, we will most likely want to retrieve the data in the same noninverted form as the data were originally saved, so even if $\overline{Q}$ outputs were added to the chip, it is unlikely that they would be used.

11-6 Serial (Shift) Registers

A second type of register is the **serial register** (usually called a **shift register**). As with parallel registers, shift registers can be constructed for any number of bits, with the most common sized being the nibble (4 bits), byte (8 bits), and 16-bit word. Figure 11-20 illustrates a

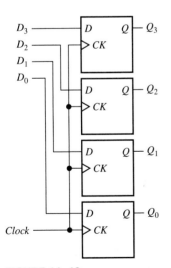

FIGURE 11–19
4-Bit parallel D-type register.

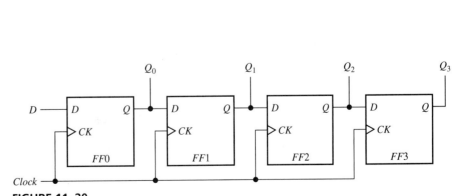

FIGURE 11–20
4-Bit serial (shift) register.

FIGURE 11–21
Shift register timing diagram.

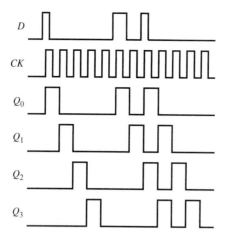

4-bit shift register. In this case, we only have one data input D, and four Q outputs Q_0 through Q_3. In operation, the register moves data to the right from the D input to the Q_3 output, with the data being moved one place on each application of the clock input. Therefore, when a clock occurs, the binary value on the D input is loaded into $FF0$. At the same time, the previous binary value in $FF0$ is loaded into $FF1$, the previous binary value in $FF1$ are loaded into $FF2$, and the previous binary value in $FF2$ is loaded into $FF3$. The result is that after the clock, the 4-bit number in the register is shifted one place to the right with new data being loaded into the leftmost bit and the previous rightmost bit being discarded. Figure 11-21 illustrates this shifting operation.

With the addition of some logic gates, it is also possible to reverse the shifting operation. More exotic shift registers are available that can be used for either parallel or serial registers, with the capability to shift either left or right. These features provide the versatility to load the register serially and unload it in parallel, or vice versa.

Shift registers have a large variety of uses, but they are mainly used in sending and receiving serial communication data. All computer modems use shift registers that allow the computer to transmit data by loading the register in parallel and then shifting the data out in serial, thereby converting parallel data to serial data. When receiving data, the shift register receives the data in serial form and then transfers the data to the computer in parallel form.

11-7 Counters

Flip flops also provide us with the capability to construct counters. Depending on their design, these counters can count in binary, decimal (BCD), or any other number base. Fundamentally, counters are classified as **asynchronous counters** or **synchronous counters**. The difference lies in the way that the individual flip flops in the counter are clocked. In general, if all of the flip flops derive their clock signal from the same source, and they are all triggered at the same time, the counter is synchronous. Asynchronous counters internally generate different clock signals for each of the flip flops, thereby causing the flip flops to be clocked at different times.

Asynchronous Counters

Asynchronous counters can count in any number base (binary, decimal, base 5, etc.) and in either direction (count up or count down). However, the most common type of asynchronous counter is the **binary ripple counter**. Figure 11-22 shows a 4-bit binary ripple counter. Notice that the clock signals for flip flops $FF1$, $FF2$, and $FF3$ are derived from the $\overline{Q}$ outputs of the previous flip flop; that is, the clock signals for the flip flops are not connected to one common clock input signal. Whenever a flip flop switches, there is an internal time delay from the time the clock occurs until the Q output actually indicates the value that was on the input. This time delay is called **propagation delay**. It is caused

FIGURE 11–22
4-Bit binary ripple counter.

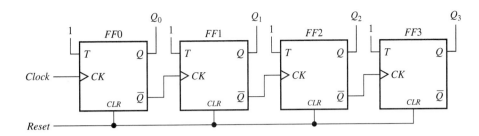

by inherent time delays in the internal gates and stray capacitance in the internal wiring of the chip. Because of the propagation delay in the flip flops in ripple counters, the flip flops are not clocked at exactly the same instant. Instead, there is a small delay (usually tens of nanoseconds) in the clocking of each flip flop. As a result, a ripple effect appears on the Q outputs of the counter. Normally, this ripple is tolerated by other circuitry connected to the counter; however, there are some applications in which ripple can cause external circuitry to read the counter value erroneously and create errors.

Assuming that the counter has been initially reset (counter output 0000_2, or state 0_{10}), when the first clock pulse occurs, $FF0$ will toggle (0001_2, or state 1_{10}). This causes the $\overline{Q}$ output of $FF0$ to switch to zero. Because $FF1$ is positive edge triggered, this has no effect on $FF1$. On the next clock pulse, $FF0$ switches to the zero state and its $\overline{Q}$ output switches high, which triggers $FF1$, causing it to switch to the one state (0010_2, or state 2_{10}). The next clock pulse again sets $FF0$ (0011_2, or state 3_{10}). When the next clock pulse occurs, $FF0$ will switch to the zero state, causing $FF1$ to also switch to the zero state. This sends a positive-going clock pulse to $FF2$ when sets the flip flop (0100_2, or state 4_{10}). This process continues until the flip flops are all set (1111_2, or state 15_{10}). On the next clock pulse, all flip flops switch back to the zero state and the process starts again. This count sequence is illustrated in timing diagram format in Figure 11-23.

Synchronous Counters

As with asynchronous counters, synchronous counters can be designed to produce any count pattern and count in any number base. Figure 11-24 shows a 4-bit synchronous binary counter. It produces the same timing diagram as that of the binary ripple counter (Figure 11-23); however, unlike the ripple counter, when the timer advances, all of the Q outputs that change state do so at the same instant because all of the flip flops in the counter are triggered by a common clock signal. In this counter, the carry from $FF1$ to $FF2$ is produced by an AND gate G_1, and the carry from $FF2$ to $FF3$ is produced by an AND gate G_2. The AND gates produce the carry signal well before the next clock signal. Because these two gates tend to "look ahead" to provide the carry signal ahead of the clock, this technique of generating a carry is sometimes called a **look-ahead carry**.

Both asynchronous and synchronous counters are available as both **up-counters** and **bidirectional counters** (also called **up/down counters**). The most common number bases

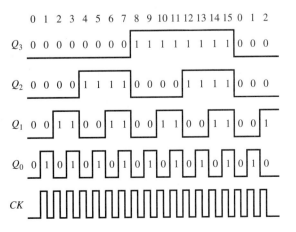

FIGURE 11–23
4-Bit binary ripple counter timing diagram.

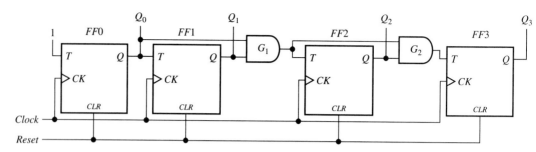

FIGURE 11–24
4-Bit synchronous binary counter.

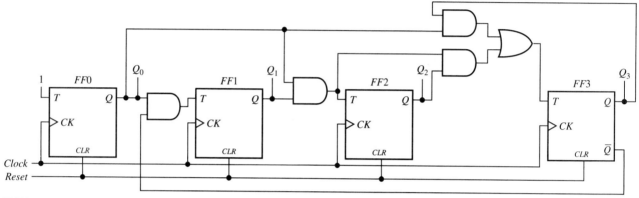

FIGURE 11–25
Synchronous BCD counter.

for counters are binary and decimal (BCD). Figure 11-25 is an **asynchronous BCD counter**. In this case, additional gates are added so that when the counter reaches a count of 9_{10} (1001_2), the appropriate carry signals are generated to load zeros into all the flip flops on the next clock pulse.

11-8 Memories

Earlier in the chapter, we investigated storage registers that can be used to store nibbles, bytes, and words of data. However, when the need arises to store large amounts of data, the use of individual flip flops or registers becomes somewhat cumbersome because it is difficult to manage the many input and output signals when large numbers of flip flops are interconnected. Therefore, for large storage applications, we generally arrange registers in arrays in which the inputs, outputs, and clock signals can be managed more easily using decoders. Additionally, by using large-scale integration fabrication techniques to construct the integrated-circuit memory, we can place extremely large amounts of storage space on single chips.

Memories are arranged in an array of bits, with each memory location containing 1, 4, 8, or 16 bits. The arrangement of the memory array generally is specified in the name of the device. For example, a 128 × 8 (pronounced, "128 by 8") memory would have 128 unique memory locations (addresses) each containing 8 bits. For this device, whenever a particular memory address is selected, the chip outputs 8 bits of parallel data. Memory sizes generally occur in multiples of 2. Any memory size above 512 is specified in "k" addresses, with 1k = 2^{10} = 1024 locations. The number of address pins on a memory chip is the $\log_2$ of the number of memory addresses. For example, a 2k × 8 memory would have $\log_2$ (2048), or 11 address pins, and 8 output data pins. Occasionally, memory chips are simply specified in the total number of individual bits. When this is done, it is up to the user to examine the number of data pins and determine the array arrangement.

EXAMPLE 11-1

A memory chip has 14 address pins and 4 data pins. Determine (a) the number of address locations, and (b) the memory array arrangement.

SOLUTION

(a) The number of address locations is $2^{14} = 16,384 = 16\,k$.

(b) This is a $16\,k \times 4$ memory.

EXAMPLE 11-2

A 64k-bit memory chip has 16 data pins. Determine (a) the number of address pins, and (b) the memory array arrangement.

SOLUTION

(a) The number of addresses in the memory is (64k locations) ÷ (16 data bits) = 4k. The number of address lines is $\log_2(4096) = 12$.

(b) This is a $4k \times 16$ memory.

Fundamentally there are three types of memory devices, the **read only memory**, or **ROM**, the **static random access memory**, or **SRAM**, and the **dynamic random access memory**, or **DRAM**. As we will see, they each have characteristics that make them uniquely suited for specific applications.

Read Only Memory (ROM)

The read only memory contains permanent (or sometimes semipermanent) data. Once it is programmed with data, the data are nonvolatile; that is, the ROM retains its data even when power is removed from the device. Since the data can only be read, ROM is ideal for storing microprocessor programs that do not need to be changed and need to be available immediately when power is applied, such as the program for a microwave oven, cellular phone, or hand-held calculator. The different types of ROMs are differentiated by the method used to store data and how easily the data can be changed.

Mask-Programmed ROM

This type of ROM is programmed to the customer's specifications by the chip manufacturer. Figure 11-26 illustrates a ROM that has four rows each having four data bits. One

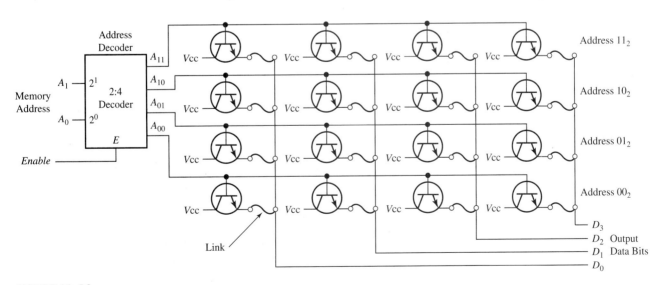

FIGURE 11–26
4×4 Read only memory (ROM).

and only one row of transistors is selected at any time by the address decoder. When the row is selected, a bias voltage is applied to all the transistors in the row, thereby switching on each of the transistors. If the emitter of any transistor has a link present, its emitter is connected to its respective output data line, which causes the data line to output a logical one. However, if the link is missing, the transistor is disconnected from the data line and it outputs a logical zero. In a mask-programmed ROM, the customer sends the manufacturer a data file containing the desired data, and the internal links are added when the memory chip is manufactured. The tooling costs to produce a mask-programmed ROM are relatively high, so this type of memory is best suited for high production volumes in which the tooling costs can be distributed over many chips. Mask-programmed ROMs cannot be reprogrammed.

Programmable ROM (PROM)

The PROM is very similar to the mask-programmed ROM except that when the chip is manufactured, all the links are included as thin fusible-link wires. Then, using a special programmer called a "PROM burner," the customer programs the chip with the desired data. The PROM burner applies a high voltage and high current to the individual fusible links in the chip and actually blows the links to create the desired zeros. Any remaining links that are not blown create the desired ones. Once programmed, PROMs cannot be reprogrammed. PROMs are excellent for low-volume applications. Because the customer can program the PROM, it is not necessary to pay a large set-up fee to the memory manufacturer, and the contents of the PROM can be changed quickly, simply by removing and discarding the chip and burning a replacement with the new data.

Electrically Programmable ROM (EPROM)

The internal construction of the EPROM is quite different from that of the ROM or PROM. Binary ones are stored in the EPROM as electrical charges (or, for the zeros, an absence of electrical charges) on internal capacitors. The insulating quality of the internal capacitors in the EPROM is very good, so the charges do not dissipate. Therefore, once the EPROM is programmed, it is capable of storing the data for several years, even 10 years or more. The main advantage of the EPROM over the ROM and PROM is that EPROMS can be easily erased and reused. EPROMs are manufactured with a clear quartz window covering the integrated circuit die inside the device. The EPROM in Figures 11-27(a) and (b) is a 256-kbit EPROM (32 k × 8) illustrating the quartz window above the die. When the die is exposed to a strong ultraviolet light, the charges stored in the capacitors are removed, and the chip is erased. Special UV EPROM erasers are available that produce a strong UV light of the proper wavelength to efficiently erase an EPROM in minutes. Because of the EPROM's sensitivity to UV light, care must be taken once a chip is programmed to prevent accidental exposure to UV light, especially that from fluorescent lights and sunlight. Normally, immediately after programming an EPROM, an opaque gummed label is placed over the quartz window. Then, if the chip needs to be erased later, the label is removed, the

FIGURE 11–27
(a) EPROM, (b) top view.   (a)   (b)

quartz window is cleaned with alcohol, and the chip is exposed to UV light. EPROMs are ideal for low-volume applications in which the data might often need to be changed. For example, they are ideal for storing the program in a prototype microprocessor system.

Electrical Erasable Programmable ROM (EEPROM)

The EEPROM functions like the EPROM except that, as the name implies, the chip can be electrically erased by applying a special voltage to certain pins of the chip. Obviously, this is much more convenient than the UV erasing method employed with EPROMs. Additionally, no special equipment is required to erase and reprogram the devices, so they can be reprogrammed in-circuit (that is, without removing them from the circuit). EEPROMS are extremely popular as memory devices in applications where it is desirable for the data to be nonvolatile but yet easily changeable. For example, EEPROMs are used to retain the favorite channel settings in a television receiver and store phone numbers in cellular phones. In these examples, when a television receiver is unplugged from the power line, or when the battery is removed from a cellular phone, the data are still available when power is restored.

Static Random Access Memory (SRAM)

SRAM memory devices internally store data in individual flip flops. Data can be written into any memory location by applying the proper binary address to the address lines, applying the data to be written to the data lines, and pulsing the "write" input pin on the device. Data can be read from the memory in any desired order simply by putting the device in read mode (that is, *not* pulsing the "write" pin) and applying the proper logical signals to the address lines. As with all devices containing flip flops, as long as power remains applied to the device, the data stored in it will be retained. However, if power is removed, all data in the chip will be lost. The flip flops are arranged in arrays so that they can be accessed in a manner similar to ROMs. However, flip flops typically require 30 to 50 transistors each. That limits the number of flip flops that can be packed onto an integrated-circuit die. For this reason, SRAMs are limited to relatively small memory sizes.

Dynamic Random Access Memory (DRAM)

DRAMs store data in a manner similar to that used by EPROMs; that is, each data bit is stored in a capacitor as a charge (a one) or lack of charge (a zero). Because a single capacitor and a few surrounding control transistors consume much less die space than a flip flop, it is possible to pack an extremely large number of bits onto a single integrated-circuit die when compared to an SRAM. For this reason, the DRAM is the memory of choice for large computer memories. However, to achieve the necessary speeds and large memory sizes required of computer memories, the capacitors in DRAMs are very small (that is, they have a very low capacitance) and are closely packed, making them more susceptible to data loss caused by leakage currents. As a result, any data stored in a DRAM will be lost because of capacitor discharge in a few tenths or hundredths of a second. Therefore, additional logical circuitry must be added to DRAM memories that will continuously refresh the stored data. This is done by a continuous process of reading the data in each memory address and then rewriting the data (called **memory refresh**). Although memory refresh seems to be an expensive and cumbersome process, the reward of having an extremely large memory in a very small space is well worth the expense and trouble.

PROBLEMS

11-1 An *RS* latch is initialized with $Q = 0$, $R = 0$, and $S = 0$. If $S = 1$, what will be the value of Q?

11-2 An *RS* latch is initialized with $Q = 0$, $R = 0$, and $S = 0$. If $R = 1$, what will be the value of Q?

11-3 A $\overline{R}\,\overline{S}$ latch is initialized with $Q = 0$, $R = 1$, and $S = 1$. If $S = 0$, what will be the value of Q?

11-4 A $\overline{R}\,\overline{S}$ latch is initialized with $Q = 0$, $R = 1$, and $S = 1$. If $R = 0$, what will be the value of Q?

11-5 In one sentence, explain the difference between a level-triggered flip flop and an edge-triggered flip flop.

11-6 In one sentence, explain the difference between a positive edge-triggered flip flop and a negative edge-triggered flip flop.

11-7 A positive edge-triggered D flip flop is initialized with $Q = 0$ and $D = 1$. If the clock input switches from a low to high level, what will be the value of Q?

11-8 A negative edge-triggered D flip flop is initialized with $Q = 0$ and $D = 1$. If the clock input switches from a low to high level, what will be the value of Q?

11-9 A positive edge-triggered T flip flop is initialized with $Q = 1$ and $T = 1$. If the clock input switches from a low to high level, what will be the value of Q?

11-10 A positive edge-triggered T flip flop is initialized with $Q = 1$ and $T = 0$. If the clock input switches from a low to high level, what will be the value of Q?

11-11 A positive edge-triggered JK flip flop is initialized with $Q = 1$, $J = 0$, and $K = 0$. If the clock input switches from a low to high level, what will be the value of Q?

11-12 A positive edge-triggered JK flip flop is initialized with $Q = 1$, $J = 1$, and $K = 0$. If the clock input switches from a low to high level, what will be the value of Q?

11-13 An 8-bit shift register is loaded with the binary number 10001001_2. If the register shifts the number left two times, what will be the binary value in the register?

11-14 An 8-bit shift register is loaded with the binary number 00110010_2, which is 137_{10}. What will be the decimal value in the register after (a) shifting left one place and (b) after shifting the original value right one place? (c) Based on your results, what general statement can you say about the effect of shifting a number left and right?

11-15 An 8-bit binary up-counter contains the number 01010011_2. What will be the result after the counter is clocked three times?

11-16 An 8-bit binary down-counter contains the number 01010011_2. What will be the result after the counter is clocked three times?

11-17 A 4-bit BCD up-counter contains the number 0111_{BCD}. What will be the result if the counter is clocked four times?

11-18 An 8-bit BCD down-counter contains the number 01_{10}. What will be the decimal result if the counter is clocked three times?

11-19 A 64k-bit ROM has 8 data pins. How many address pins does it have?

11-20 A 64k $\times$ 8 PROM has how many address pins?

11-21 State an advantage of using an SRAM when compared to a DRAM.

11-22 State an advantage of using a DRAM when compared to an SRAM.

POWER SYSTEM FUNDAMENTALS

12 Magnetic Circuits

13 Three-Phase Circuits

14 Transformers

Magnetic Circuits

OVERVIEW AND OBJECTIVES

All electrical machines (transformers, motors, and generators) operate using the principles of electromagnetics. Fundamentally, there are two ways that electromagnetics plays a part in the operation of electrical machines. First, by forcing an electric current through a coil of wire, we can produce a magnetic field, and second, by passing a magnetic field through a coil of wire (or vice versa), we can induce a voltage and corresponding current in the wire. In either case, it is this interplay between a magnetic field and the voltage (and current) in a coil of wire that is the underpinning for the operation of all electrical machines. To understand fully how electrical machines operate, what their limitations are, and how to select the appropriate machine for an application, it is vital that we first understand the fundamentals of electromagnetism and magnetic circuits.

Objectives

After completing this chapter, you will be able to

- Demonstrate an understanding of the principles of magnetism and electromagnetics.
- Show how a solenoid is constructed and how it operates.
- Calculate the magnetic flux, magnetomotive force, and flux density in a closed core.
- State the relationships among reluctance, absolute permeability, and relative permeability.
- Explain the relationship between magnetic field intensity and flux density.
- Read a *B-H* curve and find the linear and saturation regions on the curve.
- Explain why core saturation is generally detrimental to solenoid performance.

12-1 Magnetism and Electromagnetism

Although engineers and physicists have made great strides in recent years toward a further understanding of what magnetism really is and how better to use its properties, we still do not understand fully the phenomenon. However, nearly all the principles surrounding magnetism, electromagnetism, and the operation of electric machines were understood more than a century ago. It is ironic that, although in those days no one understood magnetism at the molecular and atomic levels, engineers and scientists could apply the effects of magnetism to design and construct electrical machines. Over the past few decades, we have improved upon these concepts to build smaller, more powerful, and more efficient machines, but the basic design concepts, laws, and equations developed long ago are still used today.

Magnetism is present predominantly in **ferrous** materials; that is, materials containing iron. It occurs when the atoms in the material align themselves so that the direction of electron spin of most the atoms is the same, thus creating north and south magnetic poles in the material. There are certain known conditions which can cause the magnetic alignment of the atoms to change. For example, if we pass a ferrous material through a magnetic field, it will tend to align the atoms and magnetize the material, thus creating a permanent magnet. The strength of the resulting magnet depends upon the strength of the magnetizing field, the length of time the ferrous material is exposed to the magnetizing field, and the ability of the ferrous material to retain its magnetism (called **retentivity**[1]).

We can also demagnetize materials that have been magnetized. If a magnetized ferrous material is heated to a high temperature called the Curie point (770°C for iron), or if we strike the material with a hammer, it will scramble the atomic alignment of the material and demagnetize it. It is not uncommon to accidentally "kill" a magnet by dropping it on a hard floor. We can also demagnetize a material if we expose it to an alternating magnetic field of continuously decreasing strength, a process called **degaussing**. Demagnetization is not permanent; ferrous materials can be repeatedly magnetized and demagnetized with no degradation in performance (this characteristic is used extensively in all electrical machines).

12-2 Flux

In addition to the naturally occurring magnetism in ferrous materials, it is also possible to create magnetism electrically. Whenever an electric current flows in any material (including air), magnetic field lines (called **flux**) are created in a circular fashion around the current. The field lines have a direction that is clockwise when viewed from the source of the current. To help remember this, we sometimes use what is called the **right-hand rule for flux**. If the right hand is wrapped around a conductor with the thumb pointing in the direction of the current, the four fingers will point in the direction of the magnetic flux produced by the current. The magnetization direction is important because it controls such things as the direction a motor turns and the polarity of the output voltage of a transformer.

Permanent magnets are made from high retentivity materials. However, it is also possible to make a low-retentivity ferrous material act as a magnet by winding a coil of wire around the material and connecting the wire to a direct current source. This device, shown in Figure 12-1, is called a **solenoid** or electromagnet. Solenoids have three basic characteristics that make them useful.

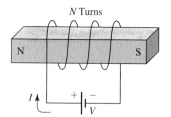

FIGURE 12–1
Solenoid (electromagnet).

- Because the material used (called the **core material**, or simply **core**) has a low retentivity, when the electrical source is disconnected from the coil, the device ceases to be an electromagnet; that is, the magnetic field becomes nearly nonexistent. Therefore, it is possible to switch an electromagnet on and off electrically. This characteristic is used in devices such as automotive electric door locks and in electric water fill valves in clothes-washing machines.

- The orientation of the north and south poles of the electromagnet is determined by the direction of current around the core. Therefore, we can reverse the north-south pole orientation of our electromagnet simply by reversing the electrical connections of the coil. If we place the electromagnet near a permanent magnet, we can have it either attract or repel the magnet as desired, simply by reversing the coil wires. This characteristic is used in reversible motors such as electric hand drills.

---

[1] The retentivity of a material is a function of the type of material. To make permanent magnets, we use core materials that have a high retentivity.

- The magnetic strength of an electromagnet is proportional to the current in the coil. Therefore, we can electrically control the mechanical force the electromagnet exerts on other surrounding ferrous metals. This principle is used to adjust the levitation height of magnetically levitated (MAGLEV) trains, to control the speed of a dc motor, and to set the deflection of pointer-style meters such as fuel gauges in automobiles. Also, if we use an alternating current in the coil with varying amplitude and frequency, we can control the magnitude and frequency of vibration that the electromagnet exerts on surrounding ferrous materials. This principle is used in audio loudspeakers and in most headphones.

Whenever the coil of a solenoid is energized, it produces magnetic field lines (or flux) in the core and sets up magnetic north and south poles at the ends of the solenoid. The flux in the core makes a complete loop; that is, it travels through the core material from south to north, exits the core at the north end, travels through the air outside the core, and re-enters the core on the south end. The number of lines of flux inside the core is constant everywhere in the core. Also, there is no loss of field lines anywhere in the magnetic circuit. We could indicate the magnitude of the magnetic flux by measuring the number of lines. However, in most electrical machines, the flux lines number in the millions or higher, so to avoid working with equations containing very large numbers, we use the unit weber (Wb), which is equivalent to 10^8 lines of magnetic flux. In mathematical equations, we represent magnetic flux with the Greek letter phi, Φ.

12-3 Flux Density

When working with electrical machines, it is also necessary to know the concentration of flux in a core material. This is called the **flux density**. It is the flux per unit of cross-sectional area. Flux density is represented by the letter B, and the unit of measure is the weber/m^2. One weber/m^2 equals one tesla (T). Therefore, we have the definition

$$B = \frac{\Phi}{A} \tag{12-1}$$

where

B = flux density in webers/m^2 (Wb/m^2) or teslas (T)

Φ = flux in webers (Wb)

A = cross-sectional area in meters^2

EXAMPLE 12-1 The core of a solenoid similar to the one shown in Figure 12-1 measures 1 cm by 1 cm by 10 cm and has a flux of 3.0 Wb. What is the flux density?

SOLUTION Because flux density is measured in Wb/m^2, we must convert the dimensions into meters. 1 cm is 0.01 m, so our cross-sectional area is

$$\begin{aligned}A &= 0.01 \times 0.01 \\ &= 0.0001 \text{ m}^2\end{aligned} \tag{12-2}$$

Now we find the flux density, which is

$$B = \frac{\Phi}{A} = \frac{3.0}{0.0001} = 30{,}000 \text{ Wb/m}^2 \tag{12-3}$$

For this problem, the length of the core (10 cm) has no bearing on the results.

12-4 Magnetomotive Force

In a solenoid, the magnitude of the flux is determined by the type of core material, the length of the magnetic circuit, the current in the coil, and the number of turns of wire in the coil. However, the coil is the device producing the magnetic field, so the magnetic force (called the **magnetomotive force**, or **mmf**) producing flux in the solenoid is determined only by the coil current and number of turns in the coil. In fact, the magnetomotive force is equal to the product of the number of turns in the coil (N) and the current (I) in the coil and is measured in the unit ampere-turns (A-t). Therefore, we can say

$$mmf = NI \tag{12-4}$$

where

mmf = magnetomotive force in ampere-turns (A-t)

N = number of turns of wire (t)

I = coil current in amperes (A)

EXAMPLE 12-2

Referring to the solenoid in Figure 12-1, if the battery voltage V is 12 V dc, the coil resistance is 30 Ω, and the coil has 350 turns of wire, what is the magnetomotive force?

SOLUTION To determine the magnetomotive force, we will first need to calculate the current using Ohm's law.

$$I = \frac{V}{R} = \frac{12}{30} = 0.4 \text{ A} \tag{12-5}$$

Next, the magnetomotive force is

$$mmf = NI = 350 \times 0.4 = 140 \text{ A-t} \tag{12-6}$$

EXAMPLE 12-3

Referring to Figure 12-1, if the 500-turn coil has a resistance of 125 Ω, what battery voltage is needed to produce a magnetomotive force of 100 A-t?

SOLUTION Rearranging the mmf equation, the required current is

$$I = \frac{mmf}{N} = \frac{100}{500} = 0.20 \text{ A} \tag{12-7}$$

Therefore, the battery voltage is

$$V = IR = 0.20 \times 125 = 25 \text{ V} \tag{12-8}$$

12-5 Reluctance

As mentioned earlier, the flux in a solenoid is determined by the type of core material, the length of the magnetic circuit, and the magnetomotive force produced by the coil. Different core materials exhibit different resistances to flux, which we call **reluctance**. Cores with high reluctances are poor magnetic conductors. In general, most electrical machines use core materials with low reluctances; that is, those that are good magnetic conductors. In equations, we use a script ($\mathcal{R}$) to designate reluctance, and the unit of measure is A-t/Wb. The flux in a core is directly proportional to the magnetomotive force and inversely proportional to the reluctance. Therefore, we can mathematically say that the flux in a core is

$$\Phi = \frac{mmf}{\mathcal{R}} = \frac{NI}{\mathcal{R}} \tag{12-9}$$

where

$$\Phi = \text{flux in webers (Wb)}$$
$$mmf = \text{magnetomotive force in ampere-turns (A-t)}$$
$$\mathcal{R} = \text{reluctance in ampere-turns/weber (A-t/Wb)}$$

■ EXAMPLE 12-4

A solenoid has a flux of 320 Wb when the magnetomotive force of its 150-turn coil is 80 A-t. (a) What is the reluctance of the core? (b) What is the coil current?

SOLUTION

(a) Rearranging the reluctance equation, the core reluctance is

$$\mathcal{R} = \frac{mmf}{\Phi} = \frac{80}{320} = 0.25 \text{ A-t/Wb} \tag{12-10}$$

(b) The coil current is

$$I = \frac{mmf}{N} = \frac{80}{150} = 0.533 \text{ A or 533 mA} \tag{12-11}$$

It is relatively easy to remember the equations that use magnetic flux, magnetomotive force, and reluctance by identifying the analogy between magnetic circuits and electrical circuits. That is, there is a correspondence between reluctance and resistance, flux and current, and magnetomotive force and voltage. Both systems are governed by Ohm's law. Therefore, we can write the Ohm's law equations for both electrical and magnetic circuits, which are

$$I = \frac{V}{R} \quad \text{and} \quad \Phi = \frac{mmf}{\mathcal{R}} \tag{12-12}$$

12-6 Determining Magnetic Field Direction of an Electromagnet

It was mentioned earlier that the north and south ends of a solenoid are determined by the direction of current around the solenoid core. There is a method for finding the north and south ends of a solenoid called the **right-hand rule for solenoids**. In this case, when the right hand is wrapped around the solenoid so that the four fingers are pointing in the direction of current flow around the core, the thumb will be pointing in the direction of the north pole of the solenoid.

12-7 Closed (Circular) Cores

Every coil of wire has inductance. The magnitude of the inductance is directly proportional to the number of turns of wire in the coil and inversely proportional to the total reluctance and the length of the magnetic circuit. Air-core inductors (simply a coil of wire with no core) have low relative inductance because the air through which the flux has to travel to make a complete circuit has an extremely high reluctance. If more inductance is needed, we can wind the coil over a ferrous bar, thus constructing a solenoid. Although this has a higher relative inductance than an air-core inductor, part of the magnetic field path still has to travel through air. We can construct an inductor with a very high relative inductance if we shape the solenoid core in a circle or rectangle so that the north and south ends meet, making a circular or closed core as shown in Figure 12-2. The closed core inductor has no north or south poles. The magnetic field makes a complete circuit inside the core material, which has a very low reluctance when compared to that of air.

FIGURE 12–2
Closed-core inductor.

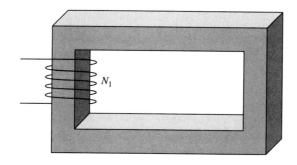

12-8 Magnetic Field Intensity

The magnetomotive force produces the flux in the core of a closed-core inductor. However, assume that we have two cores as shown in Figure 12-3 with the same cross-sectional area A and the same core material, but with different mean lengths ℓ_1 and ℓ_2 around the cores. To produce an equivalent amount of flux in the two cores, more magnetomotive force is required in core (b) than the core (a). The reason is that the longer core in (b) has more reluctance. Equation 12-9 shows that if we want to maintain the flux Φ in a core that has a higher reluctance, we also need a proportionally higher magnetomotive force. To define the strength of a magnetic field in a core, we use the term **magnetic field intensity**, H, with unit of measure ampere-turns/meter (A-t/m). The magnetic field intensity is directly proportional to the magnetomotive force and inversely proportional to the mean length of the core, or

$$H = \frac{mmf}{\ell} = \frac{NI}{\ell} \tag{12-13}$$

where

H = magnetic field intensity in ampere-turns/meter (A-t/m)

mmf = magnetomotive force in ampere-turns (A-t)

ℓ = mean length of the core in meters (m)

N = number of turns on the coil

I = coil current in amperes (A)

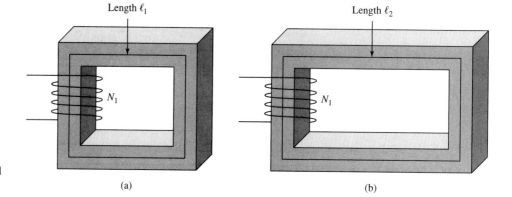

FIGURE 12–3
Two closed-core inductors with (a) short mean length, and (b) long mean length.

12-9 Permeability

We can use reluctance in our magnetic calculations to find the flux and flux densities. However, if we change the physical dimensions of the core material (the mean length or the cross sectional area), the reluctance will change. Long, thin cores have higher reluctance than short, thick ones. Therefore, it is helpful to instead define a magnetic "goodness" characteristic that is constant for a given type of core material and is independent of the material's physical dimensions. This unit is called the permeability, μ, and the unit of measure is webers/ampere-turns-meter (Wb/A-t-m). Note that the reluctance of a core is inversely proportional to the permeability of the core material, which also means that materials with low reluctance have a high permeability and are good magnetic conductors. In fact, the reluctance of any core can be calculated by knowing the permeability of the core material and the physical dimensions of the core, with the equation

$$\mathcal{R} = \frac{\ell}{\mu A} \qquad (12\text{-}14)$$

where

$\mathcal{R}$ = reluctance in ampere-turns/weber (A-t/Wb)

l = mean length of the core in meters (m)

μ = permeability in webers/ampere-turns-meter (Wb/A-t-m)

A = cross-sectional area of the core in meters^2 (m^2)

The permeability of a material is also the ratio of the flux density to the magnetic field intensity, or

$$\mu = \frac{B}{H} \qquad (12\text{-}15)$$

where

μ = permeability in webers/ampere-turns-meter (Wb/A-t-m)

B = flux density in webers/m^2 (Wb/m^2), or teslas (T)

H = magnetic field intensity in ampere-turns/meter (A-t/m)

When dealing with core materials, occasionally we encounter instances where there are intentional gaps in a core where the magnetic field must travel through air. We use these air gaps to control the reluctance and inductance of the device and occasionally to tune the resonance of circuits in which it is connected. In these cases, permeability and reluctance equations must include the permeability of the air gap. Air has a permeability of $4\pi 10^{-7}$ Wb/A-t-m. It is designated by the variable μ_0. To make our calculations easier, we generally indicate the permeability of core materials as the relative permeability μ_r, where μ_r is the ratio of the absolute permeability of the material μ with respect to μ_0. Because μ_r is a ratio of two quantities with the same unit of measure, μ_r is dimensionless.

$$\mu_r = \frac{\mu}{\mu_0} \qquad (12\text{-}16)$$

where

μ_r = relative permeability (dimensionless)

μ = absolute permeability in webers/ampere-turns-meter (Wb/A-t-m)

μ_0 = permeability of air, $4\pi 10^{-7}$ Wb/A-t-m

The permeability of nonferrous materials remains constant under all conditions of magnetic field intensities and flux densities. However, the permeability of ferrous materials is not constant. At relatively low levels of magnetic field intensity, ferrous materials exhibit the highest permeability. At these levels, a small increase in H results in a relatively large increase in the flux density B. However, as the magnetic field intensity is increased, the permeability of the material actually decreases. This trend continues until we reach a point where a large increase in H results in little or no increase in B. This point is called **core saturation**. It occurs when nearly all of the atoms in the material are magnetically aligned, leaving few unaligned atoms to provide any further increase in flux. To identify the area of

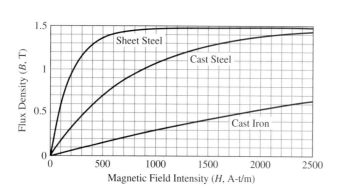

FIGURE 12–4
Saturation curves for steel.

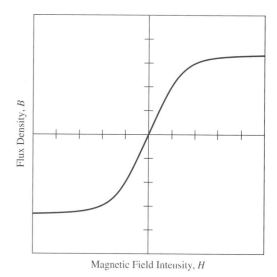

FIGURE 12–5
Bipolar saturation curve.

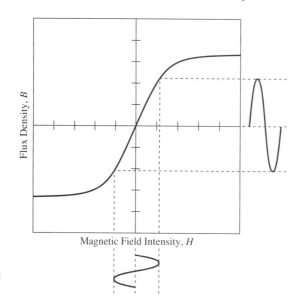

FIGURE 12–6
B-H Relationship with sinusoidal excitation.

core saturation, designers of magnetic devices often use a plot of the flux density in magnetic device design, a plot of the flux density B with respect to the magnetic field intensity H for a particular core material. This plot, as shown in Figure 12-4, is called a **B-H curve**, **saturation curve**, or **magnetization curve**. Notice that as the magnetic field intensity H increases, we gradually approach the core saturation point, the point where the flux density does not increase very much. Beyond this point the material has saturated and the permeability drops drastically. As will be discussed later in this chapter, core saturation can cause degradation of performance in dc machines, and it can have grave results in ac-operated machines. Using this graph, we can find the permeability at any value of H simply by finding the corresponding value for B and using Equation 12-15 to calculate μ.

Remember that, with respect to magnetization, core materials are bidirectional; that is, they exhibit the same B-H curve in both the positive and negative direction. Notice in Figure 12-5 that the B-H curve in the first quadrant of the graph is mirror imaged in the third quadrant. This means that if we reverse the connection of the coil of our solenoid, the magnetic field intensity and flux density will reverse. In a similar manner, Figure 12-6 shows that if we apply a sinusoidal voltage to our solenoid coil, the magnetic field intensity and

FIGURE 12–7
B-H Relationship with core saturation.

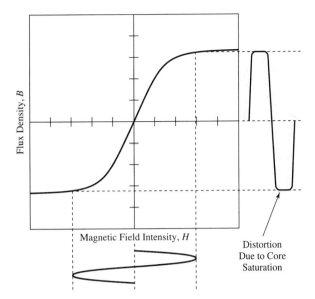

flux density also will be sinusoidal (about the origin), as long as the peak value of the magnetic field intensity does not push the core beyond the saturation point.

However, if we apply a sinusoidal voltage to the coil of the solenoid that is higher in amplitude than the rated voltage of the device, although the magnetic field intensity H still will be sinusoidal, the flux density B will not. Instead, as shown in Figure 12-7, when the core saturates, it flattens the peaks of the flux density. Generally speaking, this is an undesirable effect because when the core saturates, the inductance drops drastically. However, as we will see in our later study of transformers, in a few applications, saturating a magnetic core produces an effect that has practical applications.

EXAMPLE 12-5

A solenoid similar to the one shown in Figure 12-3(a) has a core made of cast steel (refer to Figure 12-4 for the cast steel saturation curve). It has a cross-sectional area of 4 cm² and a mean length of 20 cm. The coil has 100 turns, and the coil current is 3.2 amperes. Find the following: (a) magnetomotive force mmf, (b) magnetic field intensity H, (c) flux density B, (d) the permeability μ, (e) the relative permeability μ_r, (f) the core reluctance $\mathcal{R}$, and (g) the flux Φ.

SOLUTION
(a)
$$mmf = NI = 100 \times 3.2 = 320 \text{ A-t} \tag{12-17}$$

(b)
$$H = \frac{mmf}{\ell} = \frac{320}{0.2} = 1600 \text{ A-t/m} \tag{12-18}$$

(c) Referring to Figure 12-4 for cast steel, for $H = 1600$ A-t/m, $B = 1.3$ T.

(d)
$$\mu = \frac{B}{H} = \frac{1.3}{1600} = 8.125 \times 10^{-4} \text{ Wb/A-t-m} \tag{12-19}$$

(e)
$$\mu_r = \frac{\mu}{\mu_0} = \frac{8.125 \times 10^{-4}}{4\pi 10^{-7}} = 646.6 \tag{12-20}$$

(f)
$$\mathcal{R} = \frac{\ell}{\mu A} = \frac{0.2}{(8.125 \times 10^{-4})(0.0004)} = 615.4 \times 10^3 \text{ A-t/Wb} \quad (12\text{-}21)$$

(g)
$$\Phi = \frac{mmf}{\mathcal{R}} = \frac{NI}{\mathcal{R}} = \frac{320}{615.4 \times 10^3} = 5.2 \times 10^{-4} \text{ Wb} \quad (12\text{-}22)$$

PROBLEMS

12-1 The core of a solenoid has a magnetic flux of 2.5 Wb and a cross-sectional area of 1.5 cm². What is the flux density?

12-2 What is the flux in a solenoid that is 1.75 cm wide, 1.25 cm deep, and 20 cm long and has a flux density of 1,000 Wb/m²?

12-3 A 100-turn, 2.5-Ω coil is connected to a 10-V battery. What is its magnetomotive force?

12-4 What voltage should be applied to a 50-turn, 7-Ω coil to produce a magnetomotive force of 210 A-t?

12-5 The current in a 250-turn coil is 2.25 A, and it is wound on a core that has a reluctance of 0.5 A-t/Wb. What will be the resulting flux in the core?

12-6 A 125-turn coil is wound onto a core with a reluctance of 0.125 A-t/Wb. What coil current is required to produce a flux of 250 Wb?

12-7 A closed magnetic core has a mean length of 20 cm. When the core is excited by a magnetomotive force of 400 A-t, what will be the magnetic field intensity?

12-8 A 150-turn coil is wound onto a closed magnetic core that has a mean length of 12 inches. What coil current is required to achieve a magnetic field intensity of 2200 A-t/m?

12-9 What is the reluctance of a magnetic core that is 1.25 cm wide, 2.0 cm deep, and 15 cm long and has a permeability of 12.5×10^{-4} Wb/A-t-m?

12-10 A 1-cm-square core is 20 cm long and has a reluctance of 330×10^3 A-t/Wb. What is its permeability?

12-11 The saturation curve of a particular core material shows that at a magnetic field intensity of 1100 A-t/m the flux density is 1 Wb/m². Find (a) the absolute permeability of the core material and (b) the relative permeability of the core material.

12-12 What is the relative permeability of a magnetic core that has an absolute permeability of 2.3×10^{-3} Wb/A-t-m?

Three-Phase Circuits 13

OVERVIEW AND OBJECTIVES

All the ac circuits considered thus far have been so-called **single-phase circuits**. These are the forms most widely employed in electronic applications and are the forms most familiar to consumers. Indeed, most residential power circuits are of the single-phase type.

In large power applications such as those encountered in industrial plants, there are significant advantages in using **polyphase** power circuits; that is, a combination of several sources that have different phase angles. By far the most widely employed polyphase system is that of a **three-phase** power distribution system, which is equivalent to three separate generators, with the three phase angles equally spaced over a cycle. As we will see in this chapter and subsequent chapters, three-phase systems provide improvements in efficiency, smaller size, simplicity, and lower noise levels compared to single-phase systems.

In this chapter, we will begin a study of three-phase systems by first investigating the circuit forms and their analysis. In particular, we will show how measurements of voltage, current, and power are made. This development continues the development of ac circuit analysis considered in earlier chapters and is a necessary prerequisite to the study of ac machines in Chapter 16.

Objectives

After completing this chapter, you will be able to

- Demonstrate the phase relationship among the three voltages or currents in a three-phase system.
- State the difference between line voltage and phase voltage and convert between them.
- Show how to reverse the phase sequence in a three-phase system.
- Show why the neutral current in a four-wire balanced three-phase system is zero.
- Explain the differences between a three-phase wye and a three-phase delta system.
- Calculate power in both wye and delta three-phase systems.
- Discuss some of the advantages of three-phase systems over single-phase systems.

13-1 Introduction to Three-Phase Theory

Almost all applications of three-phase systems occur in the power system area, and as explained in Chapter 5, it is customary in that area to express voltage and current phasor forms in terms of their rms or effective values. Of course, instantaneous forms require that

FIGURE 13–1
Three-phase system using three separate voltage sources with different phase angles, connected to a common neutral.

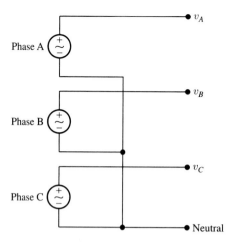

the coefficient of the sinusoidal voltage or current be a peak value. However, unless indicated otherwise, *all phasor voltage and current magnitudes will be expressed as rms values.*

Wye Configuration

Consider the circuit shown in Figure 13-1 with three ac voltage sources sharing a common ground or neutral. In power terminology, a common ground is often called the **neutral**. All three sources are assumed to have the same voltage magnitude and the same frequency, but each source has a different phase angle. These three sources could be three ac generators (which we will study later), three laboratory signal generators, or any other sinusoidal sources.

The three sources are named as Phase A, Phase B, and Phase C. The corresponding instantaneous voltages will be denoted as v_A, v_B, and v_C and the phasor voltages will be denoted as V_A, V_B, and V_C. The phase angle of Phase A will be assumed to have a reference angle of $0°$. Phase B will be assumed to lag Phase A by $120°$, and Phase C will be assumed to lag Phase A by $240°$. Based on these assumptions and with a peak value of V_p for each source, the instantaneous forms of the three voltages are

$$v_A = V_p \sin \omega t \tag{13-1}$$

$$v_B = V_p \sin(\omega t - 120°) \tag{13-2}$$

$$v_C = V_p \sin(\omega t - 240°) = V_p \sin(\omega t + 120°) \tag{13-3}$$

Note that the phase angle of $-240°$ in the complex plane is equivalent to a positive phase angle of $+120°$, and that the latter form is often more convenient to use in phasor analysis.

This connection of sources and assignment of voltages and phase angles creates a standard three-phase voltage source array and can be redrawn as a **wye configuration**. Figure 13-2 shows this form. It is the most common schematic representation of the wye circuit. If the neutral connection is momentarily ignored, the shape has the approximate form of an

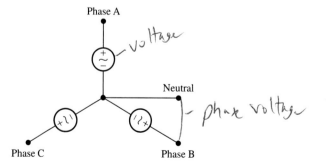

FIGURE 13–2
The three-phase system of Figure 13-1 rearranged to show that it is a wye configuration.

upside-down Y; hence, the name wye. Within the literature, there are various rotational representations of the wye, including one of the sources aligned along the *x*-axis, but they are all equivalent if the angles are displaced from each other by ±120°. The other three-phase connection is a **delta configuration**, which will be considered later in the chapter.

It is customary in three-phase power circuits to round off the values to three significant digits. Therefore, in the developments that follow, that practice will be followed in most cases.

EXAMPLE 13-1

One phase of a particular three-phase 60-Hz voltage source can be defined by the expression

$$v(t) = 170 \sin(2\pi \times 60t - 120°) = 170 \sin(377t - 120°) \qquad (13\text{-}4)$$

Find the (a) peak voltage, (b) rms voltage, (c) radian frequency, (d) cyclic frequency, and (e) phase angle.

SOLUTION This is actually a review of work covered in Chapters 4 and 5, but it is useful to discuss in the context of three-phase circuits.

(a) The peak voltage is the coefficient of the sine function and is

$$V_p = 170 \text{ V} \qquad (13\text{-}5)$$

(b) The rms voltage is

$$V_{\text{rms}} = \frac{V_p}{\sqrt{2}} = \frac{170}{\sqrt{2}} = 120 \text{ V rms} \qquad (13\text{-}6)$$

(c) The radian frequency is the factor of time in the argument of the function; i.e.,

$$\omega = 377 \text{ rad/s} \qquad (13\text{-}7)$$

(d) The cyclic frequency is

$$f = \frac{\omega}{2\pi} = \frac{2\pi \times 60}{2\pi} = 60 \text{ Hz} \qquad (13\text{-}8)$$

(e) The phase angle will be denoted as θ, and it is

$$\theta = -120° \qquad (13\text{-}9)$$

Equation 13-4 is the standard equation for phase B of a 208-V rms 60-Hz 3-phase line. The reason it is called a 208-V system will be developed in the next section.

Instantaneous Waveforms

The three voltages of Equations 13-1, 13-2, and 13-3 with $V_p = 170$ V are shown in Figure 13-3. The effective value of each voltage is 120 V rms. Notice that the graph shows each having the peak voltage of 170 V and the angular spacing of 120° between each waveform. For convenience, the abscissa of the graph is shown in units of degrees. However, if the frequency were specified, it would be simple to convert the scale to units of time.

Phasor Forms

As developed in the preceding two chapters, it is usually more convenient to work with the phasor forms for ac sinusoidal voltages. Remember that we will use rms units for the phasor magnitudes, as is customary in power applications. Therefore, the three instantaneous voltages of Equations 13-1, 13-2, and 13-3, based on a peak value of 170 V for each and

FIGURE 13–3
The three instantaneous voltage waveforms for a three-phase system based on each phase having a voltage of 120 V rms (peak value about 170 V).

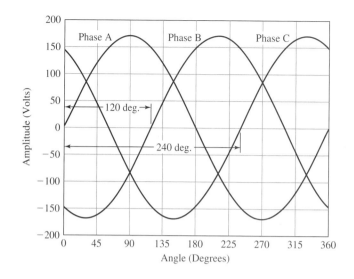

with a slight amount of roundoff, can be expressed in phasor forms as

$$\mathbf{V}_A = 120 \text{ V rms } \angle 0° \tag{13-10}$$

$$\mathbf{V}_B = 120 \text{ V rms } \angle -120° \tag{13-11}$$

$$\mathbf{V}_C = 120 \text{ V rms } \angle -240° = 120 \text{ V rms } \angle 120° \tag{13-12}$$

Avoid confusing the rms voltage of 120 V with the phase angle of 120° in expressions in which both appear. This is basically a coincidence of numbers, and there are other voltage values encountered in three-phase systems.

13-2 Phase Voltages and Line Voltages

The basic wye configuration considered in the last section is characterized by three different voltage sources, in which each voltage is measured with respect to the reference ground or **neutral**. The ideal situation corresponds to the three voltage magnitudes being equal, and that assumption has already been made and will continue to be made in subsequent developments.

Phase Voltages

The voltages of the three sources with respect to the neutral are called the **phase voltages**. The rms magnitude of each source will be denoted as V_{phase}. The three phasors can then be represented as

$$\mathbf{V}_A = V_{\text{phase}} \angle 0° \tag{13-13}$$

$$\mathbf{V}_B = V_{\text{phase}} \angle -120° \tag{13-14}$$

$$\mathbf{V}_C = V_{\text{phase}} \angle -240° = V_{\text{phase}} \angle 120° \tag{13-15}$$

The three phase voltages represented as phasors are shown in Figure 13-4.

Line Voltages

For the wye configuration, some important voltages are those measured as the difference between one phase voltage and another phase voltage. These voltages are actually line-to-line voltages, but for brevity, they are called simply **line voltages**.

There are actually three different line voltages that can be defined, and their magnitudes will be equal provided that the three phase voltages are equal and displaced apart by

FIGURE 13–4
Phasor representation of the three phase voltages.

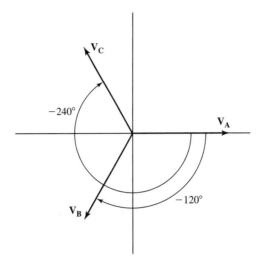

120°. Double-subscript notation will be employed with the first subscript representing the positive reference. The three values will be denoted as $\mathbf{V_{AB}}$, $\mathbf{V_{BC}}$, and $\mathbf{V_{CA}}$. They are defined as follows:

$$\mathbf{V_{AB}} = \mathbf{V_A} - \mathbf{V_B} \tag{13-16}$$
$$\mathbf{V_{BC}} = \mathbf{V_B} - \mathbf{V_C} \tag{13-17}$$
$$\mathbf{V_{CA}} = \mathbf{V_C} - \mathbf{V_A} \tag{13-18}$$

Recall that each of the voltages on the right-hand sides of the equations is measured with respect to a common neutral. Let's work out the first one to see the pattern. We will use exact fractional forms for the values to obtain an exact answer rather than one that has been rounded.

$$\begin{aligned}\mathbf{V_{AB}} &= \mathbf{V_A} - \mathbf{V_B} \\ &= V_{\text{phase}}\angle 0° - V_{\text{phase}}\angle -120° \\ &= V_{\text{phase}}(1+j0) - V_{\text{phase}}\left(-\frac{1}{2} - j\frac{\sqrt{3}}{2}\right) \\ &= V_{\text{phase}}\left(\frac{3}{2} + j\frac{\sqrt{3}}{2}\right) \\ &= V_{\text{phase}}\left(\sqrt{3}\angle 30°\right) \\ &= \sqrt{3}V_{\text{phase}}\angle 30°\end{aligned} \tag{13-19}$$

The reader can work out the values of the other two line voltages in the problems at the end of the chapter. Because of the symmetry, however, it should be clear that the rms values of the other two line voltages will be the same as that of $\mathbf{V_{AB}}$, and the only difference will be in the phase angles.

Let V_{line} represent the rms magnitude of the line voltage. We can readily see that it is related to the rms phase voltage by

$$V_{\text{line}} = \sqrt{3}V_{\text{phase}} = 1.732 V_{\text{phase}} \tag{13-20}$$

We can represent the line voltages graphically on a phasor diagram as shown in Figure 13-5. Notice that, as with the phase voltages, the line voltages are also spaced 120° apart, but they are skewed from the phase voltages by +30°. Many balanced three-phase loads are supplied using only the line voltages. When this is done, normally the neutral is

FIGURE 13–5
Phasor forms of the three line voltages formed from the phase voltages.

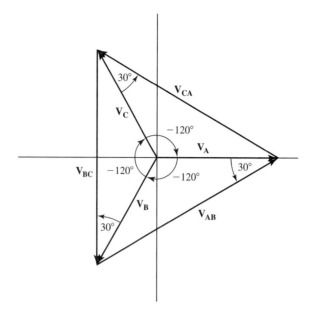

not used, creating what is called a **three-wire system**. A three-phase system that requires the neutral is called a **four-wire system**.

Several Important Conventions

At this stage, it is important to point out several standard electrical power engineering conventions regarding phase voltage and line voltage.

- When specifying the voltage of a three-phase system, we normally use line voltage, not phase voltage. This is the reason that the system of Example 13-1 was called a 208-V rms system, rather than a 120-V rms system, because $\sqrt{3} \times 120$ V rms = 208 V rms.

- Unless otherwise stated, we assume that the three line voltages in a three-phase system are equal. Likewise, we assume that all three of the phase voltages are equal. Therefore, by stating one line voltage, we imply the remaining two line voltages. This, in turn, implies the three phase voltages. For example, by stating that a system is three-phase 416 V rms, we mean that the line voltages are 416 V rms and the phase voltages are $416/\sqrt{3} = 240$ V rms.

- As previously noted, for power systems we usually round off the voltages to three significant digits. For example, a phase voltage of 120 V rms corresponds to a line voltage of 208 V rms (not 207.85 V rms), and a voltage of 38,175 V rms is generally stated as 38.2 kV rms.

▌▌ EXAMPLE 13-2

Four-wire three-phase power, rated at a phase voltage of 480 V rms, is delivered to a load using a four-conductor cable. Inside the cable, the four conductors are insulated from each other using PVC insulation. What is the highest voltage that the insulation must withstand?

SOLUTION The highest voltage between any two conductors in a three-phase four-wire system is the line voltage, not the phase voltage. The problem statement gave the phase voltage, so we must first convert it to line voltage.

$$V_{\text{line}} = \sqrt{3} V_{\text{phase}} = \sqrt{3} \times 480 = 831 \text{ V rms} \tag{13-21}$$

The highest instantaneous voltage drop between any two conductors will occur at the peak of the waveform. Therefore, we must find the peak value of the line voltage, which is

$$V_{\text{peak,line}} = \sqrt{2} V_{\text{line}} = \sqrt{2} \times 831 = 1176 \text{ V rms} \tag{13-22}$$

13-3 Three-Phase Sequencing

In most three-phase systems, the phases are connected so that B lags A, and C lags B. Therefore, the phase sequence is said to be ABC. When electrical power is connected to a three-phase motor, the direction of rotation of the motor is determined by the phase sequence. If we reverse the phase sequence to be CBA, the motor will rotate in the opposite direction.

In a three-phase system there are only two possible phase sequences, ABC and CBA. To see why this is true, consider a long sequence of the three phases . . . ABCABCAB-CABC. . . . Now choose any two of the three letters in the sequence and swap their positions. No matter which letters are chosen, reversing the positions will create the sequence . . . CBACBACBACBA. . . . Therefore, we can reverse the phase sequence of any three-phase system simply by swapping the connections of any two of the three wires.

Because the phase sequence determines the direction of rotation of three-phase motors, correct wiring of the three phases is critical. Incorrect wiring might cause a motor to turn backward, possibly damaging the motor and any machinery connected to its shaft. For this reason, a device called a **phase sequence relay** is generally installed on machines that can be damaged by reverse rotation. The phase sequence relay continuously monitors the phase sequence. If it senses an incorrect sequence, it disconnects the motor from the power line.

Although phase sequencing can be a problem in some situations, in cases where we wish to have a reversible motor, it makes the system design very convenient. By simply adding an electrical contactor that swaps two of the three power wires, we can easily make a three-phase motor reversible.

13-4 Three-Phase Wye and Delta Connections

There are two basic ways to connect three-phase power sources, the **wye** and the **delta** configurations. The **wye** connection has already been considered. Both forms will be compared in this section and the basic relationships for voltage and current will be provided.

The wye configuration is shown again in Figure 13-6(a) and the delta configuration is shown in Figure 13-6(b). As has been the practice throughout the chapter, the three voltage magnitudes will be assumed to be equal and the phase angles will be assumed to be separated by 120°. To simplify the development that follows, *only the voltage and current rms magnitudes are shown on the circuit diagrams and will be considered in the equations that follow.*

For each of the configurations, the rms magnitude of the phase voltage will be denoted as V_{phase} and the rms magnitude of the line voltage will be denoted as V_{line}. Likewise, the

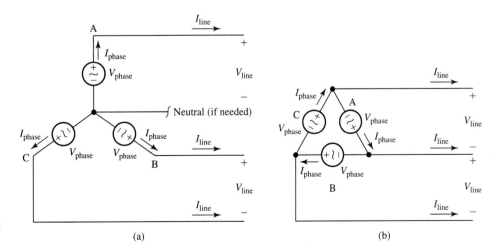

FIGURE 13–6
Three-phase source connections: (a) wye and (b) delta.

rms magnitude of the **phase current** will be denoted as I_{phase} and the rms magnitude of the **line current** will be denoted as I_{line}.

Wye Connection

Consider again the wye connection as shown in Figure 13-6(a). One terminal of each of the sources is connected to a common tie point, called the **neutral**. As we will see in later chapters, depending on how the loads are connected to these sources, the neutral might or might not be needed. Notice in the wye connection that the current flowing through each phase source (that is, the phase current) is the same as the corresponding line current. *Therefore, in the wye configuration, the line current and the phase current are the same.* In the wye connection, the relationship between the phase voltage and line voltage has already been shown to be a ratio of $\sqrt{3}$ or 1.732. In any given three-phase wye system, the line voltage is *always* higher than the phase voltage.

Delta Connection

In the delta connection, the three voltage sources are connected in a circular fashion as shown in Figure 13-6(b). In this case, the phase voltage (the voltage produced by each phase source) and the line voltage (the voltage between any two phases) for a given phase are equal. However, the reader may show with some guided exercises at the end of the chapter that the rms magnitude of the line current is $\sqrt{3}$ or 1.732 times the phase current. In the delta system, there is no neutral conductor.

Summary of Important Relationships

At this point, the relationships between the line and phase voltages, and the line and phase currents in both the wye and delta systems, are summarized as follows:

Wye System:

$$V_{\text{line}} = \sqrt{3} \times V_{\text{phase}} \qquad (13\text{-}23)$$

$$I_{\text{line}} = I_{\text{phase}} \qquad (13\text{-}24)$$

Delta System

$$V_{\text{line}} = V_{\text{phase}} \qquad (13\text{-}25)$$

$$I_{\text{line}} = \sqrt{3} \times I_{\text{phase}} \qquad (13\text{-}26)$$

Note that these equations consider only the magnitudes and not the angles. Although I_{line} and I_{phase} for the wye and V_{phase} and V_{line} for the delta have the same phase angles, the other line and phase variables have different phase angles.

13-5 Three-Phase Wye and Delta Load Connections

As mentioned earlier, it is possible to operate three-phase loads from either the phase voltages (connecting them between any phase and neutral) or the line voltages (connecting them between any two phases). If we plan to connect the loads to the phase voltages, we might need to provide a fourth wire, the neutral. When this is done, we create what is called a three-phase wye load configuration as shown in Figure 13-7(a). In this configuration, three loads, each of which is assumed to be a resistive load of value R, are connected between each of the three-phase voltages and the neutral. Each load resistor draws a current from its respective phase. These currents are called the phase currents.

As mentioned earlier, in three-phase systems we usually assume that the three voltage amplitudes are equal. If we also assume that the three load resistors are equal (called a **balanced load**), then it is intuitive that the three phase currents will also be equal and in phase with their respective voltages.

FIGURE 13–7
Three-phase balanced load configurations: (a) wye and (b) delta.

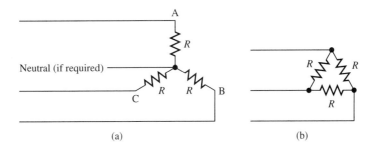

Consider a wye load configuration and define the rms magnitude of these currents to be I_{line}. Then we can write expressions for the three phasor currents as

$$\mathbf{I_A} = I_{\text{line}} \angle 0° \tag{13-27}$$

$$\mathbf{I_B} = I_{\text{line}} \angle -120° \tag{13-28}$$

$$\mathbf{I_C} = I_{\text{line}} \angle -240° = I_{\text{line}} \angle +120° \tag{13-29}$$

By KCL, the neutral current $\mathbf{I_N}$ can be expressed as the sum of the three line currents.

$$\begin{aligned}
\mathbf{I_N} &= \mathbf{I_A} + \mathbf{I_B} + \mathbf{I_C} \\
&= I_{\text{line}} \angle 0° + I_{\text{line}} \angle -120° + I_{\text{line}} \angle 120° \\
&= I_{\text{line}}(1 + j0) + I_{\text{line}}(-0.5 - j0.866) + I_{\text{line}}(-0.5 + j0.866) \\
&= I_{\text{line}} - 0.5 I_{\text{line}} - 0.5 I_{\text{line}} - j0.866 I_{\text{line}} + j0.866 I_{\text{line}} \\
&= 0
\end{aligned} \tag{13-30}$$

From the preceding development, we can draw the following conclusion: *In a three-phase wye system that has a balanced load, the neutral current will be zero, and the neutral wire is not usually needed.* (The term *usual* is used because there are some special circumstances in which it is desired even with a balanced load.) This means that in a three-phase wye configuration, it is possible to have either a three-wire or four-wire system, usually depending on whether the loads are balanced or unbalanced.

Next, consider the three-phase delta load shown in Figure 13-7(b). In this case, the voltage drop across each of the load resistors is equal to its corresponding line voltage. Although this configuration does not have a neutral, it is not necessary that the loads be balanced.

13-6 Three-Phase Power Calculations

Intuitively, one might think that the power dissipated in a balanced three-phase load depends on whether the load is wired in a wye or delta configuration. However, if we know the line voltage, line current, and power angle, or the phase voltage, phase current, and power angle, we can calculate the load power no matter how it is connected.

The power P_{phase} delivered by one phase of a three-phase system is

$$P_{\text{phase}} = V_{\text{phase}} I_{\text{phase}} \cos \theta \tag{13-31}$$

where $\cos \theta$ is the power factor.

Therefore, the total power P delivered by all three phases for a balanced system will be three times the individual phase power, or

$$P = 3 V_{\text{phase}} I_{\text{phase}} \cos \theta \tag{13-32}$$

To develop an equation for power using the line voltage and line current, we must refer to the equations for phase and line voltage and phase and line current in both the wye and delta systems.

For the wye system, we have

$$V_{\text{phase}} = \frac{V_{\text{line}}}{\sqrt{3}} \tag{13-33}$$

$$I_{\text{phase}} = I_{\text{line}} \tag{13-34}$$

Substituting Equations 13-33 and 13-34 into Equation 13-32, we obtain

$$P = \sqrt{3} V_{\text{line}} I_{\text{line}} \cos \theta \tag{13-35}$$

For the delta system, we have

$$V_{\text{phase}} = V_{\text{line}} \tag{13-36}$$

$$I_{\text{phase}} = \frac{I_{\text{line}}}{\sqrt{3}} \tag{13-37}$$

Substituting Equations 13-36 and 13-37 into Equation 13-32, we obtain

$$P = \sqrt{3} V_{\text{line}} I_{\text{line}} \cos \theta \tag{13-38}$$

Notice that the wye power equation (13-35) and the delta power equation (13-38) are the same. *The total power equation expressed in terms of line voltage for the balanced wye system is exactly the same as the corresponding equation for the balanced delta system.* Therefore, for a balanced system, if the line voltage, line current, and power factor are known, the same power equation can be used irrespective of the manner in which the load is connected.

Now let's discuss a point of possible confusion. When the power for a given phase is known for a balanced load, the total power is 3 times the power for each phase as given by Equation 13-32, and the factor 3 seems obvious. However, the factor $\sqrt{3}$ in Equations 13-35 and 13-38 might be bothersome initially, but it is because we are using line voltage and line current rather than phase voltage and phase current.

In a similar manner, we can derive equations for the total apparent power S and the total reactive power Q, which are

$$S = 3 V_{\text{phase}} I_{\text{phase}} = \sqrt{3} V_{\text{line}} I_{\text{line}} \tag{13-39}$$

and

$$Q = 3 V_{\text{phase}} I_{\text{phase}} \sin \theta = \sqrt{3} V_{\text{line}} I_{\text{line}} \sin \theta \tag{13-40}$$

EXAMPLE 13-3

A three-phase 208-V wye generator system feeds a balanced delta 30-kW load with a lagging power factor of 0.8. Assuming that line losses are negligible, determine (a) the power dissipated in each phase of the load, (b) the load phase voltage, (c) the load phase current, (d) the line current, (g) the generator phase current, and (h) the generator phase voltage.

SOLUTION

(a) The total power is the power for three phases and the power for each phase is

$$P_{\text{phase}} = \frac{P}{3} = \frac{30 \text{ kW}}{3} = 10 \text{ kW} \tag{13-41}$$

(b) Recall that the voltage specified for a three-phase system is the line voltage. Because the generator array is connected in a wye configuration and the load is connected in a delta configuration, we will add the letter L to the subscript to indicate *load*. Therefore, the voltage across each leg of the delta load is

$$V_{L\,\text{phase}} = 208 \text{ V rms} \tag{13-42}$$

(c) To determine the phase current, we begin with the relationship

$$P_{\text{phase}} = V_{L\,\text{phase}} I_{L\,\text{phase}} \cos\theta = 208 I_{L\,\text{phase}} \times 0.8 = 166.4 I_{L\,\text{phase}} = 10,000 \quad (13\text{-}43)$$

The letter L has also been added to the subscript for the current. Solving for $I_{L\text{phase}}$, we obtain

$$I_{L\,\text{phase}} = 60.1 \text{ A rms} \quad (13\text{-}44)$$

(d) The line current is related to the delta phase current by

$$I_{\text{line}} = \sqrt{3} I_{L\,\text{phase}} = \sqrt{3} \times 60.1 = 104 \text{ A rms} \quad (13\text{-}45)$$

(e) We will attach the letter G to phase subscripts for the wye-connected generator system. The phase current for each generator is

$$I_{G\,\text{phase}} = I_{\text{line}} = 104 \text{ A rms} \quad (13\text{-}46)$$

(f) The phase voltage for each generator is

$$V_{G\,\text{phase}} = \frac{V_{\text{line}}}{\sqrt{3}} = \frac{208}{\sqrt{3}} = 120 \text{ V rms}$$

▮ EXAMPLE 13-4

A three-phase balanced load has a line voltage of 408 V, a line current of 20 A, and a power factor of 0.75. Determine the total load power.

SOLUTION We are not given whether it is a wye load or a delta load; however, this information is not needed because the line voltage and line current are given. The total power is

$$P = \sqrt{3} V_{\text{line}} I_{\text{line}} \cos\theta = \sqrt{3} \times 480 \times 20 \times 0.75 = 12.5 \times 10^3 \text{ W} = 12.5 \text{ kW} \quad (13\text{-}47)$$

13-7 Comparison of Power Flow in Single- and Three-Phase Systems

It has been stated earlier that three-phase systems have significant advantages over single-phase systems with regard to power flow. In this section, a mathematical development will show why this is true for a balanced system. It is necessary to return to the instantaneous forms and use some trigonometric expansions to carry out this analysis.

Single-Phase Instantaneous Power

We begin this development by considering single-phase power and how it varies with time. For convenience, we will assume a resistive load in which the voltage and current are in phase. Let V_p represent the peak voltage and let I_p represent the peak current. The instantaneous voltage and current are given by

$$v(t) = V_p \sin\omega t \quad (13\text{-}48)$$
$$i(t) = I_p \sin\omega t \quad (13\text{-}49)$$

The instantaneous power is given by

$$p(t) = v(t)i(t) = V_p I_p \sin^2\omega t \quad (13\text{-}50)$$

A basic trigonometric identity is

$$\sin^2\phi = \tfrac{1}{2} - \tfrac{1}{2}\cos 2\phi \quad (13\text{-}51)$$

Applying this identity to the power expression, we have

$$p(t) = \frac{V_p I_p}{2} - \frac{V_p I_p}{2} \cos 2\omega t \qquad (13\text{-}52)$$

Because $V_p = \sqrt{2} V_{\text{rms}}$ and $I_p = \sqrt{2} I_{\text{rms}}$, the preceding equation can be simplified to

$$p(t) = V_{\text{rms}} I_{\text{rms}} - V_{\text{rms}} I_{\text{rms}} \cos 2\omega t \qquad (13\text{-}53)$$

Normally we are interested primarily in the average power. When the average of Equation 13-53 is formed over a cycle, the average value of the second term is zero and only the first term remains. Of course, we recognize that value as the average power for the ac voltage and current having the same phase angle in a single-phase circuit.

In contrast to the constant average power, we see that the instantaneous power actually varies with time and oscillates at a frequency equal to twice the frequency of the voltage and the current. For numerous applications, the variation of the power causes no problem, and the major focus can be on the average power, but for certain large industrial processes, particularly those involving motors, the variation is less than desirable because the pulsing power causes a large amount of hum and mechanical vibration.

Three-Phase Instantaneous Power

Let V_p represent the peak voltage for each phase of a three-phase balanced load and let I_p represent the corresponding peak current for each phase. To simplify the analysis, we will assume that the three equal loads are resistive, which means that the voltage and current for each phase will have the same phase angle.

The three voltages may be expressed as

$$v_A(t) = V_p \sin \omega t \qquad (13\text{-}54)$$

$$v_B(t) = V_p \sin(\omega t - 120°) \qquad (13\text{-}55)$$

$$v_C(t) = V_p \sin(\omega t - 240°) = V_p \sin(\omega t + 120°) \qquad (13\text{-}56)$$

The three currents may be expressed as

$$i_A(t) = I_p \sin \omega t \qquad (13\text{-}57)$$

$$i_B(t) = I_p \sin(\omega t - 120°) \qquad (13\text{-}58)$$

$$i_C(t) = V_p \sin(\omega t - 240°) = V_p \sin(\omega t + 120°) \qquad (13\text{-}59)$$

The instantaneous power in the three loads will be denoted as $p_A(t)$, $p_B(t)$, and $p_C(t)$ and can be expressed individually as

$$p_A(t) = v_A(t) i_A(t) = V_p I_p \sin \omega t \sin \omega t \qquad (13\text{-}60)$$

$$p_B(t) = v_B(t) i_B(t) = V_p I_p \sin(\omega t - 120°) \sin(\omega t - 120°) \qquad (13\text{-}61)$$

$$p_C(t) = v_C(t) i_C(t) = V_p I_p \sin(\omega t + 120°) \sin(\omega t + 120°) \qquad (13\text{-}62)$$

Our plan is eventually to add the three power values, but it is prudent to simplify each expression first. The following trigonometric identities will be used:

$$\sin A \sin B = \tfrac{1}{2} \cos(A - B) + \tfrac{1}{2} \cos(A + B) \qquad (13\text{-}63)$$

$$\cos(A - B) = \cos A \cos B + \sin A \sin B \qquad (13\text{-}64)$$

$$\cos(A + B) = \cos A \cos B - \sin A \sin B \qquad (13\text{-}65)$$

We begin the simplification with $p_A(t)$.

$$p_A(t) = V_p I_p \sin \omega t \sin \omega t = \frac{V_p I_p}{2} [\cos 0 - \cos 2\omega t] \qquad (13\text{-}66)$$

$$p_B(t) = V_p I_p \sin(\omega t - 120°) \sin(\omega t - 120°) = \frac{V_p I_p}{2} [\cos 0 - \cos(2\omega t - 240°)] \qquad (13\text{-}67)$$

$$p_C(t) = V_p I_p \sin(\omega t + 120°) \sin(\omega t + 120°) = \frac{V_p I_p}{2} [\cos 0 - \cos(2\omega t + 240°)] \qquad (13\text{-}68)$$

Next, substitute $\cos 0 = 1$ and expand the cosine terms within the brackets of the last two equations and we obtain

$$p_B(t) = \frac{V_p I_p}{2} [1 - (\cos 2\omega t \cos 240° + \sin 2\omega t \sin 240)]$$
$$= \frac{V_p I_p}{2} [1 + 0.5 \cos 2\omega t + .866 \sin 2\omega t] \quad (13\text{-}69)$$

$$p_C(t) = \frac{V_p I_p}{2} [1 - (\cos 2\omega t \cos 240° - \sin 2\omega t \sin 240)]$$
$$= \frac{V_p I_p}{2} [1 + 0.5 \cos 2\omega t - .866 \sin 2\omega t] \quad (13\text{-}70)$$

The net power is the sum of the three power values. It can be expressed as

$$p(t) = p_A(t) + p_B(t) + p_C(t) \quad (13\text{-}71)$$

We will omit some of the details because they are rather lengthy. However, when the values of power are expanded and added, quite a few terms combine or cancel in the process and the result is simply

$$p(t) = \frac{3 V_p I_p}{2} \quad (13\text{-}72)$$

Because $V_p = \sqrt{2} V_{\text{phase}}$ and $I_p = \sqrt{2} I_{\text{phase}}$, the preceding equation can be simplified to

$$p(t) = 3 V_{\text{phase}} I_{\text{phase}} \quad (13\text{-}73)$$

which has already been established in terms of the average power.

The amazing result of this development is that for a three-phase balanced system, *the instantaneous power is a constant value and is equal to the average power.* This is one of the major attributes of three-phase systems and one that can be exploited in large industrial applications, particularly those involving motors. Because the power is continuous, hum and mechanical vibration in three-phase motors are greatly reduced.

PROBLEMS

13-1 Assuming that phase A of an 832-V ac three-phase system is the reference phase (0°), and the phase sequence is ABC, write the sinusoidal expression for the phase voltage for phase C.

13-2 Assuming that phase A of a three-phase system is the reference phase (0°), and the phase sequence is ABC, what is the phase angle of the line voltage V_{CA}?

13-3 A three-phase wye connected system has a line voltage of 208 V and a line current of 5 A. Find (a) the phase voltage, and (b) the phase current.

13-4 A three-phase delta connected system has a line voltage of 416 V and a line current of 25 A. Find (a) the phase voltage, and (b) the phase current.

13-5 A wye-connected three-phase source is delivering a line voltage of 13,400 V ac at a line current of 25 amperes. Find (a) the phase voltage, (b) the phase current, and (c) the total power delivered.

13-6 A wye-connected three-phase source is delivering a line voltage of 416 V ac at a line current of 15 amperes. Find (a) the phase voltage, (b) the phase current, (c) the line voltage, (d) the line current, and (e) the total power delivered.

13-7 A delta-connected three-phase source is delivering a line voltage of 13,400 V ac at a line current of 25 amperes. Find (a) the phase voltage, (b) the phase current, and (c) the total power delivered.

13-8 A delta-connected three-phase source is delivering a line voltage of 416 V ac at a line current of 15 amperes. Find (a) the phase voltage, (b) the phase current, and (c) the total power delivered.

13-9 A wye-connected load shown in Figure 13-7(a) is dissipating 24 kW. What is the power dissipated by each resistor?

13-10 A delta-connected load shown in Figure 13-7(b) is dissipating 3.2 kW. What is the power dissipated by each resistor?

13-11 A wye-connected load shown in Figure 13-7(a) consists of three 10-Ω resistors. If the line voltage is 208 V, determine (a) the total power dissipated, and (b) the line current.

13-12 A delta-connected load shown in Figure 13-7(b) consists of three 10-Ω resistors. If the line voltage is 208 V, determine (a) the total power dissipated, and (b) the line current.

13-13 A 30-kW three-phase heater consists of three identical 10-ampere heating elements connected to a three-phase power source. As the designer, you have the choice of wiring the heating elements in either a delta or wye configuration. It is desired to connect the elements for minimum line voltage requirement. Which configuration (wye or delta) will allow for the lowest line voltage while still delivering 10 amperes to each element?

13-14 A three-phase 6-kW lighting system consists of three 2-kW 208-V lamps. The lamps can be wired either in a delta or wye configuration. Which configuration will result in the lowest line current?

Transformers

OVERVIEW AND OBJECTIVES

In Chapter 12, Magnetic Circuits, it was noted that a current in any conductor creates a magnetic field, and in a similar manner, a magnetic field passing over a coil of wire induces an electric voltage and corresponding current. In this chapter, we will study how these two phenomena can be incorporated into one device called a **transformer**. On one side of the transformer, we will have an alternating current in a coil of wire creating an alternating magnetic field. On the opposite side of the transformer, the same alternating magnetic field will induce an alternating voltage in another coil of wire. Although at first this might appear to be rather useless, as we will see, there are a seemingly endless number of potential applications for such a device.

Objectives

After completing this chapter, you will be able to

- Describe the components of an ideal transformer.
- Explain the relationship among the voltages, currents, and turns ratio in a transformer.
- Calculate the voltage, current, and power in a transformer.
- Solve problems involving transformer losses and demonstrate the effect that they have on efficiency.
- Show how to connect single-phase transformers to produce three-phase transformers.
- Apply special transformers such as autotransformers, current transformers, and potential transformers.

Whereas this chapter continues our study of electrical power systems and components, it is important to point out that in electrical power engineering, voltage and current are specified in the default units of volts rms and amperes rms. This convention will be used in this and subsequent chapters. Therefore, *in this chapter all references to ac voltages and ac currents are in rms (root mean square) unless indicated otherwise.*

14-1 Introduction

It is commonly known that a transformer is a device that can convert low voltages to high voltages (such as the transformer on an automobile engine that produces the high voltage for the ignition spark), or from a high voltage to a low voltage (such as a "wall charger" for a battery-operated electronic device). The transformer (in the form that it is used today) is a result of design and development initiated in the late 1800s by engineer George Westinghouse, and implemented by engineers William Stanley, Jr. and Nikola Tesla, both working for Westinghouse Electric Company.

In the late 1800s, Thomas Edison and George Westinghouse were working independently on a practical method of distributing electrical power from a power generation site to customers located some distance from the power source. Edison's system used direct current. In those days, without the availability of modern power switching semiconductors to transform dc voltages to higher or lower levels, Edison's dc system was limited to a single, relatively low voltage throughout the system. With a fixed voltage system, increased power demands required corresponding increases in the currents in the power lines along with associated voltage drops and power losses in the wires. Because of these voltage drops, customers who were located long distances from the power generator suffered from momentary voltage sags and consistently low line voltages. Westinghouse (and engineers Stanley and Tesla) recognized that because power is the mathematical product of voltage and current ($P = VI$), large amounts of electrical power could be distributed over a long distance by using higher voltages, thereby proportionally reducing the required currents and associated voltage drops and power losses in the power lines. They also recognized that if an alternating current system were developed, a transformer could be used to step up the voltage for transmitting the power over a long distance, and then a similar transformer could be used to step down the voltage at a distant location for use by the customers. Between the two transformers, the power line voltage could be very high. Additionally, because the line currents would be lower, smaller wire could be used for long-distance distribution, thereby reducing costs.

Because of its obvious advantages, Westinghouse's ac system was eventually adopted as the preferred power distribution system. Because the transformer has the unique ability to convert efficiently any ac voltage to any other desired ac voltage, it has become the mainstay of the electrical power distribution grid not only in the United States, but throughout the world. Additionally, the transformer has found its way into a seemingly endless number of other lower-power applications.

14-2 The Ideal Transformer

Figure 14-1 shows a simple transformer consisting of a closed core with two coils. The coils have a number of turns N_1 and N_2. When an ac voltage V_1 is applied to the coil on the left, an alternating magnetic field will be produced in the core. Because the coil on the right is wound around the core, the alternating magnetic field in the core will induce an ac voltage V_2 in the coil. The ratio of the voltages V_1 and V_2 is equal to the ratio of the number of turns on the two coils N_1 and N_2, or

$$\frac{V_1}{V_2} = \frac{N_1}{N_2} \tag{14-1}$$

The two windings are commonly referred to as the **primary winding** and **secondary winding**, where the primary is the input to the transformer, and the secondary is the output

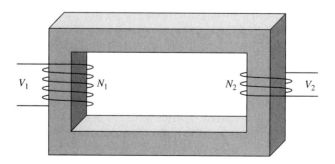

FIGURE 14–1
Simple transformer.

FIGURE 14–2
Transformer using high side and low side designation.

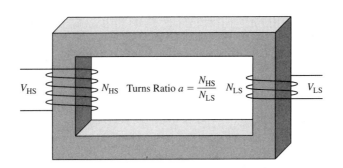

of the transformer. However, this can sometimes create confusion because the transformer is a bidirectional device; that is, we could use either winding of our transformer as the input without any change in the transformer's efficiency or performance. Therefore, if we have a new transformer that has not yet been connected into a circuit, we could ask ourselves, "Which winding is the primary and which is the secondary?" The answer depends on how the transformer is to be used. To avoid this confusion, we instead usually refer to the windings as the **high side winding** and the **low side winding**, where the high side winding is the winding with the larger number of turns.[1]

Another convenience that we use in transformer calculations is to avoid the ratio N_1/N_2 and instead use a value called the **turns ratio**, a. Turns ratio is defined as the ratio of the number of turns on the high side with respect to the number of turns on the low side, or

$$a = \frac{N_{HS}}{N_{LS}} \tag{14-2}$$

The number of turns on the high side will always be greater than or equal to the number of turns on the low side, so the turns ratio is always greater than or equal to 1. Therefore, for convenience we generally use the transformer with variable names shown in Figure 14-2. In this case, we have avoided designating any winding as a primary or secondary, but we know that the winding on the left has more turns than the one on the right because it is labeled as the high side winding.

EXAMPLE 14–1

The transformer shown in Figure 14-2 has a turns ratio of 12. If V_{HS} is 240 V ac, what is V_{LS}?

SOLUTION From Equation 14-1, we know that the voltage ratio in a transformer is the same as the turns ratio. Therefore, in Equation 14-2, we can replace N_{HS}/N_{LS} with the voltage ratio V_{HS}/V_{LS}. Rearranging the equation, we have

$$V_{LS} = \frac{V_{HS}}{a} = \frac{240}{12} = 20 \text{ V} \tag{14-3}$$

In a transformer, the placement of the coils on the core does not affect the voltages induced in the coils. No matter where on the core the coils are located, they will function the same. In fact, most transformers are constructed with the coils layered, one on top of another.

Transformers are extremely efficient devices. Typically, the power loss in a transformer is 1 percent to 5 percent of the total power transferred from the input to the output

---

[1] Unlike transformers with two windings, transformers with more than two windings will have one winding designated as the primary. This is because the primary winding must be designed to handle the total power delivered by all of the secondary windings; that is, the wire gauge of the primary must be sized sufficiently large to power the entire transformer, thereby making the primary unique.

of the transformer. For this reason, most transformer calculations simply assume that the transformer is ideal; that is, the efficiency is 100 percent, and the output power from a transformer is equal to the input power. Therefore, we can say

$$P_{\text{out}} = P_{\text{in}} \tag{14-4}$$

Because we are assuming that the input power and output power are the same, we also can assume that the volt-ampere product is the same.

$$V_{\text{HS}} I_{\text{HS}} = V_{\text{LS}} I_{\text{LS}} \tag{14-5}$$

If we rearrange Equation 14-5 and combine it with Equations 14-1 and 14-2, the result is one all-encompassing expression governing the voltage, current, and turns ratio in an ideal transformer.

$$a = \frac{N_{\text{HS}}}{N_{\text{LS}}} = \frac{V_{\text{HS}}}{V_{\text{LS}}} = \frac{I_{\text{LS}}}{I_{\text{HS}}} \tag{14-6}$$

EXAMPLE 14-2

An ideal transformer has its high side connected to a 240-V source and is consuming 400 watts of power. There is a load resistor of 2 Ω connected to the low side. Find (a) the low side voltage, (b) the low side current, (c) the turns ratio, (d) and the high side current.

SOLUTION

(a) The input power and output power of an ideal transformer are equal, so we know that the 2-Ω resistor is dissipating 400 watts. The standard power equation is

$$P = \frac{V^2}{R} \tag{14-7}$$

Rearranging the equation, we can find that the load voltage (which is the same as the low side voltage) is

$$V_{\text{LS}} = \sqrt{PR} = \sqrt{400 \times 2} = 28.3 \text{ V} \tag{14-8}$$

(b)
$$I_{\text{LS}} = \frac{V_{\text{LS}}}{R_{\text{load}}} = \frac{28.28}{2} = 14.1 \text{ A} \tag{14-9}$$

(c)
$$a = \frac{V_{\text{HS}}}{V_{\text{LS}}} = \frac{240}{28.3} = 8.49 \tag{14-10}$$

(d)
$$I_{\text{HS}} = \frac{I_{\text{LS}}}{a} = \frac{14.1}{8.49} = 1.67 \text{ A} \tag{14-11}$$

Designers of magnetically operated machines always must be mindful of the saturation point of the core material being used, because saturation causes a change in the operating characteristics of the core. This is especially critical in ac machines where the inductance (and inductive reactance) of the coil is crucial to limit current flow in the coil. As we have seen, core saturation results in an extremely low permeability. This causes the inductance of the coil that is wound around the core to become nearly zero, leaving only the resistance of the wire to limit current flow in the coil. Under this condition, the coil current increases dramatically, causing fuses in the circuit to blow or, even worse, destruction of the coil due to heat.

Because ac current flow in an inductor is proportional to the applied voltage, the flux density in a core is generally proportional to the applied coil voltage. However, as we increase the ac voltage, the increased current, magnetomotive force, and corresponding magnetic field intensity cause the core to begin to saturate, which reduces the inductive reactance of the coil. Therefore, we can see that, for a fixed frequency, the saturation of the core of an inductor generally is controlled by the amplitude of the ac voltage applied to the coil. For this reason, inductors and transformers are designed to be operated at or below specific ac voltages. Operating above the maximum design voltage will cause the core to saturate, excessive current to flow in the coil, and the coil to overheat and possibly burn out.

However, there are some instances where core saturation is desirable. Recall in our study of magnetic circuits in Chapter 12 that when a core saturates, the flux density is no longer sinusoidal but instead is clipped at the peaks. Because the secondary voltage of the transformer is induced by this flux density, the secondary voltage waveshape will also be clipped at the peaks. If we were to design the transformer so that it would normally operate in the saturation region, and we limited the primary current so that it would not overheat, our transformer could be made to regulate the output voltage and keep it somewhat constant. Granted, the output voltage would not be sinusoidal, but the peak amplitude would be limited. For electronic equipment with peak responding power supplies (standard half-wave and full-wave rectifiers), this would tend to make them more immune to large variations in the primary transformer voltage. This type of transformer is called a **constant voltage transformer**. In many cases, the secondary of the transformer is tuned by adding a capacitor, thus creating a circuit designed to resonate at the operating frequency of the transformer. This tends to remove undesirable harmonics and somewhat restore the sinusoidal waveshape of the output voltage. Such a transformer is called a **ferroresonant transformer**. Ferroresonant transformers have the added characteristic of filtering out short-term fluctuations in the primary voltage, thus adding some surge protection and voltage spike removal for the secondary load.

Another consideration when working with transformers is the effect caused by a change in frequency. Because the inductive reactance of a coil is $X_L = \omega L$, a decrease in frequency causes a proportional decrease in inductive reactance. In turn, this increases the coil current, magnetomotive force, and magnetic field intensity, which can cause core saturation.

For these reasons, all power inductors and transformers have a rated frequency, f_{rated}, and a rated voltage, V_{rated}, for each winding. Using the device in situations that would be below f_{rated} or above V_{rated} is not recommended unless other adjustments are made to the ratings. For example, it is permissible to operate a transformer at reduced frequency if we take steps to ensure that the coil current does not exceed rated current. This is done by reducing the applied ac voltage by the same proportion as the reduction in frequency, according to the expression

$$V = V_{\text{rated}} \frac{f}{f_{\text{rated}}} \quad (14\text{-}12)$$

where

V = maximum operating voltage at frequency f, in volts (V)

V_{rated} = rated voltage, in volts (V)

f = desired operating frequency, in Hertz (Hz)

f_{rated} = rated frequency, in Hertz (Hz)

14-3 Transformer Power Losses

Normally, and for most transformer calculations, we assume the transformer to be lossless; that is, we assume that the efficiency is 100 percent and that $P_{\text{out}} = P_{\text{in}}$. However, in reality there are two types of power losses in a transformer, which are called **copper loss** and **core loss**. These losses subtract from the output power of the transformer and are dissipated in the transformer itself in the form of heat.

Copper loss is simply the I^2R power loss due to the winding resistances and the corresponding winding currents. There are several ways to measure copper loss. One is simply to measure the resistance and current in each of the transformer windings, calculate the I^2R loss of each, and sum the losses. A second method (called a **short-circuit test**), which measures the total copper loss at once, is to short all of the windings except for one and apply a small ac voltage to the nonshorted winding until rated current for that winding is reached. Then, using a wattmeter, measure the power input under this condition. Because, in this case, there is no output power, the total power input to the transformer is equal to the total combined copper loss in all the windings.

Core loss is caused by two magnetic phenomena in the core of the transformer, which are **hysteresis loss** and **eddy current loss**. Both contribute to true power loss in the core and corresponding heating of the core.

Hysteresis loss is a function of the frequency and amplitude of the voltage applied to the transformer windings. Because an alternating voltage is being applied to the transformer, the magnetic field in the core changes polarity at twice the applied frequency. When this occurs, the orientation of the atoms in the core material is constantly being changed. This creates friction and heat in the core material and a corresponding power loss. The degree of magnetization increases with an increasing applied winding voltage, so the hysteresis loss is likewise a function of the transformer voltage and is greatest when the transformer is being operated at rated voltage.

Eddy current loss is also a core loss, but it is caused by electrical currents circulating in the core material. Because the core material is electrically conductive, when an alternating voltage is applied to the transformer windings, the alternating magnetic field produced in the core will also induce an electrical current in the core. Because of the ohmic resistance of the core material, there will be a corresponding I^2R power loss. Do not confuse this with the copper loss. Although both eddy current loss and copper loss are I^2R losses, the eddy current loss is proportional to the square of the magnitude of the flux in the core, which is proportional to the square of the applied primary voltage. Eddy current loss, causes heating in the core material, not the windings, and it is independent of the winding currents.

Eddy current loss can be minimized by a special core construction technique called **core lamination**. In this case, the core material is sliced into thin sheets (called laminations) and then reassembled using an insulating varnish as glue. The varnish isolates each lamination electrically and thereby minimizes the eddy currents.

Core loss can be easily measured on any transformer by performing an **open-circuit (or no-load) test**. For this test, we apply rated voltage to any winding of the transformer with all other windings open circuited and use a wattmeter to measure the real input power.

Let us summarize the types and characteristics of transformer losses.

- *Copper loss*—The I^2R power loss in the copper wire in the transformer windings. This loss is proportional to the square of the winding current. It is minimal when the transformer is operating with no load. Maximum copper loss occurs at rated load.
- *Core loss*—The power losses in the core of the transformer. There are two types:
    1. *Hysteresis loss*—frictional loss caused by the constant change in the magnetic field polarity in the core. This loss increases with an increasing voltage or frequency applied to the transformer windings.
    2. *Eddy current loss*—I^2R loss in the core material caused by circulating induced currents in the core. It is also proportional to the square of the voltage and the square of the frequency applied to the transformer windings. It can be minimized by laminating the core.

14-4 Transformer Efficiency

As mentioned earlier, most transformers are very efficient devices, typically in the 95 to 99 percent range. Transformer efficiency is the ratio of the power output to the power input. Efficiency is generally assigned the Greek letter eta (η) and the value can be expressed as either a decimal fraction or a percentage. There are two ways to calculate the efficiency of a transformer. First, if we simply measure the input and output power, we can say that the efficiency is

$$\eta = \frac{P_{out}}{P_{in}} \tag{14-13}$$

where

$\eta = $ the efficiency as a decimal fraction or as a percentage

$P_{out} = $ the output power in watts

$P_{in} = $ the input power in watts

Second, the output power of a transformer will be equal to the input power less the internal losses (copper and core loss); therefore we can say

$$P_{out} = P_{in} - P_{losses} = P_{in} - (P_{cu} + P_{core}) \tag{14-14}$$

Therefore, if we know the values of the input power and the losses, the efficiency is

$$\eta = \frac{P_{in} - P_{losses}}{P_{in}} = \frac{P_{in} - (P_{cu} + P_{core})}{P_{in}} \tag{14-15}$$

or, if we know the output power and losses,

$$\eta = \frac{P_{out}}{P_{out} + P_{losses}} = \frac{P_{out}}{P_{out} + P_{cu} + P_{core}} \tag{14-16}$$

14-5 Effect of Power Factor on Transformer Performance

We usually calculate transformer parameters assuming that the power factor of the load will be unity ($PF = 1$). However, under real conditions, a transformer might be powering a load that has a power factor less than unity. It is important to understand the effect that load power factor has on a transformer so we can properly select a transformer for an application.

Transformer power ratings are always expressed in apparent power (volt-amps), not real power (watts). If the power factor of the load is unity, then the apparent power and real power will be equal. Therefore, we can say that, for example, a 1-kVA-rated transformer is capable of delivering 1 kW to a unity power factor load. However, this is not true if the power factor is less than unity. The constraining parameters in transformer applications are the voltage and current. The maximum voltage that can be applied to a transformer winding is a function of the core material, core cross-sectional area, and the number of turns on the winding. The maximum current that a winding can tolerate is determined solely by the size (thickness) of the wire. Because of these two limitations, transformers are rated in volt-amps, not watts. The actual real power that is passing through the transformer is unimportant when sizing a transformer for an application. Instead, it is important that we do not exceed the maximum voltage or maximum current. Therefore, the product of these two determines the volt-amp rating of a transformer.

EXAMPLE 14-3

A 120 V – 120 V isolation transformer is rated at 1.2 kVA. If we apply a unity power factor load to the transformer, it can safely deliver 1.2 kW to the load. However, what is the maximum safe power (in watts) that can be delivered to the load if the power factor is 0.7?

SOLUTION The power in watts is

$$P = S \times PF = 1200 \times 0.7 = 840 \text{ watts} \tag{14-17}$$

Transformer efficiency is also adversely affected by the power factor of the load. For a fixed wattage load, as the power factor decreases, the winding current will increase, which in turn will increase the copper loss exponentially. This can be seen by examining the ac power equation.

$$P = VI \cos\theta = VI \times PF \qquad (14\text{-}18)$$

Note in this equation that, assuming our load consumes a constant amount of power, if the power factor decreases, the transformer will need to deliver more current, with a corresponding exponential ($I^2 R$) increase in copper loss.

14-6 Impedance Reflection

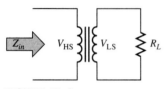

FIGURE 14–3
Impedance reflection.

When analyzing transformer circuits, it is sometimes helpful to remove the transformer from the circuit model, thus removing an additional calculation when attempting to work through a transformer's windings. To see why this is an improvement, let us examine a simple transformer with a high side winding rated at a voltage V_{HS} and a low side winding with a voltage rating V_{LS}. There is a load resistance R_L connected to the low side of the transformer, as shown in Figure 14-3. The current in the low side winding is

$$I_{LS} = \frac{V_{LS}}{R_L} \qquad (14\text{-}19)$$

The high side current is

$$I_{HS} = \frac{I_{LS}}{a} \qquad (14\text{-}20)$$

Substituting Equation 14-19 into Equation 14-20, we have

$$I_{HS} = \frac{V_{HS}}{aR_L} \qquad (14\text{-}21)$$

If we view the circuit resistance from the high side of the transformer, the resistance will appear to be

$$R_{L(HS)} = \frac{V_{HS}}{I_{HS}} \qquad (14\text{-}22)$$

Substituting Equation 14-21 into Equation 14-22, we have the low side load resistance as it appears on the high side of the transformer, which is

$$R_{L(HS)} = a^2 R_L \qquad (14\text{-}23)$$

Therefore, when performing circuit analysis on a circuit with a transformer, we can effectively move the load impedance from the low side to the high side simply by eliminating the transformer and multiplying the load impedance by the square of the turns ratio. In a similar manner, if we wish to move an impedance from the high side to the low side of a transformer, we would divide by the square of the turns ratio.

Remember that because a transformer does not pass a dc voltage or current, a resistance that is reflected through a transformer will appear as an ac impedance (not a resistance) on the opposite side. If there is a dc current flowing in the primary winding of a transformer, the load resistance will not interact with this dc current. For this reason, the term *impedance* is generally used instead of resistance when discussing impedance reflection in a transformer.

▌▌ EXAMPLE 14-4

A 60-Hz 120-V ac induction motor can be modeled as a 10.8-Ω resistance R_m in series with 25.3-millihenry inductance L_m. The motor is connected to the low side of a step-down transformer that reduces the 480-V ac line to 120 V ac. What is the effective impedance that the motor presents to the 480-V ac line?

SOLUTION The turns ratio of the transformer is

$$a = \frac{V_{HS}}{V_{LS}} = \frac{480}{120} = 4 \qquad (14\text{-}24)$$

The inductive reactance of the 25.3 mH motor inductance is

$$X_m = 2\pi f L = 2\pi(60)(0.0253) = 9.54 \ \Omega \qquad (14\text{-}25)$$

Therefore, the motor impedance is

$$\mathbf{Z_m} = R_m + jX_m = 10.8 + j9.54 = 14.41\angle 41.46° \ \Omega \qquad (14\text{-}26)$$

We can now reflect the motor impedance to the high side of the transformer.

$$\mathbf{Z_{m(HS)}} = a^2 Z_m = 4^2 \times 14.41\angle 41.46° = 230.6\angle 41.46° \ \Omega \qquad (14\text{-}27)$$

14-7 Impedance Matching Transformers

In Chapter 2, it was demonstrated that maximum power is transferred from a voltage source to a load when the load impedance is equal to the internal impedance of the source. If the load impedance is higher or lower than the source impedance, less than optimum power will be transferred.

Because impedance matching is the most desirable situation, it is often necessary to ensure that the characteristic output impedance of a signal source and the impedance of a load connected to it are the same. For example, it is desirable for a 75-Ω cable television signal (the source) to be connected to a 75-Ω input connector on a television receiver (the load). The matching of source and load impedances assures that maximum power will be transferred from the source to the load. However, often we find that the characteristic impedances of a source and load do not match. When this situation occurs, frequently a transformer is used to allow the load impedance to be matched to the signal source impedance.

An impedance connected to the low side of a transformer can be reflected to the high side of a transformer by multiplying it by the transformer turn ratio, or $Z_{HS} = a^2 Z_{LS}$. By rearranging this equation, we can find the turns ratio of a transformer that will match a high side impedance to any low side impedance.

$$Z_{HS} = a^2 Z_{LS}$$
$$a^2 = \frac{Z_{HS}}{Z_{LS}} \qquad (14\text{-}28)$$
$$a = \sqrt{\frac{Z_{HS}}{Z_{LS}}}$$

EXAMPLE 14-5

A 1-kW heating element having a resistance of 2 Ω is to be connected to an ac power line that is capable of safely delivering 2 amperes at 120 V ac, which is a maximum of 240 watts. However, if we connect the 2-Ω heater directly to the 120-V line, it will of course draw 60 amperes and overload the circuit. To use the available ac line fully, it is desired to have the heating element draw 2 amperes at 120 volts from the ac line. (a) What is the turns ratio of a matching transformer that can be used to deliver the full 240 watts to the heating element? (b) What are the voltage and volt-amp ratings of the transformer? (c) What will be the current delivered to the heating element?

SOLUTION

(a) In this case we know that the low side impedance is 2 Ω. To obtain maximum power transfer without overloading the ac line, we need to match the 2-Ω load to the ideal high side impedance, which is

$$Z_{HS} = \frac{V_{HS}}{I_{HS}} = \frac{120}{2} = 60 \ \Omega \qquad (14\text{-}29)$$

Now the transformer turns ratio can be determined, which is

$$a = \sqrt{\frac{Z_{HS}}{Z_{LS}}} = \sqrt{\frac{60}{2}} = \sqrt{30} = 5.48 \qquad (14\text{-}30)$$

(b) Because the high side of the transformer is connected to the 120-V ac line, the transformer's high side voltage rating is 120 V. The low side voltage is

$$a = \frac{V_{HS}}{V_{LS}}$$
$$V_{LS} = \frac{V_{HS}}{a} = \frac{120}{5.48} = 21.91 \ V \qquad (14\text{-}31)$$

Therefore, the transformer is rated at 120 V – 21.91 V, 240 VA.

(c) There are several ways to calculate the heating element current, the simplest of which is to use the turns ratio to find the current ratio. Equation 14-6 provides this relationship.

$$a = \frac{I_{LS}}{I_{HS}}$$
$$I_{LS} = aI_{HS} = 5.48 \times 2 = 10.95 \ A \qquad (14\text{-}32)$$

As a check of our calculations, we can also find the power delivered to the heating element, which is

$$P_{load} = I_{LS}^2 R_{load} = 10.95^2 \times 2 = 240 \ W \qquad (14\text{-}33)$$

Note in Equation 14-33 that the power delivered to the load is the same as the maximum power capacity of the ac line, thus indicating the best possible impedance match between the line and the load. Had this match been incorrect, one of two scenarios would be possible:

- If the turns ratio were too low, the load impedance reflected to the high side would be too low, causing the high side current to exceed the 2-ampere maximum.
- If the turns ratio were too high, the low side voltage would be too low, causing the power delivered to the load to be less than the ideal 240 W.

14-8 Transformer Construction

In a transformer, the physical shape of the core determines all remaining physical characteristics of the transformer. There are two fundamental transformer core construction types, the shell (or O-core) and the E-core. Each type has unique advantages and disadvantages.

Shell

The shell type transformer is shown in Figure 14-4(a). In this case, the core is either rectangular or toroidal. The windings in this type of transformer can either be separated or wound one over the other. Separating the coils has the distinct advantage of providing additional electrical isolation between windings, especially if the transformer is being designed to produce high voltages. The main disadvantages of this type of transformer are that (1) it is large, and (2) the **leakage flux** (the flux that is not contained within the core) from the coils passes through the air outside the coils and does not contribute to inducing a voltage in the other winding. As we will see, this problem is reduced in the E-core style.

FIGURE 14–4
Core types, (a) shell and (b) E-core.

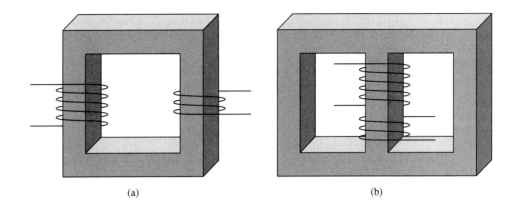

E-Core

The E-core construction style, shown in Figure 14-4(b), consists of an upper and lower bar connected by three vertical bars. All windings are located on the center vertical bar and are layered one over the other [for clarity, this layering is not shown in Figure 14-4(b)]. The flux produced by the coils travels through the vertical bar, divides equally in the top and bottom bars, and circulates through the outer vertical bars. Because the flux divides into two paths, the two outer vertical bars and the top and bottom bars will need to carry only half of the total flux and therefore can have a cross-sectional area that is one-half that of the center vertical bar. The main advantages in this type of transformer construction are that (1) is it more compact than the shell type, and (2) the leakage flux is reduced because part of the coils have core material both on the inside and outside of the coils.

14-9 Three-Phase Transformers

In addition to single-phase applications, transformers are also used in three-phase systems. As with single-phase applications, in three-phase applications the transformers are used to provide primary to secondary isolation and efficiently convert voltages to other desired values. However, as we will see, in three-phase applications, transformers also can be used to convert three-phase wye systems to delta systems, and vice versa.

Wye-to-Wye

Consider the three-transformer system shown in Figure 14-5(a), in which we have three identical single-phase transformers connected in a three-phase wye configuration for both the primary side and secondary side. On the left of the figure is shown the actual connections of the transformer windings, and on the right is a "shorthand" drawing method that is normally used to represent three-phase transformers. In this configuration, the transformer system is providing primary-to-secondary isolation and, assuming that the turns ratio of the transformers is a value other than 1, it is converting the input voltage to some other desired voltage. Because the primaries and secondaries are wired in wye configurations, this is called a wye-to-wye or wye-wye configuration. The secondary wye connection could be either three wire or four wire. The voltage and current ratios for the wye-wye configuration are

$$a = \frac{V_{\text{line(HS)}}}{V_{\text{line(LS)}}} = \frac{V_{\text{phase(HS)}}}{V_{\text{phase(LS)}}} = \frac{I_{\text{line(LS)}}}{I_{\text{line(HS)}}} = \frac{I_{\text{phase(LS)}}}{I_{\text{phase(HS)}}} \qquad (14\text{-}34)$$

FIGURE 14–5
Three-phase transformer wye and delta connection possibilities.

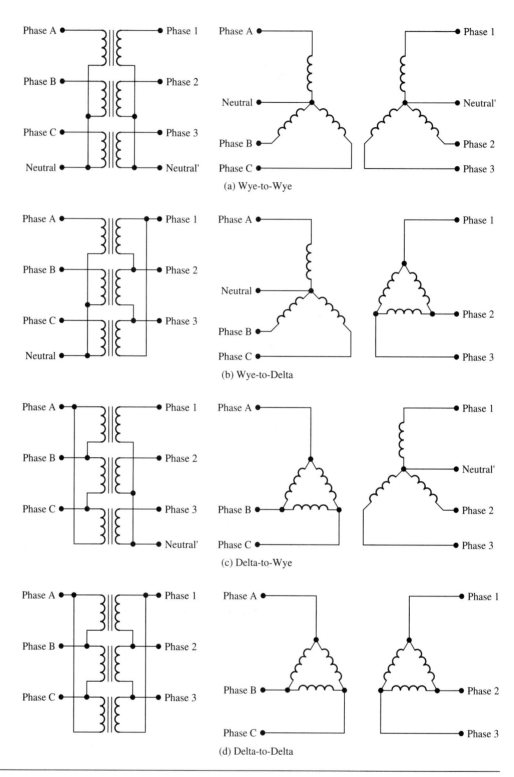

(a) Wye-to-Wye

(b) Wye-to-Delta

(c) Delta-to-Wye

(d) Delta-to-Delta

■ EXAMPLE 14-6

A wye-wye connected three-phase transformer has a turns ratio of 4 and has the high side connected to a 2240-V ac line (line voltage). What is (a) the low side line voltage, (b) low side phase voltage, and (c) high side phase voltage?

SOLUTION

(a) Using Equation 14-34, the low side line voltage is

$$V_{\text{line(LS)}} = \frac{V_{\text{line(HS)}}}{a} = \frac{2240}{4} = 560 \text{ V} \qquad (14\text{-}35)$$

(b) The low side phase voltage is

$$V_{\text{phase(LS)}} = \frac{V_{\text{line(LS)}}}{\sqrt{3}} = \frac{560}{\sqrt{3}} = 323 \text{ V} \qquad (14\text{-}36)$$

(c) The high side phase voltage is

$$V_{\text{phase(HS)}} = \frac{V_{\text{line(HS)}}}{\sqrt{3}} = \frac{2240}{\sqrt{3}} = 1293 \text{ V} \qquad (14\text{-}37)$$

Note that even though the input voltage to the transformer is 2240 V ac, because the high side of the transformer is wired in a wye configuration, the actual voltage applied to each of the high side windings is only 1293 V.

Wye-to-Delta

Next refer to the wye-to-delta, or wye-delta, connection scheme shown in Figure 14-5(b). In this case, the input or primary side of the transformer is connected in a wye configuration and the output or secondary side is connected as a delta. Remember that in this case, the line voltage and phase voltage on the secondary side are the same. Also, note that the primary and secondary phase voltages are related by the turns ratio, a.

Delta-to-Wye

The delta-to-wye or delta-wye configuration is shown in Figure 14-5(c). As with the wye-wye configuration, the secondary connection could be either three wire or four wire, as needed.

Delta-to-Delta

In the delta-to-delta configuration shown in Figure 14-5(d), because the input and output configuration are the same, the ratio of the high side and low side voltages is the same as the turns ratio of the transformer.

EXAMPLE 14-7

A delta-wye connected three-phase transformer [as shown in Figure 14-5(c)] has a turns ratio of 2 and has the high side connected to a 13,400-V ac line (line voltage). What is (a) the high side phase voltage, (b) the low side phase voltage, and (c) the low side line voltage?

SOLUTION

(a) In a delta-connected system, the line voltage and phase voltage are the same. Therefore the high side phase voltage is also 13,400 V ac.

(b) The low side phase voltage is

$$V_{\text{phase(LS)}} = \frac{V_{\text{phase(HS)}}}{a} = \frac{13{,}400}{2} = 6700 \text{ V} \qquad (14\text{-}38)$$

(c) The low side line voltage is

$$V_{\text{line(LS)}} = \sqrt{3}\, V_{\text{phase(LS)}} = \sqrt{3} \times 6700 = 11{,}605 \text{ V} \qquad (14\text{-}39)$$

14-10 Autotransformers

In many cases, we use transformers to provide electrical isolation between the primary and secondary. Transformers provide very good isolation because the secondary is not connected in any way to the primary nor to any ground. Therefore, when we design circuits using transformers, we are free to connect the secondary in any desired way. We can ground

FIGURE 14–6
Autotransformer (boost connection).

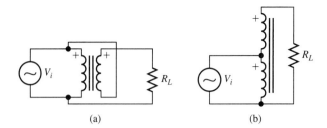

either terminal of the secondary, or connect either terminal to some other voltage, or even leave the secondary ungrounded. This creates some interesting connection possibilities.

For example, consider the transformer circuit shown in Figure 14-6(a). In this case, the primary of the transformer is connected in series with the secondary. Note that positive (+) symbols have been added to the transformer symbol. These indicate the polarity of the transformer windings. The "+" symbol indicates that when the instantaneous voltage on the "+" terminal of the primary is positive, the instantaneous voltage on the "+" terminal of all other windings on the transformer will also be positive. Therefore, in this case, we have wired the primary and secondary in series "aiding," similar to connecting two batteries in series plus-to-minus. The voltage on the load resistor R_L will be the sum of the primary and secondary voltages on the transformer. To better illustrate these connections, we have redrawn the transformer in Figure 14-6(b) with the same connections, but with the secondary positioned above the primary. This type of transformer is a special type called an **autotransformer**.

Analysis of the voltages and currents in an autotransformer might seem tricky at first, but applying some simple knowledge of how a transformer operates makes the task somewhat easier. First, there are a few fundamental rules we must remember.

- When wired as an autotransformer, the maximum current capacity of any winding of a transformer is the same as when it is used as a conventional transformer.
- The voltages, currents, and turns ratio still conform to Equation 14-6.
- Although the autotransformer circuit can deliver more apparent power (S, in volt-amperes) than the transformer alone, the transformer itself cannot safely deliver more apparent power than its rating.
- Because autotransformers can be connected in several different ways, rather than attempt to memorize the equations used to analyze each, it is easier to gain a fundamental understanding of how a basic autotransformer operates and then develop the governing equations as needed.

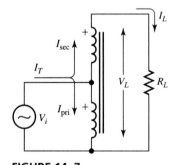

FIGURE 14–7
Autotransformer voltages and currents.

Figure 14-7 illustrates the voltages and currents in an operating autotransformer circuit. Here we have a voltage source V_i that is connected across the primary of the transformer. The primary winding current is I_{pri} and the secondary winding current is I_{sec}. Because of the way the transformer is connected, the source delivers a total input current I_T, which is the sum of the two winding currents ($I_{pri} + I_{sec}$). Because the only element connected to the top of the secondary is the load resistor R_L, the load current R_L is the same as the secondary current I_{sec}. The load resistor spans both the primary and secondary windings, so the load voltage V_L is the sum of the primary and secondary voltages ($V_{pri} + V_{sec}$).

Notice that, had we wired this transformer in a conventional manner, the total apparent power delivered from the transformer to a load would be

$$S_{load} = V_{sec} I_{sec} \qquad (14\text{-}40)$$

However, once we wire the same transformer in the autotransformer configuration in Figure 14-7, the apparent power delivered to the load becomes

$$S_{load} = \left(V_{pri} + V_{sec}\right) I_{sec} \qquad (14\text{-}41)$$

If we operate the autotransformer circuit so that the transformer is delivering its maximum rated voltages and currents, then a comparison of Equations 14-40 and 14-41 indicates that the autotransformer circuit is capable of delivering more apparent power to the load than the transformer's rated power.

This phenomenon is further illustrated in the following example.

EXAMPLE 14-8

A transformer has a primary rating of 120 V ac and a secondary rating of 24 V ac. It is wired as an autotransformer as shown in Figure 14-7 with a load resistance of 36 Ω. Find the (a) load voltage, (b) load current, (c) secondary current, (d) primary current, (e) total input current, (f) apparent power delivered to the load, and (g) minimum apparent power rating of the transformer.

SOLUTION

(a) The primary and secondary windings are connected in series aiding, so the load voltage will be their sum, which is

$$V_L = V_{\text{pri}} + V_{\text{sec}} = 120 + 24 = 144 \text{ V} \tag{14-42}$$

(b) The load current is

$$I_L = \frac{V_L}{R_L} = \frac{144}{36} = 4.0 \text{ A} \tag{14-43}$$

(c) The secondary current is the same as the load current, 4.0 A.

(d) The turns ratio of the transformer is

$$a = \frac{V_{\text{HS}}}{V_{\text{LS}}} = \frac{120}{24} = 5.0 \tag{14-44}$$

Because the primary is the high side winding, the primary current is

$$I_{\text{pri}} = \frac{I_{\text{sec}}}{a} = \frac{4.0}{5.0} = 0.8 \text{ A} \tag{14-45}$$

(e) The total input current is the sum of the primary and secondary currents, which is

$$I_T = I_{\text{pri}} + I_{\text{sec}} = 0.8 + 4.0 = 4.8 \text{ A} \tag{14-46}$$

(f) Because we are assuming the transformer to be ideal, the apparent power delivered to the load can be calculated on either the input or output side of the transformer. As an accuracy check, we will do both.

$$S_{\text{load}} = S_{\text{in}} = V_i I_T = 120 \times 4.8 = 576 \text{ VA}$$
$$S_{\text{load}} = S_{\text{out}} = V_L I_L = 144 \times 4.0 = 576 \text{ VA} \tag{14-47}$$

(g) The apparent power rating of the transformer can be calculated on the primary or secondary winding. We will do both.

$$S_{\text{rated}} = V_{\text{pri}} I_{\text{pri}} = 120 \times 0.8 = 96 \text{ VA}$$
$$S_{\text{rated}} = V_{\text{sec}} I_{\text{sec}} = 24 \times 4.0 = 96 \text{ VA} \tag{14-48}$$

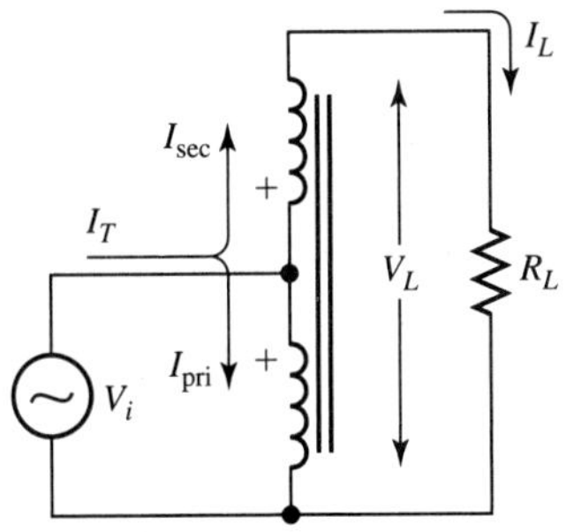

FIGURE 14–8
Autotransformer, buck configuration.

In the above example, note that we have a 96-VA transformer delivering 576 VA to the load. Although this appears to be a Herculean effort on the part of the transformer, the reason for the apparent disparity is that only 96 VA of apparent power passes through the core of the transformer (this is sometimes called the **transformed power**). The remaining apparent power, 480 VA, is conducted from the input, through the secondary winding, to the output (also called the **conducted power**). The transformer is simply giving a 24-V boost to the 120-V input voltage.

Next we will consider the same transformer connected in the same circuit, except that we will reverse the connection of the secondary as shown in Figure 14-8. In this mode, called the **buck configuration**, the secondary winding is in series opposing, or bucking, the primary winding, and the load voltage V_L is the difference between the primary voltage V_{pri} and the secondary voltage V_{sec}. Otherwise, the analysis method is the same as for the boost configuration. Therefore, this autotransformer circuit results in a voltage reduction.

EXAMPLE 14-9

A 1.2-kVA transformer has a primary rating of 240 V ac and a secondary rating of 12 V ac. It is wired as an autotransformer as shown in Figure 14-8. (a) What is the minimum value of load resistance that the transformer can safely power? (b) What is the maximum apparent power that the circuit can deliver to the load?

SOLUTION

(a) Because the primary and secondary windings are connected in series opposing, the load voltage is their difference, or 228 V ac. Therefore, the load voltage is also 228 V ac. The transformer is rated at 1.2 kVA, so the maximum secondary current is

$$I_{\text{sec(rated)}} = \frac{S_{\text{rated}}}{V_{\text{sec(rated)}}} = \frac{1{,}200}{12} = 10 \text{ A} \tag{14-49}$$

The minimum load resistance is

$$R_{L(\text{min})} = \frac{V_L}{I_{\text{sec(rated)}}} = \frac{228}{10} = 22.8 \text{ }\Omega \tag{14-50}$$

(b) The maximum apparent power that the circuit can deliver to the load is

$$S_{\text{max}} = V_L I_{\text{sec(rated)}} = 228 \times 10 = 2280 \text{ VA} \tag{14-51}$$

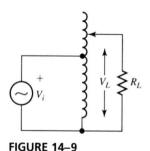

FIGURE 14–9
Variable autotransformer.

So far, we have examined autotransformers that have fixed turns ratios. However, there is a special autotransformer that has a variable turns ratio called a **variable autotransformer**, which is shown in Figure 14-9. It is often referred to by its General Radio Corp. (GenRad) trade name **Variac**©. In this device, there is only one winding. The input is connected to a portion of the winding. The output terminal is a sliding carbon brush that slides along the bare-wire turns of the winding and picks off a voltage. The winding of the transformer is wound on a cylindrical core, and the slider rotates on a shaft in a circular fashion to make contact with the windings. The variable autotransformer can efficiently provide any adjustable voltage from zero to some multiple of the input voltage.

14-11 Instrument Transformers

In addition to using transformers to deliver power to loads, transformers have some unique applications in the areas of voltage and current instrumentation. For most low-voltage, low-current applications, if we wish to monitor the voltage being applied to a load and the current through the load, we install a voltmeter and ammeter in the operator's instrument panel for the system and then connect the voltmeter terminals in parallel with the load and connect the ammeter in series with the load. However, in high-voltage, high-current applications, that is neither practical nor safe. There are electrical and fire codes (not to mention common sense!) that prevent us from routing high voltages and currents into an operator instrument panel and connecting them to meters that are potentially accessible from the outside of the panel. Therefore, for safety reasons, we need some method to reduce the voltage and current to a smaller but accurately proportional value. This can be done easily using two instrument transformers called a **potential transformer**, or **PT**, and a **current transformer**, or **CT**.

Potential Transformer

Referring to Equation 14-6, the turns ratio of an ideal transformer determines the ratio of the high side voltage to the low side voltage. Therefore, it is relatively simple to design a transformer that will reduce the ac potential of the high side to a lower (and safer) voltage on the low side. Because the low side of the transformer is powering a voltmeter, both the high and low side currents will be negligible. This means the transformer can be wound with very small gauge wire. Also, if the transformer is constructed with a relatively few number of turns, the magnetomotive force will be low, which will allow the cross-sectional area of the core to be small. These will allow us to make the transformer physically small.

In actual construction, the designer must remember that, as with all transformers, the potential transformer will have some leakage flux. This will cause the output voltage to be slightly less than that of our idealized model. Therefore, the designer will generally make some small adjustments in the turns ratio to correct for voltage losses in the transformer, thereby making the transformer deliver the rated high-side to low-side voltage ratio.

In application, the transformer is physically located in the area of the high voltage to be measured. Then the low side wires are routed from the transformer to the voltmeter in the instrument panel. The current in these wires will be very small, so the voltage drop in the wires due to wire resistance (which would cause an accompanying voltage error in the voltmeter) will be extremely low. Therefore, the length of these wires is not critical. The voltmeter in the control panel is selected so that its full scale value will match the low side voltage of the transformer under maximum high side voltage conditions. Then the indicator scale of the meter is redrawn so that the meter indicates the value of the voltage on the high side of the PT. The net result is that, even though the meter is responding to the low side voltage, its scale indicates the high side voltage.

Potential transformers are rated in two ways. Some manufacturers simply state the high side voltage and the voltage ratio (for example, 2400 V, 20 : 1). Others state the actual ratio of the PT's rated voltages. For example, if a transformer is capable of a maximum of 4800 V ac on the high side and has a ratio of 40, it will have a rating of 4800 : 120. This is done to simplify the specification, because both the voltage ratings and the voltage ratio are stated in the same specification. Most commercially available PTs have a low side rating of 120 V ac.

PTs are widely used in wattmeters to provide the voltage portion of the power calculation. It is important to note that, although a standard power transformer operates under the same electromagnetic principles as a PT, power transformers are not suitable for use as PTs. That is because PTs are designed to produce extremely accurate voltage ratios with very low phase shift. Because power in watts is a product of the voltage, current, and cosine of the angle between them, phase shift in a PT can directly affect the accuracy of a wattmeter.

EXAMPLE 14-10   A potential transformer is to be specified for a 7500-V ac instrument system. An ac voltmeter is available that has a 0 to 120 V ac full scale movement (that is, when 120 V ac is applied to the meter terminals, the pointer of the meter deflects to its maximum reading). It is decided to add some guardband by having the meter indicate 0 to 9600 V ac. What should be the rating of the potential transformer and how should the meter scale be redrawn?

SOLUTION   The specification of the PT is 9600 V, 80 : 1; or 9600 : 120. The meter scale should be redrawn so that full scale indicates 9600 V ac, or 9.6 kV ac.

Current Transformer

Referring to Equation 14-6, the ratio of the low side to high side currents of a transformer is directly proportional to the turns ratio. Therefore, if we connect a short circuit to the high side of our transformer, the current through the short circuit will be equal to the low side current divided by the turns ratio. If we connect the low side of the transformer in series with a load, the current in the high side winding will be a fraction $(1/a)$ of the load current. This method of precise current reduction allows us to sample very large load currents to produce smaller proportional currents that are safe to meter. In most CT applications, the low side winding is a single turn of wire. Note that if a single turn is used, it only needs to be a wire passing through the transformer core; it does not need to be wrapped around the core. Therefore, most CTs are constructed from a toroidal core with the high side winding wound around the core. The low side winding is added at the time the CT is installed simply by passing the high current wire through the hole in the toroid.

CTs are rated by stating the low side and high side currents. For example, a CT rated for 1000 amperes low side and 5 amperes high side would have a rating of 1000 : 5. Most current instrumentation applications use 5-ampere ac ammeters. This means that most commercially available CTs have a 5-ampere high side current rating. As with the voltmeter in PT applications, the ammeters used in CT applications usually have scale markings that indicate the low side current.

We must consider one potential safety issue when working with CTs. Take, for example, a 1000 : 5 CT that is connected into a system operating at 120 V ac. Consider the consequences of disconnecting the ammeter from the high side of the transformer while it is in operation. The ammeter on the high side serves as a short circuit for the CT. If it is removed, the high side of the transformer becomes opencircuited. This open circuit is reflected to the low side, which causes the low side impedance of the transformer to increase drastically so that the impedance is much greater than that of the load. This, in turn, causes the low side winding of the transformer to drop a majority of the applied voltage, which in our example is 120 V ac. If we multiply this low side voltage by the turns ratio of our transformer (1000 : 5 or $a = 200$), we have a high side voltage of 24,000 V ac! Naturally, this condition would cause the CT core to saturate and the high voltage would cause the insulation between the windings to break down, causing internal arcing, but the end result is that the high side voltage would still be extremely high. Under these conditions, it is very likely that the person who disconnected the ammeter would be electrocuted. For this reason, *a CT should never be disconnected while it is operating. Also, if the ammeter is removed from a CT circuit for servicing, a temporary shorting bar must be installed in its place to prevent damage to the CT and surrounding equipment should the system be switched on.* Most CTs are constructed with a shorting bar that can be rotated to short out the high side terminals when desired.

EXAMPLE 14-11

It is desired to have a 5-ampere ac ammeter measure the current in a 300-A load. With guardbanding, we have decided to design the current metering system for 500 A full scale. (a) What would be the current ratio specification for the CT for this application? (b) What would be the load current if the current through the ammeter is 1.75 A ac?

SOLUTION

(a) The CT rating would be 500 : 5.

(b) Because the CT rating is 500 : 5, the turns ratio would be $a = 100$. At 1.75 A of meter current, the load current would be

$$I_{LS} = aI_{HS} = 100 \times 1.75 = 175 \text{ A} \tag{14-52}$$

It is permissible to add extra turns of wire on the low side of a CT to reduce its turns ratio and thereby increase its sensitivity. For example, let us again consider our 500 : 5 CT. Instead of passing the low side wire through the CT, if we loop it through the CT twice (thus creating two turns), we thereby modify the characteristics of the CT to be a 250 : 5 CT.

FIGURE 14–10
Potential transformer and current transformer application.

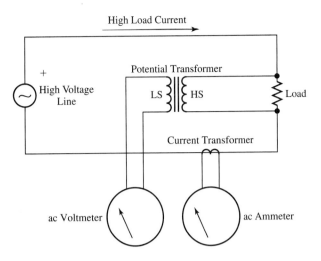

Figure 14-10 illustrates a typical application using a potential transformer and current transformer.

14-12 Useful Transformer Tips

When working with transformers of any type, a few very useful tips help us understand how a transformer will perform under certain conditions. These are summarized below.

- The core losses in a transformer are determined by the applied voltage and frequency. Winding currents have no bearing on the core loss. Reducing the load on a transformer will not reduce its core loss.
- The copper losses in a transformer are proportional to the squares of the winding currents and the square of the frequency. The applied voltage has no bearing on the copper loss.
- The maximum current that can be carried safely by a transformer winding is determined only by the wire gauge of the windings. The dimensions of the core and applied voltage have no bearing on maximum current capacity. Operating a transformer at a reduced voltage will not change its maximum current capacity.
- The real power (watts) that can be delivered safely by a transformer is the volt-amp rating of the transformer times the power factor of the load it is powering. For example, a 1-kVA transformer can deliver only 700 watts to a 0.7-power-factor load.
- Transformers are designed to be most efficient when operated at full rated load (that is, rated voltage and rated current). Operating a transformer at reduced load current or reduced voltage reduces its efficiency.
- Transformers may be operated at lower frequencies if their voltage and apparent power ratings are derated by the same ratio as the frequency reduction.

PROBLEMS

14-1 A transformer with a 2300-turn high side winding and a 900-turn low side winding has its low side connected to a 240-V ac source. Find (a) the turns ratio, and (b) the high side voltage.

14-2 The high-voltage transformer in a particular television receiver has a 25-turn primary and produces a secondary voltage of 20 kV ac when the primary voltage is 250 V ac. How many turns are on the secondary winding?

14-3 A particular transformer has a high side voltage rating of 240 V ac at 60 Hz. What is the maximum safe voltage that can be applied to the high side if the frequency is 50 Hz?

14-4 A 50-Hz transformer is purchased that will be used in North America where the line frequency is 60 Hz. The transformer is rated at 110 V – 24 V, 50 Hz. (a) What is the maximum high side voltage that can be applied to the transformer at 60 Hz? (b) What will be the low side voltage if we apply 120 V ac, 60 Hz to the high side?

14-5 The copper loss of a particular transformer is 20 W when operating at rated load. What will be the copper loss if we maintain the same applied voltage but reduce the load current to one-half of rated current?

14-6 The core loss in a transformer is 10 W when operating at rated load. What will be the core loss if we maintain the same applied voltage but reduce the load current to one-half of rated current?

14-7 A transformer has an input power of 1250 W, a copper loss of 22 W, and a core loss of 13 W. What is the efficiency of the transformer?

14-8 A transformer is delivering 475 W to a load. If the core loss is 4 W and the copper loss is 8.5 W, what is the efficiency of the transformer?

14-9 If we reflect a 120-Ω load impedance from the high side to the low side of a 120 V – 24 V transformer, what will be its reflected low side impedance?

14-10 A certain audio power amplifier has an output that is designed to operate with a 32-Ω speaker. We wish to connect a 4-Ω speaker to the amplifier and have decided that to maintain the best efficiency, we will connect an impedance matching transformer between them so that the speaker-transformer combination presents a 32-Ω impedance to the amplifier. What should be the turns ratio of the matching transformer?

14-11 A three-phase wye-to-delta transformer with a turns ratio of 3.3 has a line voltage of 1050 V ac applied to the high side. The high side line current is 15 amperes. Find (a) the high side phase voltage, (b) the high side phase current, (c) the low side phase voltage, (d) the low side line voltage, (e) the low side line current, and (f) the low side phase current.

14-12 A three-phase transformer converts a line voltage of 416 V ac delta to 208 V ac wye. What is the turns ratio of the transformer?

14-13 A 120 V – 48 V transformer is connected as an autotransformer as shown in Figure 14-7 with the 120-V side as the primary winding. The load resistance is 150 Ω. Find (a) the load voltage, (b) the load current, (c) the input current I_T, and (d) the minimum volt-ampere rating of the transformer.

14-14 A 240 V – 28 V, 1.5-kVA transformer is connected as shown in Figure 14-8 with the 240-V winding as the primary. (a) What is the maximum current that the transformer can deliver to the load? (b) What is the maximum power that can be delivered to the load?

14-15 A PT is connected to a 28,800-V line. It is to operate a 120-V meter. What specification do we need for the PT?

14-16 A PT with a turns ratio of 8 is connected to a 832-V line with a voltmeter connected to the low side winding. What would be the voltage rating of the meter so that 832 V ac on the high side line would cause full scale deflection of the meter?

14-17 A CT is used to monitor the current in a 300-A line. The meter used with the transformer is 5 amperes full scale. What is the ratio of the CT?

14-18 A CT is rated at 250 : 5. If the low side winding is passed through the toroid two times (two turns), and the low side current is 75 amperes, what is the high side current?

ELECTRICAL MACHINES

15 DC Machines

16 AC Machines

DC Machines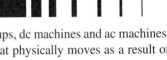

OVERVIEW AND OBJECTIVES

Electrical machines are divided into two fundamental groups, dc machines and ac machines. In this context, the term *machines* refers to any device that physically moves as a result of its interaction with an electrical circuit. This includes dc and ac generators which are designed to convert rotating mechanical power into dc and ac electrical power, and dc and ac motors which perform the exact opposite function. As we will see, dc generators and dc motors are very similar in construction, and in fact it is often possible to purchase off-the-shelf "dc machines" which can be applied as either a dc generator or dc motor.

Objectives

After completing this chapter, you will be able to

- State Faraday's Law and Lenz's Law.
- Calculate the voltage generated by passing a wire through a magnetic field.
- Sketch a simple generator and describe how it operates.
- Describe a commutator and brush assembly and state how it works.
- Find the force produced on a current-carrying wire in a magnetic field.
- State the differences between a shunt and compound dc generator and describe the performance characteristics of each.
- Sketch a simple dc motor and describe how it operates.
- State the differences among a shunt, series, and compound dc motor, and describe the performance characteristics and application examples of each.

15-1 Introduction

In our previous study of electromagnetics, we learned that a current through a wire causes a corresponding magnetic field to be induced around the wire, with the strength and direction of the magnetic field depending on the respective magnitude and direction of the current in the wire. In an opposite manner, if a magnetic field is passed through a wire, a current will be produced, with the magnitude and direction of the current dependent on the velocity of the field with respect to the wire and the flux density of the field and the direction of motion of the field with respect to the wire.

In this chapter we will be investigating the two major applications for these two electromagnetic phenomena, specifically for producing electrical power (in the case of the dc generator) and for producing mechanical power (in the case of the dc motor).

In this chapter, the reader will notice that the text often switches between SI and English units when stating force (pounds or ounces vs. Newtons), torque (foot-pounds or ounce-inches vs. Newton-meters), and distance (feet or inches vs. meters). The mixing of measurement systems is intentional. The reader will find that rotating electrical machinery specifications and nameplates often contain a mix of unit systems, and in many cases, the units of measure for a given machine depend on where the machine was designed or manufactured. Therefore, to familiarize the reader with various combinations of unit systems and the conversions between them, example problems and end-of-chapter problems will contain a mix of units.

15-2 Magnetic Induction and the DC Generator

When a wire is passed through a magnetic field, as illustrated in Figure 15-1, a voltage is induced in the wire. If the wire is connected to a complete circuit, a corresponding current will flow. **Faraday's Law** states that the magnitude of the induced current will be proportional to the number of series-connected loops (turns) of wire that are passing through the field and the perpendicular speed with which it is moving in the field, or, mathematically,

$$e = N \frac{d\Phi}{dt} \quad (15\text{-}1)$$

where

e = the induced voltage in volts (V)

N = the number of series-connected turns of wire in turns (t)

$\dfrac{d\Phi}{dt}$ = rate of change in flux in Webers/second (Wb/s)

Additionally, if the flux density is known, the induced voltage can be found using the expression

$$e = Blv \quad (15\text{-}2)$$

where

e = the induced voltage in volts (V)

B = the flux density in Teslas (T)

l = the length of the conductor that is in the magnetic field in meters (m)

v = the relative velocity between the wire and flux, in meters/second (m/s)

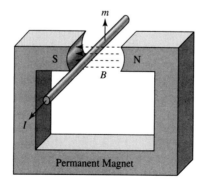

FIGURE 15–1
Magnetic induction in a wire moving in a field.

EXAMPLE 15-1

Referring to Figure 15-1, a single wire is passed through a 0.1-T field at a velocity of 0.5 m/s. (a) If the width of the field is 5 cm, what is the induced voltage? (b) For the same wire, what velocity is required to induce a voltage of 200 mV?

SOLUTION

(a) All of the given quantities are in the correct units for Equation 15-2 except for the length of the wire. Therefore, we will first convert the 5-cm length to 0.05 m. Equation 15-2 can now be used to find the voltage, which is

$$e = Blv = 0.1 \times 0.05 \times 0.5 = 2.5 \text{ mV} \tag{15-3}$$

(b) Rearranging Equation 15-2 to solve for the velocity, we have

$$v = \frac{e}{Bl} = \frac{0.2}{0.1 \times 0.05} = 40 \text{ m/s} \tag{15-4}$$

In addition to using Faraday's Law to determine the magnitude of the voltage (and current), the right-hand rule is used to determine the polarity of the induced voltage and the direction of current flow. As illustrated in Figure 15-2, if we bend the middle finger of the right hand at a right angle to the index finger and point the thumb upward, we can use the three to determine current direction. If we point the thumb in the direction of motion, M, and the index finger in the direction of flux, B, then the middle finger will be pointing in the direction of current, I. The reader is invited to try the right-hand rule shown in Figure 15-2 on the diagram in Figure 15-1.

In Figure 15-1, note that for a constant velocity of motion, the maximum voltage will be induced when the wire is moving at a right angle to the magnetic lines of flux. This is the condition that will maximize $d\Phi/dt$. If we move the wire in the horizontal direction (that is, parallel to the flux lines), $d\Phi/dt = 0$, and no voltage will be induced. In general terms, the magnitude of the induced voltage is proportional to the sine of the angle of motion with respect to the angle of the flux lines.

Figure 15-3 shows a wire that is bent into a loop, placed into a magnetic field, and rotated in the clockwise direction. The wire on the left moves upward while the wire on the right moves downward, thus creating two currents that are moving in opposite directions, thereby causing a circulating current in the wire loop. If we could monitor the current in the loop circuit with an oscilloscope while it is rotating, we would find that as the loop passes the horizontal position (the position shown in the figure), the current would be maximum,

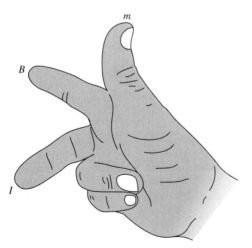

FIGURE 15–2
Right-hand rule for magnetic induction.

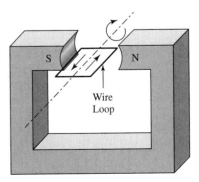

FIGURE 15–3
Wire loop rotating in a magnetic field.

and as it passes through the vertical position, the current would be zero. Also, every one-half revolution, each wire reverses its vertical direction (that is, the wire on the left that was moving upward is now on the right moving downward, and vice versa), thereby reversing the direction of current in the loop. If we rotate the loop at a constant angular velocity, we would see that the wave shape of the current is a sine wave (assuming that the flux density is uniform everywhere between the two poles).

If we rotate the wire loop at a faster rate, the increased angular velocity will cause both an increase in the induced voltage and corresponding current (because of the increase in $d\Phi/dt$) and an increase in the frequency of the generated sinusoid. As an example of this phenomenon, when we wish to charge an automobile's battery faster, we generally "rev" the engine slightly. The increased rotating speed of the alternator produces a higher voltage and higher current, thereby increasing the battery charge rate.

If we could somehow open the wire loop in Figure 15-3, we could insert a load into the circuit and have our sinusoidal generator deliver power to the load. However, this presents a problem because the wire loop rotates while the load remains fixed in position. This can be done with devices called **slip rings** and **brushes**. Figure 15-4 shows our generator with slip rings and brushes installed. The slip rings are durable metal rings with smooth, polished surfaces. They are installed on the generator with each one electrically connected to opposite ends of our wire loop and mounted so that they rotate with the wire loop. Each brush is made of a carbon composition that will easily slide against the slip ring while making good electrical contact. They are held snug against the slip ring with springs (not shown). In operation the slip rings rotate while the brushes remain fixed in position. Therefore, we can connect an electrical load to the brushes, as shown in Figure 15-4, to extract power from the wire loop as it rotates. In operation, there is a small voltage drop across the contact made by the brush and the commutator segment called the **brush drop**, which is 0.5 to 1 volt per brush.

If we wish for our generator to produce dc instead of ac, we will need an additional device to perform rectification of the ac waveform. One method of performing this rectification is to use a bridge of rectifier diodes. However, the most common method used is to rework the slip rings in our generator as shown in Figure 15-5. Here we have removed one of the slip rings and split the other one into two segments that are separated by a thin insulator. Each of the two segments is connected to one end of our rotating wire loop. Next, we add the second brush positioned on the left side so that each brush touches one of the two segments. This assembly is called a **commutator and brush assembly**. Recall that earlier we determined that as the wire loop rotates, the current direction reverses every half-cycle of rotation. However, note that with our commutator and brush assembly, the connections between the commutator segments and the brushes also reverse every half-cycle of revolution. Therefore, the resulting signal delivered to the load is a rectified sinusoid as shown in Figure 15-6.

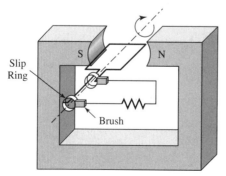

FIGURE 15–4
AC Generator with slip rings and brushes.

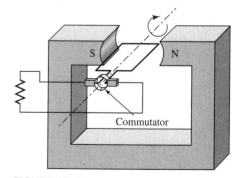

FIGURE 15–5
DC Generator with commutator and brushes.

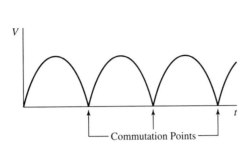

FIGURE 15–6
DC Generator output waveform.

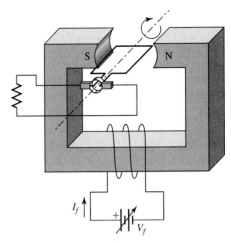

FIGURE 15–7
DC Generator with field control.

An unusual phenomenon occurs as the wire loop rotates that should be explored. Because of the width of the brushes, for a short period of time during rotation, the brush actually touches both commutator segments (called commutation), thereby creating a short. One could conclude that this short circuit would cause an extreme amount of arcing and therefore burn the brush and commutator. However, notice that commutation occurs when the wire loop is in the vertical position with the conductors moving parallel to the flux. Therefore, at the time of commutation, the voltage and current in the loop is zero, thereby rendering the problem insignificant.

Earlier, we learned that we can increase or decrease the output voltage of our generator by changing the rotating speed of the wire loop. However, there are occasions where the mechanical device (called a **prime mover**) that is rotating the wire loop operates at a constant speed, thereby making it necessary to find another method to adjust the output voltage. Figure 15-7 shows such a method. In this case, we have replaced the permanent magnet with an electromagnet. A dc power supply is connected so that the electromagnet produces the same north-south field polarity, but because the power supply is adjustable, we can now control the magnetomotive force of the coil and the corresponding flux density. Therefore, the power supply voltage (called the **field voltage**, V_f) directly controls the output voltage of the generator. This arrangement allows us to vary the output voltage of the generator from near zero[1] to the point where the core of the electromagnet saturates.

With the arrangement shown in Figure 15-7, one might ask why we need the generator at all. Why not eliminate the generator and use the power supply to power the load? To answer, we must understand the source of most of the power delivered by the generator. With a large number of turns on the field coil, we can make the field current and corresponding power input to the field very small while maintaining a large magnetomotive force. The actual power delivered by the generator comes from the prime mover, not the field. The power input to the field is not delivered to the load but instead is dissipated as heat in the field windings. In large generators, the output power can be thousands of times more than the power input to the field. In fact, as we will see later, many generators actually supply their own field current (called **self-excited generators**).

There are three methods to reverse the output voltage polarity of our generator. Naturally, the most obvious is to reverse the connections to the brushes. However, on large generators, the conductors can be extremely large, making it difficult or impossible to

---

[1] Because the magnetic core retains some small amount of residual magnetism, it is impossible to decrease the generator output voltage to zero without stopping the rotor. Even with the field power supply switched off, there will still be a small generated output voltage.

rewire the connections. If we reverse the direction of rotation of the wire loop, the output polarity will reverse. However, if the generator is being operated from a turbine or a reciprocating engine, reversing the direction of rotation is not possible. The third method involves reversing the connections to the field power supply. This will reverse the direction of current in the field coil, the north-south polarity of the magnet, and the polarity of the generated voltage.

Naturally, one loop of wire in our generator rotor will produce only a very small voltage. According to Faraday's Law, to increase the output voltage we could increase $d\Phi/dt$ either by increasing the rotor speed or increasing the flux density. However, there are limitations to the maximum speed that we can spin the rotor because centrifugal force will tend to make the rotor fly apart. We could also increase $d\Phi/dt$ by increasing the field flux density, but this too has limitations because of core saturation. Therefore, we usually increase the number of turns on the rotor to construct a generator with a higher output voltage.

Another method to increase output voltage without increasing the rotor speed or field flux density is to add additional poles to the field. Figure 15-8 illustrates a four-pole field for a dc generator (the rotor is not shown). The field windings are connected in series so that the magnetic polarity alternates from pole to pole (that is, north, south, north, south). When the rotor is rotated in this field system, the rotor wires will pass through twice as many magnetic polarity changes for each revolution. Assuming that the flux in each pole is the same as that for our two-pole generator, we can achieve the same output voltage by spinning the rotor at only half speed. Therefore, for the same rotor speed and pole flux, the four-pole generator can produce twice the voltage of a two-pole generator. Practically speaking, the only limitation to the numbers of poles in a generator is in the space required to house the windings. We can add any number of poles that we wish, as long as we have them in pairs.

Another improvement that can be made to our generator is in the rotor windings. Notice in Figure 15-6 that when the rotor passes through a commutation point, the voltage is near zero. If we add an additional coil to the rotor, as shown in Figure 15-9, and physically orient it at 90° to the original coil, then we will always have at least one of the two coils generating a usable voltage. We now have four commutator connections to make, so we must split the commutator into four segments. Figure 15-10 illustrates the output voltage from a two-coil rotor. Notice that the commutation point now occurs at such a point that the output voltage never drops to zero. By adding more coils and more commutator segments, we can achieve a generator with an output voltage that is very smooth, thereby approaching a continuous dc voltage. Figure 15-11 is a photograph of a rotor with several windings, with the commutator divided into many segments. Also, notice in the photograph that the rotor core on which the coils are wound is laminated. Although this is a dc machine,

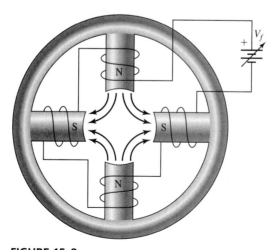

FIGURE 15–8
DC Generator four-pole field.

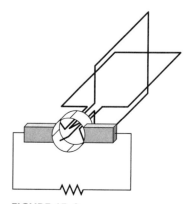

FIGURE 15–9
DC Generator rotor with two coils.

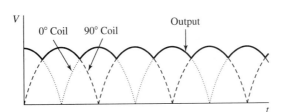

FIGURE 15–10
Coil and output waveforms for a two-winding rotor.

FIGURE 15–11
Rotor with several rotor coils and commutator segments.

the rotor spins within a magnetic field, so the rotating motion will induce eddy currents in the rotor core material, which are minimized by core lamination.

As indicated in Equation 15-1, the induced voltage in a generator is directly proportional to the speed at which the rotor rotates. If the field flux density is held constant (as with a permanent magnet), the ratio of output voltage to rotating speed is extremely linear and repeatable. For this reason, permanent magnet generators are often used as inexpensive speed transducers (called **tachogenerators**). By connecting a voltmeter to the generator output and calibrating the meter in rpm, we can construct a very inexpensive, simple, and reliable tachometer.

EXAMPLE 15-2

A permanent magnet dc generator is connected to the crankshaft of an automobile engine and the output of the generator is connected to a voltmeter. With the engine running at 4000 rpm, the generator outputs 13 volts. What is the engine speed when the generator output is (a) 10 volts, (b) 3 volts, (c) 0 volts?

SOLUTION The ratio of rpm to output voltage for the generator is 4000 rpm/13 V = 333.3 rpm/V. Because the generator is linear over the entire range, multiply each of the output voltages by 333.3 rpm/V to find the engine speed.

(a) 10 V × 333.3 rpm/V = 3333 rpm
(b) 3 V × 333.3 rpm/V = 1000 rpm
(c) 0 V × 333.3 rpm/V = 0 rpm

15-3 Shunt and Compound DC Generator

Depending on the performance characteristics we desire from our dc generator, there are several different ways to connect the armature and the field. Generally speaking, there are only two types of field windings for a dc generator,[2] the **shunt field** and the **series field**.

The shunt field is designed to be connected in parallel with the rotor (called the **armature**) of the generator so that current generated by the armature can power the field. Alternately, the shunt field can also be powered from a separate dc power source. In either case, it is desirable to build the shunt field so that its resistance is high enough to keep the

---

[2] Most large dc generators also have interpole and compensating windings which correct for electromagnetic phenomena within the generator. These additional windings do not affect the basic understanding of generator operation and therefore are not within the scope of this text.

field excitation current low while delivering a strong magnetic field. Therefore, the shunt field consists of many turns of wire to produce a high magnetomotive force, and the wire is small so that the field coil will have a high resistance.

The series field is designed to be connected in series with the generator load and therefore must be capable of carrying the large currents produced by the generator. As a result, the series field is constructed of fewer turns of large wire. The combination of high current and fewer turns results in a series field with a magnetomotive force roughly equivalent to that of the shunt field.

In both the self-excited shunt generator and the compound generator (which uses both the series and shunt fields and is always self-excited), the power for the field excitation comes from the generator itself. However, this could leave us to wonder how it starts itself, since we have a, "Which comes first?" scenario. The field coil needs power to produce a magnetic field, and the rotor needs a magnetic field to produce that power. Actually, all generators are designed using a core material that has some small amount of retentivity. In doing so, even though there is no field coil excitation, there is still enough residual flux to allow the armature to produce a small voltage. This small voltage excites the field coil, which adds additional flux to the residual field, thereby strengthening the field. This, in turn, produces more voltage. The process (called **voltage buildup**) continues to "snowball" until the generator reaches full rated output voltage. Therefore, in self-excited generators, it is important that the residual field is oriented in the same direction as the field produced by the field coil; otherwise, the flux produced by the field coil will cancel the residual magnetism, causing no voltage buildup. The residual flux can be oriented in the desired direction by a simple process called **flashing the field**. This is done by simply connecting rated voltage of the desired polarity from a dc power supply to the shunt field coil, thereby remagnetizing the core in the proper direction.

Shunt Generator Model

When analyzing generators for electrical performance, the various generator components are schematically pictured as voltage sources and coils with resistances. Figure 15-12 shows a simplified schematic of a self-excited shunt generator. In this model, we have the armature which generates the armature voltage V_a and delivers an armature current I_a. Connected in parallel with the armature is the field coil with resistance R_f and field current I_f. The output voltage of the generator (also called the **terminal voltage**) is V_t, and it delivers an output current I_t. If a load resistance R_L is connected to the generator, the load current will be I_L. From the schematic, we can write several governing equations, which are

$$I_a = I_f + I_t \tag{15-5}$$

$$I_f = \frac{V_t}{R_f} \tag{15-6}$$

$$V_t = V_a \tag{15-7}$$

$$I_L = I_t = \frac{V_t}{R_L} \tag{15-8}$$

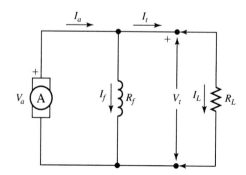

FIGURE 15–12
DC Shunt generator model.

EXAMPLE 15-3

A dc generator modeled as shown in Figure 15-12 delivers a terminal voltage V_t of 125 V dc to a load resistance of 25 Ω. The generator has a field resistance of 250 Ω. Determine (a) the load current, (b) the field current, (c) the field power, (d) the armature current, and (e) the total power delivered to the load.

SOLUTION

(a) Solve for the load current using Equation 15-8.

$$I_L = I_t = \frac{V_t}{R_L} = \frac{125}{25} = 5 \text{ A} \tag{15-9}$$

(b) The field current is found using Equation 15-6.

$$I_f = \frac{V_t}{R_f} = \frac{125}{250} = 0.5 \text{ A} \tag{15-10}$$

(c) The field power is

$$P_f = V_f I_f = 125 \times 0.5 = 62.5 \text{ W} \tag{15-11}$$

(d) The armature provides current to both the field and the load; therefore, the armature current is

$$I_a = I_f + I_L = 0.5 + 5 = 5.5 \text{ A} \tag{15-12}$$

(e) The load power is

$$P_L = V_t I_L = 125 \times 5 = 625 \text{ W} \tag{15-13}$$

Although Equations 15-5 through 15-8 can be used for rough estimates of shunt generator performance, there are components missing from the model of Figure 15-12 which create errors in our calculations. We are aware that because the armature windings are made of wire (which has resistance), when the armature delivers current to the field and load, the armature wire will incur an $I \times R$ voltage drop. The drop will reduce the output voltage of the generator so that the output voltage now becomes dependent to some extent on the load current. Additionally, we also know that there will be a small brush drop which is also going to reduce the terminal voltage. Therefore, Figure 15-13 illustrates a more accurate representation of an actual shunt generator. Note that the armature resistance R_a and the brush drop V_b have been added in series with the armature. Also, we have renamed the armature voltage E_a because this voltage is now an internal induced voltage that cannot be

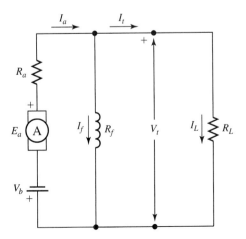

FIGURE 15–13
More precise dc shunt generator model.

measured directly.[3] For this model, Equations 15-5, 15-6, and 15-8 are still valid. However, we now need to rewrite Equation 15-7, which becomes

$$V_t = E_a - V_b - I_a R_a \qquad (15\text{-}14)$$

■ EXAMPLE 15-4

A dc shunt generator is rated at 120 V dc and 15 amperes. It has an armature resistance of 0.1 Ω, a total brush drop of 2 volts, and a field resistance of 240 Ω. Assuming that the field flux remains constant, what is the no-load output voltage of the generator?

SOLUTION For any load current up to full rated load, the induced armature voltage E_a remains directly proportional to the field flux. The flux is assumed to be constant, so E_a will not change. Therefore, our solution method will be to find E_a under full load, and then use it to find the output voltage when the load is removed. We begin by finding the armature current I_a.

$$I_a = I_L + I_f = I_L + \frac{V_t}{R_f} = 15 + \frac{120}{240} = 15.5 \text{ A} \qquad (15\text{-}15)$$

Next, rearrange Equation 15-14 to solve for E_a.

$$E_a = V_t + V_b + I_a R_a = 120 + 2 + 15.5 \times 0.1 = 123.55 \text{ V} \qquad (15\text{-}16)$$

Now, with E_a known, we can perform a no-load analysis. In this case, I_L becomes zero. However, because the shunt field is still connected, we must include the field current in the analysis. The load current is zero, so the armature current I_a is now

$$I_{a(\text{NL})} = \frac{E_a - V_b}{R_a + R_f} = \frac{123.55 - 2}{0.1 + 240} = 0.51 \text{ A} \qquad (15\text{-}17)$$

Now we find the no-load output voltage, which is

$$V_{t(\text{NL})} = E_a - V_b - I_{a(\text{NL})} R_a = 123.55 - 2 - 0.51 \times 0.1 = 121.50 \text{ V} \qquad (15\text{-}18)$$

The **efficiency** of a generator is the ratio of its output power to its input power. The output power of the generator is the product of the voltage and current. However, because the input power of the generator is mechanical power, we must calculate the power (in watts) based on the rotational speed and shaft torque. The input power is

$$P_{\text{in}} = \frac{T n_r}{7.04} \qquad (15\text{-}19)$$

where

P_{in} = the input power in watts (W)

T = the input shaft torque in foot-pounds (ft-lb)

n_r = the rotation shaft speed in revolutions per minute (rpm)

Therefore, the shunt generator efficiency is

$$\eta = \frac{P_{\text{out}}}{P_{\text{in}}} = \frac{V_t I_t}{(T n_r / 7.04)} \qquad (15\text{-}20)$$

where

η = the efficiency (dimensionless)

V_t = the generator terminal voltage in volts (V)

I_t = the generator output current in amperes (A)

---

[3] Some texts use the variable E_o instead of E_a to designate the internally generated armature voltage.

Rotor Copper Loss

This is the I^2R loss in the rotor due to the resistance of the wire. This loss varies with the square of rotor current.

Rotor Core Loss

Because the rotor core (the iron upon which the rotor windings are wound) is rotating inside a magnetic field, there will be eddy current and hysteresis losses in the rotor core. These losses vary with the field flux and the rotor speed.

Field Copper Loss

The I^2R loss in the field windings dues to the resistance of the wire. This loss varies with the square of the field current.

Brush Loss

There is power loss in the brush-commutator interface. This loss is proportional to the rotor current and brush drop and is $V_b I_a$.

Friction

These are losses due to mechanical friction. They include the friction of the shaft bearings and the friction created by the commutator and brush assembly.

Windage

These are losses due to wind resistance of the rotor. In most generators, cooling fins are attached to the rotor to circulate air through the generator, thus promoting cooling and allowing the generator to be operated at higher output currents. These cooling fins increase the windage loss.

Notice that there are rotor core losses, but there are no core losses in the stationary field core (called the **stator**). There are no stator core losses because the magnetic field in the stator core is constant. Therefore, there are no eddy current or hysteresis losses.

■ EXAMPLE 15-5

A shunt dc generator rotating at 1800 has an output voltage of 200 V dc. The input torque is 6.17 ft-lb. If the efficiency is 71 percent, what is the output current?

SOLUTION We rearrange Equation 15-20 to solve for I_t, which is

$$I_t = \frac{\eta(Tn_r/7.04)}{V_t} = \frac{0.71(6.17 \times 1800/7.04)}{200} = 5.6 \text{ A} \qquad (15\text{-}21)$$

One additional problem exists with our self-excited shunt generator. Its output voltage is not adjustable. We can add voltage adjustment by controlling the field current. Figure 15-14 shows our self-excited generator with a rheostat installed in series with the shunt field. By increasing the resistance of the rheostat, we reduce the field current, field flux, and, consequently, the generator's output voltage. A field rheostat provides the capability to adjust the field current from near zero (depending on the maximum resistance of the rheostat) to full rated current. Therefore, with the field rheostat, we can adjust the generator output voltage from near zero to full rated voltage.

Example 15-4 illustrated that the output voltage of a shunt generator sags somewhat as the load current increases. In addition, because the generator is supplying its own field current, as the output voltage sags, the field current sags by the same ratio. This, in turn,

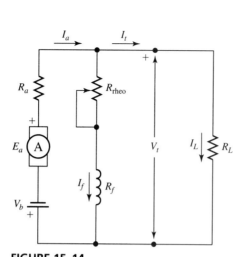

FIGURE 15–14
Shunt dc generator with field rheostat.

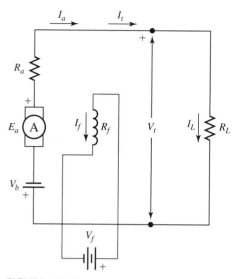

FIGURE 15–15
Separately excited shunt dc generator.

causes a further reduction in the overall voltage sag. We can overcome some of the voltage sag by disconnecting the shunt field from the armature and powering the field from an external power source V_f, thus constructing the separately excited shunt generator illustrated in Figure 15-15. Note that for this generator configuration, the armature current I_a, the generator output current I_t, and the load current I_L are all the same current. Also, because the generator no longer needs to supply current to the shunt field, this generator configuration can deliver more current to the load than the self-excited generator.

When calculating the efficiency of the separately excited shunt generator, the power input to the field must be added to the mechanical input power. Therefore, the efficiency is

$$\eta = \frac{P_{\text{out}}}{P_{\text{in}}} = \frac{V_t I_t}{(T n_r / 7.04) + V_f I_f} \tag{15-22}$$

Compound Generator Model

Figure 15-16 shows the schematics of two compound generators. In this configuration, the shunt and series fields are connected. In Figure 15-16(a) the shunt field is connected across the armature, thus creating the **short shunt compound generator**. Figure 15-16(b) shows

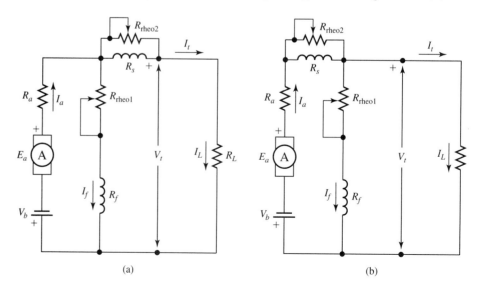

FIGURE 15–16
Compound generator, (a) short shunt and (b) long shunt.

the shunt field connected in parallel with the load, which is called a **long shunt compound generator**. In both cases, the series field has a parallel rheostat (called a **diverter**) to divert current around the field, thereby providing adjustment of the series field current. We assume that the series field is connected, so that it enhances the field flux. If so, then for both generators as the load current increases, the shunt field provides more flux to the field. Increased flux results in increased output voltage, so this has the effect of increasing the output voltage of the generator as the load current increases. By carefully adjusting the shunt field and diverter rheostats, it is possible to "tune" the generator so that there is no sag in the output voltage from no-load to full-load current, thereby creating a near-perfect dc source.

There is little difference between the performance characteristics of the short shunt and long shunt generators. However, the short shunt is the most common configuration because the shunt field current does not flow through the series field and diverter. Therefore, it is slightly more efficient than the long shunt configuration. There are other more exotic combinations of connections of this type of generator, such as reversing the series field (called a **differentially compounded generator**). Further information on these types of connections can be found in any text on electric machines.

15-4 Motor Action and the DC Motor

When we operate a generator, we are converting mechanical power into electrical power. As the generator produces more power, we must input more mechanical power to the generator (in the form of torque). According to **Lenz's Law**, any voltage, current, flux, or force that is produced by any transformer action will oppose the action that caused it. For example, in our generator, the current and voltage induced by rotating the armature will cause a force to be created that opposes the rotation of the armature.

In a similar application of Lenz's Law, if we pass a current through a conductor that is in a stationary position within a magnetic field, a force will be produced. The force will be at a right angle to the magnetic flux lines, and the magnitude of the force will be proportional to the current in the conductor, the number of turns of wire within the field, and the flux density of the magnetic field. Stated mathematically, this is

$$F = BlI \qquad (15\text{-}23)$$

where

F = the resulting mechanical force in newtons (N)

B = the flux density in teslas (T)

l = the effective length of the wire (meters) in the field multiplied by the number of turns

I = the current in the conductor in amperes (A)

Figure 15-17 illustrates this phenomenon. The combination of current in the wire and the flux in the permanent magnet provides an upward force on the wire. The force occurs

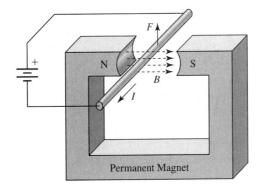

FIGURE 15–17
Force on a current-carrying wire in a magnetic field.

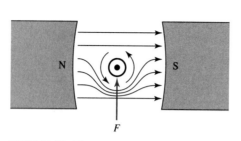

FIGURE 15–18
Flux compression and resulting force.

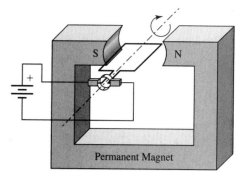

FIGURE 15–19
Simple dc motor.

because of the interaction between the magnetic field around the wire resulting from the current through it, and the magnetic field produced by the permanent magnet. Figure 15-18 shows a cross section of how this interaction produces force. In this illustration, the direction of current flow through the conductor is "out of the page" (the dot in the center of the conductor can be considered to be the tip of an arrowhead indicating current direction). Using the right-hand rule, we can easily determine that this will produce a counterclockwise magnetic field around the conductor. This circular magnetic field around the conductor opposes the flux above the wire and enhances the flux below the wire. As a result, the field bends the flux lines produced by the permanent magnet and forces them to pass underneath the conductor, a phenomenon called **flux compression** or **flux bunching**, which, in turn, produces a counteractive force on the wire, pushing it in the upward direction.

We can use the force produced by this phenomenon to produce torque. If we place a loop of wire in a magnetic field as illustrated in Figure 15-19 and pass a current through the wire, it will produce forces that tend to rotate the wire loop. In our illustration, the current direction will be into the page on the left side and out of the page on the right. This will cause the wire on the left to be pushed upward and the wire on the right to be pushed downward. If the wire loop is suspended on a shaft, the shaft will rotate in the clockwise direction. When the rotor reaches the commutation point, the commutator and brush assembly will reverse the current direction in the wire loop, thereby reversing the forces on the wires and continuing the clockwise motion. Note in Figure 15-19 that there is no difference between this motor illustration and one of the earlier dc generator figures. Actually, there is little difference between most dc motors and dc generators. In many cases they can be used interchangeably. For example, many electric-start lawn mowers have a starter motor that doubles as a dc generator (called a motor generator). When starting the engine, electrical power is consumed from the battery to operate the motor-generator, which in turn produces mechanical torque to start the engine. Once the engine starts, the motor-generator becomes a generator, consuming mechanical power from the engine through the shaft and delivering resulting electrical power to recharge the battery.

There is a potential problem in our motor design of Figure 15-19. At the point where commutation occurs, the wire loop will be in the vertical position. Because forces produced by the rotor will always be at a right angle to the field flux (in this case, vertical), none of the force produced will contribute to rotating motion. If the rotor is started with the wire loop in the horizontal position, it is likely that inertia will carry the rotor through the vertical position and therefore keep the rotor spinning. However, if the rotor is in the vertical position when power is applied, there will be no torque produced and the motor will fail to start. This problem can be solved by additional loops of wire at various angles with respect to the original loop, with additional commutator segments to connect them. In this manner, there will always be at least one loop of wire near the horizontal position which can provide torque. Therefore, no matter what the position of the rotor when power is applied, it will always start.

When the rotor of the dc motor spins within the magnetic field, not only is it producing usable torque, but, like the generator, it is also generating a voltage. This voltage will always be opposite to the applied voltage (again, Lenz's Law), and is called a **counter-emf**, **cemf**, or **back-emf**. When the motor is started and the rotor is not spinning, the cemf will be zero. Without a cemf, the rotor current will simply be

$$I_{a(\text{start})} = \frac{V_t - V_b}{R_a} \quad (15\text{-}24)$$

where

$I_{a(\text{start})} =$ the armature starting current in amperes (A)

$V_t =$ the applied voltage in volts (V)

$V_b =$ the brush drop in volts (V)

$R_a =$ the armature resistance in ohms (Ω)

The armature resistance is generally on the order of a few ohms or less, so the starting current is very high. In fact, the starting current is so high that if sustained, it will likely burn out the armature windings. For this reason, dc motor installations always include circuit fuses or circuit breakers that will remove power from the motor if it fails to start.

Once the motor starts, the rotor windings rotating in the magnetic field begin generating a cemf. Because the cemf opposes the applied voltage, the actual effective voltage applied to the rotor is the difference between the applied voltage and the cemf. Therefore, as the rotor picks up speed and the cemf increases, the rotor current will decrease to a sustainable level that does not cause motor overheating. With the motor running, the armature current is

$$I_a = \frac{V_t - V_b - V_{\text{cemf}}}{R_a} \quad (15\text{-}25)$$

where

$V_{\text{cemf}} =$ the induced counter emf in the armature windings in volts (V)

If we increase the mechanical load on the motor, the rotor speed will decrease, causing a corresponding decrease in the cemf. Referring to Equation 15-25, note that a decrease in the cemf causes an increase in the armature current. In a dc motor, the torque produced is directly proportional to the armature current. In fact, it is relatively easy to add instrumentation to monitor a dc motor's relative output torque by simply connecting an ammeter in series with the rotor circuit.

The motor illustrated in Figure 15-19 is a **permanent magnet dc motor**. Most small instrumentation and hobby motors are permanent magnet motors. They are simple, inexpensive, and very reliable. Their speed can be controlled by varying the applied armature voltage. Their direction of rotation can be reversed by reversing the polarity of the applied voltage. They exhibit good starting torque and excellent speed regulation under varying loads. However, for larger motors, it is difficult and expensive to construct the large high-flux permanent magnets required to operate efficiently. Therefore, large motors use electromagnetic fields as shown in Figure 15-20. In this case, the field coil is powered from the same voltage source that is powering the armature.

It is important to fully understand the relationship between the field flux, cemf, armature current, and motor speed. For the dc motor in Figure 15-20, consider the scenario in which the field coil accidentally becomes disconnected. Intuitively, one would think that in this case the flux density would fall to a very low value, causing a corresponding loss in developed torque and speed. However, this is not the case. When the field coil is disconnected, the only remaining flux is the residual flux in the core. This causes a drastic reduction in the cemf and a large increase in the armature current. If the rotor is not mechanically loaded, the increase in armature current and torque causes an increase in the rotor speed. Without intervention, it is possible that the rotor speed can increase to a point where the centrifugal force causes disintegration of the rotor. Fortunately, in most cases a catastrophe

FIGURE 15–20
DC Motor with electromagnetic field.

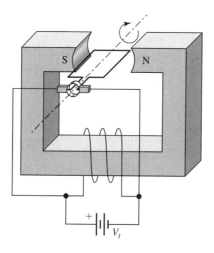

such as this will be prevented by the proper selection and installation of fuses or circuit breakers in the motor circuit.

15-5 Shunt, Series, and Compound DC Motor

As with dc generators, dc motors are available with both shunt and series field windings. The windings are similar to those in the dc generator, with the shunt field coil made from many turns of small wire and the series field coil made from fewer turns of large wire. When using these two windings, there are fundamentally three ways a dc motor can be connected: as a **shunt motor**, a **series motor**, and a **compound motor**. Each connection scheme results in a motor with very different performance characteristics and corresponding applications.

Shunt Motor

The shunt motor shown in Figure 15-21 is connected the same as a shunt generator; that is, with the shunt field connected in parallel with the armature. Therefore, the voltage applied to both the field and the armature is the same. The shunt motor exhibits good starting torque and speed regulation. It is an excellent connection method for general-purpose applications.

For this connection method, the field current is

$$I_f = \frac{V_t}{R_f} \tag{15-26}$$

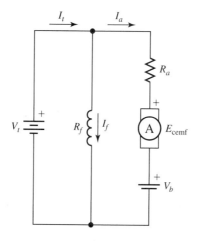

FIGURE 15–21
Shunt dc motor.

the armature current is

$$I_a = \frac{V_t - V_b - V_{\text{cemf}}}{R_a} \tag{15-27}$$

and the total current is

$$I_t = I_a + I_f \tag{15-28}$$

EXAMPLE 15-6

A shunt dc motor like that shown in Figure 15-21 has an applied voltage of 140 V dc, a field resistance of 184 Ω, an armature resistance of 0.3 Ω, and a total brush drop of 2 volts. When operating under a certain mechanical load, the motor runs at 1600 rpm and the total motor current (armature plus field) is 5.6 amperes. If the load is increased and the motor speed drops to 1550 rpm, what will be the total current?

SOLUTION Intuitively, when the load on any motor is increased, the motor will draw more current. Therefore, we know that the answer to this problem will be greater than 5.6 amperes.

First, we know that because the applied voltage remains constant, the field current will also remain constant. Using Equation 15-26, the field current is

$$I_f = \frac{V_t}{R_f} = \frac{140}{184} = 761 \text{ mA} \tag{15-29}$$

Under the initial load condition (before the load is increased), the armature current is

$$I_a = I_t - I_f = 5.6 - 0.761 = 4.839 \text{ A} \tag{15-30}$$

and the cemf is

$$V_{\text{cemf}} = V_t - V_b - I_a R_a = 140 - 2 - (4.839)(0.3) = 136.55 \text{ V} \tag{15-31}$$

Because the cemf in a motor is directly proportional to rotor speed, we can find the cemf under the increased load by the simple ratio

$$\frac{n_r}{n_r'} = \frac{V_{\text{cemf}}}{V_{\text{cemf}}'} \tag{15-32}$$

$$V_{\text{cemf}}' = \frac{n_r' V_{\text{cemf}}}{n_r} = \frac{(1550)(136.55)}{1600} = 132.28 \text{ V}$$

Next, we find the armature current under the increased load, which is

$$I_a = \frac{V_t - V_b - V_{\text{cemf}}}{R_a} = \frac{140 - 2 - 132.28}{0.3} = 19.06 \text{ A} \tag{15-33}$$

Because the field current is constant, we add the field current from Equation 15-29.

$$I_t = I_a + I_f = 19.06 + 0.761 = 19.82 \text{ A} \tag{15-34}$$

Notice the extremely large increase in armature current (nearly fourfold), indicating a fourfold increase in load torque. Because the speed decreased only 3.1 percent with this large change in load, it is easy to conclude that this type of motor has very good speed regulation, a common characteristic of shunt motors.

The speed of a shunt motor can be controlled in one of two ways, armature control or field control. In either case, the motor must be connected so that the field and armature are excited by separate power supplies so that the voltage of each can be independently controlled. With armature control, we apply rated voltage to the field and vary the armature voltage from zero to rated voltage. The result is that the motor speed can be varied from zero to rated speed. With field control, we apply rated voltage to the armature and vary the field from rated voltage to near zero. This results in a rotor speed that varies from somewhat below rated speed to a speed that is above rated speed. Under this method, it is important to monitor the speed of the motor to avoid excessive speed and potential damage to the rotor.

The direction of rotation of a shunt motor can be reversed by reversing either the shunt field connections or the armature connections. Reversing the polarity of the applied voltage will not reverse the direction of rotation because it reverses the current direction in both the shunt field and armature, resulting in no change in rotational direction.

Series Motor

In the series-connected dc motor, the series field is used instead of the shunt field. As shown in Figure 15-22, the series field is connected in series with the armature. Analysis of the series motor is somewhat simpler because the motor circuit is a simple series circuit. The armature current is

$$I_a = I_f = I_t = \frac{V_t - V_b - V_{\text{cemf}}}{R_s + R_a} \qquad (15\text{-}35)$$

When the series motor is energized, assuming that the rotor is not rotating, the cemf will be zero. Therefore, the armature current is limited only by the resistance of the series field, the resistance of the armature, and the brush drop. Therefore, the starting current is

$$I_{t(\text{start})} = \frac{V_t - V_b}{R_s + R_a} \qquad (15\text{-}36)$$

The brush drop is in the order of a few volts, and the armature and series field resistances are very low values, in many cases less than 1 Ω. Therefore, the starting current for a series motor is extremely high. Because the starting current flows through the series field, the flux produced by the field increases directly proportional to the starting current. Additionally, as with the shunt motor, the torque produced by the armature is directly proportional to the armature current. Therefore, the overall starting torque in a series motor is proportional to *the square of the starting current*. For this reason, series dc motors are used only in applications requiring a large starting torque, where it is very likely that the motor must start with a heavy mechanical load, such as cranes, elevators, and diesel-electric trains.

Additionally, note that, because the series field and the armature are connected in series, as the armature current varies, the series field current also varies. As a result, the series motor has poor speed regulation. In fact, if the motor is completely unloaded, its speed will increase without limit until the motor self-destructs. Because of this, series motor applications are designed so that it is very unlikely that the motor will become disconnected from the load. Therefore, they are not connected to the load with drive belts, chains, or

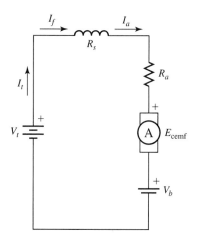

FIGURE 15–22
Series dc motor.

clutches. Instead, they are usually used in a direct-drive configuration where the motor is connected directly to a gearbox or drive mechanism.

EXAMPLE 15-7

A 200-V series dc motor like that shown in Figure 15-22 has an armature resistance of 0.5 Ω and a series field resistance of 0.25 Ω and a total brush drop of 1.5 V. When it is operating at rated load, the cemf is 190 V dc. Find (a) the rated load current, and (b) the starting current.

SOLUTION

(a) The rated load current can be found using Equation 15-35.

$$I_t = \frac{V_t - V_b - V_{cemf}}{R_s + R_a} = \frac{200 - 1.5 - 190}{0.25 + 0.5} = 11.33 \text{ A} \quad (15\text{-}37)$$

(b) Applying Equation 15-36, the starting current is

$$I_{t(\text{start})} = \frac{V_t - V_b}{R_s + R_a} = \frac{200 - 1.5}{0.25 + 0.5} = 264.7 \text{ A} \quad (15\text{-}38)$$

The speed of a series motor can be controlled only by varying the applied voltage. The direction of rotation can be reversed by reversing either the field connections or the armature connections. Reversing the polarity of the applied voltage will *not* reverse the direction of rotation because it reverses the current direction in both the field and armature.

Compound Motor

The compound dc motor shown in Figure 15-23 is a combination of the shunt and series motors, with the two fields connected so that their flux is additive. The motor exhibits the performance characteristics of both the shunt and series motors. The series field causes the motor to have extremely high starting torque, whereas the shunt field gives it excellent speed regulation. Care must be taken when connecting the motor fields, to ensure that the flux produced by the fields is additive. Reversing of one of the fields results in unpredictable and potentially dangerous results, including reversed direction of rotation and runaway speed.

Reversing the direction of rotation of a compound motor is possible using one of two methods: (1) Reverse the armature connections, or (2) reverse the connections of both the series and shunt fields. Reversing the polarity of the applied voltage will *not* reverse the direction of rotation because it will reverse the currents in both the armature and the fields.

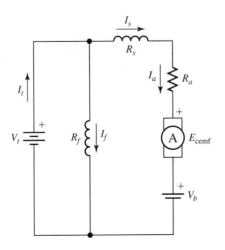

FIGURE 15-23
Compound dc motor.

Motor Efficiency

As with all electrical or mechanical devices, the efficiency is the ratio of the output power to the input power. For a dc motor, the input power is the electrical power in watts, and the output power is the shaft power in watts. Mathematically, this is

$$\eta = \frac{P_{out}}{P_{in}} = \frac{(Tn_r/7.04)}{V_t I_t} \tag{15-39}$$

where

η = the efficiency (dimensionless)
P_{out} = the output power in watts (W)
P_{in} = the input power in watts (W)
T = the shaft output torque in foot pounds (ft-lb)
n_r = the rotor speed in revolutions per minute (rpm)
V_t = the applied input voltage in volts (V)
I_t = the applied input current in amperes (A)

If the motor is separately excited, the field power must be included in the input power. Therefore, the efficiency equation becomes

$$\eta = \frac{P_{out}}{P_{in}} = \frac{(Tn_r/7.04)}{V_t I_t + V_f I_f} \tag{15-40}$$

where

V_f = the field voltage in volts (V)
I_f = the field current in amperes (A)

15-6 Dynamic Braking of DC Motors

There is little difference between a dc generator and a dc motor, so whenever power is removed from the motor's armature, it becomes a generator as the motor speed decreases (assuming that we maintain field excitation). If we connect the armature to a resistive load after removing power, the energy stored in the rotor in the form of angular momentum will be transferred to the resistive load, thereby rapidly decreasing the rotor speed. This method, called **dynamic braking**, can be used to bring the motor to a stop quickly. In some cases we use dynamic braking to transfer stored energy from a motor into a storage device, such as a battery, for later use.

Another method to decelerate or stop a dc motor rapidly is called **plugging**. When plugging a dc motor, we momentarily reconnect the motor in such a way as to reverse the direction of rotation. For example, in a shunt motor we could momentarily reverse the connections of either the shunt field or the armature. Because plugging would result in excessive line currents and motor torque, a current-limiting device, usually a resistor, is inserted in series with the motor during plugging. Without the resistor, the line currents would likely overload the voltage source, or trip circuit breakers; or the excessive torque could loosen or destroy armature windings.

PROBLEMS

15-1 A wire is passed through a 0.15-T magnetic field that is 0.01 m wide at a rate of 1 m/s. What is the generated voltage?

15-2 A wire is passed through a 0.3-T magnetic field that is 1″ wide at a rate of 1 ft/s. What is the generated voltage?

15-3 A four-turn coil of wire is passed through a 0.6-T magnetic field that is 20 mm wide. At what rate does it need to be moved to generate 0.05 volts?

15-4 A coil of wire is passed through a 0.1-T, 35-mm-wide magnetic field at a rate of 1.14 m/s. How many turns of wire are needed on the coil to produce 1 volt?

15-5 A self-excited shunt dc generator is delivering a rated voltage and current of 140 V and 5 A to a load. If the field resistance is 175 Ω, determine (a) the field current, and (b) the armature current.

15-6 When a self-excited shunt generator with a 150-Ω field is connected to a 10-Ω load, the output voltage is 200 V. To operate under these conditions, a field current of 650 mA is required through the field. What value of field rheostat is required?

15-7 A dc shunt generator is rated at 120 V dc and 20 amperes. It has an armature resistance of 0.08 Ω, a total brush drop of 1.5 volts, and a field resistance of 200 Ω. Assuming constant field flux, what is the no-load output voltage of the generator?

15-8 A self-excited dc shunt generator has an induced armature voltage of 190 V dc. It has an armature resistance of 0.15 Ω, a total brush drop of 2 volts, and a field resistance of 275 Ω. It is connected to a 25-Ω load. Determine (a) the voltage delivered to the load, (b) the field current, and (c) the armature current.

15-9 A self-excited shunt generator is operating at 1450 rpm with a shaft input torque of 13.39 ft-lb. It is delivering 15 A at 125 V dc. What is the efficiency?

15-10 A self-excited dc shunt generator is operating at 750 rpm. It is 65 percent efficient when it delivers 35 A at 240 V dc. What is the input shaft torque?

15-11 A separately excited shunt generator is operating at 1500 rpm and a shaft torque of 17.72 ft-lb. The field voltage is 140 V dc and the field current is 775 mA. If the generator is delivering 18 A at 145 V dc, what is the overall efficiency?

15-12 A separately excited shunt generator is delivering 35 A to a 4-Ω load. The 165-Ω field is excited with 145 V dc. It the rotor is operated at 900 rpm with a shaft torque of 50.42 ft-lb, what is the overall efficiency?

15-13 A short shunt compound generator has the following operating parameters: $I_a = 12.5$ A, $I_f = 950$ mA, $R_L = 36$ Ω. What is the output voltage V_t?

15-14 A short shunt compound generator has the following operating parameters: $R_a = 0.12$ Ω, $E_a = 150$ V dc, $V_b = 1.75$ V dc, $R_s = 0.11$ Ω, $R_f = 185$ Ω, $R_L = 10$ Ω. What is the output voltage V_t?

15-15 A wire is resting in a 20-mm-wide magnetic field that has a flux density of 0.4 T. If a current of 3.5 amperes passes through the wire, what is the force created?

15-16 A wire is resting in a magnetic field that is 1″ wide and has a flux density of 1.4 T. What current is needed in the wire to produce a force of 1 N?

15-17 A 3-turn coil of wire is located in a 0.75-T magnetic field that is 15 mm wide. If the current in the coil is 2.2 A, what is the force that is created?

15-18 A 50-turn coil of wire is located in a 0.5-T magnetic field that is 1.25″ wide. What current is needed to produce a force of 3.5 ounces?

15-19 A 120-V shunt motor has an armature resistance of 2.5 Ω, a brush drop of 2 volts, and a field resistance of 150 Ω. When the motor runs at rated load, the cemf is 115 V dc. Determine (a) the starting current, and (b) the running current at rated load.

15-20 A 200-V shunt motor has an armature resistance of 0.8 Ω, a brush drop of 1.5 volts, and a field resistance of 125 Ω. When the motor runs at rated load, the total current is 12.5 A. Determine the cemf when it is running at rated load.

15-21 A 175-V shunt dc motor draws 15 A of current when it delivers 9.7 ft-lb of torque at 1250 rpm. What is the efficiency?

15-22 A 150-V shunt motor is 68 percent efficient when it delivers 5.5 ft-lb of torque at 1800 rpm. Under these operating conditions, what is the input current to the motor?

15-23 For each of the applications listed below, choose the best type of dc motor (shunt, series, compound, or permanent magnet).

(a) compressor (must have high starting torque and run at a constant speed)

(b) crane

(c) conveyor

(d) elevator

(e) ventilation fan (very low starting torque)

(f) subway

(g) radio-controlled model racer

AC Machines

OVERVIEW AND OBJECTIVES

As mentioned in Chapter 15, electrical machines can be divided into two fundamental groups, dc machines and ac machines. Earlier chapters have explored the principles of operation and applications of dc machines, including both dc generators and dc motors. This chapter will explore the principles of operation of ac machines (both generators and motors), their operational characteristics, and their practical applications.

Objectives

After completing this chapter, you will be able to

- Describe the theory of operation of the three-phase alternator, and calculate an alternator's input power, output power, losses, output frequency, and efficiency.
- Discuss the process to connect an alternator to a power grid and remove an alternator from a power grid.
- Describe the principles of operation of the three-phase synchronous motor.
- State the parts of a three-phase induction motor and describe its principles of operation. Perform calculations of three-phase induction motor synchronous speed, rotor speed, slip, and efficiency.
- Describe the operating principles of the single-phase induction motor, define the operating characteristics of various types, and perform calculation of speed, slip, and efficiency.
- State the operating principles of the reluctance motor.
- Describe a universal motor and state the differences and similarities between the universal motor and the series dc motor.

16-1 Introduction

When an ac voltage is applied to an electromagnet, an alternating magnetic field is produced; that is, the north and south poles of the electromagnet continually reverse. The rate at which the pole reversal occurs depends upon the frequency of the applied ac voltage. In this chapter we will investigate ways in which we can turn this continuously changing magnetic field into a *rotating* magnetic field so that we can produce torque, rotating speed, and the resulting rotating mechanical power.

Although some single-phase electrical machines will be investigated in this chapter, a majority of the ac machines will either use or generate three-phase electrical power. Therefore, the reader is encouraged to review Chapter 13 before beginning this chapter.

As in previous chapters, in this chapter all references to ac voltages and currents are in rms (root mean square) unless indicated otherwise.

16-2 AC Generator (Alternator)

In our discussion of dc generators, we first explored a fundamental ac generator that uses slip rings and brushes, as illustrated in Figure 16-1. In this figure there is an armature, which is the rotating part (rotor) of the generator, and a field, which is the stationary part (stator) of the generator. Although the generator in Figure 16-1 is a usable design, it suffers from one disadvantage in a high-current application. When the armature delivers high currents, these currents must pass through the slip rings and brushes, thereby necessitating heavy-duty slip rings and brushes and the arcing and wear associated with the high currents.

This shortcoming can be overcome by exchanging the roles of the rotor and stator. Figure 16-2 shows an ac generator (called an **alternator**) in which the field is the rotor (the rotating part) and the armature is the stator (the stationary part). In this illustration, the field coil is part of an electromagnet (instead of a permanent magnet) excited by a dc power supply V_f. Because the field coil is rotating between the pole pieces of the core, it produces an alternating magnetic field in the core, which in turn induces a corresponding alternating voltage in the armature coil. If the field coil is wound with many turns of small wire, the resistance will be sufficient to keep the field current I_f small and thereby prevent excessive wear on the slip rings and brushes. This generator can then be made to produce a high current by simply winding the armature with heavy-gauge wire.

The output voltage of the alternator can be controlled in one of two ways. By Faraday's Law, the alternator's output voltage is

$$e = N \frac{d\Phi}{dt} \tag{16-1}$$

where

e = the induced voltage in volts (V)

N = the number of series-connected turns of wire in turns (t)

$\dfrac{d\Phi}{dt}$ = the rate of change in flux in Webers/second (Wb/s)

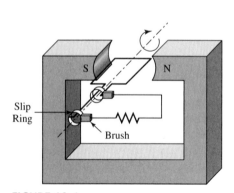

FIGURE 16–1
Fundamental ac generator.

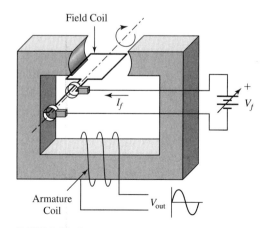

FIGURE 16–2
AC Generator with rotating field and stationary armature.

Once the alternator is constructed, it is not practical to adjust the output voltage by changing the number of turns of wire N in Equation 16-1. However, it is possible to vary the rate of change in the flux, $d\Phi/dt$, by either adjusting the field current, I_f (which controls the flux density), or by adjusting the rotor speed, n_r. In applications where the alternator must produce a constant output frequency, the rotor is operated at a constant rotating speed and the output voltage amplitude is controlled by varying the field current.

At this point it is useful to recognize some important characteristics of alternators.

- The output frequency of an alternator can be varied only by changing the rotor speed.
- The output voltage of the alternator can be varied by changing either the field current or the rotor speed.
- The power input to the field ($V_f \times I_f$) is dissipated entirely inside the alternator in the form of heat. None of the field power is converted to output power.
- Because the brushes and slip rings are in the field circuit, and because the field current is low, the power loss in the brushes and slip rings is relatively small.
- Although the output current capacity of the alternator can be increased simply by increasing the size of the stator wire, remember that a higher output current will cause a proportional increase in the input shaft torque and subsequent increase in the mechanical stress on the field windings.
- Because the rotating field causes an alternating magnetic field in the stator core, the core will suffer hysteresis and eddy current losses. Therefore, the stator core must be laminated to minimize eddy currents. However, the magnetic field in the rotor core is constant, so it does not need to be laminated.

Although the alternator in Figure 16-2 is a usable design, because it has a single-phase ac output, the input shaft torque increases and decreases with every revolution of the rotor. This is apparent by noting that when the field coil is in the horizontal position, it induces a peak voltage and corresponding current in the armature coil. Because this causes the armature to deliver its maximum peak power to the load, the torque at this point will be high. However, when the field coil is in the vertical position, no voltage is induced in the armature coil, and the output power is zero. At this rotor position, the input shaft torque is very small. Therefore, as the field is rotated, the input shaft torque encounters one maximum and one minimum value every half-revolution. This constant change in shaft torque causes the alternator to be noisy and to vibrate mechanically.

A common method used to create a smoother and quieter running alternator is to make it a three-phase alternator as illustrated in Figure 16-3(a). Here we have three armature coils A, B, and C that are wound on pole pieces that are physically positioned 120° apart. The field of this alternator is shown as a permanent magnet for clarity, but in practice, it is usually an electromagnet. As the field is rotated, the north and south poles of the field align with each of the three pole pieces at 120° intervals, causing positive and negative voltages to be generated in each armature coil. In the position shown in Figure 16-3(a), the north

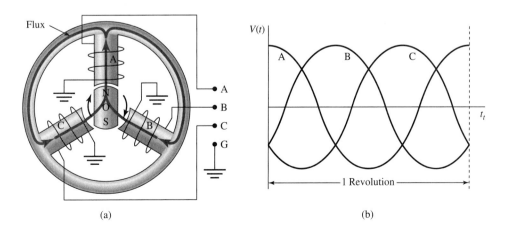

FIGURE 16–3
(a) Simple three-phase ac alternator and (b) resulting output waveform.

pole of the field is aligned with stator pole A. The flux passing through stator pole A divides equally at the top, passes through the outside frame of the alternator, and returns to the south pole of the rotor through stator poles B and C. Using the right-hand rule, note that this will induce a maximum positive voltage in stator coil A and negative voltages of one-half of the maximum amplitude in stator coils B and C. This is illustrated in the plot in Figure 16-3(b). As the field is rotated clockwise 120°, the north of the rotor aligns with stator pole B (which produces a positive maximum voltage in stator coil B). Then when the field is rotated clockwise to 240°, the north pole of the field aligns with stator pole C, producing a maximum voltage in coil C. Notice in Figure 16-3(b) that the waveforms generated by the three stator coils make up a three-phase ac signal. If a balanced three-phase load (either delta or wye) is connected to the three output terminals, the total power delivered to the load is constant over one complete revolution of the rotor. Therefore, the torque input required by the rotor of the alternator is constant, and the vibration and noise are reduced drastically when compared to that of the single-phase alternator.

The efficiency of this alternator can be further increased by decreasing the reluctance between the rotor core and the armature core. This can be done by constructing the alternator with *pairs* of pole pieces, with one-half of each of the armature coils wound around each of the pair. Figure 16-4 shows an armature for a three-phase alternator in which each of the stator coils is wound on two pole pieces. For example, one-half of coil A is wound on stator pole A and the other half is wound on stator pole A′, which is physically located at 180° with respect to pole A. The two portions of coil A are wound in a direction so that flux flowing out of pole A′ and into pole A will cause additive voltages to be induced in the two halves of the coil.

When a field coil is place in the center of the pole pieces and rotated, the magnetic flux flows smoothly as the rotor poles align with the stator poles. This rotation is illustrated for a two-pole three-phase alternator in Figure 16-5. This illustration shows that even when the rotor poles are not aligned with the stator poles, the total field flux is distributed proportionally between the poles to produce a corresponding proportional voltage in each of the coils. Notice also that anytime that the field flux is perpendicular to any pole pair, the voltage induced in that coil is zero. For example, in Figure 16-5(a), the field flux is perpendicular to pole pieces A and A′. Therefore, at that instant, the voltage for phase A is zero.

The alternator illustrated in Figure 16-5 is called a **two-pole three-phase alternator**. *When naming the alternator type, the number of poles refers to the number of poles per output phase, not the total number of poles.* It is common to have alternators with more than two poles per phase. However, poles are always in pairs.

The number of poles and the rotor speed determine the output frequency of the alternator, which is

$$f = \frac{P n_r}{120} \tag{16-2}$$

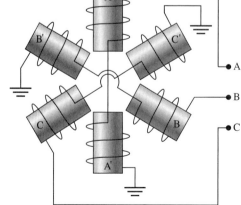

FIGURE 16–4

Two-pole three-phase alternator armature.

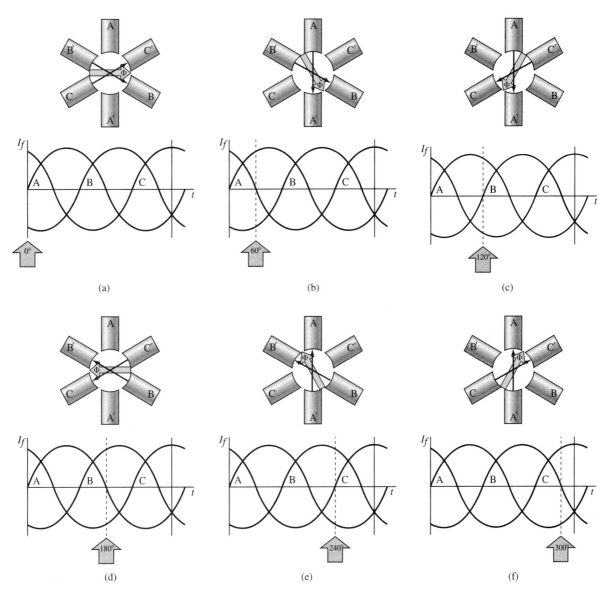

FIGURE 16–5
Two-pole three-phase alternator, and output voltage waveforms.

where

f = the frequency in hertz (Hz)
P = the number of poles per output phase (dimensionless)
n_r = the rotor speed in revolutions per minute (rpm)

■ EXAMPLE 16-1 A 60-Hz diesel-operated shipboard generator is a six-pole three-phase alternator. However, because of a diesel engine governor problem, it is rotating at 1225 rpm and the alternator is producing the wrong frequency. (a) What is the output frequency? (b) What engine speed is needed for the alternator to produce 60 Hz?

SOLUTION

(a) The frequency is

$$f = \frac{Pn_r}{120} = \frac{6 \times 1225}{120} = 61.25 \text{ Hz} \qquad (16\text{-}3)$$

(b) Rearranging Equation 16-2, the ideal diesel engine speed is

$$n_r = \frac{120f}{P} = \frac{120 \times 60}{6} = 1200 \text{ rpm} \qquad (16\text{-}4)$$

■ **EXAMPLE 16-2**

An alternator used in a hydroelectric plant has 240 poles and is connected to a 60-Hz national power grid. What is the rotor speed?

SOLUTION Rearranging Equation 16-2, the rotor speed is

$$n_r = \frac{120f}{P} = \frac{120 \times 60}{240} = 30 \text{ rpm} \qquad (16\text{-}5)$$

■

Note in Example 16-2 that to minimize water cavitation and pipe erosion, hydroelectric turbines are designed to rotate at relatively low speeds. Therefore, the alternators for large hydroelectric plants must be designed to generate the correct frequency when the field is being rotated slowly. This is achieved by constructing the alternators with a large number of pole pairs. Because of the physical space required to house a large number of poles, hydroelectric alternators are very large in diameter. In contrast, alternators used with steam turbine prime movers are designed for high rotating speeds, have few pole pairs, and are relatively small in diameter.

Alternators that are used on large power grids must be connected to the grid using a carefully monitored process. Connecting an alternator to a large grid with the alternator generating an incorrect voltage, phase, or frequency can have dangerous and catastrophic electrical and mechanical consequences, which include burning out circuits, physically loosening and destroying field windings and armature windings within the alternator, or even shearing the alternator's rotor shaft. Because of the potential for damage caused by incorrectly connecting an alternator to a grid, the process is generally performed by an automated industrial controller or digital computer. The correct procedure is outlined below.

1. With the alternator offline and field excitation set to a typical value, increase the speed of the alternator and prime mover until the output frequency of the alternator is the same as the power grid frequency.

2. Adjust the field excitation until the output voltage of the alternator is the same as the voltage of the power grid.

3. Fine-tune the rotor speed until the output phase of the alternator is the same as the phase of the power grid.

4. Connect the alternator to the grid.

Once the alternator is online, the mechanical output power of the prime mover can be increased. This causes the alternator to introduce power to the grid. Note that this does not cause an increase in the alternator's rotor speed. Instead, it causes the alternator's output phase to slightly lead that of the power grid, thereby causing current and power to flow from the alternator into the power grid.

The procedure to take an alternator offline is as follows.

1. Reduce the power output of the prime mover until the electrical power being transferred from the alternator to the grid is near zero (this is normally monitored using wattmeters).

2. Open the circuit connecting the alternator to the grid.

3. Switch off the prime mover and allow the alternator to come to a stop.

Alternators are most commonly used as electrical power generators for automotive electrical systems. However, automotive electrical systems are primarily 12-volt dc systems; therefore the output of the alternator must be rectified (converted to dc) and regulated to maintain the output voltage within a 13-volt to 16-volt dc range. Figure 16-6 is a schematic

FIGURE 16–6

A 12-V dc automotive alternator system.

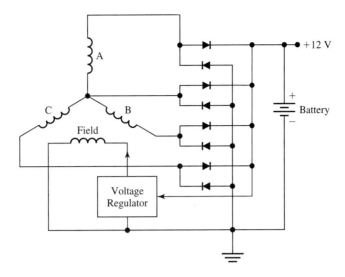

diagram for a typical automotive alternator system. The alternator is connected in a wye configuration, although some systems use a delta connection scheme. The three-phase ac output of the alternator is converted to dc by a three-phase full-wave bridge rectifier consisting of eight rectifier diodes. The dc output voltage of the rectifiers (which is also the battery voltage) is monitored by an integrated-circuit voltage regulator which directly controls the alternator field current. If the battery voltage is too low, the voltage regulator increases the field current, which in turn increases the alternator output voltage. The regulator also indirectly controls the charge maintenance of the battery. If the battery charge is low, the battery voltage will also be low. When the engine is started and the alternator begins producing power, the regulator will detect this low voltage and increase the alternator output voltage, which in turn increases the charge current into the battery. As the battery charges, its internal voltage will increase, thereby demanding less current from the alternator. In a steady-state scenario, the battery voltage is maintained slightly below the alternator output voltage, which keeps a trickle-charge current flowing from the alternator into the battery. In most automotive systems, the regulator maintains the alternator output voltage in the range of 13 to 16 volts dc, depending on the engine speed, the condition of the battery charge, and the other power demands of the automotive electrical system.

The efficiency of an alternator is the ratio of the output electrical power to the sum of the input mechanical power and the field power, or, stated mathematically,

$$\eta = \frac{P_{out}}{P_{in}} = \frac{P_{out}}{P_{shaft} + P_{field}} = \frac{3V_{phase}I_{phase}\cos\theta}{Tn_r/7.04 + V_f I_f} \tag{16-6}$$

where

η = the efficiency (dimensionless)

P_{out} = the output power in watts (W)

P_{in} = the input power in watts (W)

P_{shaft} = the mechanical input power in watts (W)

P_{field} = the electrical power input to the field in watts (W)

V_{phase} = the alternator output phase voltage in volts (V)

I_{phase} = the alternator output phase current in amperes (A)

$\cos\theta$ = the load power factor (dimensionless)

T = the shaft input torque in foot-pounds (ft-lb)

n_r = the rotor speed in revolutions per minute (rpm)

V_f = the field voltage in volts (V)

I_f = the field current in amperes (A)

The internal power loss in the alternator is the difference between the input power (the shaft power plus the field power) and the electrical output power, or

$$P_{\text{loss}} = P_{\text{in}} - P_{\text{out}}$$
$$= P_{\text{shaft}} + P_{\text{field}} - P_{\text{out}} \quad (16\text{-}7)$$
$$= (Tn_r/7.04) + V_f I_f - 3V_{\text{phase}} I_{\text{phase}} \cos\theta$$

where

P_{loss} = the total internal power losses in the alternator in watts (W)

The losses in the alternator are as follows.

- *Field loss*—The $V_f I_f$ power loss in the field caused by the resistance of the wire and brush loss.
- *Stator copper loss*—The $I^2 R$ power loss in the stator windings caused by resistance of the wire.
- *Stator core loss*—Hysteresis and eddy current loss in the stator core.
- *Friction and windage loss*—Rotor loss caused by bearing friction and wind resistance.
- *Stray loss*—Minor losses not included in the above categories. For approximation calculations, this loss can be ignored.

■ EXAMPLE 16-3

A six-pole alternator is delivering 1 kVA, 60 Hz, at 0.9 lagging power factor. The field is excited with 120 V dc at 0.6 A. The input torque is 7.96 ft-lb. Determine (a) the rotor speed, (b) the total input power, (c) the output power, (d) the efficiency, and (e) the losses.

SOLUTION

(a) A six-pole alternator that generates 60 Hz has a rotor speed of

$$n_r = \frac{120f}{P} = \frac{120 \times 60}{6} = 1200 \text{ rpm} \quad (16\text{-}8)$$

(b) The total input power is the sum of the mechanical input power plus the field power, or

$$P_{\text{in}} = P_{\text{shaft}} + P_{\text{field}} = \frac{Tn_r}{7.04} + V_f I_f = \frac{7.96 \times 1200}{7.04} + 120 \times 0.6 = 1429 \text{ W} \quad (16\text{-}9)$$

(c) The output power is

$$P_{\text{out}} = S \cos\theta = 1000 \times 0.9 = 900 \text{ W} \quad (16\text{-}10)$$

(d) The efficiency is

$$\eta = \frac{P_{\text{out}}}{P_{\text{in}}} = \frac{900}{1429} = 0.63 \quad (16\text{-}11)$$

(e) The internal power loss is

$$P_{\text{loss}} = P_{\text{in}} - P_{\text{out}} = 1429 - 900 = 529 \text{ W} \quad (16\text{-}12)$$

Synchronous Motor

If mechanical power is disconnected from an alternator's rotor shaft while it is connected to a power line, the alternator will become a **synchronous motor** and will continue to operate. This is because the power line excites the stator windings, thereby creating a rotating magnetic field in the alternator core. Because the field excitation is still applied, the electromagnet created by the field coil follows the rotating magnetic field created by the stator windings. The rotating speed of the shaft is determined by the line frequency and the number of stator poles per phase. It is

$$n_r = \frac{120f}{P} \qquad (16\text{-}13)$$

Note that the alternator cannot be started as a synchronous motor. Because of the inertia of the rotor, the rotating field that is created by the stator windings moves too fast for the rotor to instantaneously "catch up" and synchronize. Instead of rotating, the alternator remains stalled and loudly "growls." The alternator rotor must be started using some other method, and then, once the speed is near the synchronous speed as defined by Equation 16-13, it can be switched into synchronization.

Although the alternator acts as a synchronous motor when mechanical power is removed from the input shaft, alternators are not specifically designed to be used as synchronous motors. However, synchronous motors are available that can provide mechanical power output and operate at synchronous speeds. Although they can be used for timing applications, instead they are generally used for power factor correction. The power factor of a synchronous motor can be adjusted by varying the field current, and, in fact, the power factor can be made either leading (capacitive) or lagging (inductive). Because the power factor can be made leading, the synchronous motor is sometimes called a **synchronous capacitor** or **synchronous condenser**.

16-3 Three-Phase Induction Motor

The most popular ac motor for applications exceeding a few horsepower is the three-phase induction motor. It is simple, quiet, extremely reliable, powerful for its size, and has few moving parts. The field of the induction motor is in the stator, and the armature is on the rotor. The field is much like the field of a synchronous motor; that is, it has pairs of pole pieces onto which stator coils are wound. The simplest three-phase induction motor is the two-pole motor, which has three pairs of poles, one pair for each phase. The stator windings can be connected in either a delta-, or three- or four-wire wye scheme.

Because the three-phase induction motor field is constructed in much the same way as the synchronous motor, when three-phase power is applied to the windings, a rotating magnetic field is produced in the rotor. This field is of constant flux but rotates at a speed which is determined by the synchronous motor speed, which is

$$n_s = \frac{120f}{P} \qquad (16\text{-}14)$$

where

n_s = the synchronous speed of the rotating field in revolutions per minute (rpm)

However, as we will see, although the field produced by the stator coils rotates at a synchronous speed, the rotor does not.

The key to understanding the operational principle of a three-phase induction motor lies in knowing how the armature operates both electrically and magnetically. As shown in Figure 16-7, the induction motor armature consists of two conductive end rings connected by many conductive rotor bars. The assembly is mounted securely on the rotor shaft, usually by core laminations that fill the space between the end rings. This type of rotor construction is called a **squirrel-cage rotor** because of its physical appearance. When a magnetic field (that is, the armature field) passes through the rotor bars, a current is induced in the bars. Because one side of the rotor will be exposed to a north magnetic field while the other side is exposed to a south field, the current flows through the bars in one direction

FIGURE 16–7
Squirrel-cage rotors.

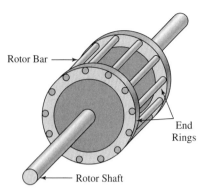

on one side of the rotor, through the end rings, and back through the rotor bars on the opposite side of the rotor, thus making a complete circuit. The bars have an extremely low resistance, so the rotor bar currents are very large. By Lenz's Law, these heavy currents produce a high-flux magnetic field that opposes the armature field that induced the currents. This opposition between the armature field and the rotor field produces the force that spins the rotor. In effect, the rotating armature field pushes the rotor in a circular fashion, thus producing torque. Note that there are no electrical connections to the rotor circuit and that the armature currents are produced by magnetic induction. That is why this type of motor is called an **induction motor**. The motor can be electrically analyzed by using standard transformer analysis techniques, with the armature field coils as the primary and the rotor bars as the secondary. Because there are no electrical connections to the rotor, there is no need for brushes or slip rings. The only moving part in the motor is the rotor, and the only parts that wear out are the bearings that support the rotor on both ends. Because of this simplicity, the three-phase squirrel-cage induction motor is the most reliable of all motors.

By Faraday's Law, the magnitude of the voltage and resulting current (and flux) produced in the rotor bars is proportional to the rate of change of armature flux. When the rotor is not rotating, this flux is alternating at the synchronous speed defined by Equation 16-14. This produces a high current and opposing flux, resulting in a high starting torque. However, when the rotor is rotating in the same direction as the armature field, the relative motion between the armature field and the rotor bars decreases. In fact, the relative alternating frequency of the flux is proportional to the difference between the synchronous speed of the stator field and the speed of the rotor. It is called the **slip speed**, n. Mathematically, this is

$$n = n_s - n_r \tag{16-15}$$

where

n = the slip speed in revolutions per minute (rpm)

n_s = the synchronous speed in revolutions per minute (rpm)

n_r = the rotor speed in revolutions per minute (rpm)

Lower slip speed causes a decreasing amount of induced voltage, current, and opposing flux in the rotor, which also decreases the produced torque. Theoretically, if the rotor were to operate at the synchronous speed ($n_r = n_s$), the relative slip speed would be zero; the voltage, current, and field produced by the rotor would be zero; and no torque would be produced. Because the motor has internal mechanical losses (friction and windage), even with no mechanical load on the shaft, torque will be needed to operate the rotor. Therefore, induction motors are not capable of operating at synchronous speed, even when unloaded. In actuality, when an induction motor is started, the rotor speed increases until an equilibrium is reached in which the slip speed produces a rotor field and torque that exactly offsets the mechanical load on the motor shaft. At this point, rotor acceleration ceases and the motor runs at that speed.

If the mechanical load on the motor shaft is increased, the increased load torque causes the rotor speed to decrease slightly, which increases the slip speed, thereby increasing the output torque to offset the increased mechanical load. Conversely, if the mechanical load on the motor is decreased, the excess torque causes the rotor speed to increase slightly until equilibrium is again reached and the rotor speed "levels off" at a new value.

Slip speed can also be stated as a percentage of the synchronous speed. This is called the **slip**, s. It is the ratio of the slip speed to the synchronous speed, or

$$s = \frac{n}{n_s} = \frac{n_s - n_r}{n_s} \qquad (16\text{-}16)$$

where

$$s = \text{the slip (dimensionless), often given in percent (\%)}$$

The slip ranges from near zero to 1, with a slip of near-zero meaning that the rotor is operating near synchronous speed, and a slip of 1 indicating that the rotor is stalled.

The torque produced by an induction motor depends upon the slip and the electrical characteristics of the rotor. For most squirrel-cage rotors, the maximum mechanical torque is not produced when the rotor is stalled. Instead, maximum torque is produced at a speed called the **breakdown speed**. Therefore, if the mechanical load on the motor is increased so that the rotor speed drops below the breakdown speed, the motor will stall.

A nameplate is attached to all motors which defines the electrical and mechanical characteristics of the motor. This allows engineers and electricians to determine quickly if the motor is operating correctly, and, in the event of a failure, to locate a replacement motor that has matching characteristics. In addition to the voltage and current ratings, the nameplate indicates the shaft output horsepower, full load speed, environmental operating parameters, and physical dimensions. When analyzing nameplate data, it is sometimes necessary to convert the output shaft horsepower to watts. This is done using the conversion expression

$$P_{\text{watts}} = P_{\text{hp}} \times 746 \qquad (16\text{-}17)$$

where

$$P_{\text{watts}} = \text{the power in watts (W)}$$
$$P_{\text{hp}} = \text{the power in horsepower (hp)}$$

▌ EXAMPLE 16-4

A 10-hp four-pole three-phase squirrel-cage induction motor is operating at rated load. The nameplate ratings are as follows: line frequency 60 Hz, line voltage 208 V ac, line current 28.7 A, power factor 0.88 lagging, and slip 3 percent. Determine (a) the input power, (b) the output power (in watts), (c) the motor efficiency, and (d) the operating rotor speed.

SOLUTION

(a) The input power is

$$P_{\text{in}} = \sqrt{3} \, V_{\text{line}} I_{\text{line}} \cos \theta = \sqrt{3} \times 208 \times 28.7 \times 0.88 = 9099 \text{ W} \qquad (16\text{-}18)$$

(b) The output power is

$$P_{\text{watts}} = P_{\text{hp}} \times 746 = 10 \times 746 = 7460 \text{ W} \qquad (16\text{-}19)$$

(c) The efficiency is

$$\eta = \frac{P_{\text{out}}}{P_{\text{in}}} = \frac{7460}{9099} = 0.82 \qquad (16\text{-}20)$$

(d) To find the rotor speed, we must first use Equation 16-14 to calculate the synchronous speed, which is

$$n_s = \frac{120 f}{P} = \frac{120 \times 60}{4} = 1800 \text{ rpm} \qquad (16\text{-}21)$$

Now the rotor speed can be determined by rearranging Equation 16-16 to solve for the rotor speed.

$$n_r = n_s(1 - s) = 1800 \times (1 - 0.03) = 1746 \text{ rpm} \quad (16\text{-}22)$$

Induction motors typically operate at slip values between 0.3 percent and 3 percent. Therefore, as long as the motor is operating within the range of no load to rated load, the number of poles in the motor can be determined using the equation

$$P \cong \frac{120f}{n_r} \quad (16\text{-}23)$$

The number of poles must be an integer value, so we simply use Equation 16-23 and then truncate the result to find the value of P.

EXAMPLE 16-5

A 60-Hz induction motor operating at rated load is rotating at 1150 rpm. (a) Find the number of poles in the motor. (b) What is the synchronous speed of the motor? (c) If the mechanical load is removed from the motor, what will be its approximate rotating speed?

SOLUTION

(a) Using Equation 16-23, we have

$$P \cong \frac{120f}{n_r} = \frac{120 \times 60}{1150} = 6.26 \quad (16\text{-}24)$$

Truncating the result, the number of poles is 6.

(b) The synchronous speed is

$$n_s = \frac{120f}{P} = \frac{120 \times 60}{6} = 1200 \text{ rpm} \quad (16\text{-}25)$$

(c) For an unloaded induction motor, the slip is approximately 0.3 percent, or 0.003. Therefore, we use Equation 16-16 to find the approximate unloaded speed, which is

$$n_r = n_s(1 - s) = 1{,}200(1 - 0.003) = 1196 \text{ rpm} \quad (16\text{-}26)$$

Although the most popular rotor type for three-phase induction motors is the squirrel-cage rotor, some are constructed with a wound armature instead of using rotor bars. This type of motor is called a **wound rotor three-phase induction motor**. In this case, three sets of windings are placed on the rotor and connected in parallel so that armature currents can circulate as they do in the squirrel-cage rotor. Because the additional wire adds resistance to the rotor circuit, the result is that, although it operates using the same principles as any three-phase induction motor, it has different starting and running torque characteristics. The main effect on the motor's characteristics is that the breakdown speed of the motor can be varied, even to the extent that the breakdown point can be adjusted to be zero rpm, thereby creating a motor that produces its highest torque at startup. The tradeoff in adding resistance to the rotor circuit is that the motor exhibits poorer speed regulation than the squirrel-cage motor does.

Additionally, some wound rotor induction motors have the rotor coil connections brought out using slip rings and brushes so that additional resistance can be added to or removed from the rotor circuit while the motor is operating, thereby resulting in a motor with starting and running characters that can be adjusted "on the fly." In this case, rheostats can be added so that resistance can be added at startup to produce very high starting torque, and then once the motor reaches running speed, the resistance can be removed to exhibit the good speed regulation characteristic of squirrel-cage motors. Of course, the tradeoff in having such a versatile motor is that it has brushes, slip rings, and rotor windings, which increase the complexity, lower the reliability, and increase the required maintenance of the motor.

16-4 Single-Phase Induction Motor

Although the three-phase induction motor is the motor of choice for a vast majority of industrial applications, it cannot be used in residential and small-business applications where three-phase power is not available. In three-phase motors, the rotating magnetic field is produced by the 120° phase angle between the three phases. However, with single-phase systems, because phased power is not available, a different method has been devised to create a rotating magnetic field even though the applied power is single phase.

Interestingly, single-phase induction motors use squirrel-cage rotors that are of the same design as those used in three-phase induction motors. However, because the motor is operating from single-phase power, the principle of operation of the stator is quite different. Figure 16-8 illustrates a simplified stator arrangement for a two-pole single-phase induction motor. Notice that the field coils are connected to produce complementary magnetic fields; that is, when one of the pole pieces produces a north magnetic pole, the other produces a south pole, and vice versa. Because alternating current is applied to the coils, vertically oriented alternating magnetic flux will be produced. This is called the **main flux** or **direct flux**.

When a rotating squirrel-cage rotor is placed in this field, as shown in Figure 16-9(a), the main flux induces circulating currents in the rotor bars. At the instant in time that is shown in Figure 16-9, the main magnetic field direction is from top to bottom. The rotor is moving in

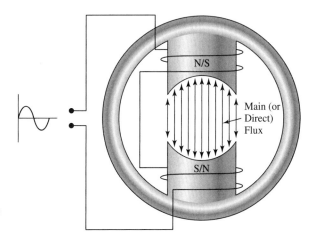

FIGURE 16–8
Stator of a two-pole single-phase induction motor.

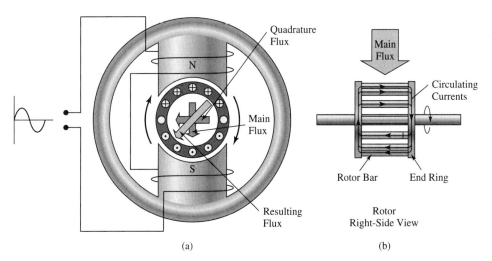

FIGURE 16–9
Single phase motor (a) main and quadrature flux, and (b) rotor currents.

the clockwise direction, so the rotor bars at the top will be cutting through the main flux from left to right, and the bottom rotor bars from right to left. Therefore, currents will be induced in the top rotor bars with a direction into the page, and in the bottom rotor bars with a direction out of the page, as indicated in Figure 16-9(a) (the reader is invited to use the right-hand rule to verify these current directions). As illustrated in Figure 16-9(b), because the rotor bars are connected by the end rings, the rotor bar currents will produce a net circulating current that travels into the top bars, down the back end ring, out of the bottom bars, and up through the front end ring. This circulating current, in turn, produces a second magnetic field through the rotor in the direction from right to left as illustrated in Figure 16-9(a). Because the second magnetic field is perpendicular to the main field, the second field is called the **quadrature field**, and the flux is called the **quadrature flux**.

Because the rotor current is a product of magnetic induction, the phase angle of the rotor current will lag the stator current by 90°. Therefore, whereas the main and quadrature fields are a direct result of the stator and rotor currents, respectively, the quadrature field will lag the main field by approximately 90°. The two magnetic fields are perpendicular, so we can add their vector values to determine the resulting combined magnetic field, as shown in Figure 16-10. This resulting magnetic vector not only will have magnitude, but because it is composed of two perpendicular components, it also will have a direction which rotates with respect to time, thus producing a rotating magnetic field. Note that the main and quadrature fields do not rotate. The main field is always a vertical field and the quadrature field is always a horizontal field. The fields change intensity and polarity with the amplitude and polarity of the alternating current in the armature field. However, when the fields are combined, collectively they create a rotating magnetic field. This rotating field excites the rotor and produces a counteractive force that mechanically drives the rotor to spin.

Recognize that the quadrature field is a result of the rotation of the rotor. Therefore, if the rotor is not turning, there is no quadrature field. Without a quadrature field, there is no resulting rotating field and no rotor torque. In short, a single-phase induction motor needs a substitute quadrature field to produce starting torque. This is provided by adding another pair of magnetic poles and a coil (called the **auxiliary winding**) to the motor that are orthogonal to the main field coils and pole pieces. For the auxiliary winding to produce a magnetic field that lags the main field, an electrical phase shift circuit must be added to the auxiliary winding circuit, as illustrated in Figure 16-11. The phase shifting is done in a variety of ways. This will be explained in more detail later. In some single-phase motors, the auxiliary winding remains always connected, thereby providing the quadrature field needed to start the motor in addition to enhancing the rotor-produced quadrature field after the motor begins to run. In other motors, the auxiliary winding is connected only during starting. It is disconnected by a centrifugal switch which opens when the motor reaches some predetermined percentage of running speed. On these types of motors, because the auxiliary winding is only used for starting, the auxiliary winding is also called the **start winding**.

The magnitude of the starting torque is determined by the flux density of the auxiliary winding and the phase angle of the auxiliary winding current with respect to the main winding current. Auxiliary windings that produce a flux density that is approximately equal to the main flux density, and are excited by a current that is shifted nearly 90°, will produce

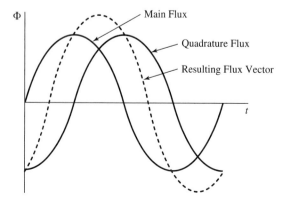

FIGURE 16–10
Main and quadrature flux waveforms and the resulting sum.

FIGURE 16–11
Single-phase motor with auxiliary winding.

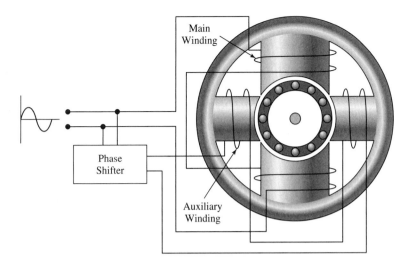

the highest starting torque. However, it is not always necessary to design a motor to have high starting torque. For example, a motor that operates a ventilation fan does not need to have a high starting torque; therefore it can be constructed with a less expensive phase shifter and an auxiliary winding that produces a lower phase shift and requires a lower winding current. Therefore, single-phase induction motors are available with a wide variety of construction methods, each with different starting and running characteristics and each uniquely suited for various applications.

In general, a single-phase induction motor is not as quiet as its three-phase counterpart. Because the rotating magnetic field is produced by a combination of two phases (main and quadrature), it is difficult to maintain a constant flux density throughout the entire 360° rotation of the field, thus causing the motor to hum. Additionally, because single-phase induction motors are less efficient than three-phase induction motors, they are usually available only in lower-horsepower sizes, typically 10 horsepower or less, with the most popular sizes being in the $\frac{1}{4}$-hp to 1-hp range.

All of the equations used for three-phase motor calculations, including those for synchronous speed, slip, rotor speed, and efficiency, also apply to single-phase motors. For equations requiring the number of poles P, the number of poles in a single-phase induction motor is the number of main winding poles. The auxiliary winding poles are not included in the count. Some single-phase induction motors have multiple pole pairs with windings that can be reconnected to provide changeable speeds. For example, a 60-Hz motor can be rewired for four-pole, six-pole, or eight-pole operation to provide synchronous speeds of 1800 rpm, 1200 rpm, or 900 rpm, respectively. As an example of a multispeed motor application, some automatic clothes washers have a delicate cycle in which the motor is rewired by a selector switch to operate at a slower speed.

The direction of rotation of a single-phase induction motor is determined only by the phase relationship between the main winding and the auxiliary winding. Once the motor is started, it will run in either direction using the main winding only. Therefore, the direction of rotation can be reversed by reversing the two auxiliary winding connections, or by reversing the two main winding connections, but not both.

Selecting a single-phase induction motor for a particular application requires knowledge of the characteristics of the various types of single-phase induction motors. All single-phase induction motors have basically the same type of main winding and squirrel-cage rotor. The difference among the various types is in the construction and excitation of the auxiliary winding. Fundamentally, there are four types of motors, which are described below.

Resistance-Start Split-Phase Induction Motor

The **resistance-start split-phase induction motor** has an auxiliary winding which serves as a start winding only. Once the motor reaches approximately 75 percent of rated speed, a centrifugal switch disconnects the auxiliary winding, allowing the motor to run on the main winding only. If the motor speed drops to below 75 percent, the centrifugal switch reconnects the auxiliary winding. The phase shift between the main and auxiliary winding is provided by the selection of the wire size and number of turns on the windings. The auxiliary winding consists of smaller wire with more turns than the main winding. Therefore, the main winding has more inductive reactance than resistance ($X_L > R$), whereas the auxiliary winding has more resistance than inductive reactance ($X_L < R$). This causes the currents through the two windings to be out of phase, thereby providing the required phase shift for starting. The resistance-start split-phase motor provides moderate starting torque and good speed regulation. Because the only moving electrical part in the motor is the centrifugal switch, this is the most reliable of all moderate-horsepower single-phase induction motors. It is best suited for household appliances and centrifugal pumps. Because it produces a moderate amount of noise, it is usually not well suited for ventilation fans.

Capacitor-Start Split-Phase Induction Motor

The **capacitor-start split-phase induction motor** uses a capacitor in series with the auxiliary winding to provide the required phase shift for starting. Once started, the auxiliary winding is disconnected by a centrifugal switch. The motor is designed with heavy wire in the auxiliary winding, which is capable of carrying large starting currents, and because the capacitor provides nearly 90° of phase shift, this motor has the highest starting torque of all single-phase induction motors, as much as 300 percent of the rated running torque. Once started, the motor has running characteristics that are identical to the split-phase motor. It is best suited for applications requiring high starting torques, such as reciprocating pumps and air-conditioner compressors.

Permanent Split Capacitor, or PSC Motor

Like the capacitor-start motor, the **permanent split-capacitor induction motor** (or **PSC motor**) also has a capacitor in series with the auxiliary winding to provide phase shift for starting. However, the capacitance value is much smaller, thereby limiting the current through the auxiliary winding. Because this prevents overheating of the winding, it remains always connected. There is no centrifugal switch. Because the capacitor value is small, the phase shift angle is smaller, resulting in lower starting torque. However, the capacitor is always connected, so the motor runs much more smoothly and quietly than other single-phase motors. It is best suited for applications requiring low starting torque and quiet operation, such as ventilation fans.

Two-Value Capacitor; Capacitor-Start, Capacitor-Run; or CSCR Motor

The **CSCR induction motor** (also called a **two-value capacitor induction motor**, or a **capacitor-start, capacitor-run induction motor**) is a combination of the capacitor-start motor and the PSC motor. When started, two capacitors are connected in parallel to provide the starting current for the auxiliary winding. One of the capacitors is large, providing the bulk of the starting current. Once started, a centrifugal switch opens and disconnects the larger capacitor, leaving a smaller run-capacitor connected. Therefore, the CSCR motor has the high starting torque characteristics of a capacitor-start motor and the quiet running characteristics of the PSC motor.

Shaded-Pole Induction Motor

The **shaded-pole induction motor** uses one main winding only. It does not have an auxiliary winding. In this type of motor, the rotating magnetic field is produced by copper bars (called **shading bars**) which are placed into slots in the pole piece and are wrapped around one side, as illustrated in Figure 16-12. In doing so, part of the field flux passes through the nonshaded part of the pole as usual, whereas the remainder of the flux passes through the shaded portion. Because the copper bar makes one complete turn around the shaded

FIGURE 16–12
Shaded-pole induction motor.

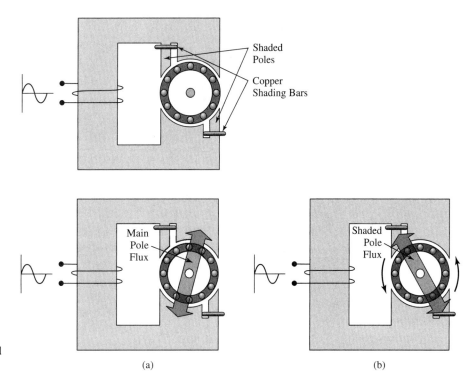

FIGURE 16–13
Shaded-pole motor, (a) main pole flux followed by (b) shaded pole flux.

portion of the pole, it creates a short circuit. When the field is excited by the armature winding, a high alternating current is induced in the shading bar. By Lenz's Law, this high current produces another magnetic field that opposes the armature field. However, because the field produced by the shading bar is the product of an induced current, it will lag the main field by an angle in the range of $0°–90°$. This causes the field in the shaded portion of the pole piece to lag the main field, thereby creating a sweeping motion of the overall magnetic field as illustrated in Figure 16-13. In the figure, this sweeping action will be from right to left on the top pole and from left to right on the bottom pole. The end result will be that this squirrel-cage rotor will rotate in the counterclockwise direction.

The direction of rotation for the shaded-pole motor is fixed. It is determined by the placement of the shading bars. Because the angle between the main flux and the shaded-pole flux is small (typically $20°$), the motor has very low starting torque and is very inefficient. However, because the motor has no capacitor or centrifugal switch and the only moving part is the squirrel-cage rotor, it is an extremely reliable motor. Shaded-pole motors typically are used in low-torque applications such as refrigeration and circulation fans. Many shaded-pole motors are available with **impedance-protected** armature coils. The armature coils in these types of motors are wound of many turns of small wire, giving the coil a relatively high resistance. Therefore, should the motor stall, the armature current is limited by the coil resistance and the coil is thereby prevented from overheating and burnout. An impedance-protected motor can be stalled indefinitely without damage.

Additionally, some shaded-pole motors are constructed with two or more shading bars on each pole. These provide a smoother flux sweep and a larger angle to the sweep, thereby making the motor run more quietly and providing more torque. A shaded-pole motor with two pairs of shading bars is shown in Figure 16-14.

Practical Aspects of Single-Phase Induction Motors

Single-phase inductions motors are somewhat more prone to failure than comparable three-phase motors. Therefore, it is helpful to understand some of the practical aspects of these types of motors and common trouble symptoms.

1. With the exception of the PSC and shaded-pole motor, all single-phase induction motors use a centrifugal switch connected in series with the auxiliary winding. When power is

FIGURE 16–14
Shaded-pole motor.

applied to the motor, if the switch has failed to close, the motor will not start and will simply hum. This condition can be easily diagnosed by removing the load from the motor shaft and spinning the shaft by hand. If the motor starts, the problem is in the auxiliary winding. Although the centrifugal switch could be the problem, on motors that have capacitors in the auxiliary winding circuit, the problem is more likely to be the capacitor.

2. The capacitor-start, PSC, and CSCR motors contain one or two capacitors. With age, motor capacitors tend to dry out, causing the capacitance value to decrease. Eventually, the low capacitance (and corresponding high reactance) of the aging capacitor will restrict the auxiliary winding current to a point where insufficient starting torque will be developed and the motor will fail to start. Because the auxiliary winding current will be limited by the capacitor, it is not likely the motor will be damaged. If the capacitor value is degraded (reduced), the motor will start under no mechanical load but will not start when loaded. On the other hand, if the capacitor is open (zero capacitance), the motor will not self-start under any condition. Starting problems can be diagnosed easily by removing the load from the motor shaft and spinning the shaft by hand. If the motor starts, the problem is in the auxiliary winding and is likely either the centrifugal switch or the capacitor. Because an open capacitor and an open centrifugal switch exhibit the same symptom, it is difficult to diagnose which component is the problem without removing and measuring the capacitor and checking the switch for continuity.

3. If a centrifugal switch sticks in the closed position, it will fail to disconnect the auxiliary winding when the motor reaches full speed. The motor will start normally, but because the auxiliary winding current is much higher than the main winding current, the motor will hum loudly upon reaching running speed, and it will overheat. If the motor is equipped with a thermal protection switch, the switch soon opens, thereby removing power from the motor. If the motor does not contain a thermal protection switch, it is likely that the start winding will burn out, evidenced by smoke coming from the ventilation holes in the motor.

16-5 Reluctance Motor

The single-phase **reluctance motor** combines the characteristics of a squirrel-cage induction motor and a synchronous motor. As a result of its construction, the reluctance motor starts as an induction motor, but once it reaches a speed at which the slip is very low, the rotor locks into synchronization with the rotating stator field and runs as a synchronous motor.

The motor is constructed much like a standard single-phase squirrel-cage induction motor, except that the rotor core has holes in it that create various values of reluctance depending on the relative angle between the rotor and the magnetic field produced by the stator. Figure 16-15(a) shows a cross-sectional view of a reluctance motor rotor. Because the holes in the core create a high-reluctance path for field flux, there are paths within the rotor that have a lower reluctance, namely the paths containing continuous core material as illustrated in Figure 16-15(b). The fact that the reluctance of the rotor varies with the flux angle is the key to the motor running synchronously.

FIGURE 16–15
(a) Reluctance motor rotor construction and (b) low-reluctance paths.

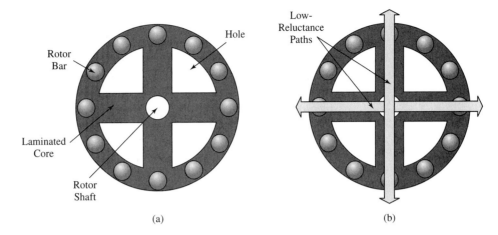

When the reluctance motor is started, it starts as an induction motor. The rotating magnetic stator field induces a current in the rotor bars, which, in turn, produces an opposing magnetic field, which pushes the rotor ahead of the field and brings the rotor up to operating speed. At induction-motor operating speed, the low value of slip will cause the relative angle between the rotor position and the stator field to change continuously. However, as soon as this relative angle is such that the stator field aligns with one of the low-reluctance paths in the rotor, the low reluctance causes a large increase in the flux passing through the rotor core. The increased flux creates a magnetic attraction between the stator field and the rotor, which will increase the rotor speed to match the rotating speed of the stator flux, thereby locking the rotor at synchronous speed. Once lockup occurs, the slip becomes zero, there is no relative motion between the stator flux and rotor bars, and the current in the rotor bars becomes zero. At this point, the rotor bars cease to contribute to the operation of the motor. As long as the rotor continues to operate in synchronization, all torque produced by the rotor is provided by the magnetic attraction between the stator flux and the rotor core. While the reluctance motor is operating at synchronous speed, if the load torque is increased so that the rotor drops out of sync, the rotor slip again creates induced currents in the rotor bars, and the motor reverts to operating as an induction motor.

Like induction motors, reluctance motors can be constructed to operate from single-phase or three-phase power. Reluctance motors are used in timing applications that must operate from ac power, such as clocks and appliance timers. They are most commonly found in washing machine and clothes dryer timers.

EXAMPLE 16-6

A four-pole reluctance motor is operating from a 60-Hz power line. It is operating synchronously and is producing 0.1 ft-lb of torque. Find (a) the rotor speed, and (b) the output horsepower.

SOLUTION

(a) Use Equation 16-14 to calculate the synchronous speed, which is

$$n_s = \frac{120f}{P} = \frac{120 \times 60}{4} = 1800 \text{ rpm} \tag{16-27}$$

(b) The output power in watts is

$$P_{\text{out}} = \frac{T n_r}{7.04} = \frac{0.1 \times 1800}{7.04} = 25.57 \text{ W} \tag{16-28}$$

and the output power in horsepower is

$$P_{\text{hp}} = \frac{P_{\text{watts}}}{746} = \frac{25.57}{746} = 0.034 \text{ hp} \tag{16-29}$$

16-6 Universal Motor

In Chapter 15, we explored the characteristics and applications of the series dc motor. Recall that to reverse the direction of rotation of a series dc motor, the connections to the series field or the armature should be reversed, but not both. In other words, the direction of rotation cannot be reversed simply by reversing the polarity of the applied voltage. Because power can be applied to the series dc motor in either polarity without a change in the direction of rotation, this would imply that a voltage source that constantly changes polarity (that is, an ac source) could be connected to a series motor and the motor would operate normally. A series connected motor that is designed for ac applications is called a **universal motor**. It consists of the same components as a series dc motor, including a series field, an armature, and a brush and commutator assembly.

The universal motor exhibits many of the same operating characteristics of the series dc motor. It has very high starting torque, poor speed regulation, and operates at a very high speed which is only limited by friction, windage, and the mechanical load on the rotor. Universal motors are most commonly used in portable power tools such as hand drills, sanders, and routers, and small kitchen appliances such as food processors, mixers, and blenders. Although universal motors can be operated from dc power also, the proper applied dc voltage is not the same as the rated ac voltage because the rated ac voltage takes into account the inductive reactance of the series field. For example, operating a 120-V ac universal motor on 120 V dc would likely damage the motor.

PROBLEMS

16-1 A six-pole three-phase alternator is rotating at 850 rpm. What is the output frequency?

16-2 A 180-pole alternator used in a hydroelectric power station generates 60 Hz. At what speed does it rotate?

16-3 A three-phase alternator is generating 1 kW of power. The shaft speed is 1800 rpm, and the input torque is 6.31 ft-lb. What is the overall efficiency of the alternator?

16-4 A turbine-driven two-pole 21-megawatt shipboard propulsion generator (alternator) produces 4160-volt, three-phase, 60-Hz power. The rotor rotates at 3600 rpm and the shaft torque delivered from the turbine to the alternator is 42,337 ft-lb. Determine (a) the number of poles in the alternator, and (b) the efficiency of the alternator.

16-5 A four-pole three-phase synchronous motor is connected to a 50-Hz power line. What is the rotor speed?

16-6 An eight-pole synchronous motor is connected to a variable-frequency power source. It is desired to operate the motor at 925 rpm. What frequency must be applied to the motor?

16-7 A four-pole three-phase 60-Hz induction motor operates at 1725 rpm. What is the percent slip?

16-8 A six-pole three-phase 60-Hz induction motor is operating at rated load. The nameplate rated slip is 2.5 percent. What is the rotor speed?

16-9 A three-phase 60-Hz induction motor is being selected for an application requiring approximately 900 rpm. How many poles will be required?

16-10 A three-phase 60-Hz induction motor is operating under rated load at 595 rpm. How many poles does it have?

16-11 A single-phase induction motor operates at 120 V ac, 60 Hz, 3.5 amperes, 0.65 lagging power factor. It delivers 0.69 ft-lb of torque at 1750 rpm. What is the motor's efficiency?

16-12 A single-phase induction motor has the following nameplate ratings: 1 hp, 240 V ac, 60 Hz, 1175 rpm, 7.15 amperes, 0.62 lagging power factor. When the motor is operating at rated load, (a) what is the internal power loss in the motor, (b) what is the efficiency of the motor, and (c) what is the shaft torque?

16-13 A single-phase induction motor is needed to operate a ventilation fan. The starting torque is very low, but the motor must operate as quietly as possible. Which type of motor is best suited for the application?

16-14 A single-phase induction motor is needed to operate a reciprocating pump. There is a possibility that the pump must start on a compression stroke which requires a large amount of torque. Noise is not a consideration. Which type of motor is best suited for the application?

16-15 What is the rated operating speed of a six-pole 60-Hz reluctance motor?

16-16 An eight-pole reluctance motor is connected to a variable-frequency power source. What frequency is needed to operate the motor at 750 rpm?

CURRENT ADDITIONAL TOPICS

17 Programmable Logic Controllers

18 Digital Communications

Programmable Logic Controllers

OVERVIEW AND OBJECTIVES

Since the late 1960s, the methods used to control large manufacturing machines, shipboard electrical systems, and manufacturing processes have undergone a major revolution. In this time period, we have seen machine-control circuitry evolve from cumbersome hard-wired logical circuits to more sophisticated and versatile programmable machine controllers that are much more capable and cost-effective than their hard-wired predecessors. In this chapter, we will explore both the older hard-wired control circuits and the newer machine controllers called **programmable logic controllers (PLCs)**.

Objectives

After completing this chapter, you will be able to

- Read and analyze electrical controls ladder diagrams.
- Identify various electrical controls and schematic symbols.
- State and describe various parts of a PLC.
- Describe how a PLC operates when it executes a program.
- Write simple ladder logic programs for a PLC.
- Design and draw simple PLC-to-machine wiring diagrams.

17-1 Introduction

As the name implies, a programmable logic controller (PLC) is a system that is designed to control machines by performing logical operations that are programmed into the system's memory. Although a PLC typically contains one or more microprocessors, it is not considered to be a computer. Instead it is a special-purpose device designed to perform controlling and monitoring operations in an industrial environment. As we will see, the PLC was designed to take the place of more rudimentary and expensive control circuitry. We will explore both this older method of controlling machines and the newer method that has replaced it, the PLC.

17-2 Introduction to Machine Controls

A machine control system is fundamentally a closed-loop system. For inputs, the controller has feedback sensors, which are used for monitoring the status or state of a manufacturing machine, and switches, which are used for operator input. The outputs of the control system are connected to motors and solenoids, which directly control the operation of the machine, and to lamps on the control panel, which inform the operator of the present state of the machine. In operation, the controller continuously monitors the machine's performance with respect to the desired performance that is input by the operator and makes adjustments to the machine's operating parameters so that the machine performs as desired. Programmable machine controls have an internally stored program that executes when the controller is powered. By programming the controller, it is possible to achieve extremely sophisticated and precise machine performance.

Before the 1960s, machine controls were constructed using electromechanical relays which were wired to perform the fundamental logical operations AND, OR, and NOT. Each relay was approximately the size of a D-cell battery (or larger) and performed the equivalent of one logic gate. By wiring the relays together, low-level logical operations were combined to form higher-level operations. As one can imagine, using this method to achieve even a moderate level of sophistication required a complex maze of wiring, a large number of relays, and massive electrical panels to house the relays.

Equally as formidable as the design and wiring of these relay-logic control systems was the engineering effort required when performance changes were made to the control systems. Such changes were commonplace in industries that changed products on a regular basis, such as the automobile industry. In the 1960s when minicomputers became available, production engineers at General Motors recognized that the logical operations that were being performed by relays also could be performed as logical software operations in a minicomputer. By doing so, the state of each relay instead could be a stored bit, thereby reducing the size of a control system drastically. This also had the very desirable effect of making performance changes (that formerly required rewiring of the relays) into software revisions.

Over the years following the invention of this first PLC, the engineers recognized that although this was a vast improvement over the previous hard-wired relay logic systems, the programming of these new controllers was a complex task. The electricians who formerly had rewired the relay logic control systems found that they needed to program computers, a task for which they had not been trained. Therefore, the engineers who designed the first PLC systems began to modify the systems to make PLC programming easier. The result was a graphical programming language in which the actual programming was done by manipulating relay components on a computer screen. Then, when the relay logic appeared as desired, the computer translated the graphical presentation into the complex computer language needed to mimic the corresponding relay circuits. This graphical language was named **ladder logic**. Once perfected, electricians found that they could enter the programs into the PLC because the circuits being graphically represented on the computer monitor appeared the same as the relay logic circuits to which they were accustomed.

Because of this similarity between machine control circuits and PLC programming techniques, we will begin our study of PLCs by familiarizing ourselves with conventional hard-wired relay logic components and circuit diagramming.

17-3 Machine Control Components

Reference Designators

When designing the circuitry for a control system, it is important to be able to identify and locate each of the components. For this reason, each component is assigned a **reference designator** consisting of a letter followed by a number. The letter indicates the type of component (fuse, transformer, relay, switch, and so on), and the number is unique to the component. For example, if our system has four fuses in it, we may wish to assign to them the reference designators F1, F2, F3, and F4. In a similar manner, control relays could be labeled CR1, CR2, CR3, and so forth. The reference designator number assigned to each component is done in sequence without skipping numbers. For this reason, if we see a transformer T3 in a schematic, we can assume that transformers T1 and T2 are also somewhere in the system. In addition to writing each component's reference designator on the electrical schematic diagram, we also indicate where the component is physically located in the system. This can be done by either actually writing the reference designator on the physical component or providing a component layout drawing (sometimes called a **roadmap**), which shows the physical location of each component with its reference designator.

Sometimes we make modifications to machine control circuits which involve the removal of some components. When this is done, we do not renumber all subsequent components. Instead, we simply make a notation on the drawing called "Unused Reference Designators" followed by a list of the components that have been removed. This is generally done as a courtesy to the engineers, technicians, and electricians who use the drawings regularly so that they do not waste time looking for components that no longer exist.

Similarly, modifications sometimes require the addition of components. Normally, we would assign the next sequential unused number. However, it is sometimes difficult and time-consuming to determine this number. To avoid this dilemma, we also make a notation on a schematic drawing under the heading "Last Used Reference Designator." Under this heading is a list of the highest numbered component of each type. For example, the list may be

Last Used Reference Designator:

CR23, T2, F6, LS5, PB5, SS2

Then, for example, if our modification adds a control relay and a selector switch, they will be assigned the reference designators CR24 and SS3, respectively. After making the modifications, it is important to update the "Last Used Reference Designator" list on the drawing.

Wire Numbering

Another method used to make the design, construction, and maintenance of control circuitry easier is to use wire numbers. When drawing the electrical wiring diagram of the control circuit, the designer assigns a number to each electrical node in the circuit. When the equipment is wired, the wires that make up each node are numbered by attaching wire markers to the ends of the wires that indicate the node number. Figure 17-1 shows a wire with a wire marker attached. Both ends of every wire attached to a node are numbered (and obviously both numbers will be the same). The result is that when completed, every wire in the system will be marked with the number of the node to which it is attached, thereby making the circuit much easier to maintain and troubleshoot. Additionally, because all the wires are numbered, it is acceptable to wire nearly all of the control circuit with the same color of wire,[1] thereby reducing the cost of the system and eliminating the need to stock spools of different-colored wire.

FIGURE 17–1
Wire number marker.

---

[1] Electrical code requires that all safety ground wires in a control system be green or green with a yellow stripe (a stripe is called a **tracer**). The colors of all other wires in the system can be chosen by the designer but cannot be green.

Control Transformer

Generally, industrial machines are large and powerful devices requiring a large amount of electrical power for operation. To keep current demand of the machines to a minimum, the machines are generally powered by three-phase power with a line voltage of 208 V ac or more. However, it is considered unsafe to connect any voltage higher than 120 volts (ac or dc) to an operator's control panel. Therefore, we must use a step-down transformer to reduce the incoming line voltage to a safe level. Such a transformer is called a **control transformer**, and the reduced output voltage of the transformer is called the **control voltage**.

Up to this point in our studies of electrical circuits, diagrams have always been read from left to right, then top to bottom. However, as we will see, machine controls diagrams are read from top to bottom, and then left to right. For this reason, when we draw a control transformer, instead of having the primary winding(s) on the left and secondary winding(s) on the right, we will draw it rotated 90° clockwise with the primary winding(s) on the top and secondary winding(s) on the bottom as shown in Figure 17-2.

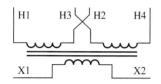

FIGURE 17–2
Control transformer symbol.

The control transformer in Figure 17-2 has two primary windings for the purpose of making the transformer more versatile by giving it the ability to operate from two different line voltages. Let's say, for example, that our transformer has primary windings rated at 208 V ac each. Therefore, if we need for it to operate from a 208-volt line, we could simply connect the two primary windings of our transformer in parallel by connecting H1 to H3, and H2 to H4. However, if our transformer is to be connected to a 416-volt line, we could wire our two primary windings in series by connecting H2 to H3. By making the transformers with two matching primary windings, the transformer can be used in a wider variety of applications, thus requiring transformer distributors to stock fewer transformer types and allowing end users to keep fewer spare parts on hand.

The secondary winding of the transformer produces the control voltage, 120 V ac or less. The secondary is generally connected so that the right side (X2) is grounded to the frame of the machine and the left side (X1) becomes the output voltage. These are distributed throughout the control circuit and are called the **ground rail** (or neutral rail) and the **hot rail**.

Fuses

Fuses are used in control circuits to protect both the personnel using and maintaining the equipment and the equipment itself, should an electrical fault occur in the control system. Fuses used in control systems are generally cartridge-type fuses, that is, an insulating cylinder with a metal cap on each end. In electrical controls diagrams, the fuse symbol resembles a cartridge fuse as shown in Figure 17-3. Fuses can be mounted in special fuse blocks that are available from electrical suppliers; however, most control transformers are available with one or more fuse blocks mounted on the top of the transformer as shown in Figure 17-4.

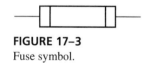

FIGURE 17–3
Fuse symbol.

Switches

Although there are a very large number of switch types, the most popular switches used in control systems are the **pushbutton switch, selector switch**, and **limit switch**. Pushbutton and selector switches are used on control panels where operator input is needed, and limit switches are mounted on the machine to sense the positions of mechanical parts of the machine. For example, a limit switch is mounted in an automobile door frame to sense when the door is ajar.

Pushbutton switches can be either **momentary** or **maintained**. A momentary pushbutton switch is activated when it is pressed. When it is released, a spring returns the switch to the deactivated position. The reset button on a computer and a doorbell button are momentary pushbuttons. Maintained pushbuttons latch in the activated position when pressed once, and then unlatch when pressed again. The ON-OFF switch on a computer monitor is a maintained pushbutton. The electrical contacts on switches can be either **normally open (NO)** or **normally closed (NC)**. The naming of a contact is determined by its state when

FIGURE 17–4
Control transformer with fuse blocks.

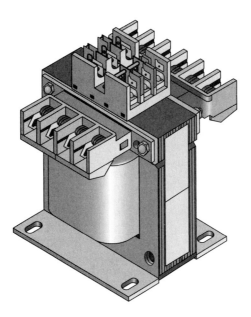

the switch is deactivated. For example, if we wish to have a momentary switch contact that switches ON (conducts) when the button is pressed and switches OFF when it is released, we would need a normally open contact. Conversely, normally closed contacts are ON when the button is released and OFF when the button is pressed. All large machines have an **emergency-stop** pushbutton, which is a maintained pushbutton with a normally closed contact. In operation, when the button is pressed, the switch latches in the activated position, but the electrical contact opens, thereby shutting off power to the machine. Figures 17-5(a) and (b) show the schematic symbols for normally open (NO) and normally closed (NC) pushbutton switches, and Figure 17-5(c) shows a pushbutton with two sets of contacts, one NO and one NC.

Selector switches operate in much the same manner as pushbuttons, but the actuation of the contacts is done with a cam, so that the operator turns the switch actuator instead of pressing it. The switch actuator has two or more detent positions. For example, in most automobiles, the speed of the fan for the heater and air conditioner is controlled by a selector switch. The schematic symbol for the selector switch is much the same as that for the pushbutton switch except that a "handle" is drawn on the top of the actuator as shown in Figure 17-6. Figure 17-6(a) shows a selector switch with a contact that is open when the selector is to the left and closed when the selector is turned clockwise. The contact of the switch in Figure 17-6(b) is closed when the selector is to the left and open when it is turned clockwise. The selector switch in Figure 17-6(c) is a combination of the other two switch types.

Limit switches operate much like pushbutton switches, except that the actuator is designed to be moved by a mechanical part of the machine on which it is mounted. For this reason, many limit switches have a lever added so that the switch can be actuated with a camming action. Figure 17-7 shows a limit switch with a lever and wheel that is designed for such a camming action. Figure 17-8 shows the schematic symbols for the NO and NC limit switches.

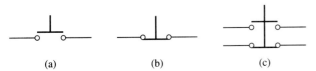

FIGURE 17–5
Pushbutton switch symbols, (a) normally open, (b) normally closed, (c) both.

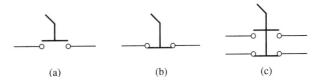

FIGURE 17–6
Selector switch symbols.

FIGURE 17–7
Limit switch.

FIGURE 17–8
Limit switch symbols, (a) NO, (b) NC.

The reference designator for a switch depends on the switch type. For example, we have "PB" for pushbuttons, "SS" for selector switches, and "LS" for limit switches.

Lamps

In machine control systems, lamps are used primarily on operator control panels to indicate the machine's state of operation or to serve as warning indicators. They can be of any color, but the color is generally determined by a colored lens cap covering the bulb. The lamp bulb itself is usually a clear incandescent lamp. The schematic symbol for a lamp is shown in Figure 17-9. On most schematics, the lamp color is written next to the lamp symbol. The lamp in Figure 17-9 is red.

FIGURE 17–9
Lamp.

Relays

Relays are the most versatile components in control circuits because not only do they allow us to control motors, valves, and lamps, they also are used to perform logical operations. A relay is an electromagnetically operated switch consisting of a wire coil and one or more electrical contacts of either NO or NC configuration. Relays that are designed to switch higher currents are also called **contactors**.

Figure 17-10(a) shows a cross-sectional view of a relay. It consists of a magnetic core onto which is wound a coil of copper wire. The core is open at the top with a cap that is spring loaded so that the cap is pushed upward and away from the core. This also pushes a moving contact strip that is mounted on the cap upward against two stationary contacts

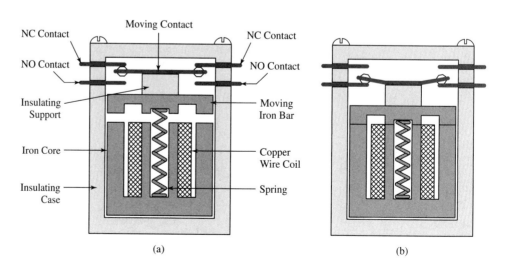

FIGURE 17–10
Relay cross section,
(a) de-energized and
(b) energized.

called the normally closed (or NC) contacts. This is called the **de-energized** condition. In the de-energized condition, the two NC contacts of the relay will be electrically connected by the moving contact.

When power is applied to the relay coil, the current in the coil produces a magnetic field in the core which pulls the cap downward. This, in turn, causes the moving contact to move downward and make contact with the two normally open (or NO) contacts as shown in Figure 17-10(b). This is called the **energized** condition. The moving contact is made of a flexible metal so that the contacts bend slightly to keep mechanical pressure on the contact, thereby assuring a good connection and providing a wiping action to keep the contacts clean. In the energized condition, the two NO contacts of the relay will be electrically connected.

When beginning a study of relays and contactors, a common question is why a relay is needed at all. Because an electric current is operating the relay coil and the relay contacts are switching the current to another device, why not remove the relay and have the input current control the device directly? The answer is twofold.

- Each relay requires a minimum amount of flux and corresponding magnetomotive force to move the contacts mechanically. If we wind a large number of turns on the coil, we can produce this required magnetomotive force with a very small current, in some cases only a few milliamperes. At the same time, we can mount heavy-duty contacts on the relay capable of switching tens or even hundreds of amperes of current. In this case, the relay becomes somewhat of an amplifying switch; that is, a small amount of input current controls a large amount of output current. Relays are available that are capable of switching several amperes of current when the relay coil is operated by a TTL logic signal.

- Because a relay can have normally closed contacts, we can use the relay as a logical inverter. That is, when the input current to the coil is zero (OFF), the normally closed contacts will be ON, and vice versa.

When a relay is drawn in an electrical schematic, the symbols shown in Figure 17-11 are used. Notice that in Figure 17-11(a), the symbol for a NO relay contact looks similar to that for a capacitor. It is important not to confuse these two symbols. In electrical controls diagrams, the capacitor will have one flat plate and one curved plate, whereas the NO relay contact will have two flat plates, as shown in Figure 17-11(a). Additionally, because the components of a single relay might be physically separated on the schematic, the contacts and coil of a relay are always marked with the reference designator for that relay. For example, if a relay is CR9, the coil will be labeled CR9, and every contact of the relay, no matter where it is located on the schematic diagram, will also be labeled CR9. If a relay has more than one contact, we usually append a letter, a, b, c, and so on, to the end of the reference designator for each contact. For example, if relay CR17 has four contacts, the coil will be CR17, and the contacts will be CR17a, CR17b, CR17c, and CR17d.

Time-Delay Relays

When designing control circuitry, at times it is necessary for a relay to wait for a short period after its coil is energized before it closes its contacts. As an example, consider a high-pressure hydraulic pump that is powered by an electric motor. Because of the high pressure involved, when we start the pump, we usually open a valve (called a dump valve) that connects the high-pressure (outlet) side to the low-pressure (inlet) side of the pump, thereby relieving the pressure on the pump. This reduces the amount of torque required to start the pump, which in turn reduces the starting current for the electric motor. After the pump

FIGURE 17–11
Relay symbols, (a) NO contact, (b) NC contact, (c) coil.

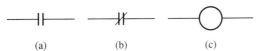

FIGURE 17–12
Delay-on time-delay relay,
(a) timing diagram,
(b) schematic symbols.

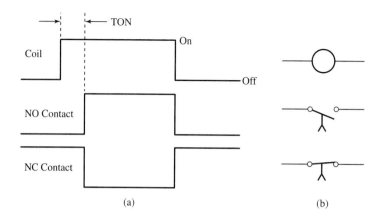

reaches full speed, we can then open the dump valve, thus allowing the pump to charge the hydraulic system. If we install a dump valve that is electrically operated, we will need to apply power to it a few seconds after energizing the pump motor. This function can be done using a **time-delay relay** (also called **TDR**). There are two fundamental types of TDRs, **delay-on** and **delay-off**.

Delay-On

Delay-on relays are also called TON (pronounced "tee-on") relays. When the coil of a delay-on relay is energized, its contacts will actuate a preset time later. The amount of delay is usually adjustable using a control knob on the relay. When the delay-on relay is switched off, the contacts deactivate immediately—there is no time delay. Figure 17-12(a) shows a timing diagram of a delay-on time-delay relay. Note that the delay occurs when the relay is switched on but not when it is switched off. Figure 17-12(b) shows the schematic symbols for the relay components corresponding to the timing diagram. Note that because this type of time-delay relay has the time delay occur when it is energized, the arrows on the contacts point toward the contacts.

Delay-Off

When the coil of a delay-off relay (also called TOF, and pronounced "tee-off") is energized, its contacts will actuate immediately. However, when the coil is de-energized, the contacts do not immediately switch to their de-energized position. Instead, there is a time delay in the de-activation of the contacts. Figure 17-13(a) shows a timing diagram of a delay-off time-delay relay. Note that the delay occurs when the relay is switched off but not when it is switched on. Figure 17-13(b) shows the schematic symbols for the relay components corresponding to

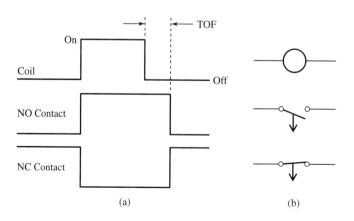

FIGURE 17–13
Delay-off time-delay relay,
(a) timing diagram,
(b) schematic symbols.

the timing diagram. Note that because this type of time-delay relay has the time delay occur when it is de-energized, the arrows on the contacts point away from the contacts.

17-4 Ladder Diagrams

Earlier, we learned that the engineers who designed the first PLCs also developed a graphical PLC programming language called ladder logic. Because electrical controls schematics also use ladder diagramming techniques, to design and analyze ladder logic programs, we must first understand how control diagrams are constructed.

Electrical controls diagrams look quite different from electronic schematic diagrams. First, and as mentioned earlier, electrical circuit diagrams are always read from left to right, then top to bottom, whereas machine controls diagrams are read from top to bottom, and then left to right. Second, in electronic diagrams, the power and ground buses are generally not shown, whereas in controls diagrams they are drawn as two parallel vertical rails which form the fundamental structure for the entire diagram. There are also several less prominent differences, which we will see later.

We begin at the top of the diagram with the power source for the control circuit. If the controls are to operate from ac power, there will be a control transformer at the top of the diagram. (If the controls are dc-powered, there will be a dc power supply at the top.) In addition to the transformer, a fuse must be included to protect the transformer in case of a short in the controls circuit. Also, for safety purposes, a maintained NC pushbutton called an "emergency stop" switch is also connected in series with the transformer secondary. In practice, the emergency stop (also called E-stop) switch is red and is located in a prominent place on the machine where it can be easily reached by the operator. Also, for additional safety reasons, one terminal of the transformer secondary must be grounded. Figure 17-14 shows our circuit up to this point. Note that because the right side of the transformer secondary is grounded (called the **ground rail**), the left side will be the "hot" side (called the **hot rail**). Any circuitry that is connected between the hot and ground rails will be powered whenever the transformer is powered and the E-stop switch is on. The primary of the transformer will be powered by the incoming line voltage. On a controls schematic, these transformer primary connections will also be shown; however, on our diagram we will omit these connections for brevity. Also in Figure 17-14, notice that we have begun numbering the nodes in our circuit, with the ground rail as wire 1, the wire connecting the fuse and E-stop switch as wire 2, and the hot rail as wire 3. When the circuit is constructed, the wires will have wire markers with these node numbers.

Now that the basic framework of our diagram is completed, we can begin adding control circuitry. Let's begin by adding an example circuit in which a NO pushbutton and a NO limit switch operate a lamp. The logic of our circuit will be that the lamp will illuminate when both the pushbutton and the limit switch are on. The circuit is shown in Figure 17-15. As we add more circuits to our diagram, it will begin to look like a ladder with the hot and ground rails as the uprights and the control circuits as the rungs. For this reason, each circuit in the diagram is called a **rung**.

Control circuit rungs have some unique characteristics. First, a rung consists of two basic types of components, the components that control and the components that are controlled, called the **loads**. Controlling components can be connected in series, parallel, or

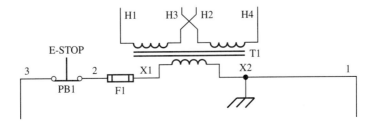

FIGURE 17–14
Top of an electrical controls diagram.

FIGURE 17-15

Control circuit with one rung.

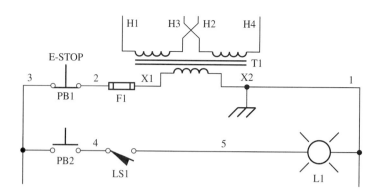

any combination. Loads are always connected in parallel. For safety reasons, the controlling components are *always* on the left side of the rung, and the loads are *always* on the right side. This is because if any of the wires in a rung become accidentally shorted to ground, as soon as the rung is activated, it will blow the fuse, thereby shutting down the control circuit. Second, when analyzing the rung logic, we always analyze from left to right. For the rung in Figure 17-15, we can say that when PB2 is on AND (notice the logical operation, AND) LS1 is on, L1 will be on. In ladder diagrams, whenever controlling components are connected in series, we have a logical AND operation.

Next, let's add an example rung to our circuit in which a relay coil CR1 is energized whenever either of two selector switches, SS1 or SS2, is on. Figure 17-16 shows our ladder diagram with the additional rung. Notice also that because we will be concentrating on the rung logic, we have eliminated the power source from the diagram. In future diagrams, the power source will be implied. In our second rung, when controlling components are connected in parallel, we will have a logical OR operation. Also, whenever there are parallel current paths, each of the parallel paths is called a **branch**. In rung 2, SS1 forms one branch and SS2 forms a second branch. Additionally, note that we have not shown any of the contacts for relay CR1. It is presently assumed that the CR1 contacts will be used in other rungs that are not yet shown or that the contacts of CR1 may be controlling the line voltage being applied to an electrical machine (such as a motor) which may be shown on a separate diagram.

Figure 17-17 shows our ladder diagram with a third rung added. This rung has two branches with two components in series in each branch. Notice also that we have specified two contact sets for relay CR1, one NO and one NC. Both of these contact sets are controlled by the CR1 coil, which, in turn, is controlled by selector switch SS1 or SS2. The logic for rung three is as follows: Lamp L2 will be illuminated when (CR1 is ON AND PB3 is ON) OR (when CR1 is OFF and PB4 is ON). Parentheses have been added to the previous sentence for clarity and to make it appear more like a Boolean logic expression. When verbally discussing ladder logic, we often use the words "enabled" and "disabled". A device is enabled when all components in series with it are ON, and a device is disabled when any one of the components in series with it is OFF. Therefore, in the third rung, we

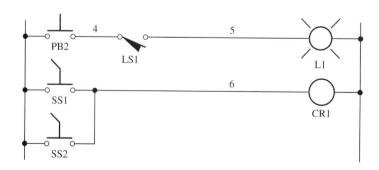

FIGURE 17-16

Control circuit with two rungs.

FIGURE 17–17
Control circuit with three rungs.

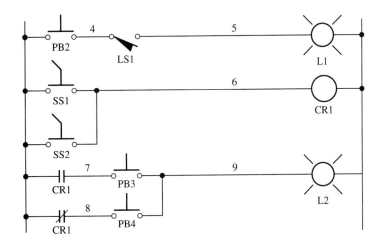

could say that when CR1 is OFF, PB3 is disabled and PB4 is enabled. Conversely, we could say that when CR1 is ON, PB3 is enabled and PB4 is disabled.

17-5 Latch Circuit

In addition to performing logical operations such as AND, OR, and Invert, ladder logic can also perform bit storage operations. Consider the ladder rung in Figure 17-18. This circuit employs a form of digital feedback. The pushbuttons PB1 and PB2 control relay CR1; however, notice that a NO contact of CR1 is connected in parallel with PB1. When power is applied to the circuit and PB1 is open, no power is applied to CR1 and L1, which in turn means that the NO contact of CR1 is open. When PB1 is pressed, power is applied through NC PB2 to CR1 and L1. CR1 energizes, closing the NO contact in parallel with PB1. At this point, PB1 is no longer needed to keep CR1 energized because the NO CR1 contact shunts current around PB1. Therefore, the operator can release PB1 and the rung will remain energized, or latched. The rung can be de-energized in one of two ways: (1) by removing power from the control circuit, or (2) by pressing NC pushbutton PB2. In either case, the rung will return to its initial de-energized condition. This type of circuit is called a **latch circuit**, and the bridging relay contact CR1 is called a **sealing contact** or **seal-in contact**.

For safety reasons, on most heavy machinery the circuit that controls the starting and stopping of the machine is required to be a latch circuit. To see why this is required, consider a scenario in which a factory under normal operation experiences a power failure. In the course of the power failure it is likely that the machine operators will take advantage of the power outage to leave their work stations and take a short break. If the machines are controlled by a latch circuit, then when power is restored, all the machines must be manually restarted by the operators when they return to their work stations. However, if the machines are controlled by simple on/off selector switches, when power is restored, the machines will abruptly (and autonomously) restart, possibly without an operator at the controls.

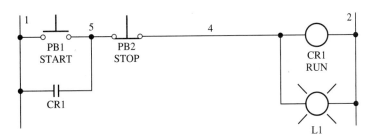

FIGURE 17–18
Latch circuit.

In addition to simple 1-bit storage operations, additional circuitry can be added to the latch circuit to make it into a T, D, or JK flip flop. With the latch circuit in conjunction with the AND, OR, and Invert operations, it is theoretically possible to construct a digital computer using ladder logic circuitry. The only limitations to this would be size, complexity, and the large power requirement. Although we will not be constructing a digital computer as our course of study, we will later see how ladder logic can be used to construct logical circuitry that is increasingly complex and powerful.

17-6 PLC Types

Although all PLCs operate in much the same manner, they are available in various physical configurations. Fundamentally, these different configurations determine the flexibility and cost of the PLC.

Open Frame

The open-frame PLC is also sometimes called a **single-board PLC**. It is usually the lowest cost of all PLCs, and in most cases it also offers the fewest features. It is completely contained on one printed circuit card (with the exception of a power supply) and is generally not inside its own enclosure, hence the name *open frame*. Open-frame PLCs are purchased with a fixed number of input and output (I/O) bits, which in most cases cannot be later increased. For this reason, the designer must be careful to plan accurately the number of I/O bits needed before specifying the PLC. Most designers order a PLC with more I/O than will be needed in order to have "spares" to cover any unforeseen design changes or subsequent modifications. Although the open-frame PLC is very inexpensive and easy to use, it is disadvantaged because if an I/O bit fails, the entire PLC must be taken offline for repair. It is possible to move the defective input or output to one of the spare I/O pins on the PLC, but this also requires reprogramming the PLC to move the I/O bit in the software.

Shoe Box

The shoe-box PLC is also sometimes called a **brick** simply because of its size and shape. Bricks are more expensive than open-frame PLCs but less expensive than modular PLCs. Like the open-frame PLC, the shoe-box PLC comes with a fixed number of I/O bits. However, most brands have available I/O expansion modules that can be either plugged directly into the PLC or connected using an interface cable. This allows the designer to expand the I/O capacity of the PLC later, usually in blocks of 8 or 16 bits. Although the shoe box is more expandable and flexible than the open-frame PLC, it is still somewhat limited by a slower CPU processing speed and a reduced instruction set. Like the open-frame PLC, the main disadvantage in the shoe-box PLC is that if an I/O bit fails, the entire PLC must be taken offline for repair.

Modular

The modular PLC is designed for both flexibility and speed. It is used in applications requiring control of a large number of I/O bits, making large numbers of numeric calculations, or executing large programs. The PLC consists of a backplane containing electrical sockets into which are plugged the various PLC modules. The modules can be the CPU, power supply, input modules, output modules, communications modules, and a large variety of other module types. The designer purchases a backplane and then chooses the appropriate modules needed for the particular application. With most modular PLCs, the modules can be arranged on the backplane in any order, making it easy to customize the PLC for any application. Unlike the open-frame and shoe-box PLCs, if any part of a modular PLC should fail, the problem can be remedied easily by swapping the defective module with a spare module.

17-7 PLC Configuration

All PLCs consist of similar functional parts that make up the complete PLC system. In some cases the parts are required, and in other cases they are optional. The parts include the following:

CPU

The CPU is the processor for the PLC. Generally it is a microprocessor or microcontroller with associated support chips and circuitry. In modular PLCs, there may be several CPUs from which to choose which vary in speed, instruction set, and amount of associated memory. Some of the more powerful modular PLCs have distributed CPUs in which some of the computing power resides in other modules. For example, a module that handles high-speed Ethernet communications with other PLCs may contain a CPU dedicated to communications, thereby freeing the main CPU of that task.

Power Supply

The type and location of the PLC power supply depends on the PLC type. For open-frame PLCs, the power supply is usually external to the PLC board. For these, the designer would select and purchase a power supply that has the appropriate input and output voltages. For shoe-box PLCs, the power supply is contained within the brick. In this case, the designer must select a PLC that is capable of operating from the machine's control voltage. For modular PLCs, the power supply is a module that is plugged into the PLC backplane. In this case, the designer selects a power supply module with an input voltage that is the same as the machine's control voltage. The power supplies contained in shoe-box and modular PLCs sometimes also have output terminals that can supply low voltage, low current (usually $+24$ V dc or $+12$ V dc) that is suitable to power sensors, signal conditioners, and output circuitry.

Input

Most inputs to PLCs fall into the categories of individual bits (called **discrete inputs**) or **register input**. Discrete inputs are suitable for connections to individual input devices such as pushbutton switches, sensors, or limit switches. Each discrete input has one screw terminal on the PLC for attaching the wire from the input device. Register inputs are designed for parallel communication of bytes or words of data. They are suitable for connection to any devices that output numeric information in parallel digital form such as keypads, encoders, or thumbwheel switches. Although discrete and register inputs are the most commonly used, most PLC manufacturers offer many other varieties of input types.

Output

The most commonly used PLC output types are discrete **relay outputs** and discrete **solid-state outputs**. Relay outputs can be used to switch either ac or dc signals and are available in either NO, NC, or both NO and NC contact arrangements. They are designed so that whenever the PLC program switches on a relay output, the PLC switches on the coil of the output relay. The designer decides how each relay output is connected and what external devices it will operate. Solid-state outputs function much the same as relay outputs, except that they can only switch dc signals. Each solid-state output consists of a transistor (either PNP or NPN) which is controlled by the PLC. When the transistor is switched on by the PLC, the transistor conducts current to the output terminal (PNP) or from the output terminal (NPN).

Programming Unit

In most cases, PLC program development is done on a desktop or notebook computer using special development software for the particular brand and model of PLC that is being used. However, after the PLC-operated system is completed and operating, if program modifications are needed, it is not feasible to remove the PLC from the machine to take the PLC to an office or lab area for reprogramming. Therefore, most PLC manufacturers offer a

programming unit, which is a calculator-size device that plugs into the PLC via an interface cable. The programming unit has a keypad and display which allow personnel to view and modify the PLC program without disconnecting the PLC from the machine. The programming unit also gives maintenance personnel the capability to view the status of the PLC's inputs, outputs, and internal registers while the PLC is running, thus providing a valuable troubleshooting tool.

Internally, the CPU of the PLC contains three distinct areas of memory that serve very different purposes. First, a **read-only memory** (ROM) area contains the CPU's firmware. This is a program installed by the manufacturer that provides the CPU with basic information on how to interpret a user program, how to read inputs and write outputs, and how to perform the many other functions that the manufacturer provides with the PLC as features. ROM is not accessible by the user. The second memory area is EEPROM (electrically erasable programmable read only memory). The user stores programs in EEPROM and has the capability to read, write, and modify this area of memory. Although EEPROM can be modified by the user, it is nonvolatile; that is, if power is lost, the data stored in the EEPROM are retained and are available once power is restored. In many PLCs, additional program space can be provided by purchasing an EEPROM memory expansion module. The third PLC memory type is **random access memory** (RAM). RAM is used by the PLC to temporarily store data produced by the execution of the program. For example, the status of the discrete inputs and outputs, the results of mathematical calculations, and the elapsed time values of internal timers are stored in RAM. RAM is volatile; therefore, the PLC does not use RAM for the storage of any critical information.

17-8 PLC Operation

When a PLC is initially powered up, the CPU first performs a predefined series of initializations and self-tests to verify that the internal circuitry is functioning properly. Then, the PLC begins a continuous loop of two distinct operations:

1. **Perform I/O update**—In the I/O update process, the PLC outputs its most recently computed output bit values and then captures the binary values of all the inputs. In a PLC, the inputs are captured only once for each execution loop; they are not continuously monitored. Therefore, if an input signal from an external device changes while the PLC is executing the program, the input change is ignored until the next I/O update. Also, note that the outputs that appear on the PLC output terminals are updated only once for each execution loop. Even if it is determined during program execution that an output state should change, the PLC does not immediately change the output. Instead it waits until I/O update, at which time it updates all the outputs simultaneously.

2. **Solve the program**—During this part of the execution loop, the PLC scans the user program, calculating the result of each rung (on or off). The rung calculations are based on the information stored in RAM, consisting of the most recently captured inputs, calculated output states, timer values, counter values, and many other variables. Because the user program looks very much like an electrical controls diagram, the PLC executes the program linearly; that is, from top to bottom and left to right, in much the same way that you would read an electrical controls diagram. Once a rung has been solved, the PLC moves on to the next rung. At no time does the PLC "look back" at a previously solved rung. For each scan, each rung is solved only once.

Once the PLC has performed I/O update and solved the program, it returns to I/O update and begins again. Therefore, the sequence is as follows: Perform I/O update, solve the program, perform I/O update, solve the program, and so on.

The time that it takes a PLC to perform one I/O update plus one program solution is called the PLC's **scan time**. PLC scan times range from a few hundreds of microseconds to a few hundreds of milliseconds. They depend on the microprocessor clock frequency, the efficiency of the firmware program, and the size of the user program. For a particular

PLC, because the microprocessor clock frequency and firmware are fixed, the scan time is dependent solely on the size of the user program. PLC scan time is a critical parameter in any machine controls design because a PLC that scans slowly might not "see" an input event that has an on-time that is shorter than the scan time. For example, a PLC that has a 10-ms scan time might miss an input event that is less than 10 ms in duration because the I/O updates are occurring only once every 10 ms. All PLC manufacturers publish a scan time formula in the PLC user's manual that can be used by the designer to calculate the scan time based on the size of the user program that has been loaded into the PLC.

17-9 Fundamental PLC Programming

PLCs are programmed using ladder logic language, which is a graphic language consisting of NO and NC contacts and coils. When inputting and editing a program, the program is displayed on the computer screen in much the same way as an electrical controls diagram appears, with relay contacts and coils arranged in a ladder arrangement. All PLCs require that program rungs be arranged so that contacts are on the left side of the rungs and coils are on the right side. Generally coils may not be connected in series (there are exceptions), but some PLCs allow coils to be connected in parallel.

External inputs to the PLC are available in the program as relay contacts, and they generally have reference designators that are determined by the PLC manufacturer. One of the most common input reference designators is to use the letters "I" or "IN" followed by the number of the input, starting with input IN0 (or I0). Although there is physically only one IN0 on the PLC, we are allowed to use as many IN0 contacts as we need in our program because the PLC's CPU duplicates the value of IN0 for us. Additionally, the CPU inverts IN0 so that we can also have available NC IN0 contacts to use in our program. *Therefore, in all PLCs, we are allowed to have as many NO and NC contacts of any relay as needed.* The maximum number allowed is only limited by the amount of memory space available in the CPU to store our program. When an input is energized (ON), all NO contacts of the input that are contained in the program will be closed, and all NC contacts of it will be open.

Outputs from our program to the actual physical outputs on the PLC appear in the program as relay coils. We will only have one coil in our program for each output that we wish to control.[2] When an output coil is energized in the program, the corresponding PLC output will be ON (after I/O update, of course). As with inputs, PLC outputs each have a reference designator. Some of the more common reference designators for outputs are OUT0, OUT1, OUT2, and so on, or Q0, Q1, Q2, and so on.

When learning to program a PLC, it is always best to start by writing an extremely small one-rung program, loading it into the PLC, and testing it. Then, once the programmer is satisfied that the entire program development process is understood, more complex programs can be attempted. For most PLCs, the smallest program is a NO input IN0 controlling an output OUT0 and is shown in Figure 17-19. When this program is run, as long as there is no signal applied to input terminal IN0 of the PLC, the NO contact IN0 in the program will be open, thereby keeping output OUT0 off. However, when an input signal is applied to terminal IN0 of the PLC, when the next I/O update occurs, IN0 is stored in the PLC as an ON bit, which causes the software contact IN0 in our program to close and energize coil OUT0. On the next I/O update, the output terminal OUT0 of the PLC is switched on. Therefore the result of our simple program is to allow the PLC input IN0 to

FIGURE 17–19
Simple PLC program.

---

[2]Most PLCs allow programmers to use a particular relay coil multiple times in a program. However, when learning PLC programming, to avoid confusion, this technique is not advisable. The technique of using multiple coils of the same relay should be attempted only by experienced PLC programmers.

366 CHAPTER 17 • Programmable Logic Controllers

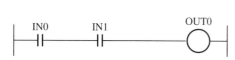

FIGURE 17–20
Series contact arrangement.

FIGURE 17–21
Parallel contact arrangement.

control the PLC output OUT0. The time delay between the time IN0 is switched on and OUT0 switches on is determined by the scan time of the PLC and where in the PLC cycle the input is switched on. If IN0 is switched on immediately after I/O update, it is possible to have a delay of two scan times—one scan to capture the input value and one additional scan to switch on the output.

PLC programs also give us the ability to "wire" devices in series or parallel in software instead of actually physically connecting the devices with wires. For example, assume that we are constructing the control circuit for a large machine that is operated by two persons. We will mount two pushbuttons on the machine, one for each operator, and we wish to design the control circuit so that the machine will run only when both operators press their respective pushbuttons at the same time. We could physically wire the pushbuttons in series and connect the series combination to one input of the PLC. However, we are given more flexibility if we wire each pushbutton to a separate input (IN0 and IN1, for example) and then connect them in series in the PLC program, as shown in Figure 17-20. Then, if at some later time we wish to reconfigure the switches so that they are in parallel or so they operate other functions within the program, no wiring change is required. Instead, we "rewire" the switches within our program. Figure 17-21 shows the same two inputs IN0 and IN1 connected in parallel instead of series. For this contact arrangement, OUT0 will be switched on when either IN0 or IN1 is on.

We also can use our program to invert the logical operation of an input. For example, the contact arrangement in Figure 17-22 has one NO contact of IN0 and one NC contact of IN1 connected in series. For this program, output OUT0 will be on when IN0 is on and IN1 is off. In this case, a NO pushbutton switch connected to IN1 would act as a NC switch because we inverted it in the program.

Earlier in this chapter we constructed a latch circuit (shown in Figure 17-18) which has two pushbuttons, a NO START and a NC STOP, controlling a relay coil RUN. We can construct this circuit in software by first connecting our switches to the PLC inputs and the relay coil to a PLC output. Let us assume we wire START to IN0, STOP to IN1, and RUN to OUT0. We will also assume that both the START and STOP switches have NO contacts. The program to create the latch (in Figure 17-23) appears very much like the hardware version of the circuit in Figure 17-18. Notice in this program that we have placed a NO contact of OUT0 in parallel with IN0. In PLC programs not only do we have contacts of the inputs available to us, but we can also have contacts of the output "relays" at our disposal. These contacts switch on and off whenever the output changes state. Our OUT0 contact is of the NO type, so it will be on whenever OUT0 is on. An additional advantage of this implementation of the latch circuit is that both the START and STOP switches have NO contacts,

FIGURE 17–22
Series contact arrangement with one NC contact.

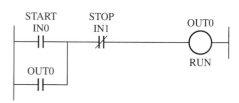

FIGURE 17–23
Latch program.

thereby allowing us to purchase two identical switches instead of one NO and one NC. If our implementation is for an actual manufacturing machine, when we purchase spare parts, we only need to stock one spare NO switch that can be used for either of these switches.

■ EXAMPLE 17-1

Write a PLC program that has two inputs, IN0 and IN1, and one output OUT0. The output will be on whenever the two inputs disagree. When the two inputs are in the same state, OUT0 is to be off.

SOLUTION For this program, there are two conditions that will cause OUT0 to be on, which are IN0 off and IN1 on, or IN0 on and IN1 off. This will require two branches, each having one contact of IN0 and IN1 in series. In the first branch we will have a NO IN0 and a NC IN1, and in the second branch we will have a NC IN0 and a NO IN1. The resulting program is illustrated in Figure 17-24.

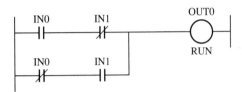

FIGURE 17-24
Example 17-1 solution.

■ EXAMPLE 17-2

A factory has three tanks of ammonia used for a manufacturing process. You are asked to design a warning system so that when any two or more of the three tanks are empty, a light illuminates, thereby warning the factory personnel to place an order for more ammonia. A switch is installed in the bottom of each tank that switches on when the tank is empty. The switches are connected to PLC inputs IN0, IN1, and IN2, and the warning light is connected to PLC output OUT0. Write the PLC program to switch the light on when any two or all three of the tanks are empty.

SOLUTION There are four possible conditions that can cause a warning light. They are

1. IN0 on and IN1 on
2. IN0 on and IN2 on
3. IN1 on and IN2 on
4. IN0 on, IN1 on, and IN2 on

However, notice that if condition 4 is true, conditions 1, 2, and 3 will also be true. Therefore, condition 4 is redundant and does not need to be considered as part of the solution. The solution will consist of three branches, with each branch satisfying one of the conditions 1, 2, and 3 above. The resulting program is shown in Figure 17-25.

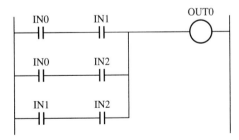

FIGURE 17-25
Example 17-2 solution.

■ EXAMPLE 17-3

For Example 17-2 above, after the switches are wired to the PLC and the system is powered for testing, it is discovered that the wrong fluid-level sensor switches have been purchased and that instead of switching on when the tanks are empty, these sensors switch *off* when the tanks are empty. You decide that it is not necessary to exchange the switches and instead you will correct the problem by modifying your PLC program. Draw the ladder diagram of the program to work with these switch types.

SOLUTION Programmable logic has one main advantage. It can be easily modified, and in much less time and expense than would be involved in changing the switches. Because the logic of the purchased switches is opposite of that in the original design, we will simply reinvert the logic of each switch inside our program by replacing all occurrences of the inputs with their logical inverse. Figure 17-26 shows the solution.

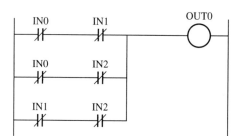

FIGURE 17-26
Example 17-3 solution.

In addition to having available inputs and outputs that the PLC programmer can use in programs, all PLCs also have **internal relays**. Internal relays are software relays that can be used for any general-purpose application within a program. An internal relay has one coil and as many NO and NC contacts as needed. Internal relays are not accessible from the inputs or outputs of the PLC, so they can be controlled only from within a program, and the contacts can operate only within the rungs of a program. The number of internal relays allowed in a program depends on the PLC brand and model and the amount of memory in the CPU. Internal relays have their own numbering system, such as CR0, CR1, CR2, and so on, or M1, M2, M3, and so on.

Internal relays are best used to reduce the number of contacts within a program. The maximum size of a PLC program that will fit into a CPU memory is determined by the total number of contacts, coils, and other elements in a program. If we can reduce the number of elements in a program, we can fit larger, more powerful programs into our PLC, and the program will execute more quickly because the CPU evaluates fewer elements when solving the program.

To see how internal relays help reduce the number of elements in a program, we will write a program in which outputs OUT0, OUT1, and OUT2 are to be switched on any time inputs IN0, IN1, and IN2 are all on. Figure 17-27 illustrates one method to implement the program. This method uses 12 elements (the number of contacts, plus the number of coils). However, if we use an internal relay CR0, we can reduce the number of elements to 10 by implementing the program as shown in Figure 17-28. In this program, we have inputs IN0, IN1, and IN2 operating the coil of internal relay CR0. Then in the remainder of the rungs, we have a NO coil of CR0 operating the outputs. The performance of the program is the

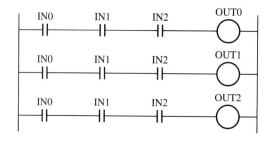

FIGURE 17-27
Program with three inputs and three outputs.

FIGURE 17–28
Simpler program with three inputs and three outputs.

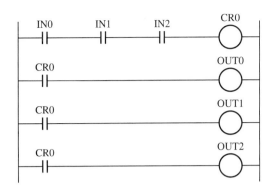

same as that of Figure 17-27, although Figure 17-28 will execute slightly more quickly because of the reduced number of elements.

17-10 PLC Programs with Timers

All PLCs also have available internal programmable timers that the programmer can use in programs. The number of allowable timers in a program depends on the PLC brand and model and the amount of memory in the CPU. The most common type of timer available is the nonretentive on-timer or TON. The TON timer works in the same way as the hardware version TON; that is, when power is applied to the timer coil, the timer begins timing. When the timer times out, it switches on and its contacts change state. If power is interrupted to the coil of the timer, it de-energizes and resets its internal timer to zero.

Depending on the PLC brand and model, timer coils are represented in the PLC program as either a standard coil symbol or as a rectangular box as shown in Figure 17-29. In most PLCs, when timers are placed in a program rung, they are placed on the right side of the rung. Timers have two variables associated with them, the **accumulator value** and the **preset value**. The accumulator value is the timed value. When the timer's coil is energized, the accumulator value will advance. When the coil is de-energized, the accumulator value will reset to zero.[3] The timer's preset value is the time value at which the timer will time out and switch its contacts. The preset value is a constant. It is entered when the timer is entered into the program. In PLCs, standard timers increment in 0.1-second increments; however, the preset value shown in the program omits the decimal point. Therefore, the programmer must multiply the desired time by 10 and enter that value as the preset value. The timers shown in Figure 17-29 will time out in 2.0 seconds.

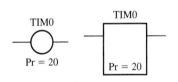

FIGURE 17–29
Timer symbols.

■ **EXAMPLE 17-4**

When a machine is switched on, it takes a few seconds for the pump to reach operating speed. Therefore, we need a READY light on the operator's control panel that illuminates 5 seconds after the machine has been switched on to tell the operator when the machine is ready to use. In addition to other inputs and outputs, our PLC has an input IN0 which is on when the machine is switched on and an output OUT1 that controls the READY lamp. Write the portion of the PLC program that will control the READY lamp.

SOLUTION Another advantage in using PLCs is that a rung in a PLC program can be a stand-alone entity that is not affected by anything else in the program. Therefore, we can write a PLC program in parts, test the parts, and then assemble them later into the final version of the program. Figure 17-30 shows the part of the program that will control the ready lamp. In the first rung, the input IN0 controls the coil of the timer TIM0, which has

---
[3] Most PLCs have timers that advance starting from zero. Some PLCs start the accumulator value at the timer's preset value and have it decrement to zero. Consult the PLC user's manual to determine how the timers advance for your particular PLC.

FIGURE 17–30
Solution for Example 17-4.

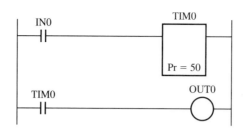

a preset of 5.0 seconds. When IN0 switches on, TIM0 begins timing. When the accumulated value of TIM0 reaches and exceeds its preset value (in this case, 5.0 seconds) the timer switches on. In the second rung, this closes the NO TIM0 contact, which, in turn, switches on OUT0 and the READY lamp.

EXAMPLE 17-5

An oiler is installed on a machine which periodically applies oil to a mechanism in the machine. The oiler is operated electrically, and when given a 1-second trigger, it squirts a predetermined amount of oil into the machine. It is determined that when the machine is operating, the oiler needs to be triggered every 15 seconds. Write a ladder logic program that will produce an output OUT4 that is on for 1 second every 15 seconds. The oiler timing circuit is to be active whenever input IN3 is on.

SOLUTION In this example, we have two timing operations, a 15-second timer that times the interval between oiler triggers and a 1-second timer that produces the electrical pulse to trigger the oiler. The program is shown in Figure 17-31. In rung 1, when input IN3 switches on, timer TIM0 starts timing. Fifteen seconds later, TIM0 times out. At this time, in rung 2, the TIM0 contact closes and timer TIM1 starts timing. At the same time, in rung 3, the TIM0 contact closes. Because TIM1 has not yet timed out, the NC TIM1 contact is presently closed, thereby switching on output OUT4. One second later, TIM1 times out. In rung 3, this switches off OUT4, and in rung 1, the NC TIM1 contact opens and switches off TIM0. When TIM0 switches off, the NO contact of TIM0 in rung 2 opens and switches off TIM1. At this point we have returned to the starting condition, and TIM0 again starts timing. The program will continuously provide a 1-second duration output on OUT4 every 15 seconds, for as long as IN3 remains on.

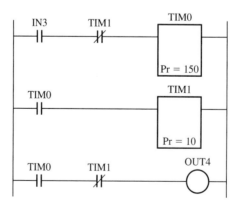

FIGURE 17–31
Solution for Example 17-5.

In addition to nonretentive TON timers, many PLCs offer non-retentive TOF timers, and retentive TON and TOF timer. Although these additional types of timers are convenient elements to have available when programming, many smaller PLCs offer only nonretentive TON timers. For experienced programmers, having only TON timers available is a minor inconvenience because the other types of timers can be constructed from TON timers.

17-11 Advanced PLC Features

In addition to standard NO and NC contacts, coils, and timers, PLCs also have programming features that perform more sophisticated tasks. Although these features are not within the scope of this text, a brief overview of some of the more useful ones will be presented.

Counters

As the name implies, **counters** provide a variety of counting features. Generally counters are available as either unidirectional counters or bidirectional counters. Unidirectional counters have two input control lines, which are **count** and **reset**. When the count line is switched on, the counter advances one count. To count again, the count line must be switched off and then on again. When the reset line is on, the counter resets to zero. The reset is overriding; that is, it takes precedence over the count line, thereby disabling any counting while the reset line is on. Bidirectional timers also have a reset line, but to operate the timer in both directions, they will have either a count line and a direction line, or a count-up line and a count-down line.

Like timers, counters also have a preset value. As the counter advances, its contacts are de-energized until the count value equals the preset value. When the count and preset values are equal, all contacts of the counter are energized. Counters allow the programmer to track the number of times an operation has occurred. For example, they can be used to count the number of times a lamp has been flashed or the number of products placed in a shipping container.

Sequencers

Sequencers are programming elements that count in much the same manner as counters. However, they have the added feature of allowing the programmer to enter contacts of the sequencer that switch on at defined count values. Therefore, it is possible to have a contact switch on when count = 1, or count = 5, or any other count value. As a result, the PLC can control a sequence of operations. For example, on count = 1 we can have the PLC load a piece of raw material, on count = 2 it can punch the product, on count = 3 it can bend the product, and on count = 4 it can eject the finished piece.

Mathematics

PLCs that have register I/O can input and output parallel bytes or words of data. The PLC's register I/O can be connected to keypads, digital displays, or other computers. Once data have been input to the PLC in byte or word form, the PLC needs the capability to perform mathematical operations on the data. To do this, most PLCs also have programming elements that perform mathematical operations. When a rung containing a math operation is energized, the math operation is performed, and when the rung is de-energized, the math operation is skipped. The math operation can be as simple as data move, data copy, compare values, add, subtract, multiply, or divide, or as complex as trigonometric, logarithm, or exponential functions. The math operations are performed on bytes or words of memory within the PLC. Math operations give the PLC the capability to analyze digital values and make control decisions based on the results.

17-12 PLC Wiring

In addition to programming the PLC, the PLC system designer must determine how to wire the PLC into the system so that the PLC has proper power applied to it and the inputs and outputs function properly. An understanding of how PLC inputs and outputs function is important in ensuring that the PLC I/O will function correctly when the system is powered.

Power Connections

When purchasing a PLC, the designer must specify the type of power that will be available to operate the PLC. This can be either ac or dc power. For ac, the most common voltages are 120 V ac and 240 V ac, although many PLCs can operate from any ac voltage from 85 volts to 250 volts. Generally, for ac operation, the PLC can be connected directly to the power line or to the secondary of a control transformer. For dc operation, the most common voltages are 12 V dc and 24 V dc, which must be provided by an external dc power supply of the proper voltage.

Discrete Inputs

Depending on the design of the control circuitry for a particular application, the PLC may be required to operate with either dc or ac inputs. Therefore, PLCs are available with inputs that can be used with either dc or ac control voltages.

Figure 17-32 shows a PLC input circuit for dc input operation. The circuit consists of a current-limiting resistor connected to an optical isolator (opto-isolator). When a voltage is applied to the PLC input terminal, current flows through the resistor and the LED in the opto-isolator and out to the common terminal. The light produced by the LED illuminates an optically sensitive transistor which switches on. The transistor's output is connected to the CPU of the PLC. The purpose of the opto-isolator is to electrically isolate the CPU from the input terminals, thereby protecting the CPU from damage in the event of an electrical surge on the input terminals. The value of the current limiting resistor depends on the dc voltage rating of the PLC input.

The ac input circuit is illustrated in Figure 17-33. It is much the same as the dc input circuit except that the opto-isolator contains two LEDs that are facing in opposite directions. In doing so, the LEDs can operate from a current that is alternating in polarity. It is important to note that the ac input circuit in Figure 17-33 can also be operated from dc control signals.

The input circuit is connected to the control signals so that when the control signal is on, current flows through the opto-isolator. Figure 17-34 illustrates a four-point PLC input connected to several different types of switches. Notice that the switches are connected so that when any of the switch contacts are closed, current will flow from the dc power supply, into the input terminal, through the resistor and LED for that input, and out of the common terminal to return to the power supply, thereby completing the circuit. For input IN0, we have used a NC pushbutton switch. Therefore, IN0 will be normally on. It will switch off when PB3 is pressed. Also, although we have connected the negative terminal of our power supply to the common terminal on the PLC, because we are using a PLC with bipolar inputs (that is, having two LEDs in the opto-isolator) we could reverse the power supply connections or even replace the dc power supply with an ac source.

Discrete Outputs

Most PLC discrete outputs can be classified as either **relay** or **solid state**. Of these, the most common and versatile output type is the relay. With the relay output, each output of the PLC

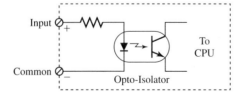

FIGURE 17–32
PLC dc input circuit.

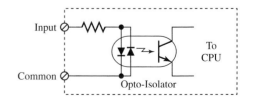

FIGURE 17–33
PLC ac input circuit.

FIGURE 17–34
Connection of input devices to a four-point PLC input.

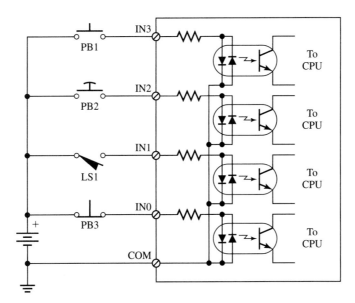

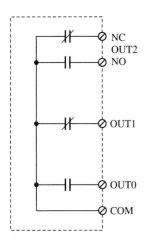

FIGURE 17–35
Various PLC relay output configurations.

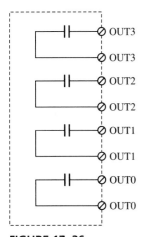

FIGURE 17–36
Four-point form A relay output.

consists of one relay, with the coil of the relay controlled by the CPU. The contacts of the relay are brought out to screw terminals on the PLC. The relay contacts available can be normally open (called **form A**), normally closed (called **form B**), or a combination of one form A and one form B with a shared common terminal (called **form C**). Figure 17-35 illustrates these three relay configurations. Output OUT0 is a form A output, OUT1 is form B, and OUT2 is form C.

Although Figure 17-35 shows the relays connected to a single common, relay outputs are also available with independently connected relays. For example, Figure 17-36 shows a four-point form A relay output with independently connected relays. This type of output is the most versatile of all the relay output types because each of the relays can be connected to a different voltage or even to a mix of ac and dc circuits.

Another very popular output type is the **solid-state output**. In this type of output, the controlling devices are opto-isolators, similar to those used on discrete PLC inputs. However, in this case, the PLC's CPU controls the LED in the opti-isolator and the transistor is connected to the output terminals. The load current switching is done with a transistor, so there are no moving components to wear out. Therefore, one of the main advantages in using solid-state outputs is reliability. The main disadvantage in using solid-state outputs is that the switching devices are transistors, so they cannot switch ac loads.

Figure 17-37 illustrates a four-point solid-state output consisting of NPN transistors with various loads connected to the output terminals. Note that because NPN transistors are used, current must flow *into* the output terminals.[4] Therefore, one terminal of each of the loads must be connected to the positive terminal of the power supply so that when any output is on, the NPN transistor in the PLC grounds the other load terminal, thereby completing the circuit. The four loads connected to our four-point solid-state output system are as follows: OUT0, indicator lamp L1; OUT1, indicator lamp L2; OUT2, electrically operated pneumatic valve V1; and OUT3, relay coil M1. The M1 relay connected to output OUT3 is a motor control relay (on most controls diagrams, the "M" reference designator is reserved for motor control relays). Motor control relays are designed to carry large currents required by motors.

The PNP solid-state output is similar to the NPN solid state output, except that the output transistors are PNP type. For these types of outputs, current flows out of the output terminal.[5] To arrange the circuit so that current will flow out of the output terminals, we must

---

[4] Because current flows *into* NPN outputs, they are also called **sinking outputs**.
[5] Because current flows *out of* PNP outputs, they are also called **sourcing outputs**.

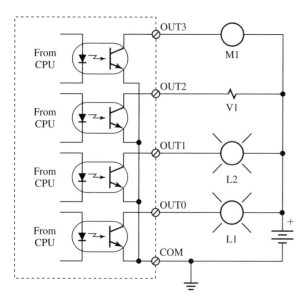

FIGURE 17–37
Four-point NPN solid-state output with various loads.

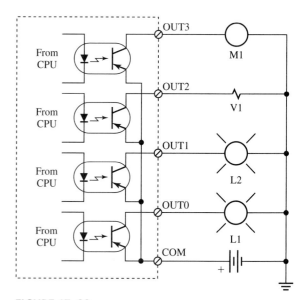

FIGURE 17–38
Four-point PNP solid-state output with various loads.

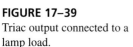

FIGURE 17–39
Triac output connected to a lamp load.

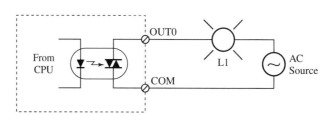

apply a positive voltage to the common terminal as illustrated in Figure 17-38. This will reverse the direction of current flow through the loads, as compared to the NPN output circuit in Figure 17-37.

A third type of output that is very popular with PLC system designers is the triac output. The triac is fundamentally a solid-state switch that can be electronically controlled (in this case, the control comes from the PLC's CPU). The device performs much like a relay, except that it can be used to switch only ac currents (that is, it will not perform properly when connected into a dc circuit). To isolate the PLC's CPU from the output circuit, the optically coupled triac is used in all PLC triac outputs, as illustrated in Figure 17-39.

When selecting PLC output types, there are some important points that should be considered.

- Although the examples previously shown for relay, solid-state, and triac outputs illustrate only a few outputs in a group, most PLC manufacturers provide output systems in groups of 8 or 16 output points.
- If it is uncertain whether the loads will be ac or dc loads, or if there is uncertainty as to the voltages that will be switched, purchase relay outputs.
- When relays switch on, their contacts will bounce slightly. If a relay output is being connected to a digital logic circuit, be aware that the digital circuitry will consider each bounce of a relay contact as a pulse. This can create strange and unrepeatable results.
- If the output circuits will all be dc within the 10- to 30-volt range, use solid-state outputs. They are more reliable than relays and they do not incur contact bounce when they switch on.

- If all loads are ac operated, use triac outputs. They are more reliable than relays.
- When selecting output types, *always* read the manufacturer's specifications for the PLC being purchased to ensure that the selected output types are capable of switching the voltages and currents that will be used in the system.

PROBLEMS

Draw ladder logic diagrams of circuits that will perform the following operations.

17-1 Switch on output OUT0 when both inputs IN0 and IN1 are on.

17-2 Switch on output OUT1 when either input IN2 is on or input IN3 is on.

17-3 Switch on output OUT2 when either input IN3 is off or input IN4 is off.

17-4 Switch on output OUT3 when both inputs IN5 and IN6 are off.

17-5 Switch on output OUT4 when inputs IN7 and IN8 agree (that is, both on or both off).

17-6 Switch on output OUT5 when inputs IN9 and IN10 are on, or when IN11 is on and IN12 is off.

17-7 Switch on output OUT6 when any two (and only two) of inputs IN13, IN14, IN15, and IN16 are on.

17-8 Switch off output OUT7 when both IN17 and IN18 are on. (*Hint*: Draw a truth table.)

17-9 Switch on OUT8 when IN20, IN21, and IN22 are either all off or all on.

17-10 Draw a ladder logic circuit that will switch on output OUT0 3.5 seconds after either IN0 or IN1 is switched on. OUT0 should switch off immediately when both IN0 and IN1 are off.

17-11 Draw a ladder logic circuit that will switch on OUT0 for exactly 2 seconds every time IN3 is switched on. Assume that each time IN3 is switched on, it will remain on for more than 2 seconds.

17-12 Inputs IN0, IN1, IN2, and IN3 of a PLC are connected to the binary coded decimal (BCD) output of a numeric keypad, with IN0 being the least significant bit and IN3 being the most significant bit. Write a ladder logic program that will switch on output OUT2 when the operator presses key 5 (0101_2) or key 9 (1001_2). OUT2 should be off for all other keys.

17-13 Design and draw a ladder logic program that will continuously flash a lamp connected to OUT3. The lamp should flash on for 1 second and off for 0.5 second.

17-14 Design and draw a ladder logic program that will detect IN0 switching on *before* IN1. If IN0 switches on before IN1, output OUT0 will switch on. If IN1 switches on before IN0, OUT0 will switch off. (*Hint*: Remember, the PLC solves the program from top to bottom and left to right.)

Digital Communications

OVERVIEW AND OBJECTIVES

This chapter will provide an introduction to a topic that has made a tremendous impact on the world in all walks of life in recent years, namely the broad area of digital communication. Most readers will likely be familiar with the term through the use of computers and the Internet, but it is also becoming widespread in television, consumer electronics, the telephone system, and data collection equipment.

Historically, some forms of digital communication were in usage even before the advent of radio. The Morse code, in which letters and numbers are encoded by combinations of short pulses (dots) and long pulses (dashes), dates back to the nineteenth century. Teletype transmission has also been in existence for many years.

Nothing has affected the area of digital communication more than the **Internet** and the personal computer. The rapid growth of this area has brought digital communication into millions of households, and the future looks very bright for many new innovations and applications of this technology.

This chapter will be devoted to an introduction to the concepts of digital communication. The emphasis will be directed toward the principles of encoding, transmission, and decoding of signals that are represented by digital forms.

Objectives

After completing this chapter, you should be able to

- Define various terminology associated with digital communications.
- Discuss the form of a **pulse code modulated (PCM)** signal.
- Define the strategies involved in **unipolar** and **bipolar offset encoding** of A/D converters.
- State and apply the relationships for step size and error for both unipolar and bipolar offset encoding.
- Explain the concept of **time-division multiplexing (TDM)** as it relates to PCM.
- Define **amplitude-shift keying (ASK)**, and discuss its properties.
- Define **frequency-shift keying (FSK)**, and discuss its properties.
- Define **binary phase-shift keying (BPSK)**, and discuss its properties.

18-1 Introduction and Terminology

Chapters 9, 10, and 11 were all devoted to digital circuits. A fundamental topic associated with such systems is the **binary number system**. The most basic form of digital communication involves transmission of signals from one point to another using a series of distinct pulses or other suitable electronic forms. Each group of pulses utilizes the binary number system to represent a character or level of some type.

Binary Communication versus *M*-ary Communication

Historically, digital communication was first performed using binary encoding, in which each symbol could assume only two levels. This form of transmission can be referred to as **binary communication**. It is still widely used today. However, the need to send more data in a given time interval has led to much research and development of so-called ***M*-ary communication** systems, in which more than two levels are transmitted, with M representing the number of levels. This process will be explained in more detail later. However, the development will begin with the assumption of a binary system, and until indicated otherwise, that assumption will apply in the developments that follow.

Bits and Bytes

The fundamental element of digital data is the **bit**, which has only two possible states, usually designated as 1 and 0. In various contexts, the states of the bit can also be considered *on* and *off* or *true* and *false*. For purposes of computer storage and transmission, bits are usually combined in groups of eight, called **bytes** (or sometimes **octets**).

Relationship between Bits and Levels

A **word** will be defined as any combination of N bits representing some character or level. Let us first consider the most elementary possibility in which a word consists of only 1 bit. This means that the signal can assume only two possible values, indicated as 0 and 1, so it is very limited. Next, assume that each word consists of 2 bits. The signal can now assume four possible values or states, indicated as 00, 01, 10, and 11. If each word contains 3 bits, the signal can assume eight possible values, and so on.

Let L represent the number of distinct possible levels, values, or states and let N represent the numbers of bits per word. The discussion of the preceding paragraph is readily generalized to be

$$L = 2^N \qquad (18\text{-}1)$$

The inverse relationship is

$$N = \log_2 L \qquad (18\text{-}2)$$

The relationship of Equation 18-2 results in an integer only if L is an integer power of 2. Therefore, if the relationship is being used to determine the number of bits in each word required to encode a certain number of levels, the result could be a noninteger value, in which case the value would need to be rounded up the next higher integer.

Because most calculators do not have logarithms to the base 2, it is possible to determine a value based on the following relationship:

$$\log_2 x = 3.32 \log_{10} x \qquad (18\text{-}3)$$

Synchronous versus Asynchronous

Digital transmission may be characterized as either **synchronous** or **asynchronous**. Synchronous transmission is achieved by means of a master clock or timed reference, which is either transmitted with the signal or available at both ends through some other means.

Asynchronous (also called **start-stop**) data transmission systems have the capability of initiating or terminating transmissions in a much more flexible manner. This is achieved by provided start and stop information with each transmitted data unit, usually a byte, so it is essentially self-clocking. The advantages of asynchronous transmission are obvious for such applications as computer terminals and Internet transmission, in which the beginnings and endings of data exchange are random.

Parallel versus Serial

Digital transmission may also be characterized as either **parallel** or **serial**. In parallel transmission, each bit of the transmitted data unit has a separate transmission path, and all bits are sent simultaneously. An example of parallel transmission is the connection between the parallel port of a computer and a printer or the transmission of data between a computer's **central processing unit (CPU)** and **memory**.

In serial transmission, only one channel is used and the bits are transmitted in sequence. In general, parallel transmission is faster but is usually limited to short distances because of the requirement for multiple channels.

Textual versus Binary Data

The transmission of alphanumeric characters is usually referred to as **textual data**. However, digitized audio and video are often treated in much the same manner in computer and Internet communication. These data are referred to as **binary data** in this particular context.

Simplex versus Duplex

The transmission channel and connection between two data devices such as computers or terminals may be characterized in three different ways: (1) **simplex**, (2) **half duplex**, and (3) **full duplex**.

- In a simplex connection, transmission is always in one direction only, and that direction is fixed.
- In a half-duplex connection, transmission can take place in either direction, but not simultaneously. Thus, the sending device and receiving device must take turns.
- In a full-duplex connection, transmission can take place in both directions simultaneously.

Marks and Spaces

Although the normal digital terms of *ones* and *zeros* are perfectly correct, some of the earlier terminology uses the terms *mark* and *space*. These terms may still be found in many references and on some equipment. A **mark** is equivalent to a logical 1, and a **space** is equivalent to a logical 0.

Unipolar versus Bipolar

The two states of a **unipolar** binary signal are a positive or negative voltage or current level and a value of 0. Thus, the polarity of the signal is always in one direction or zero, and it never assumes the opposite direction. For example, a logical 1 might be represented by 5 V and a logical 0 might be represented by 0 V.

The two states of a **bipolar** binary signal are a positive voltage or current and a negative voltage or current. Thus, the signal assumes both polarities, and the level of 0 occurs only in the transition phase between two bits.

The terms *unipolar* and *bipolar* as given here apply to the form of the data signal. However, we will later see that analog-to-digital and digital-to-analog converters use the terms *unipolar* and *bipolar* in reference to the analog signal being sampled, so the meanings are somewhat different.

Data Rate

Information in digital communication is measured in bits. The data rate R is defined as the information transmitted per unit time and is measured in bits/second (b/s), with prefixes added as required. In dealing with computers and file transmission, it is easy to get confused between bits and bytes because memory allocations on a computer usually are specified in bytes. Just remember that a byte is equivalent to 8 bits and the data rate is usually specified in bits/second. See Example 18-4 at the end of this section.

EXAMPLE 18-1

How many distinct characters can be formed with a word length equal to 1 byte?

SOLUTION Because 1 byte is equivalent to 8 bits, we have

$$L = 2^8 = 256 \tag{18-4}$$

EXAMPLE 18-2

Commercial music compact discs (CDs) use 16-bit words to represent the signal samples. Determine the number of distinct levels that can be created.

SOLUTION The number of levels is

$$L = 2^{16} = 65{,}536 \tag{18-5}$$

Comparing the results of these examples, one can see how quickly the number of possible levels increases as the number of bits per word increases. This is a good illustration of why music stored on compact discs has such high quality.

EXAMPLE 18-3

It is desired to encode a signal into no fewer than 100 levels. Determine the number of required bits per word.

SOLUTION Although a trial-and-error approach could lead to a quick solution, we will illustrate a more formal approach by using Equations 18-2 and 18-3.

$$N = \log_2 L = \log_2 100 = 3.32 \log_{10} 100 = 3.32 \times 2 = 6.64 \tag{18-6}$$

This value is not an integer, so we must round upward and express the value as

$$N = 7 \text{ bits} \tag{18-7}$$

This result indicates that 128 possible levels could actually be obtained with the number of bits required.

EXAMPLE 18-4

Assume that a particular data file on a computer has a size of 3 Mbytes and that it is to be uploaded at a speed of 200 kbits/s. How long will it take the file to transfer?

SOLUTION Let M represent the total number of bits in the file, and the value is

$$N_{\text{total}} = 3 \times 10^6 \text{ bytes} \times 8 \text{ bits/byte} = 24 \times 10^6 \text{ bits} \tag{18-8}$$

Let T_{tran} represent the transmission time. It can be determined as follows:

$$T_{\text{tran}} = \frac{\text{bits}}{\text{bits/s}} = \frac{M}{R} = \frac{24 \times 10^6}{200 \times 10^3} = 120 \text{ s} = 2 \text{ minutes} \tag{18-9}$$

18-2 Encoding and Transmission of Textual Data

We will begin the study of digital communication by considering the process of representing textual data with binary words. When you type a character on a computer keyboard, it is generating a unique code word that enters computer memory. Thus, each word represents a particular character. Most characters of interest for this purpose are **alpha-numeric** characters, meaning that they are either letters or number. This area of digital communication is often referred to as **data communication**.

The applications of data communication include transmission between two computers, between a computer and a remote terminal, between a business terminal and a central accounting system, and most of the communication on the Internet. The key factor here is that the input and output data employ digital codes corresponding to characters, rather than samples of an analog signal.

A basic consideration in all digital data communication systems is that each of the possible characters must have a unique code word. A means must be provided at the data source for converting each possible character to its associated code word, and a similar means must be provided at the destination for converting the code word back to its proper character.

Over the years, several different codes have been developed. There is little reason to list them in this limited treatment. Rather, we will focus on the particular code that has the widest usage, namely the ASCII code.

ASCII

The most widely used encoding process in modern digital data systems is the **American Standard Code for Information Interchange**, hereafter referred to as the **ASCII** code. Originally, the ASCII code was a 7-bit code. With 7 bits, the number of possible distinct levels or characters is $2^7 = 128$. This value is sufficient to encode all required letters (upper-case and lower-case), numerals, punctuation marks, and several special characters used on keyboards and in data transmission. Many applications use the eighth bit in a byte as a **parity check** bit.

The **Extended ASCII Code** incorporates 8 bits for data, which provides a total of 256 characters. It may also have an additional parity bit. The Extended ASCII Code is commonly used in modern computers.

A complete summary of the basic 7-bit ASCII code is provided in Table 18-1. The 128 possible binary codes, the corresponding characters they represent, and the definitions of special characters are listed there. Many of the special characters employ terminology peculiar to data processing procedures and will not be discussed here.

Bit 7 is considered as the **most significant bit (MSB)**, and bit 1 is considered as the **least significant bit (LSB)**. As we will see shortly, however, the usual manner of serial transmission is with the *LSB transmitted first and the MSB transmitted last*. One could argue that because ASCII words represent characters rather than values, it is immaterial which bits are considered MSB and LSB as long as the order of transmission is understood. However, certain properties have a common pattern with respect to basic binary encoding, so it is desirable to have a defined order based on the MSB to the LSB.

Asynchronous Serial Data Transmission

The manner in which a given character is encoded for asynchronous serial transmission in many systems will now be illustrated. Refer to Figure 18-1, in which the upper-case letter W is used as the example. From Table 18-1 or from Example 18-1, the form of the binary code with the MSB first is 1010111. For data transmission, the normal order of transmission will be *with the LSB first*; that is, as 1110101. However, more bits are required, as we will see shortly.

382　CHAPTER 18　• Digital Communications

Table 18–1 American Standard Code for Information Interchange (ASCII Code).

Bits 7	0	0	0	0	1	1	1	1
6	0	0	1	1	0	0	1	1
5	0	1	0	1	0	1	0	1
Bits 4321								
0000	NUL	DLE	SP	0	@	P	`	p
0001	SOH	DC1	!	1	A	Q	a	q
0010	STX	DC2	"	2	B	R	b	r
0011	ETX	DC3	#	3	C	S	c	s
0100	EOT	DC4	$	4	D	T	d	t
0101	ENQ	NAK	%	5	E	U	e	u
0110	ACK	SYN	&	6	F	V	f	v
0111	BEL	ETB	'	7	G	W	g	w
1000	BS	CAN	(	8	H	X	h	x
1001	HT	EM	)	9	I	Y	i	y
1010	LF	SUB	*	:	J	Z	j	z
1011	VT	ESC	+	;	K	[	k	{
1100	FF	FS	,	<	L	\	l	|
1101	CR	GS	-	=	M	]	m	}
1110	SO	RS	.	>	N	^	n	~
1111	SI	US	/	?	O	_	o	DEL

NUL	Null		DLE	Data Link Escape
SOH	Start of Heading		DC1	Device Control 1
STX	Start of Text		DC2	Device Control 2
ETX	End of Text		DC3	Device Control 3
EOT	End of Transmission		DC4	Device Control 4
ENQ	Enquiry		NAK	Negative Acknowledge
ACK	Acknowledge		SYN	Synchronous Idle
BEL	Bell		ETB	End of Transmission Block
BS	Backspace		CAN	Cancel
HT	Horizontal Tabulation		EM	End of Medium
LF	Line Feed		SUB	Substitute
VT	Vertical Tabulation		ESC	Escape
FF	Form Feed		FS	File Separator
CR	Carriage Return		GS	Group Separator
SO	Shift Out		RS	Record Separator
SI	Shift In		US	Unit Separator
			DEL	Delete

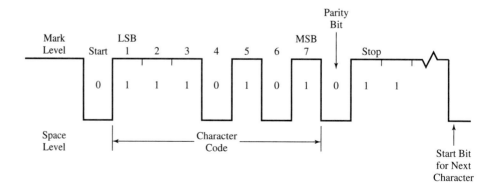

FIGURE 18–1
Asynchronous serial transmission of the letter W in ASCII form.

For asynchronous transmission, the normal level of the signal before the beginning of the word is the mark or "1" level. (There is an old expression "marking time" that can help jog the memory.) The actual level associated with the mark level can vary with the type of system and might even be a negative voltage in some standards. However, this figure shows the mark level as the upper level in accordance with the simplest type of interpretation.

The beginning of a new character is indicated by a transition between the *mark* level and the *space* level as shown. This transition starts the timing process at the receiver for the given word. Assuming normal tolerances of components, the synchronization certainly should last through the remainder of the word transmission.

Following the start bit, the 7 bits representing the character W are now transmitted, with the LSB transmitted first and MSB last. From Table 11-1, the 7 bits arranged in the order of transmission from left to right are 1110101.

After the completion of the data bits, a **parity** bit is inserted. This bit can be used to determine if there is 1 bit in the preceding 7 bits that was received erroneously. It can be based on either **odd parity** or **even parity**. With odd parity, the parity bit is either a 0 or 1 as required to make the net number of data bits plus the parity bit an odd number of 1s. With even parity, the strategy is to create an even number of 1s.

In this example, odd parity is employed. The character W has five 1s, so already there is an odd number of 1s, and the parity bit is a 0.

Following the parity bit, a *stop* signal is now transmitted. The stop signal is always at the mark level. As previously mentioned, it must have a minimum length, which in this example is assumed to be 1 bit interval. The signal then remains at the mark level until the next start bit is received. This could represent any arbitrary length of time provided that the minimum stop element length has been satisfied.

Textual Data Conclusion

Textual characters are characterized by the fact that as long as there are enough bits in each word to represent all possible characters, and as long as there are no bit errors in transmission, the transmitted result would be 100 percent correct. However, that is not the case with analog data converted to digital forms, as will be discussed in subsequent sections.

EXAMPLE 18-5

Write the binary codes for upper-case W and lower-case w and compare them. What can you conclude?

SOLUTION The two characters beginning with the MSB (bit 7) are obtained from Table 18-1 and are compared as follows:

W   1010111

w   1110111

The two codes differ only in bit 6 (the one following the MSB). This pattern is true for all of the letters. An upper-case letter may be converted to a lower-case letter by complementing bit 6.

EXAMPLE 18-6

Write the ASCII code for the number 9 and indicate any pattern that is observed.

SOLUTION From Table 18-1, the ASCII code for the number 9 is 0111001.

The last four digits of the binary form represent the corresponding 4-bit binary representation of the number 9. This pattern is true for all the numbers from 0 through 9.

18-3 Sampling an Analog Signal

The term **analog** is widely used to describe signals and systems that involve functions that are defined at any arbitrary level and with the time scale continuous. Most standard radio and television broadcasting has traditionally been achieved by analog methods, although the newer trends in many segments of the industry involve digital technology. In a sense, the terms *analog* and *digital* are almost considered the opposite of each other, at least in a casual comparison sense. The terms **continuous time** and **discrete time** are more formal ways of describing analog and digital, respectively, but we will continue to employ the simpler terms.

The question now arises as to how we can take an analog signal and convert it into a digital signal. We have already seen that a digital signal can assume only a finite number of levels and an analog signal could be considered as having an infinite number of levels. The answer lies in selecting a sufficient number of bits to represent each sample within a specified accuracy. The number of possible levels increases exponentially with the number of bits per word, so the error can be made as small as one chooses, within limits.

The more difficult question is related to the sampling process. Consider the analog signal illustrated in Figure 18-2(a). Assume that it is sampled at intervals of T seconds and represented by the samples shown in Figure 18-2(b). At this point, the signal is known as a **sampled-data signal**. It eventually will be converted into a sequence of digital words, but for the discussion at hand, the form shown is convenient.

Obviously, we cannot see the entire original signal, so a natural question is whether or not we lose information. Said differently, could something happen between samples that might cause us to miss some important information? This important question requires that we introduce and discuss the concept of **bandwidth**.

Bandwidth

All signals that are processed in a communication system can be expressed in terms of sinusoidal components. The relative strength of the sinusoidal components that make up the signal is called the **frequency spectrum**. For example, it is known that the human voice can be represented by sinusoidal components whose frequencies range from well under 100 hertz (Hz) to several kilohertz (kHz). In contrast, the video portion of a television signal has frequency components that extend above 4 megahertz (MHz).

Knowledge of the frequency content of the signal is very important in estimating the bandwidth required in the appropriate communications channel. Conversely, the bandwidth of a given channel can limit the bandwidth of possible signals that could be transmitted over the channel. For example, assume that a given channel is an unequalized telephone line having a frequency response from a few hundred hertz to several kilohertz. The channel might be perfectly adequate for simple voice data but would hardly suffice for high-fidelity sound or for high-quality video. Thus, the frequency response of the channel must be adequate over the frequency range encompassing the spectrum of the signal, or severe distortion and degradation of the signal will result.

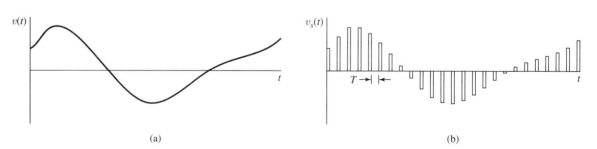

FIGURE 18–2
(a) An analog or continuous-time signal and (b) a sampled version of it.

Signals such as voice and music forms, in which most of the spectral components start near dc and reach some maximum frequency, are referred to as **baseband** signals. The bandwidth in this case is usually indicated as the highest frequency in the signal. The mathematical technique used to determine the spectrum of a signal is called **Fourier analysis** or, in common operational terms, **spectral analysis**.

In some cases, it is possible to transmit information over a narrower channel by slowing down the information rate to a level compatible with the channel. For example, in some space communications systems, the bandwidth has been kept quite low to keep the total noise power to a minimum. In some cases, higher-frequency data can be first recorded and then played back at a slower speed, which minimizes the required bandwidth. In this case, a longer processing time is being substituted for a wider bandwidth.

Sampling Theorem

The fundamental key to representing a signal in terms of samples is the sampling theorem. It can be stated as follows:

A signal bandlimited from dc to some maximum frequency f_m can be reproduced by taking no less than $2f_m$ samples/second or, equivalently, by taking samples no greater that $1/(2f_m)$ seconds apart.

Let f_s represent the sampling rate measured in samples/second and let T represent the time duration between the beginnings of successive samples. Within the context of this development, samples/second and hertz will be considered as equivalent. The preceding theorem may be stated in either of the two forms that follow:

$$f_s \geq 2f_m \qquad (18\text{-}10)$$

or

$$T \leq \frac{1}{2f_m} \qquad (18\text{-}11)$$

The sampling frequency is related to the sampling time interval by

$$f_s = \frac{1}{T} \qquad (18\text{-}12)$$

In various theoretical developments, it is convenient to use the forms of Equations 18-10 and 18-11 with equal signs. However, in practice, the sampling rate is required to be somewhat greater than the theoretical minimum, to deal with the real-life limitations of the signal restoration process.

Aliasing

If a signal is sampled at a slower rate than required from the sampling theorem, a phenomenon known as **aliasing** occurs. This results in distortion of the signal, and the original signal cannot be converted back without some error. The reader might have observed this phenomenon while watching westerns in a theater or on television. Often wagon wheels will appear to be turning at a slow rate or even backward while the wagon is moving forward. Because movies are filmed using a standard sampling rate for the frames, the rotation of the spokes might assume a frequency that is too high for the given sampling rate, and the aliasing effect occurs.

▮▮ EXAMPLE 18-7   The highest frequency for voice-grade telephone signals is about 3.4 kHz. Determine the theoretical minimum sampling rate and the corresponding sampling time interval.

SOLUTION   The theoretical minimum sampling rate is

$$f_s = 2f_m = 2 \times 3400 = 6800 \text{ Hz or samples/s} \qquad (18\text{-}13)$$

The corresponding sampling interval would be

$$T = \frac{1}{f_m} = \frac{1}{6.8 \times 10^3} = 147 \times 10^{-6} \text{ s} \qquad (18\text{-}14)$$

The actual sampling rate used in many commercial voice-grade telephone systems is 8 kHz, which is somewhat higher than the theoretical minimum as required in practical systems.

III EXAMPLE 18-8

A signal having a baseband bandwidth from near dc to 16 kHz is to be sampled and converted to digital form. The sampling rate is chosen to be 25 percent greater than the theoretical minimum. Determine the (a) sampling rate and (b) sampling time interval.

SOLUTION

(a) The sampling rate is chosen to be 25 percent greater than the theoretical minimum so it can be determined as

$$f_s = 1.25 \times 2 \times f_m = 2.5 \times 16 \times 10^3 = 40 \times 10^3 = 40 \text{ kHz} \qquad (18\text{-}15)$$

Thus, it will be necessary to take 40,000 samples of the signal per second to meet the specifications.

(b) The sampling time interval is

$$T = \frac{1}{f_s} = \frac{1}{40 \times 10^3} = 0.025 \times 10^{-3} \text{ s} = 25 \text{ μs} \qquad (18\text{-}16)$$

18-4 Pulse Code Modulation

Once the sampling theorem is understood and accepted, we can continue the study of digital communication by assuming that an analog signal is to be approximated by a sequence of samples, each of which is converted to a binary word. The basic form of the resulting digital signal is referred to as **pulse code modulation (PCM)**.

The block diagram shown in Figure 18-3 illustrates the concept of PCM generation. The elements in the block are identified in a form to illustrate the processes involved rather than to show an actual implementation. In practice, several of these operations may occur in the same circuit.

Assume an analog signal $v(t)$, band limited from dc to f_m hertz, which is to be converted to PCM form. The signal is first filtered by an **anti-aliasing** analog filter whose function is to remove any superfluous frequency components above f_m that might appear at the input and be shifted into the data band by the aliasing effect. The signal is then sampled at a rate $f_s > 2f_m$. At this point, the signal can be considered as having the same form as the sampled signal in Figure 18-2.

Quantization

A process unique to all digital representation of analog signals is now performed. Each sampled-data pulse is converted into (or replaced by) one of a finite number of possible values. This process is called **quantization**, and the quantizer block in the diagram of

FIGURE 18–3
Block diagram illustrating the steps involved in generating a digital PCM signal. (In actual practice, several of the steps may be accomplished in the same circuit.)

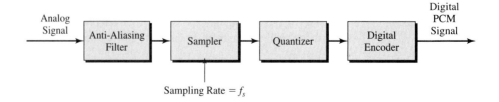

Figure 18-1 is assumed to accomplish this purpose. Following the quantization process, each of the standard values generated is encoded into the proper form for transmission or subsequent signal processing. Each of the composite encoded samples appearing at the output of the encoder is called a **PCM word**. Thus, the infinite number of possible amplitude levels of the sampled signal is converted to a finite number of possible PCM words.

The end result of the basic sampling, quantization, and encoding process in a binary PCM system then is a sequence of binary words, each having N bits. The value of each binary word represents in some predetermined way the amplitude of the analog signal at the point of sampling. In effect, the basic process is an **analog-to-digital (A/D)** conversion. Depending on the system requirements and other factors, some of the readily available A/D converters could be used for the PCM transmitter encoding process.

Decoding

Skipping for the moment all the detailed steps between the encoding process and the receiver, an inverse decoding process must be employed at the receiver to convert the received digital words into a usable analog form if the desired output is an analog signal. This process is a form of **digital-to-analog (D/A)** conversion, and some of the available D/A converters could be used for the receiver decoding purpose.

Signal Restoration

With analog communication processes, there are various ways to degrade a signal or change it in form in the transmission medium. With binary digital or PCM systems, *as long as the signal level is sufficiently large that it is possible to tell the difference between the two values transmitted, the signal theoretically can be reconstructed to the accuracy inherent in the sampling and quantization processes.* This is one of the outstanding advantages of digital systems.

Time-Division Multiplexing

One of the distinct advantages of sampling is that several signals from different information sources may be combined together and transmitted over a common channel by means of **time-division multiplexing (TDM)**. This process is illustrated for PCM in Figure 18-4. In this particular simplified system, each of the individual data channels is sampled at the same rate, and a given word representing the encoded value of the particular sample is transmitted between the point of sampling that channel and the next channel in the sequence.

To ensure that the electronic commutation at the receiver is exactly in step with that at the source, some means of synchronization must be employed. One way this can be

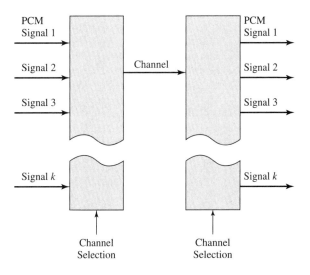

FIGURE 18–4
Multiplexing and demultiplexing of PCM signals.

achieved is by means of a special synchronization word or bit combination that is inserted at the beginning of a frame. The synchronized word is chosen to have a different pattern (or value) than other possible words that could be transmitted. At the receiver, all incoming words are sensed by digital circuitry, which establishes the beginning of a new frame when the synchronization word is received.

18-5 Analog-to-Digital and Digital-to-Analog Conversion

In this section, we will consider the manner in which a PCM signal is encoded in its basic form and the corresponding quantization characteristics. By "basic" encoding, we refer to the natural process of representing numbers in the conventional or *natural binary* number system form, and conversions or interpretations of those numbers in normalized or fractional forms. Some systems encode or convert the samples to special data forms.

To simplify the discussion in this section, $N = 4$ bits will be used for illustration. This choice results in $L = 2^4 = 16$ words, which is sufficiently large that the general trend can be deduced; and yet it is sufficiently small that the results can be readily shown in graphical and tabular form.

The possible 16 natural binary words attainable with 4 bits and their corresponding decimal values are shown in the two left-hand columns of Table 18-2. (The two right-hand columns will be explained later.) Note that the smallest value is at the bottom of the table, and the largest value is at the top. For each binary number, the bit farthest to the left is called the **most significant bit (MSB)**, and the bit farthest to the right is called the **least significant bit (LSB)**.

Depending on the exact circuitry involved, the dynamic range of the signal, and other factors, the 16 possible levels could be made to correspond to particular values of the analog signal as desired. However, some standardization has been achieved with many of the commercially available A/D and D/A converters, and the discussion here will concentrate on some of these standard forms. Typically, A/D and D/A converters are designed to operate with maximum analog voltage levels of 2.5, 5, 10, or 20 V, although means are often provided for changing to other values if desired.

Table 18–2 Natural binary numbers, decimal values, and unipolar and bipolar offset values for 4 bits.

Natural Binary Number	Decimal Value[1]	Unipolar Normalized Decimal Value[2]	Bipolar Offset Normalized Decimal Value[2]
1111	15	$15/16 = 0.9375$	$7/8 = 0.875$
1110	14	$14/16 = 0.875$	$6/8 = 0.75$
1101	13	$13/16 = 0.8125$	$5/8 = 0.625$
1100	12	$12/16 = 0.75$	$4/8 = 0.5$
1011	11	$11/16 = 0.6875$	$3/8 = 0.375$
1010	10	$10/16 = 0.625$	$2/8 = 0.25$
1001	9	$9/16 = 0.5625$	$1/8 = 0.125$
1000	8	$8/16 = 0.5$	0
0111	7	$7/16 = 0.4375$	$-1/8 = -0.125$
0110	6	$6/16 = 0.375$	$-2/8 = -0.25$
0101	5	$5/16 = 0.3125$	$-3/8 = -0.375$
0100	4	$4/16 = 0.25$	$-4/8 = -0.5$
0011	3	$3/16 = 0.1875$	$-5/8 = -0.625$
0010	2	$2/16 = 0.125$	$-6/8 = -0.75$
0001	1	$1/16 = 0.0625$	$-7/8 = -0.875$
0000	0	0	$-8/8 = -1$

[1] Decimal value for integer representation of binary number.
[2] Decimal value with binary point understood on left-hand side of binary number.

Normalization

Because of these different voltage levels and the widely different decimal values of the binary number system as the number of bits is changed, it is frequently desirable to *normalize* the levels of both the analog signal and digital words so that the maximum magnitudes of both forms are unity. The *normalized input analog voltage* is defined as

$$\text{normalized input analog voltage} = \frac{\text{actual input analog voltage}}{\text{full-scale voltage of A/D converter}} \qquad (18\text{-}17)$$

where the full-scale voltage of the A/D converter is typically 2.5, 5, 10, or 20 V. At the D/A converter in the receiver, the output voltage is

$$\begin{pmatrix}\text{actual output}\\ \text{analog voltage}\end{pmatrix} = \begin{pmatrix}\text{normalized value}\\ \text{of digital word}\end{pmatrix} \times \begin{pmatrix}\text{full-scale voltage}\\ \text{of D/A converter}\end{pmatrix} \qquad (18\text{-}18)$$

Most of the subsequent discussions in this section will be based on the normalized forms of both the analog and digital values. The normalized voltage will be denoted as $x(t)$ with subscripts added as appropriate. The full-scale voltage will be denoted as V_{fs}.

Normalization of the values of the binary words in Table 18-2 to a range less than 1 is achieved by adding a binary point to the left of the values given in the table. To simplify the notation, the binary point will usually be omitted, but it will be understood in all discussions in which the normalized form is assumed.

We will now consider the manner in which the normalized analog level can be related to the normalized digital level. The two most common forms employed in A/D conversion are (1) the **unipolar representation** and (2) the **bipolar offset representation**. The subscript u will be used for *unipolar* parameters, and the subscript b will be used for *bipolar* parameters. Don't confuse the use of the terms *unipolar* and *bipolar* as used here with the use of the same terms to describe a binary data train. Here the reference relates to the polarity of the analog signal.

Unipolar Encoding

The **unipolar** representation is most appropriate when the analog signal is always of one polarity (including zero). If the signal is negative, it can be inverted before sampling, so assume that the range of the normalized analog signal $x(t)$ is $0 \leq x < 1$. Let X_u represent the unipolar normalized quantized decimal representation of x following the A/D conversion. The 16 possible values of X_u for the case of 4 bits are shown in the third column of Table 18-2. Note that 0000 in binary corresponds to true decimal 0 and that the binary value 1000 corresponds to the exact decimal midrange value 0.5. However, the upper value of binary 1111 does not actually reach decimal 1 but instead has the decimal value $15/16 = 0.9375$.

Let ΔX_u represent the normalized step size, which represents on a decimal basis the difference between successive levels. This value is also the decimal value corresponding to 1 LSB and is in the general case

$$\Delta X_u = 2^{-N} \qquad (18\text{-}19)$$

The actual step size in volts Δv_u is obtained by multiplying the normalized step size by the full-scale voltage.

$$\Delta v_u = \Delta X_u V_{\text{fs}} = 2^{-N} V_{\text{fs}} \qquad (18\text{-}20)$$

for the **unipolar** system. The largest normalized quantized decimal value attainable $X_u(\max)$ differs from unity by the value of 1 LSB and is

$$X_u(\max) = 1 - \Delta X_u = 1 - 2^{-N} \qquad (18\text{-}21)$$

for the **unipolar** system. Thus, unity can be approached to an arbitrarily close value, but it can never be completely reached. The resulting distribution about the midpoint is not completely symmetrical, but it has the advantages that the absolute zero levels of both number systems are identical, and the exact midpoint of the analog voltage corresponds to the binary value whose MSB is 1 and all other bits are 0.

Bipolar Offset Encoding

The **bipolar offset representation** is most appropriate when the analog signal has both polarities. Specifically, it assumes that the normalized range of the analog signal is $-1 \le x < 1$. Let X_b represent the bipolar normalized quantized decimal representation of x following the A/D conversion. The 16 possible values of X_b for the case of 4 bits are shown in the fourth column of Table 18-2. Note that 0000 in binary corresponds exactly to the decimal value of -1, and the binary value 1000 corresponds to the exact decimal value of 0. However, the upper value of binary 1111 does not actually reach the decimal value 1 but instead has the decimal value $7/8 = 0.875$.

Let ΔX_b represent the normalized step size or value corresponding to 1 LSB for the bipolar case. This value is

$$\Delta X_b = 2^{-(N-1)} = 2^{-N+1} \qquad (18\text{-}22)$$

The actual step size Δv_b is

$$\Delta v_b = \Delta X_b V_{\text{fs}} = 2^{-N+1} V_{\text{fs}} \qquad (18\text{-}23)$$

for the **bipolar offset** case, which is twice as large as for the unipolar case. This value on a normalized basis appears to be twice as great as for the unipolar case, but that is because the normalized peak-to-peak value is twice as great as for the unipolar case. The largest normalized quantized decimal value attainable $X_b(\max)$ is

$$X_b(\max) = 1 - \Delta X_b = 1 - 2^{-N+1} \qquad (18\text{-}24)$$

for the bipolar offset case.

Observe from the fourth column of Table 18-2 that for 4 bits there are 8 binary words corresponding to negative decimal quantized levels, 7 binary words corresponding to positive decimal levels, and 1 binary word (1000) corresponding to a decimal level of 0. In general, there are $M/2 = 2^{N-1}$ binary words corresponding to negative quantized decimal signal levels, $M/2 - 1 = 2^{N-1} - 1$ binary words corresponding to positive decimal values, and one word corresponding to decimal 0. This last level is always represented by a binary word whose MSB is 1 and whose other bits are 0.

As an additional point of interest, the twos-complement representation of a given binary number can be obtained by replacing the MSB in the bipolar representation by its logical complement. In fact, the bipolar offset representation is sometimes referred to as a "modified twos-complement" representation.

Quantization Curves

The quantized decimal values and their binary representations have now been defined for two possible forms. The exact manner in which the quantized decimal values represent different ranges of the analog signal is defined by means of the **quantization characteristic curve**. Both **rounding** and **truncation** strategies may be employed. In *rounding*, the sampled value of the analog signal is assigned to the *nearest* quantized level. In *truncation*, the sampled value is rounded down to the next lowest quantized level. For example, assume that two successive quantization levels correspond to 6.2 and 6.4 V, respectively. With rounding, a sample of value 6.29 V would be set at 6.2 V, and a sample of 6.31 V would be set at 6.4 V. With truncation, both samples would be set to 6.2 V. As might be expected, the average error associated with truncation is greater than for rounding. Nevertheless, there are applications in which truncation is used.

From this point on, we will assume the more common process of *rounding*. The quantization characteristic curve for an ideal 4-bit A/D converter employing *rounding* and *unipolar* encoding is shown in Figure 18-5. The horizontal scale represents the decimal

FIGURE 18–5
Unipolar quantization characteristic for a 4-bit A/D converter.

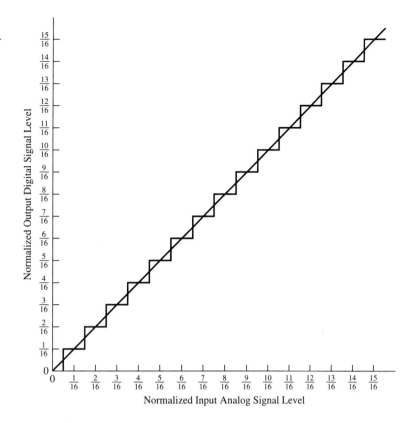

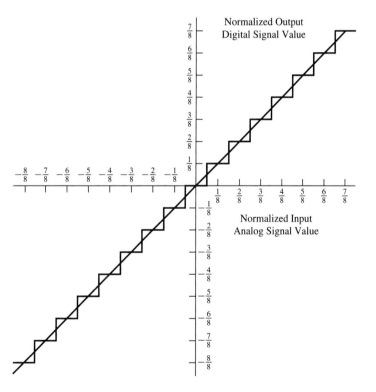

FIGURE 18–6
Bipolar offset quantization characteristic for a 4-bit A/D converter.

analog signal level on a *normalized* basis. The vertical scale represents the decimal representation on a *normalized unipolar* basis. The straight line represents the true analog ideal characteristic in which the output would equal the input.

The corresponding quantization curve for *bipolar offset* encoding is shown in Figure 18-6. As in the previous case, rounding is assumed.

Quantization Error

With the assumption of rounding, the theoretical peak quantization error is one-half of one LSB. Let E_u represent the peak unipolar normalized error, and let E_b represent the peak bipolar normalized error. These values are

$$E_u = \frac{\Delta X_u}{2} = 2^{-(N+1)} \qquad (18\text{-}25)$$

and

$$E_b = \frac{\Delta X_b}{2} = 2^{-N} \qquad (18\text{-}26)$$

Let e_u represent the actual peak unipolar error, and let e_b represent the actual peak bipolar offset error. These values are obtained by multiplying the preceding two values by the full-scale voltage. Hence,

$$e_u = E_u V_{\text{fs}} = 2^{-(N+1)} V_{\text{fs}} \qquad (18\text{-}27)$$

and

$$e_b = E_b V_{\text{fs}} = 2^{-N} V_{\text{fs}} \qquad (18\text{-}28)$$

EXAMPLE 18-9

The normalized analog waveform shown as the smooth curve in Figure 18-7 is to be converted to a PCM signal by a 4-bit A/D converter whose input-output normalized characteristic curve is given by Figure 18-6. Sampling will occur at $t = 0$ and at 1-ms intervals thereafter. Over the time interval shown, construct on the same scale the quantized form of the signal that is actually encoded and subsequently reconstructed at the receiver. In addition, list the corresponding binary words that are transmitted.

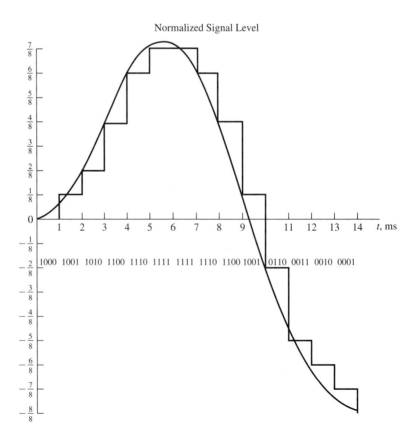

FIGURE 18–7
Analog and quantized signals used in Example 18-9.

Table 18–3 Data supporting Example 18-9.

Time, ms	Closest Quantization Value	Binary Representation
0	0	1000
1	1/8	1001
2	2/8	1010
3	4/8	1100
4	6/8	1110
5	7/8	1111
6	7/8	1111
7	6/8	1110
8	4/8	1100
9	1/8	1001
10	−2/8	0110
11	−5/8	0011
12	−6/8	0010
13	−7/8	0001
14	−8/8	0000

SOLUTION The analog signal is represented in normalized form, so the values may be used directly on the quantization characteristic scales of Figure 18-6. If the actual voltage level had been given instead, it would be convenient to divide the actual voltage level by the full-scale voltage so that the results could be used directly on the normalized scale. The quantized signal is constructed by reading the exact analog voltage at sampling points, noting which quantization level is nearest, and indicating the corresponding binary word using Table 18-2 if necessary.

At $t = 0$, the exact analog voltage is 0, so the quantized value is also 0. The corresponding binary bipolar offset value is 1000. At $t = 1$ ms, the analog voltage is closer to 1/8 than 0, so the quantized level is 1/8. The corresponding binary value is 1001. This process is continued for the entire duration of the signal, and the results are summarized in Table 18-3 and partially in Figure 18-7. The reader should find it instructive to verify the results given.

The "staircase" function shown represents the form of the converted signal following a digital-to-analog conversion provided that no further processing is used. Obviously, there are big gaps in the signal, but remember that we are using only 4 bits per word to illustrate the process. This resolution would have only limited application in practice.

■ **EXAMPLE 18-10** An 8-bit A/D converter with a full-scale voltage of 20 V is to be employed in a binary PCM system. The input analog signal is adjusted to cover the range from zero to slightly under 20 V, and the converter is connected for unipolar encoding. Rounding is employed in the quantization strategy. Determine the following quantities: (a) normalized step size, (b) actual step size in volts, (c) normalized maximum quantized analog level, (d) actual maximum quantized level in volts, (e) normalized peak error, and (f) actual peak error in volts.

SOLUTION The various quantities desired may be readily calculated from the relationships developed in this section. The results, with some practical rounding, are summarized in the following:

(a) The normalized unipolar step size is

$$\Delta X_u = 2^{-N} = 2^{-8} = 0.003906 \qquad (18\text{-}29)$$

(b) The actual step size is

$$\Delta v_u = \Delta X_u V_{\text{fs}} = 0.003906 \times 20 = 78.12 \text{ mV} \qquad (18\text{-}30)$$

(c) The normalized maximum quantized level is

$$X_u(\text{max}) = 1 - \Delta X_u = 1 - 0.003906 = 0.9961 \quad (18\text{-}31)$$

(d) The actual maximum quantized level is

$$v_u(\text{max}) = X_u(\text{max})V_{\text{fs}} = 0.9961 \times 20 = 19.92 \text{ V} \quad (18\text{-}32)$$

Alternately, the actual step size could have been subtracted from 20 V to yield the same result.

(e) The normalized peak error is

$$E_u = \frac{\Delta X_u}{2} = \frac{0.003906}{2} = 0.001953 \quad (18\text{-}33)$$

(f) The actual peak error is

$$e_u = E_u V_{\text{fs}} = 0.001953 \times 20 = 39.06 \text{ mV} \quad (18\text{-}34)$$

EXAMPLE 18-11 The 8-bit A/D converter of Example 18-10, while maintaining the same peak-to-peak range, is converted to bipolar offset form so that it can be used with an analog signal having a range from −10 V to just under 10 V. Repeat all the calculations of Example 18-10.

SOLUTION Notice that the peak-to-peak voltage is still 20 V as in Example 18-10. Many A/D converters use a unipolar internal encoding process but accomplish the bipolar operation by adding a dc bias equal to half the peak-to-peak voltage. In this case, the bias level would be 10 V. Some manufacturers would still specify the full-scale voltage as 20 V, but with the approach given here, the full-scale voltage will be assumed as the peak positive level, that is, $V_{\text{fs}} = 10$ V.

(a) The normalized bipolar step size is

$$\Delta X_b = 2^{-N+1} = 2^{-7} = 0.007812 \quad (18\text{-}35)$$

(b) The actual step size is

$$\Delta v_b = \Delta X_b V_{\text{fs}} = 0.007812 \times 10 = 78.12 \text{ mV} \quad (18\text{-}36)$$

Note that although the normalized step size appears to be twice as great as for unipolar, the actual step size in volts is the same because the peak-to-peak voltage is the same.

(c) The normalized maximum quantized level is

$$X_b(\text{max}) = 1 - \Delta X_b = 1 - 0.007812 = 0.9922 \quad (18\text{-}37)$$

(d) The actual maximum quantized level is

$$v_b(\text{max}) = X_b(\text{max})V_{\text{fs}} = 0.9922 \times 10 = 9.922 \text{ V} \quad (18\text{-}38)$$

Alternately, the step size could have been subtracted from 10 V to yield the same result.

(e) The normalized peak error is

$$E_b = \frac{\Delta X_b}{2} = \frac{0.007812}{2} = 0.003906 \quad (18\text{-}39)$$

(f) The actual peak error is

$$e_b = E_b V_{\text{fs}} = 0.003906 \times 10 = 39.06 \text{ mV} \quad (18\text{-}40)$$

which is the same as for unipolar encoding because the peak-to-peak range is the same.

18-6 Modulation Methods

Some forms of digital communication are performed directly with the baseband pulse forms consisting of the two binary levels representing ones and zeros. However, in the more general case, it might be necessary to provide **modulation** of the signal to transfer the information to a higher frequency to ensure more efficient transmission.

In this section, we will investigate the means by which digital signals can be shifted to a higher frequency range for transmission. This could be the radiofrequency range when electromagnetic wave propagation is desired, or it could be the audio frequency range for transmission over telephone lines and the Internet.

Amplitude-Shift Keying

The process of **amplitude-shift keying (ASK)** is probably the simplest and most intuitive form of higher-frequency encoding. In this process, turning the carrier on for the bit duration represents a 1, and leaving the carrier off for the bit duration represents a 0. Thus, a fixed-frequency RF carrier is simply gated off and on in accordance with the bit value.

Consider the arbitrary baseband bit stream shown in Figure 18-8(a). The form of an ASK signal is illustrated in Figure 18-8(b). Bear in mind that in this figure and others that will be shown later, it is not feasible to show more than a few cycles of the higher-frequency sinusoid. In audio frequency situations such as encountered on telephone lines, there might be only a few cycles within the width of a pulse, but in the RF range, there might be hundreds or thousands of cycles in a pulse width.

The ASK signal is sometimes referred to as **on-off keying (OOK)**. Classical telegraphy can be thought of as a form of ASK although that medium employs both short (**dot**) and longer (**dash**) intervals in which the carrier is turned on.

Frequency-Shift Keying

The process of binary **frequency-shift keying (FSK)** involves the transmission of one or the other of two distinct frequencies. A 0 is represented by the transmission of a frequency f_0 and a 1 is represented by the transmission of a frequency f_1. The transmitter oscillator is thus required to switch back and forth between two separate frequencies at the data rate.

For the same data pattern used in explaining ASK, the form of the corresponding FSK signal is shown in Figure 18-9.

Binary Phase-Shift Keying

In many references, **binary phase-shift keying (BPSK)** is referred to simply as **phase-shift keying (PSK)**. However, to distinguish it from forms using more than two levels, we will use the first reference in this text.

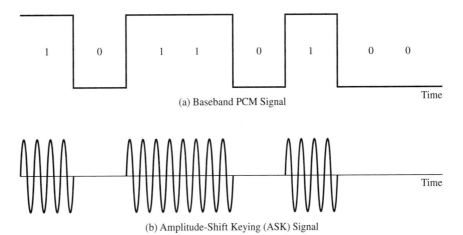

FIGURE 18–8
Baseband PCM signal and the amplitude-shift keying (ASK) modulated signal.

FIGURE 18–9
Baseband PCM signal and the frequency-shift keying (FSK) modulated signal.

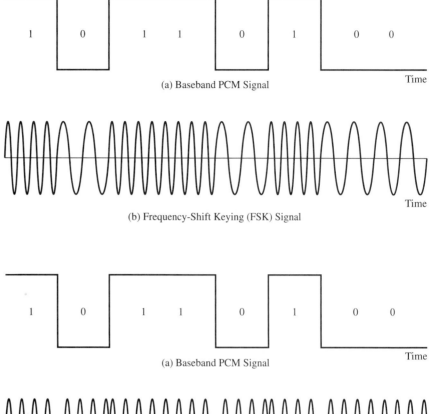

FIGURE 18–10
Baseband PCM signal and the binary phase-shift keying (BPSK) modulated signal.

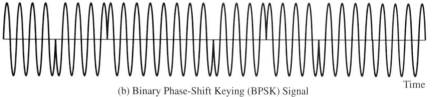

The process of BPSK uses a fixed-frequency sinusoid whose relative phase shift can assume one of two possible values. The values assumed will be 0° and 180°. For the same data stream used in explaining ASK and FSK, the form of a BPSK signal is shown in Figure 18-10. Note how a change in a bit value results in a sudden change in the phase of the sinusoid.

18-7 *M*-ary Encoding

Early in the chapter, the concept of ***M*-ary encoding** was mentioned briefly. It is based on the transmission of more than two levels. Consider, for example, the process of transmitting four possible levels. In this case, each unit received accomplishes the same process as if 2 bits were transmitted because 2 bits provide four possible outcomes. Likewise, if eight possible levels could be interpreted at the receiver, each unit would correspond to 3 bits.

M-ary Relationship

Let *M* represent the possible number of possible levels that can be used in a transmission system and let *k* represent the number of bits associated with the process. We have

$$k = \log_2 M \tag{18-41}$$

or

$$M = 2^k \tag{18-42}$$

These relationships have already been used with different notation and with a different purpose. Early in the chapter, the number of possible distinct levels or words that could be

FIGURE 18–11
Relative phase sequence of four states (dibits) used in QPSK.

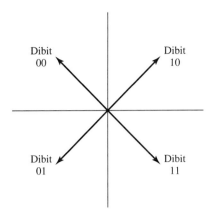

achieved with binary encoding involved these forms. Here we are determining how many bits are equivalent to a multilevel encoding scheme. In most practical systems, M is selected as an integer power of 2, so k will represent that integer value.

Symbol Rate and Data Rate

In M-ary systems, it is necessary to consider both the **symbol rate** s and the **data** or **information rate** R. The symbol rate is the number of signal elements transmitted per second. However, the data rate is the number of equivalent bits transmitted per second. The relationship is

$$R = s \log_2 M = sk \tag{18-43}$$

The symbol rate is also called the **baud rate**. The integer k is then the number of bits per symbol or bits/baud.

QPSK

One of the earliest forms of M-ary encoding and transmission was that of 4-ary, and the modulated format is known as **quadriphase shift keying (QPSK)**, which is widely used. It is achieved by using four different phase angles equally spaced in the phasor form as shown in Figure 18-11. This form provides two bits per symbol, so for a given bandwidth, the data rate is twice as great. Each one of the four possible values is called a **dibit**.

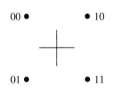

FIGURE 18–12
Constellation diagram for QPSK.

Constellation Diagrams

Figure 18-11 displays the four states of QPSK in the form of a phasor diagram. As systems become increasingly complex, it is convenient to change the phasor diagram into a simpler figure showing the points representing the tips of the phasors. In that manner, variable amplitudes may be shown more easily. This type of figure is called a **constellation diagram**. The constellation diagram for QPSK is shown in Figure 18-12.

QAM

Many developments over the past few years have resulted in very high data rates over limited bandwidth channels. This is possible only by special encoding schemes in which many levels can be perceived at the receiver, providing the equivalent of many bits per symbol.

A general approach for modulated waveforms having many different levels is that of **quadrature amplitude modulation (QAM)**. The result is a combination of amplitude levels and phase shifts that create the possible number of outcomes. A constellation diagram for a typical QAM system is shown in Figure 18-13. This system has 16 possible levels.

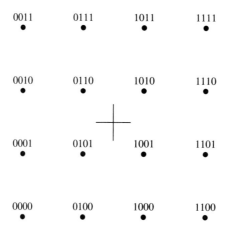

FIGURE 18–13 Constellation diagram for 16-level QAM.

Signal-to-Noise Ratio

When signal levels are relatively large, electrical noise can often be ignored. However, in communications systems, the signal levels at the input of receiving equipment are often on the order of microvolts and smaller. Electrical noise can then be a very disturbing factor and can completely obscure a signal. Everyone has likely encountered a problem of trying to monitor a signal when the noise effects are troublesome.

Communication processes in the presence of noise are usually described by a signal power to noise power ratio, usually denoted as (S/N). Let P_r represent the received signal power and let N_r represent the noise power. The signal-to-noise ratio is then expressed as

$$S/N = \frac{P_r}{N_r} \qquad (18\text{-}44)$$

The value is usually expressed in decibel form, which is

$$(S/N)_{dB} = 10 \log_{10}(S/N) = 10 \log_{10}\left(\frac{P_r}{N_r}\right) \qquad (18\text{-}45)$$

A metric for measuring the performance of digital communication systems is the **bit error rate (BER)**, that is, the number of bits read incorrectly per unit of time. These errors are caused by the presence of noise. For a given digital system, the BER increases as the signal-to-noise ratio decreases. Moreover, as the number of levels M is increased, the required signal-to-noise ratio generally increases. Many digital systems have a fallback position involving lower data rates when the signal-to-noise ratio decreases.

PROBLEMS

18-1 Determine the number of possible PCM values that can be encoded for (a) 6 bits and (b) 10 bits.

18-2 Determine the number of possible PCM values that can be encoded for (a) 12 bits and (b) 14 bits.

18-3 It is desired to represent an analog signal in no less than 500 values. Determine the minimum number of bits required for each word.

18-4 It is desired to represent an analog signal in no less than 5000 values. Determine the minimum number of bits required for each word.

18-5 Assume that a particular data file of size 4 Mbytes is to be downloaded at a speed through a dial-up connection at a rate of 50 kbits/s. How long will it take the file to download?

18-6 A file is uploaded from a computer at a rate of 500 kb/s in a time duration of 6 minutes. How large is the file in bytes?

18-7 A unipolar analog signal is sampled at intervals of 1 ms by a 4-bit A/D converter whose normalized input-output characteristic is given by Figure 18-5. The voltage values at the sampling points are provided in the table that follows and the full-scale voltage is 10 V. Determine the 4-bit digital word that would be generated at each point.

Time, ms	0	1	2	3	4	5	6
Value, V	0	1.5	2	3.5	5.5	7.7	9.4
Digital word							

18-8 A unipolar analog signal is sampled at intervals of 1 ms by a 4-bit A/D converter whose normalized input-output characteristic is given by Figure 18-5. The voltage values at the sampling points are provided in the table that follows and the full-scale voltage is 5 V. Determine the 4-bit digital word that would be generated at each point.

Time, ms	0	1	2	3	4	5	6
Value, V	0	1.5	2.4	3.5	4.5	4	0.5
Digital word							

18-9 A bipolar analog signal is sampled at intervals of 1 ms by a 4-bit A/D converter whose input-output characteristic is given by Figure 18-6. The voltage values at the sampling points are provided in the table that follows and the full-scale voltage is 5 V (based on a peak-to-peak range of 10 V). Determine the 4-bit digital word that would be generated at each point.

Time, ms	0	1	2	3	4	5	6
Value, V	3.8	2.7	2.3	0.25	−2	−3.5	−4.8
Digital word							

18-10 A bipolar analog signal is sampled at intervals of 1 ms by a 4-bit A/D converter whose input-output characteristic is given by Figure 18-6. The voltage values at the sampling points are provided in the table that follows and the full-scale voltage is 10 V (based on a peak-to-peak range of 20 V). Determine the 4-bit digital word that would be generated at each point.

Time, ms	0	1	2	3	4	5	6
Value, V	−9	−6	−1	0	4	6	8
Digital word							

18-11 A 16-bit A/D converter with a full-scale voltage of 20 V is to be employed in a binary PCM system. The input analog signal is adjusted to cover the range from zero to slightly under 20 V, and the converter is connected for unipolar encoding. Rounding is employed in the quantization strategy. Determine the following quantities: (a) normalized step size, (b) actual step size in volts, (c) normalized maximum quantized analog level, (d) actual maximum quantized level in volts, (e) normalized peak error, and (f) actual peak error in volts.

18-12 A 6-bit A/D converter with a full-scale voltage of 20 V is to be employed in a binary PCM system. The input analog signal is adjusted to cover the range from zero to slightly under 20 V, and the converter is connected for unipolar encoding. Rounding is employed in the quantization strategy. Determine the following quantities: (a) normalized step size, (b) actual step size in volts, (c) normalized maximum quantized analog level, (d) actual maximum quantized level in volts, (e) normalized peak error, and (f) actual peak error in volts.

18-13 The 16-bit A/D converter of Problem 18-11, while maintaining the same peak-to-peak range, is converted to bipolar offset form so that it can be used with an analog signal having a range from −10 V to just under 10 V. Repeat all the calculations of Example 18-11.

18-14 The 6-bit A/D converter of Problem 18-12, while maintaining the same peak-to-peak range, is converted to bipolar offset form so that it can be used with an analog signal having a range from −10 V to just under 10 V. Repeat all the calculations of Example 18-12.

18-15 The commercial Bell T1 PCM TDM system uses a sampling rate of 8 kHz and each sample is represented by 8 bits. A total of 24 signals are sampled within each sampling interval (frame) and one extra bit is added to each frame. Determine the total data rate.

18-16 A 12-bit PCM TDM system must be designed to process six data channels. Channels 1 to 4 each are to be sampled at a rate of 1 kHz whereas channels 5 and 6 are each to be sampled at a rate of 2 kHz by taking two samples of each per frame. To provide guard band, a sampling rate 25 percent above the theoretical minimum is to be employed. Determine the total data rate.

18-17 A baseband signal has frequency components from dc to 8 kHz. Determine the (a) theoretical minimum sampling rate and (b) maximum time interval between successive samples.

18-18 A baseband signal has frequency components from dc to 20 kHz. Determine the (a) theoretical minimum sampling rate and (b) maximum time interval between successive samples.

18-19 Assume that the signal of Example 18-17 is sampled at a rate 25 percent above the theoretical minimum. Determine the (a) sampling rate and (b) maximum time interval between successive samples.

18-20 Assume that the signal of Example 18-18 is sampled at a rate 40 percent above the theoretical minimum. Determine the (a) sampling rate and (b) maximum time interval between successive samples.

18-21 The constellation diagram for a particular QAM system has four different amplitudes and each amplitude has eight different possible phase shifts. Determine the number of bits per symbol.

18-22 The constellation diagram for a particular QAM system has 16 different amplitudes and each amplitude has 8 different possible phase shifts. Determine the number of bits per symbol.

MULTISIM LABORATORY

This section of the text is devoted to the use of the program **MultiSIM**® to provide simulation experiments that may be performed on a computer. These experiments represent the types of activities that might be performed in a real educational electrical/electronics laboratory. We don't claim that the experiences with a computer are exactly the same, but we do claim that similar educational objectives can be achieved.

Notational Differences

The notational forms appearing on a MultiSIM circuit schematic and the output data screens differ somewhat from the forms appearing in the text. For example, a variable such as R_1 in the text will appear as R1 in MultiSIM. Hopefully, this will not cause any serious problems in interpretation, and it is certainly not a result of the editing process.

The major dilemma occurs when an analysis is performed within a MultiSIM example. Should the MultiSIM forms for the variables be used or should the standard text forms be used? Frankly, in each case involved, an arbitrary decision was made as to the best way to show the results. In some cases the MultiSIM form is used while in other cases the equations revert back to the text style. In each case, the decision was based on the form that would likely be easier for the reader to interpret.

Finally, some of the MultiSIM examples stray somewhat from the text notation whenever the choice makes a result easier to interpret. For example, a symbol such as Vout can be readily interpreted as the output voltage, and this choice may convey more meaning to the reader than the original symbol in a text example. Overall, we don't feel that anyone will be misled by the differences in notation, but just be aware of the differences as they appear in various examples.

Introduction

This section of the text is devoted to the instructions and applications associated with the use of MultiSIM. It is not a prerequisite to anything else in the text, but it serves as a significant supplement to all of the theory and applications covered in other chapters.

MultiSIM is comprehensive circuit analysis software that permits the modeling and simulation of a wide variety of electrical and electronic circuits. It offers a very large component database, schematic entry, analog/digital circuit simulation, and many other features, including seamless transfer to printed circuit board (PCB) layout packages. The program is a product of **Interactive Image Technologies Ltd.**[*] It has evolved from the company's earlier **Electronics Workbench (EWB)** program. The circuit simulation portion of the program is based on the popular **SPICE** program (**S**imulation **P**rogram with **I**ntegrated **C**ircuit **E**mphasis). SPICE was developed at the University of California at Berkeley and was a batch-oriented program. However, MultiSIM is interactive and offers many user-friendly features. **National Instruments, Inc.** now owns the company.

The authors believe that, in the broadest sense of modern technology, MultiSIM is a type of laboratory that can be used to serve virtually all of the same educational functions as a traditional teaching laboratory. In fact, we will take that statement one step further and say that there are important operations that can be studied with MultiSIM that are very difficult to implement in a traditional laboratory. One example is the study of the effect of worst-case parameter variations in a circuit design, a task readily implemented with the program, but one that might be difficult in a traditional laboratory.

The circuit components parallel those of actual laboratory units and must be wired into the circuit in essentially the same fashion as in an actual laboratory. Thus, many laboratory skills can be taught with a computer and software without damaging any components or instruments.

Following this introduction is a section called **Primer**. It provides general information about MultiSIM, including instructions for selecting the parts, wiring, and so on. Numerous exercises then follow with a format that ties the exercises directly to the text chapters. For example, the section entitled **MultiSIM Examples for Chapter 1** has various examples that link to Chapter 1. The examples have titles such as **MultiSIM Example MSM-1-1, MultiSIM Example MSM-1-2**, and so on. The examples linking to Chapter 2 have titles such as **MultiSIM Example MSM-2-1, MultiSIM Example MSM-2-2**, and so on. Figure numbers will have a similar numbering format, and this pattern will become clearer as the figures are considered. The only exceptions to the format are the figures in the PRIMER, which are numbered as Figures **MSM-A, MSM-B**, and so on.

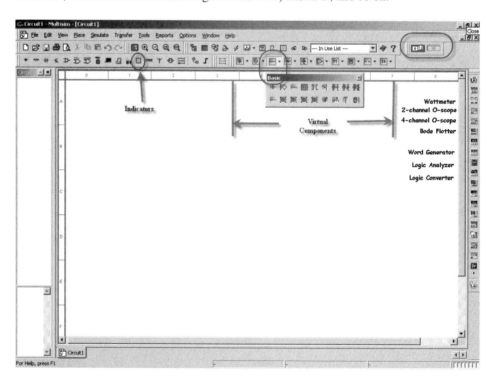

FIGURE MSM-A
Initial screen of MultiSIM version 9.

---

[*] Interactive Image Technologies, Ltd.
111 Peter Street, Suite 801
Toronto, Ontario, Canada
M5V 2H1

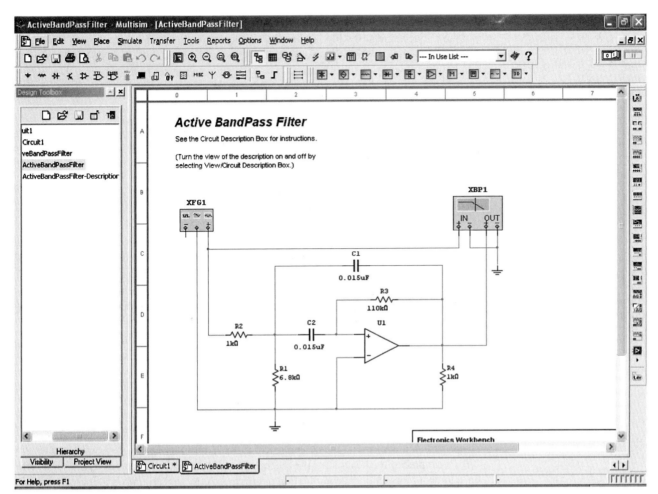

FIGURE MSM-B
Typical circuit schematic provided with MultiSIM.

The MultiSIM processes that will be considered here are (1) the construction of a circuit schematic on a computer screen that serves as a representation of an actual circuit and (2) the dynamic response of the circuit using computer **simulation**. The schematic diagram is created using a mouse and various window options. The type of analysis desired (dc, ac, transient, and so on) is then applied to the circuit to perform the simulation and the output data (voltages, currents, and so on) are obtained in various formats.

The major instructions in this primer are based on Version 9, which had just been released at the time the book was being prepared for production. However, in previous updates, utilities have been provided for converting circuits developed in earlier versions to the latest version. If the circuit files are used with a later version, there might be a need to perform this conversion. Like most software, there will likely be revisions and updates that appear in later versions. This might mean that there are some minor changes in some of the instructions or windows in future updates. When in doubt, use the **Help** file.

Primer

Upon activating MultiSIM, the initial screen of MultiSIM Version 9 will probably have a form similar to that shown in Figure MSM-A. However, there are numerous options for displaying and moving different utilities on the screen, and the previous settings might have been left in a different form than those in the figure. Therefore, don't panic if the screen is a little different, but as we discuss various options, you will see some of the possible ways that the display can be adjusted.

The initial blank space occupying most of the screen is called the **Circuit Window**, and the initial name at the top is **Circuit1**. If this space does not appear, left-click on **File** and then left-click on **New**. If this space has been minimized, left-click on the **Maximize** button in the upper right-hand area.

Toolbars

Various rows of options called **Toolbars** appear at the top and side of the screen. The various toolbars may be activated or deactivated on the screen. Left-click on **View** and then hold the cursor over **Toolbars**. A window will open containing the names of all the toolbars. A check mark is beside each one that is visible. Left-click on a given toolbar name to toggle the choice between a check mark and the absence of a check mark.

Some of the menu choices and the options within menus are those encountered in most Windows-based applications, and some are peculiar to MultiSIM. When the mouse arrow is moved over most of the buttons, their names appear on the screen. If in doubt about the function of a given operation, refer to the **Help** file. You do not need to understand all of the options available to get started with MultiSIM. Experience will lead to familiarity with those features most appropriate to your needs. We will concentrate here on the ones that are essential to get started, and as experience with the program is acquired, more options will become apparent.

After you eventually create a circuit, you can save it under a different name by left-clicking on **File** and then left-clicking on **Save As**. You can then name the file. You can also navigate through your file system according to your personal preference for file storage. All circuit files in Version 9 will automatically be assigned the extension **ms9** when you save them.

The various buttons in the row just above the Circuit Window represent the **Components Toolbars**. When the arrow is placed above a given button or **bin**, the name assigned to the particular parts bin appears on the screen. Most of these bins have either the verb **Place** or the verb **Show** preceding the name of the bin, but we will refer to the bin names without the verb. Each toolbar representing a combination of bins may be moved around on the screen for convenience if desired. To move a bar, left-click on the area at the top, hold down the left mouse button, and drag it to any desired location. Release the mouse button when you reach the desired location.

The various buttons arranged in a column on the right-hand side of the screen form the **Instruments Toolbar**. When the arrow is placed above a given button, the name of the instrument appears on the screen. The instruments in this toolbar tend to be the more sophisticated and multifunctional in form. There is also a component bin with the name **Indicator**. This bin contains some of the simpler instruments such as a dc voltmeter, dc ammeter, and so on, and we will employ these instruments extensively.

In the large blank area on the screen, the **Circuit Window**, you will create the circuit schematic. A schematic of a typical circuit containing instrumentation is shown in Figure MSM-B. This particular circuit was used as an example on the MultiSIM software disk and has the title **ActiveBandPassFilter.ms9**.

Creating a Circuit Schematic

The process of creating a circuit diagram or schematic consists of selecting and dragging the components from a parts bin and connecting the components using wire. In some cases, it might be desirable to drag all the components out to the circuit window first and then wire them together. However, the procedures may be interchanged at any time, and parts may be moved around to make room for other parts as necessary.

Using a Grid for Layout

A grid is very helpful in laying out the components. If it does not initially appear on the screen, the grid can be toggled on and off if desired from an option reached with the **View** button.

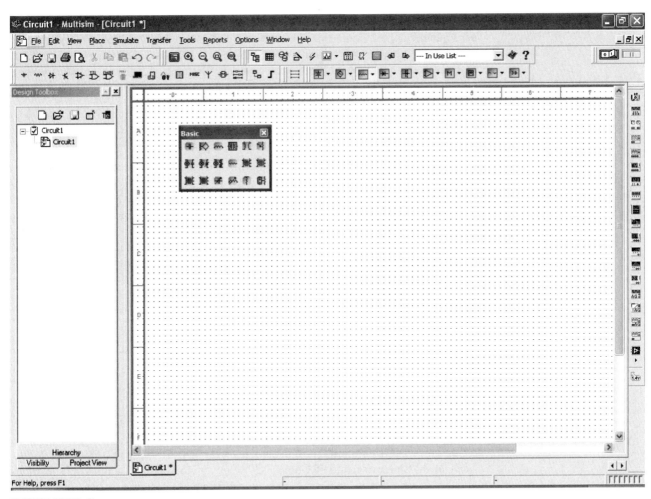

FIGURE MSM-C
Screen with basic components toolbar open.

The option is called **Show Grid**. It can be toggled on and off by a check mark or the absence of a check mark.

Parts Bin Toolbars

As previously noted, the **Components Parts Toolbars** are represented initially by the rows of buttons above the Circuit Window. Each button identifies a particular parts bin based on a common grouping. As you move the arrow across each button, the name is identified.

To open a given parts bin, left-click on its button. In Figure MSM-C, the **Basic Family** parts bin containing passive components (R, L, and C) and a number of other parts appear. A parts bin may be closed by left-clicking on the **X** in the upper right-hand corner.

As you look through various bins, don't worry if there seems to be an overwhelming number of parts at this point. You will quickly learn to find those that you use often, but it is unlikely that you will ever need more than a small percentage of the available parts.

Selecting and Dragging a Part from a Bin

To obtain a part for the circuit being created, first left-click on the bin button containing the part. For certain parts, you might need to use a little trial and error before you find the right bin. After the part is identified in the appropriate bin window, move the arrow to the correct button and left-click on it. You can then release the mouse button because that component has now been locked to the arrow. With many parts, an additional **Components Browser** window will open with a title such as **Select a Component**. You can then select the value or parts number of the component. When the choice is made, click **OK**. The arrow will appear with a marker attached to it, and it may be moved to the position at which it will be placed. Don't be overly concerned about its exact location at this time. It can be changed later. When you reach the approximate location at which the component is to be placed, left-click the mouse and the component will be released at that point on the screen.

Selecting and Moving Components on the Screen

It is frequently desirable to select and move a component that is already in the schematic workplace. Move the arrow to the component position and place it in contact with the component. At the point where contact is established, left-click once and hold the mouse button down. A rectangle now appears as sort of a box around the component. While holding the left mouse button down, you can then drag the component to a new location on the screen. The labels will move with the part. When the proper location is reached, release the left button and the component will be detached. The rectangles may be eliminated by left-clicking outside of the area of the component.

The selection process just described may also be used to move the label or value of the component or the node number or name. Bring the arrow in contact with the label or value instead of the component itself, and the remainder of the procedure is the same.

Rotating a Component

Components always have a fixed orientation when they come out of the parts bin, and this might or might not be the desired final orientation. All components may be rotated by integer multiples of $90°$. First, select the component by left-clicking on it.

Rotation can be achieved by means of a left-click on the **Edit** menu, followed by moving the arrow to **Orientation**. The options that appear to the right are **Flip Horizontal, Flip Vertical, 90 Clockwise, and 90 CounterCW** (counterclockwise). The titles are somewhat self-explanatory, but the reader may experiment with these operations when an actual circuit is constructed. Two successive steps of either one of the flip operations or four successive steps of either one of the rotate operations will restore the original orientation. Another way to rotate by $+90°$ is by depressing **Ctrl** followed by **R** (for rotation by $90°$).

Deleting a Component

To delete a component, first select it. The simplest way to delete it is to press the **Delete** key on the keyboard, although there is a **Delete** option under the **Edit** menu items. If the component is a two-terminal device in which external wires have been connected, the component will be replaced by a short between the terminals.

Ground

All circuits must have at least one ground point. The quickest way to obtain a ground symbol is to use the **Power Source Family** button, which is identified by the battery symbol. You can also obtain a ground from the **Source** bin. The ground establishes the point from which all voltages are measured. To eliminate extra wiring, ground points may be used at various points in the circuit. Again, however, there must be at least one ground symbol, and it is automatically established within the program as node **0**.

Wiring the Components Together

At some point, you will be ready to start wiring the components together. You may have all the parts that you need in the circuit window or you may prefer to wire some portion as you go along. For this purpose, you will need wire.

A basic rule to follow is that *all terminals that are to be connected must have a section of wire between them*. There is a temptation in some cases to simply bring component terminals together, in which case they may appear to be connected. Unless there is a section of wire between them, however, they will not connect. Think of the wire as having solder at each end; without it, there will not be a connection. This rule applies to the ground symbol and to all components.

The program automatically establishes node numbers from **1** upward in the order in which you wire them together, although all ground symbols assume node **0**. Therefore, to establish any kind of numbering system, connect the nodes in the desired numerical order. However, this is not absolutely necessary, because you can perform some renumbering later.

To start a wire from a given terminal, move the arrow as closely as possible to the tip of the terminal. At the point where a connection is possible, the arrow will then change to a set of crosshairs. Left-click at this point, and the process creates a piece of wire that can be moved to some other terminal. The wire will initially appear as a dashed line until the final connection is made. It is recommended that you move in straight lines parallel to either the horizontal or vertical sides of the screen. You can make one $90°$ turn without any further clicking, but if you make more than one turn, you will need to left-click at the point of the second and subsequent turns. When you reach the desired connection point, left-click at the terminal. The dashed line will now appear as a solid line, meaning that you have wired the two terminals together. The program creates a node number at this point, but at first it will not be visible. It will take a little practice to become fully proficient in the wiring process.

To delete a section of wire, first select it by left-clicking on it. Then press the **Delete** key.

Opening a Component Properties Window

In addition to providing initial part numbers and values in the selection process as described earlier, many components will require further parameter values and properties to be established before running a circuit simulation. Moreover, you may change some of the parameters without having to go back to the parts bin. This process can be performed at any time after the component is obtained from the parts bin. One of the great virtues of circuit simulation is the ease with which parameter values may be changed throughout a simulation study.

Parameter values and properties are specified in the **Component Properties** window. To open such a window, proceed as if you were planning to select the given component and place the arrow on the component. Double left-click on the component, and the window should appear. The name of the window depends on the particular type of component. A wide variety of properties may be specified. For the simplest types of component, only one or two values need to be specified. Others require a multitude of entries. In the examples that follow, many of these windows are discussed in detail.

After you have entered all the data required to specify the component, left-click **OK**. If for any reason, you do not change any values or if you wish to return to the previous state, left-click **Cancel**. The component may now be deselected by moving the arrow away from the component and left-clicking.

Because a section of wire is treated essentially the same as a component, the preceding process may be used to open a properties window for changing its characteristics. This concept will be discussed in the next paragraph as it relates to node numbering.

Node Numbering

Each section of wire connecting two terminals constitutes a node. Starting with the number 1, each node is numbered by the program in the order in which the connections are made except, of course, any connection to a ground, which is numbered as 0. The default condition for MultiSIM 9 initially does not display this information on the screen.

To change a node number or to view the existing number, first double left-click on the wire section as discussed in the last section. In this case, the window has the title **Net**. To change the number or use a name, type the desired node number or name in the **Net name** slot. To have the number or name show on the schematic, click on the slot next to **Show**. After either or both changes, left-click on **OK**.

One caution: If you are renumbering, the program will not allow you to use a number that is already active. To use a fixed node numbering order, try to connect the nodes in the order planned as previously suggested. However, you can always label nodes temporarily with numbers outside of the range, to get access to a desired number. Remember that you can also label nodes with names such as **IN** and **OUT**.

Analysis and Simulation

There is a button resembling a switch with the numbers 0 and 1. Use the mouse to move the arrow over this switch and left-click to initiate a simulation run. When all the results are obtained, it can be switched again to deactivate the simulation. The same procedure can be used with a button that looks like a lightning flash. It has the name **Run/stop simulation (F5)**. From that information, it can be deduced that pressing **F5** will also activate a simulation.

There are some other simulation options, but they are not heavily emphasized in the work here. Search for a button that looks like a triangular waveform. When the cursor is brought over it, the name **Grapher/Analyses List** will appear. The square button is used to activate the grapher, which is the screen for displaying a curve of some circuit variable. The small rectangular button beside it can be used to set the various forms of analysis used to simulate circuits. The latter operation can also be performed by a left-click on the **Simulate** button followed by a left-click on **Analyses**.

Some readers may explore this alternate procedure, but it is only used in a few places in the text. This is mostly because the procedure becomes a little more mathematical in form and less related to the manner in which a real laboratory procedure is conducted. We want the procedures developed here to parallel very closely the real-world procedure of a teaching laboratory environment, so the earlier approach will be primarily emphasized.

Notational Conventions

Please note again the discussion of the first page of Part VIII explaining the need to have different notational forms on the schematic diagrams as compared with text equations.

Standard Components versus Virtual Components

Figure MSM-A earlier provided a typical view of the MultiSIM screen. As discussed then, the Component Toolbars appear above the Circuit Window. The one farthest to the left represents *standard components* and the one to its right represents *virtual components*. For detailed design work where standard values are required, there is an obvious need in many cases to use standard values. However, for much of the work in this book, the virtual components are somewhat easier to use because they can be set to any arbitrary values and they will be employed as appropriate.

Reference Designation

As components are placed on a schematic, they will each be named. For example, the first resistor pulled from a resistor bin will be denoted as **R1**, and the second one will be denoted by **R2**, and so on. However, the names can be changed by the procedure that follows.

Double left-click on the device, and the properties window for the device will open. Then left-click on the **Label** tab. Type the desired name in the **RefDes** slot and left-click **OK**.

One caution: You cannot use a name that is currently assigned to another component. Suppose you wish to swap the names of two resistors **R1** and **R2**. If you tried to change either one as they are, the program would generate an error message because the other one is already assigned the desired name. Therefore, you would need to assign one temporarily to a name not currently on the schematic. Then you can perform the desired steps. For that reason, it is convenient when possible to extract components to be partially identified by numbers in the order for desired labeling.

The Virtual Oscilloscope

The oscilloscope (o-scope) is an extremely versatile and useful laboratory measurement device. An actual o-scope displays a plot of voltage amplitude versus time on a cathode-ray tube (CRT) screen with the vertical axis representing amplitude and the horizontal axis representing time. MultiSIM offers a choice between a two-channel and a four-channel o-scope.

The controls for the o-scope are as follows:

- The **Time base** scale control adjusts the time scale on the horizontal axis in time per division. In this book the Y/T selection must remain selected. In Figure MSM-D, the time base scale is 200 μs/div. Note that the displayed sine wave requires 5 divisions to complete one complete cycle, therefore the period (T) of the waveform is:

$$\mathbf{T} = \left(200\ \frac{\mu s}{\mathbf{div}}\right) \times 5\ \mathbf{div} = 1000\ \mu s = 1\ \mathbf{ms}$$

- The Channel A (Channel B) scale control adjusts the volts per division on the vertical axis for the selected channel. The waveform in Figure MSM-D requires 4 divisions to

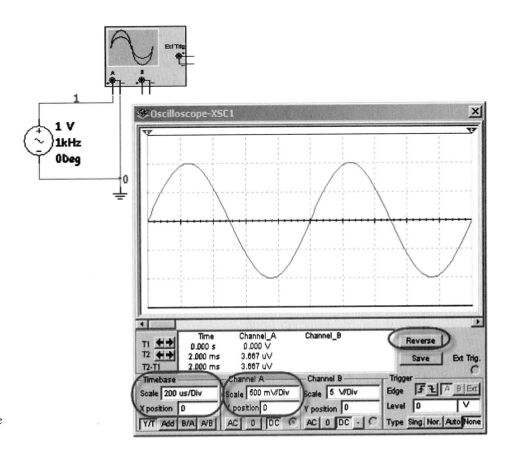

FIGURE MSM-D
MultiSIM dual-channel o-scope screen.

go from the negative peak to the positive peak. The vertical scale is 500 mV/div, therefore the peak-to-peak and peak values of the waveform are determined as follows:

$$V_{\text{peak-peak}} = \left(500 \, \frac{\text{mV}}{\text{div}}\right) \times 4 \, \textbf{div} = 2000 \, \textbf{mV} = 2 \, \textbf{V}$$

$$V_{\text{peak}} = \frac{V_{\text{peak-peak}}}{2} = \frac{2 \, \text{V}}{2} = 1 \, \text{V}$$

The **y-pos** control adjusts the DC level (vertical position) of the waveform. This is used primarily to adjust waveforms so that they can be observed without interference from other displayed waveforms. The default background color of the display is black, but it can be switched to white with the **Reverse** button.

The **AC/0/DC** control deserves special mention.

When **0** is selected, the input is grounded out and the trace can be adjusted to a useful zero reference point.

When **AC** is selected, only the time-varying or ac portion of the waveform is displayed. The dc portion of the waveform is blocked out from the display. This means that a symmetrical waveform will be centered around zero volts.

When **DC** is selected, all portions of the waveform are displayed.

MultiSIM Examples for Chapter 1

EXAMPLE MSM-1-1 Construct a MultiSIM model for the circuit of Example 1-11 (Figure 1-24). Use an ammeter to measure the current, and several voltmeters to measure resistive branch voltages. Compare the simulation results with the calculated results.

SOLUTION The three resistors in the circuit of Figure 1-24 were denoted as R_1, R_2, and R_3. The Reference ID (**RefDes**) names on the schematic will be labeled as **R1, R2**, and **R3** by default provided that they are taken from the parts bin in that order. However, the names can be changed as discussed in the Primer. The initial reference ID names of the two voltage sources are **V1** and **V2**, but they are changed to the names **Vs1** and **Vs2**, respectively, to be compatible with the names in the circuit example.

Using the procedure in the Primer, the circuit schematic is created as shown in Figure MSM-1-1(a). The dc sources are obtained by left-clicking on the ground symbol on the farthest left parts bin as shown in Figure MSM-1-1(b). Once the window opens, select **POWER_SOURCES/DC_POWER** and then **OK**. A battery will attach to the arrow. Drag it to the approximate location where it is to be placed and left-click again. It will then be released and be available for modification. If it were necessary to change the orientation (which is not the case for either source in this example), you would select it by left-clicking on it, then left-clicking on **Edit** on the **Menu Row**. You should then follow the procedure discussed in the Primer. A circuit ground is also available in the same parts bin as the dc source. Drag it to the desired location and left-click to detach it.

An alternate method for locating each of these devices is to right-click inside the circuit area and then click on select **Place Component**. The same window as shown in Figure MSM-1-1(b) will open. Select **Sources/POWER_SOURCES/DC_POWER (Group/FAMILY/COMPONENT)**, then select the components as discussed above.

The default value of a dc voltage source is 12 V. To set any other value, double left-click on the source and a window with the title **POWER_SOURCES** will open. The tab will likely be in the default **Value** position, but if not, left-click on that tab. The value is then set in the slot provided. Note that there is a second slot that may be set for **V, mV**, and so on. In this example, only one of the two sources will need to be reset because the value of the other is the default value. Leave the other slots at the default

FIGURE MSM-1–1(a)
MultiSIM circuit used for Example MSM-1-1.

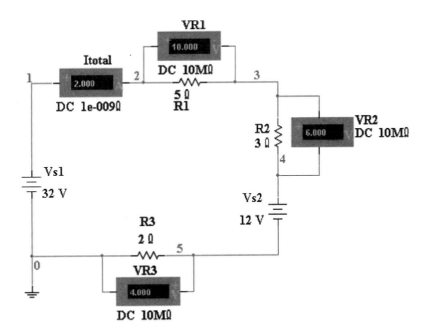

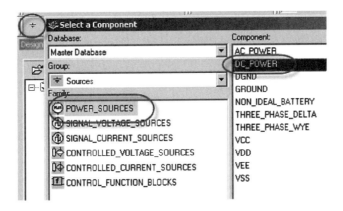

FIGURE MSM-1–1(b)
MultiSIM power sources component window.

values of **0** because they are not used in this example. Left-click on **OK** and the new value will be retained. You can then change the Reference ID names to **Vs1** and **Vs2**.

We won't discuss all of the options underneath the **Label** and **Display** tabs at this point because they are more cosmetic than operational, but the reader may experiment with them as more experience with the program is acquired.

The resistors are obtained by left-clicking on the parts bin with the resistor and the blue background. The **Basic Virtual Family** menu will appear as shown in Figure MSM-1-1(c). Select **Place Virtual Resistor**, at which point a resistor will attach to the arrow. Drag it to the approximate location where it is to be placed and left-click again. It will then be released and is available for further modification. If it is necessary to change the orientation, select the resistor to be rotated by right-clicking on it and then selecting 90 Clockwise or 90 CounterCW.

The default value of a virtual resistance comes out as **1 kΩ**. To set any other value, double left-click on the resistor and a window with the title **BASIC_VIRTUAL** will open. The tab will likely be in the **Value** position, but if not, left-click on that tab. The value is then set in the slot provided. Note that there is a second slot that may be set for **Ohm, kOhm, Mohm,** and so on. Leave all the blocks from **Tolerance** on down at their default values. Left-click on **OK** and the value will be set. Again, an alternate method of selection would be to right-click inside the circuit area, select **Place component**, and then when the window in

FIGURE MSM-1–1(c)
MultiSIM basic virtual family window.

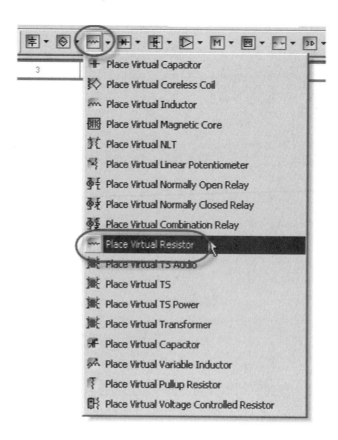

Figure MSM-1-1(b) appears, select **Basic/BASIC_VIRTUAL/RESISTOR_VIRTUAL** followed by **OK**.

As in the case of the dc source window, some other tabs on the resistor window are the **Label** and **Display** tabs. Many other components will have similar options on the properties windows, and once you begin developing experience using the program, you may alter the labels and displays as desired.

The ammeter and voltmeters are all obtained from the **Indicator** bin. This bin's icon has a small figure 8 on it (see Figure MSM-A). Figure MSM-1-1(d) represents a portion of the window which will open when this icon is selected. Select either **VOLTMETER** or **AMMETER** and you will note four variations of each. They are denoted as

AMMETER_H	VOLTMETER_H
AMMETER_HR	VOLTMETER_HR
AMMETER_V	VOLTMETER_V
AMMETER_VR	VOLTMETER_VR

This is a little confusing, but the variations save you the trouble of changing polarities and terminals in the circuit window, although it can be done within the circuit if you later

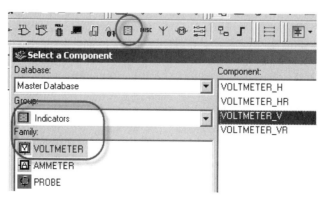

FIGURE MSM-1–1(d)
MultiSIM indicators window.

change the orientation. The differences are reflected in the extensions in the end and the codes for both the ammeter and voltmeter follow.

- **H**   The meter has **horizontal** connections and the positive terminal is on the **left**.
- **HR**   The meter has **horizontal** connections and the positive terminal is **reversed** (meaning on the **right**).
- **V**   The meter has **vertical** connections and the positive terminal is on the **top**.
- **VR**   The meter has **vertical** connections and the positive terminal is **reversed** (meaning on the **bottom**).

For this particular circuit, the one ammeter used has the form **AMMETER_H**. The voltmeter on the top row has the form **VOLTMETER_H** and the two voltmeters on the right have the form **VOLTMETER_V**. Finally, the voltmeter on the bottom row has the form **VOLTMETER_VR**. Don't spend too much time trying to get all of these designations straight, because you can always perform some simple checks when you need to determine the forms, and you can change them in the circuit window if necessary by performing a rotation or a flip.

The circuit is energized by either left-clicking on the switch or one of the other procedures discussed in the Primer. After a short settling time, the instruments will display the various values of the circuit variables. When the circuit is deenergized, the instruments maintain the measured values. However, upon reactivating the circuit, the instruments will initially assume zero values, but will then attain the final values again. If any of the parameter values were changed during the period when the circuit was deactivated, the new values could differ from those obtained earlier.

In this example, the MultiSIM values are identical to those obtained in Example 1-11 as is shown in the table that follows.

	Calculated	Simulated
Itotal	2 A	2 A
VR1	10 V	10 V
VR2	6 V	6 V
VR3	4 V	4 V

■ EXAMPLE MSM-1-2

Construct a MultiSIM model for the circuit of Example 1-13 (Figure 1-28) and use a voltmeter and several ammeters to measure the branch current and the common branch voltage.

SOLUTION The construction of the circuit schematic, shown in Figure MSM-1-2, follows the same procedure as in the previous example. The discussion here is necessarily

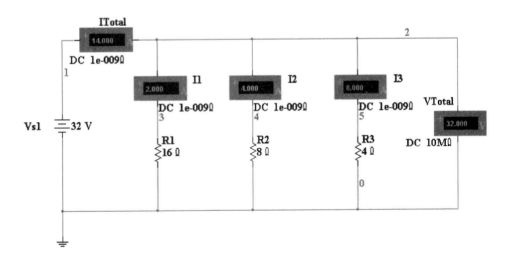

FIGURE MSM-1–2
MultiSIM circuit for Example MSM-1-2.

brief because no new types of components or analysis are involved. Subsequent examples will provide detailed discussion when new concepts are introduced.

Each of the ammeters and the single voltmeter had their **reference IDs** modified to indicate the name of the specific current or voltage. To save space, the meter internal resistance values were moved by selecting the text and moving it. The user also might delete the display of the internal resistance value of each meter by double left-clicking on each meter, selecting the **Display** tab, deselecting **Use Schematic Global Setting**, and then deselecting **Show values**.

Activation of the circuit produces results that are identical to those of Example 1-13 as is shown in the table that follows.

	Calculated	Simulated
Itotal	14 A	14 A
I1	2 A	2 A
I2	4 A	4 A
I3	8 A	8 A
Vtotal	32 V	32 V

■ EXAMPLE MSM-1-3

Modify the circuit constructed in Example MSM-1-2, to measure power. Use several wattmeters to measure the total circuit power and the power dissipated by each resistor.

SOLUTION The construction of the circuit schematic follows the same procedure as in the previous example and is shown in Figure MSM-1-3. The wattmeter may be found with the **Virtual Instruments** on the right side of the MultiSIM workspace (see Figure MSM-A). When you place the mouse over the third virtual instrument from the top, the word Wattmeter will appear.

A wattmeter is a four-terminal device which measures both the voltage across a device (the first two terminals marked with a V) and the current through a device (the second two terminals marked with an I) and then displays the power measurement in watts.

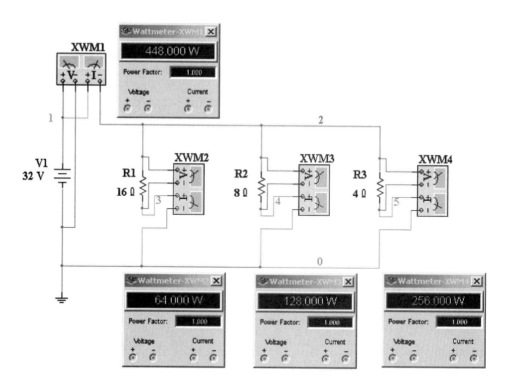

FIGURE MSM-1–3
MultiSIM circuit for Example MSM-1-3.

Note how the wattmeters in Figure MSM-1-3 are connected. Note that a branch had to be broken and connected so that the current would enter the positive current terminal and flow out of the negative current terminal. Three of the wattmeters have been rotated to fit in the circuit better.

Activation of the circuit produces power results which are identical to those of Example 1-13 as is shown in the table that follows. To read a wattmeter, double left-click on it and a display providing a reading opens as shown in Figure MSM-1-3.

	Calculated	Simulated
PTotal	448 W	448 W
PR1	64 W	64 W
PR2	128 W	128 W
PR3	256 W	256 W

MultiSIM Examples for Chapter 2

EXAMPLE MSM-2-1

Construct the MultiSIM circuit shown in Figure MSM-2-1. Use the branch current measurements to calculate the mesh currents, and then compare the results with the predicted values.

SOLUTION Build the circuit shown in Figure MSM-2-1. Note that the indicated branch current directions are based on the polarities of the installed ammeters, not on any understanding of the expected direction of the branch currents.

Simulate the circuit and note the measured branch current values:

	Simulated
Ia	−1.000 mA
Ib	−1.002 mA
Ic	+2.000 mA

Note that the simulated results indicated negative current values for branch currents **Ia** and **Ib**. This is an indication that the actual currents for these two branches are actually flowing in the opposite direction. In other words, branch currents **Ia** and **Ib** are flowing upward into node 5, while branch current **Ic** is flowing as assumed, out of node 5.

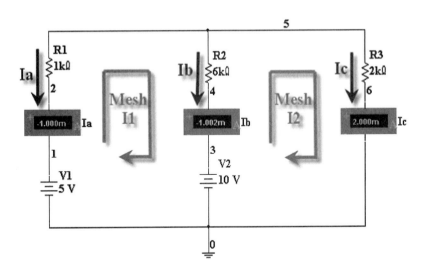

FIGURE MSM-2-1
MultiSIM circuit for Example MSM-2-1.

If a mesh current analysis is performed on the circuit of Figure MSM-2-1, the resulting mesh currents are:

	Calculated (clockwise)
I1	1 mA
I2	2 mA

Observations of the branch and mesh currents (as drawn) result in mesh currents:

$$\text{Mesh } I1 = -I_a = 1 \text{ mA}$$
$$\text{Mesh } I2 = I_c = 2 \text{ mA}$$

The direction in which a current is flowing may be just as important as the magnitude of the current. As was pointed out earlier, each ammeter in Figure MSM-2-1 is connected with the positive terminal on top and the negative terminal on the bottom. If the resulting current measurement were positive, the current would be known to be flowing downward (from + to −) as can be seen in the I3 measurement. In the case of the **Ia** and **Ib** measurements, however, the measured current value was negative; therefore, the measured currents were flowing upward.

The ammeters in Figure MSM-2-1 have been set to hide the internal resistance values and the **Reference ID** names. Each ammeter has had a label attached indicating the branch current name. The label name is applied in the same window as the **Reference ID**.

Note that the ammeter result for **Ib** is slightly different from the calculated value. This sometimes happens in simulations (and in an actual laboratory setting). The difference is minimal and is a result of the input resistance of the meters. In the case of ammeters, this resistance is ideally 0 ohms, whereas in reality there is some resistance, which will cause a small inaccuracy in some measurements. The default input resistance of a MultiSIM ammeter is 1×10^{-9} ohms. Although it is possible to double-click on the meter and modify the resistance value, this is normally not necessary. Just keep the issue in mind when you analyze your simulation results.

While the subject of meter inaccuracy is being discussed, it should be noted that voltmeter input resistance is ideally infinite; in reality the meter will have something less than ideal (the default internal resistance in MultiSIM is 10×10^6 ohms). Again, this causes a small amount of inaccuracy, but it is not necessary to modify this setting as long as the cause is kept in mind while analyzing simulation results. Don't forget that meters in actual laboratories will have these same issues, and the option of changing the input resistances will not be available.

■ EXAMPLE MSM-2-2

Construct the MultiSIM circuit shown in Figure MSM-2-2. Calculate the two node voltages, **V1** and **V2**, and compare the calculated values with the simulation results.

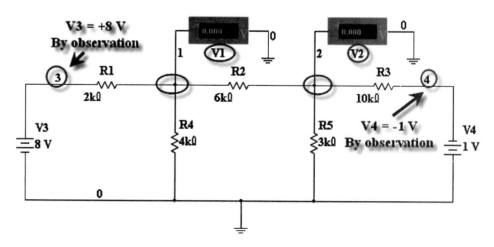

FIGURE MSM-2-2
MultiSIM circuit for Example MSM-2-2 showing nodal voltage values.

SOLUTION Node voltage analysis performed on the circuit of Figure MSM-2-2 yields the following results:

	Calculated	Simulated
V_1	4.564 V	4.563 V
V_2	1.101 V	1.101 V

Remember from the discussion in the last example that the meters might result in slightly different values than the calculated values, as seen in the simulated values above.

There are three other node voltages in the circuit of Figure MSM-2-2, but their values may be determined by inspection. As seen in the figure,

	Observed
V_3	+8 volts
V_4	−1 volts
V_0	0 volts

As seen in the placement of the voltmeters, node voltages are always referenced to some circuit common point, quite often circuit ground. Once the node voltages are known, the voltages across each component can be found simply by subtracting the node voltages on either side of the device. For instance, the voltage across **R1**, **VR1**, is:

$$VR1 = V3 - V1$$
$$= +8.000 - 4.563$$
$$= +3.437 \text{ V}$$

▌▌ EXAMPLE MSM-2-3

Resistor **R2** in Figure MSM-2-3(a) is considered as the circuit load resistor. Calculate the Thevenin equivalent circuit from the viewpoint of **R1** and Nodes A and B. Build the circuit in MultiSIM and use a virtual multimeter to measure the simulated Thevenin resistance and the open-circuit voltage, **VAB**. Compare the calculated values with the simulation results.

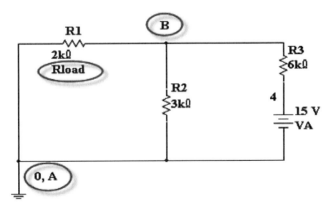

FIGURE MSM-2–3(a)
MultiSIM circuit for Example MSM2-3.

SOLUTION When calculating the Thevenin circuit, the load resistor is removed and the open-circuit voltage **VAB** is calculated by using the voltage divider rule. Then the battery, **VA**, is replaced by a short circuit and the resistance from the viewpoint of nodes A and B is then calculated. The calculated Thevenin circuit values are:

	Calculated	Simulated
VTH = VAB	−5 V	−5 V
RTH	2 kΩ	2 kΩ

FIGURE MSM-2–3(b)
MultiSIM circuit showing RThevenin measurement.

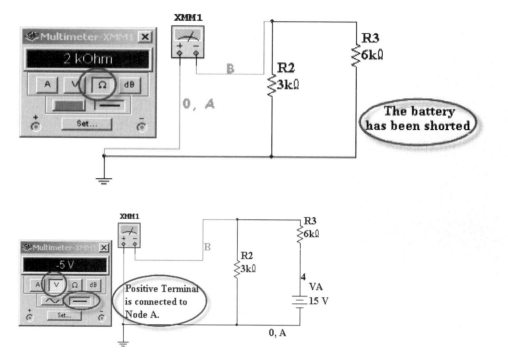

FIGURE MSM-2–3(c)
MultiSIM circuit showing VThevenin (Vopen-circuit) measurement.

The top node in Figure MSM-2-3(a) has been changed from node 1 to node B by left-clicking on the node and then changing the node name. The ground node (node 0) cannot be renamed, but that node has the ability to have a label attached to it, in this case A.

Figure MSM-2-3(b) represents the MultiSIM circuit used to measure the Thevenin resistance from the viewpoint of nodes A and B. The **virtual multimeter** is the top virtual instrument on the right side of the workspace. It can be set to measure current, voltage, resistance, or decibels (dB). In this figure, the multimeter is set to measure resistance and results in a value of 2 kOhms as supported by the calculations.

Figure MSM-2-3(c) represents the MultiSIM circuit used to measure the open-circuit voltage **VAB**, which is the Thevenin voltage. Note that measurements are being taken from the viewpoint of nodes A and then B, so the positive terminal of the multimeter must be connected to node A. The multimeter must also be set to measure DC volts by clicking on the straight line in the device window. The resulting measurement is –5 V as supported by the calculations.

▌▌ EXAMPLE MSM-2-4

Circuit MSM-2-4(a) is the Thevenin equivalent circuit obtained in Example MSM-2-3 with the load resistor returned to the circuit. Measure the load voltage and current. Then perform a source transformation to convert this circuit to a Norton equivalent circuit and use MultiSIM to show that the load voltage and current remain the same. Finally, reconnect the original circuit shown in Figure MSM-2-3(a) and use a voltmeter and ammeter to measure the load voltage and current. Compare the three results of the three circuit simulations.

SOLUTION Connect the circuit of Figure MSM-2-4(a) in MultiSIM and determine the simulated values for the load voltage and current. Note that the positive terminal of the battery is connected to node A just as the positive terminal of the voltmeter in Example MSM-2-3 was connected to node A. Again, this is because the voltage was being determined from the viewpoint of nodes A and B, **VAB** instead of nodes B and A, **VBA**. The simulation results are shown in the table on page 419. (Even though the values are positive, the polarities have the negative terminal at the top, making the readings negative.)

Connect the Norton equivalent circuit (Figure MSM-2-4(b)) in MultiSIM and determine the values for the load voltage and current. Note that the tip of the arrow of the current source is placed where the positive terminal of the battery had been in

FIGURE MSM-2–4(a)
Thevenin equivalent circuit used to measure Vload and Iload.

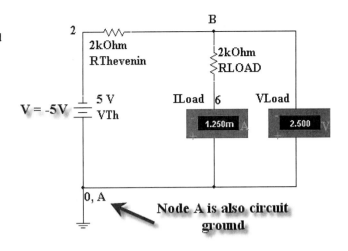

FIGURE MSM-2–4(b)
Norton equivalent circuit used to measure Vload and Iload.

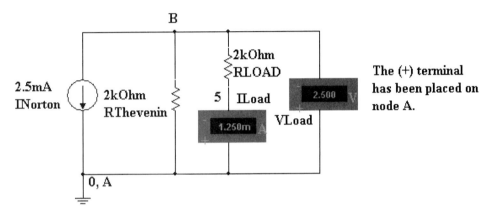

FIGURE MSM-2–4(c)
The original MultiSIM circuit for Example MSM-2-4 showing the Vload and Iload measurements.

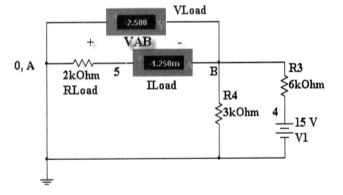

Figure MSM-2-4(a). Note that the simulated results are the same as with the Thevenin equivalent circuit discussed previously.

Finally, connect the original circuit (Figure MSM-2-4(c)) in MultiSIM and determine the load voltage and current of the original circuit. Note that the simulation values are the same as for the Thevenin and Norton equivalent circuits.

	Calculated	Simulated Thevenin	Simulated Norton	Simulated Original
VLoad	−2.5 V	−2.5 V	−2.5 V	−2.5 V
ILoad	−1.25 mA	−1.25 mA	−1.25 mA	−1.25 mA

EXAMPLE MSM-2-5

The circuit in Figure MSM-2-5(a) is the Thevenin equivalent circuit from Example MSM-2-4. Use MultiSIM to develop a dissipated power versus load resistance plot. From this plot determine the value of load resistance at which the maximum power is transferred from the circuit to the load.

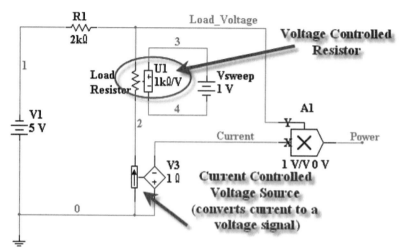

FIGURE MSM-2-5(a)
MultiSIM circuit for Example MSM-2-5.

SOLUTION One of the available analysis options in MultiSIM is called **Parameter Sweep**. This option will sweep the desired parameter, such as resistance, and output the selected circuit parameter. The problem with this analysis option is that it creates a table of values, not a plot. It would still be possible to create a plot using this option by exporting its results into a spreadsheet and creating a plot inside the spreadsheet.

To create the plot directly within MultiSIM, it is necessary to add a few more components to the circuit.

Introduction to the Current Controlled Voltage Source (ICVS)

The first part is a **Current Controlled Voltage Source (ICVS)**, which is found in the **Sources/CONTROLLED_VOLTAGE_SCOURCES/CURRENT_CONTROLLED_VOLTAGE_SOURCE** parts bin. This device will produce a voltage waveform which is proportional to the current in the branch where the device is connected as shown in Figure MSM-2-5(a). The device is placed so that the current arrow is pointing in the direction of the current flow. For this example, there is no advantage in scaling the resulting voltage waveform; therefore the scale factor is left at its 1 to 1 ratio default value.

Introduction to the Voltage Controlled Resistor (VCR)

The next device is a **Voltage Controlled Resistor (VCR)** found in the **Basic/BASIC_VIRTUAL/VOLTAGE_CONTROLLED_RESISTOR-VITUAL** parts bin. As shown in Figure MSM-2-5(a), a battery is connected to one set of terminals while the resistor portion of the VCR is placed where the load resistor is located. The value of the battery is unimportant because it will be swept by the **DC Sweep Analysis** option between two different values. For this example, we want to have the resistance axis of the plot scaled in kΩ. For this reason, the scaling factor of the **VCR** is set to 1 kΩ/V (the default value).

Introduction to the Multiplier Control Block

The last new device is a **Multiplier Control Block** found in the **Sources/CONTROL_FUNCTION_BLOCKS/MUTIPLIER** parts bin. This device will multiply the two input waveforms and output the product. The **multiplier** in this example will multiply the load voltage and the load current to produce the power waveform. Note that the associated wires have had their **Reference IDs** renamed.

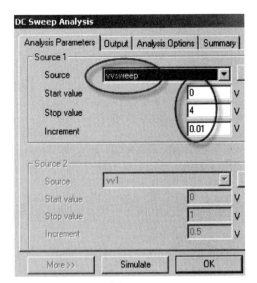

FIGURE MSM-2-5(b)
DC Sweep Analysis parameters window.

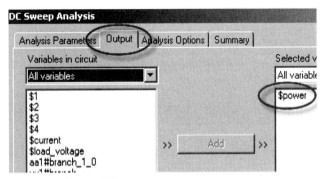

FIGURE MSM-2-5(c)
DC Sweep Analysis output window.

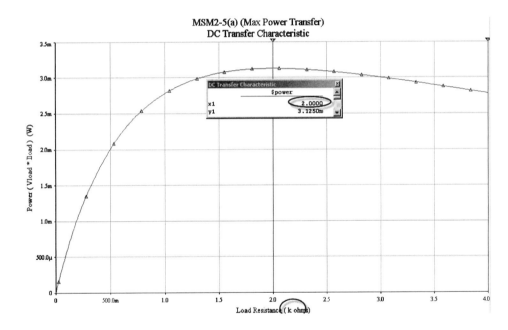

FIGURE MSM-2-5(d)
MultiSIM DC Sweep Analysis plot for Example MSM-2-5.

Introduction to the DC Sweep Analysis

The **DC Sweep Analysis** option (Figure MSM-2-5(b) and (c)) is set up to sweep a dc voltage over a selected range of voltages. In this example, the Reference ID of the battery driving the **VCR** was changed to **Vsweep**. The DC Sweep is set to sweep VVsweep from 0 V to 4 V in 0.01-V increments. The VCR's scale is set to 1 kΩ/V. This means that the load resistor will be essentially swept from 0 Ω to 4 kΩ in 10-ohm increments. The **Power** node was selected as the output node in the analysis **Output window** (Figure MSM-2-5(c)).

Once the **simulate** button was selected, a plot like the one in Figure MSM-2-5(d) was created. The names for the vertical and horizontal axis, and the title of the plot, were changed accordingly.

The maximum value of the resistance was found to be 2 kΩ by using the **Grapher cursor** (found in the **Grapher toolbar**). This result supports the fact that the maximum power will be transferred to the load if the value of that load is equal to the circuit's Thevenin resistance.

MultiSIM Examples for Chapter 3

Introduction to MultiSIM's Piecewise Linear Voltage Source (PWLVS)

It is often necessary to supply input voltage signals which are not available as a source. MultiSIM has a device called a Piecewise Linear Voltage Source, which can be programmed to provide any voltage vs. time signal that can be represented by straight-line connections. It can either be programmed directly into the device or can be provided via a .txt file. This device can be found in the **Sources** parts bin. The device in Figure MSM-3-1(a) has been programmed to provide the voltage signal as shown in Figure 3-4. This signal is able to be programmed with just three data points, so it can be programmed directly into the device window. If a signal is much more than six or seven points, it is best to program the device via a **.txt** file. This device will be demonstrated in the next few examples.

■ **EXAMPLE MSM-3-1**

Figure MSM-3-1(a) is the MultiSIM circuit for Example 3-1 (Figure 3-4). Provide a MultiSIM **Grapher** output of the capacitor's voltage and current waveforms.

SOLUTION Build the circuit in MSM-3-1(a) in MultiSIM. Program the **Piecewise Linear Voltage Source** to provide the voltage waveform shown in Figure 3-4. This program is shown in Figure MSM3-1(b). The Reference IDs of two of the wires in the circuit file have been changed to **VC** and **IC**.

When the circuit is simulated, set the settings on the o-scope to the following:

Timebase = 2 ms/div   **Channel A = 20 V/div**   **Channel B = 5 mV/div**

The resulting o-scope display is shown in Figure MSM-3-1(c). Use the o-scope measurement cursors to measure the peak current values. The positive peak of the rectangular current wave is measured to be 6 mA and the negative peak is -3 mA. Remember that the current controlled voltage source created a voltage waveform which was proportional to the circuit current. Therefore, even though the o-scope units were mV, the true units are mA.

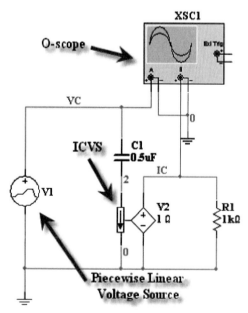

FIGURE MSM-3-1(a)
MultiSIM circuit used for Example MSM-3-1(a).

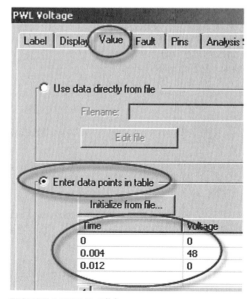

FIGURE MSM-3-1(b)
Piecewise linear voltage source setup window.

FIGURE MSM-3–1(c)
Measurement for peak current values.

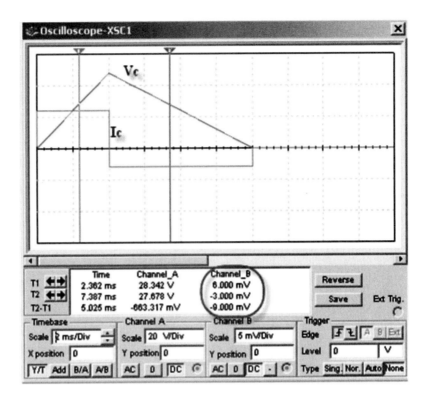

EXAMPLE MSM-3-2 Add the necessary devices to the circuit in Figure MSM-3-1 so that the **Power** waveform as shown in Figure 3-5 can be observed and validated.

SOLUTION Construct the circuit in Figure MSM-3-2(a) in MultiSIM using the circuit created in Example MSM3-1 and adding a **Multiplier Control Function Block**. Be sure to rename the output and input nodes of the **Multiplier** as shown.

The **Piecewise Linear Voltage Source** is programmed in the same manner as in Example MSM-3-1. Simulate the circuit and monitor the current and the power waveforms.

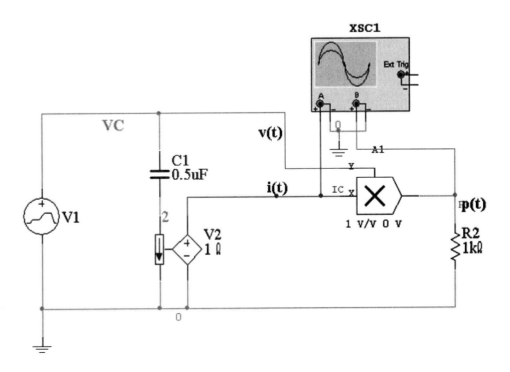

FIGURE MSM-3–2(a)
MultiSIM circuit for Example MSM-3-2.

FIGURE MSM-3–2(b)
O-scope output demonstrating the P(t) measurement.

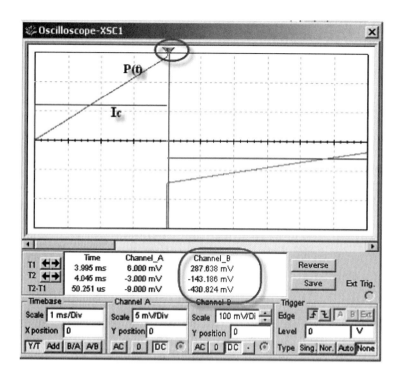

Use the cursors to measure the peak values. Remember that the actual units for the power measurements are in mW instead of mV. Thus, the positive peak power is about 288 mW and the negative peak is about −143 mW.

EXAMPLE MSM-3-3

Build a MultiSIM circuit that will allow you to simulate the circuit in Example 3-9 (Figure 3-16) and observe and validate the voltage and current waveforms shown in Figure 3-17. Although the waveforms in Figure 3-17 only show the charge waveforms, include the discharge waveforms in your output.

SOLUTION Build the circuit in Figure MSM-3-3(a) in MultiSIM. The original circuit in Figure 3-16 included a 12-V battery and a switch. A better way to accomplish the same thing and also include the discharge phase is to use a 12-V **Clock Voltage Source**. This source provides a rectangular wave which cycles between 0 V and some settable voltage (usually 5 V). This device is accessed as a 5-V **Clock Voltage Source** in the **Sources** parts bin. To set the clock up for this problem, use the time constant of 2 ms calculated in Equation 3-62.

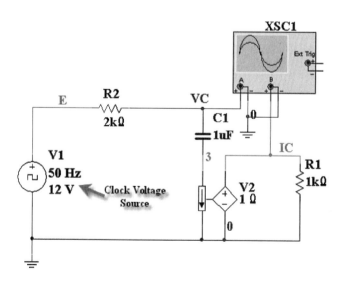

FIGURE MSM-3–3(a)
MultiSIM circuit for Example MSM-3-3.

FIGURE MSM-3-3(b)
Clock waveform setup.

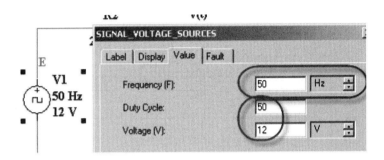

Because it takes about 5 time constants to totally charge or discharge, the **Time High** (T_H) and **Time Low** (T_L) of the clock should be:

$$T_H = T_L = 5\tau = 5\,(2\text{ ms}) = 10\text{ ms}$$

The rectangular waveforms **Period (T)** is:

$$T = T_H + T_L = 10\text{ ms} + 10\text{ ms} = 20\text{ ms}$$

The rectangular waveforms **frequency (f)** is:

$$\text{Frequency} = \frac{1}{T} = \frac{1}{20\text{ ms}} = 50\text{ Hz}$$

And finally the rectangular waveforms **duty cycle (D)** is:

$$\text{Duty Cycle }(D) = \frac{T_H}{T} \cdot 100\% = \frac{10\text{ ms}}{20\text{ ms}} \cdot 100\% = 50\%$$

All the necessary parameters for setting up the **Clock Voltage source** to use as both a voltage source and switch are now known. During the Time High portion of the cycle a 12-V dc voltage will be applied to the circuit and the capacitor will charge. During the Time Low portion of the cycle there will be 0 volts applied and the capacitor will discharge. Figure MSM-3-3(b) shows how the clock is set up.

Simulate the circuit and measure the voltage and current waveforms. The output curves are shown in Figure MSM-3-3(c). The voltage waveform peaks at the expected value of 12 V. The current waveform cycles between +6 mA and −6 mA.

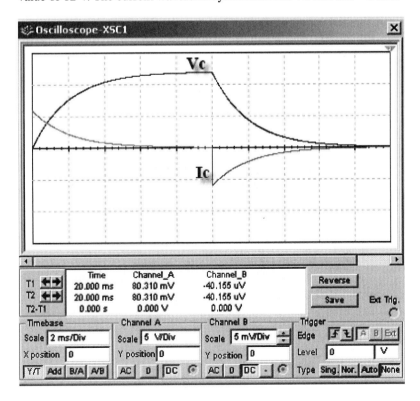

FIGURE MSM-3-3(c)
Current and voltage waveforms for Example MSM-3-3.

■ EXAMPLE MSM-3-4   Build a MultiSIM circuit to simulate the circuit in Example 3-10 (Figure 3-18) and observe and validate the voltage and current waveforms shown in Figure 3-19. While the waveforms in Figure 3-19 only show the charge waveforms, include the discharge waveforms in your output. Show two complete cycles.

SOLUTION Build the circuit in Figure MSM-3-4(a) in MultiSIM. This circuit is exactly the same circuit as in Example MSM-3-3 except for the initial charge on the capacitor. This is accomplished in MultiSIM by making two changes to the circuit.

Double left-click on the capacitor and you will see the window shown in Figure MSM-3-4(b). Change the **Initial Conditions** setting to −8 V.

Select the Simulate pull-down menu and select **Interactive Simulation Settings**. Modify the **Initial conditions** block from the default value to **User-defined** as shown in Figure MSM-3-4(c), then select OK.

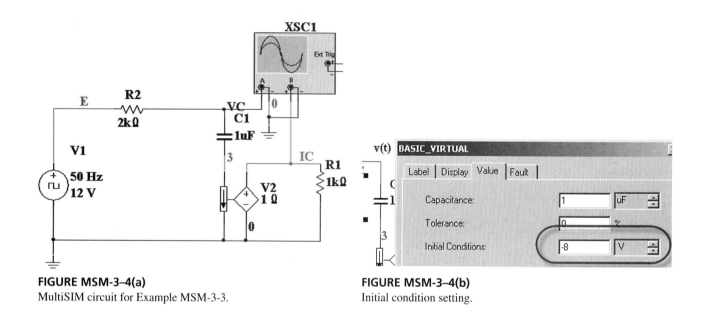

FIGURE MSM-3-4(a)
MultiSIM circuit for Example MSM-3-3.

FIGURE MSM-3-4(b)
Initial condition setting.

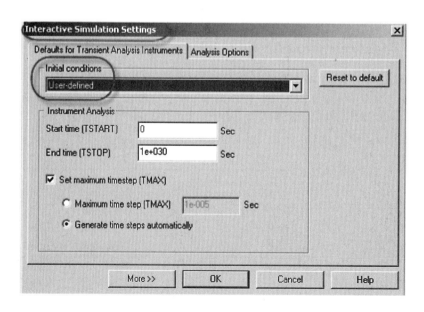

FIGURE MSM-3-4(c)
Change initial condition parameter to 'user defined' in the interactive simulation settings window.

FIGURE MSM-3–4(d)
Voltage and current waveforms with initial condition.

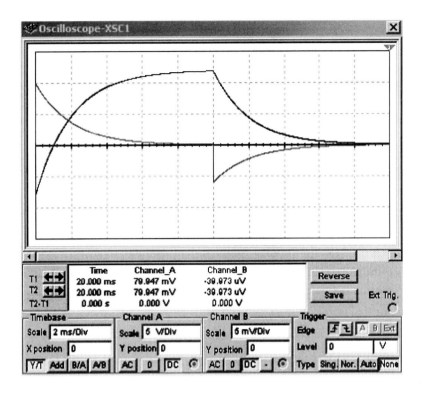

Simulate the circuit. Note that the results as shown in Figure MSM-3-4(d) indicate that the voltage starts at the given initial voltage of -8 V and then returns to the waveform found in Example MSM-3-3 in the next interval.

MultiSIM Examples for Chapter 4

EXAMPLE MSM-4-1

Build a MultiSIM circuit to simulate the circuit in Example 4-6 (Figure 4-15) and observe the voltage and current waveforms.

SOLUTION The circuit is shown in Figure MSM-4-1(a). If you were required to measure the voltage across the resistor, you might be tempted to place the measurement device so that the positive terminal was on the left side and the negative terminal was on the right side. This might work as long as the device was floating, that is, not internally grounded. If it had been internally grounded, as a lot of oscilloscopes are, then the measurement device would have just shorted out the inductor and all measurements made would have been incorrect.

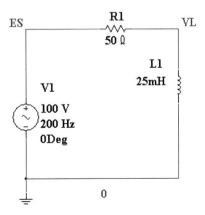

FIGURE MSM-4–1(a)
MultiSIM circuit for Example MSM-4-1.

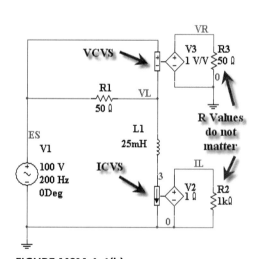

FIGURE MSM-4–1(b)
MultiSIM file for Example MSM-4-1 with controlled sources used to plot VR and IL.

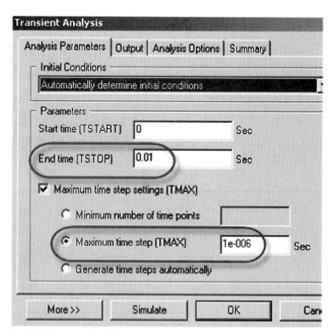

FIGURE MSM-4–1(c)
Transient analysis window.

Luckily, MultiSIM has another way to do this so that the two measurements (**VR** and **VL**) can both be taken at the same time and compared with each other. A **Voltage Controlled Voltage Source (VCVS)** is a source which will output a voltage which is proportional to another voltage but which is isolated from that voltage. This device can be found in the **Controlled Voltage Sources** parts bin along with the previously mentioned ICVS. Figure MSM-4-1(b) shows the circuit with both the VCVS and the ICVS added and the node names changed to represent the signal that would be measured at each node.

Build the circuit in MSM-4-1(b) in MultiSIM. In an earlier example the Current Controlled Voltage Source (ICVS) was used to allow the current in a circuit to be plotted as a voltage waveform. Use an ICVS and plot the inductor current versus time and a VCVS to plot the resistor voltage.

Before simulating the circuit, the simulation time required to produce two cycles needs to be determined. The frequency of the source is 200 Hz, therefore the period (T) of the source will be:

$$T = \frac{1}{f} = \frac{1}{200\text{ Hz}} = 5\text{ ms}$$

This time the o-scope will not be used to measure the signals. Instead, a **Transient Analysis** will be performed. This mode will provide the same type of waveform as the o-scope but will provide more user control of the results. Select the **Simulate** pull-down menu and then select **Analysis/Transient Analysis**.

The resulting Transient Analysis window as shown in Figure MSM-4-1(c) will open. To observe the phase angles between different signals, it helps to observe two complete cycles; therefore the simulation should be set to last for two times the period or 10 ms (**TSTOP**). Change **TMAX** from its default of **1e-005** to **1e-006** to produce smoother output traces (or set the **Minimum number of time points** to 5000).

Modify the **TSTOP** value to 0.01. Select the Output tab and select the required output variables as shown in Figure MSM-4-1(d). Simulate the circuit and format the resulting **Grapher** output so that all three waveforms can be observed. This means that the current waveform, which will be on a significantly different scale than the voltage waveforms, will have to be reassigned to the right axis and scaled separately.

FIGURE MSM-4–1(d)
Transient analysis output tab.

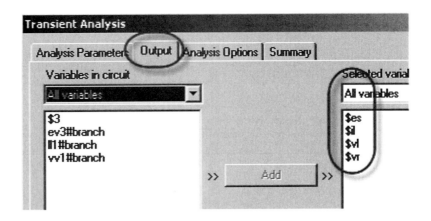

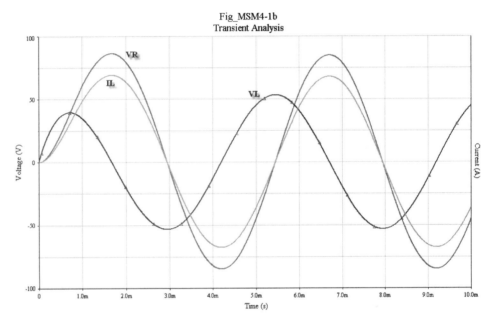

FIGURE MSM-4–1(e)
Transient analysis plot for Example MSM-4-1.

Figure MSM-4-1(e) shows the three output waveforms and their phase relationships with each other. Note that there is a transient interval before the waveforms reach steady-state conditions.

EXAMPLE MSM-4-2 The circuit in Figure MSM-4-1(a) is the circuit from Example MSM-4-2. Use MultiSIM to measure the necessary information to verify the value of the circuit total impedance.

SOLUTION The impedance of the circuit in Figure MSM-4-2(a) is determined in the steps that follow.

$$X_L = 2\pi f L = 2\pi(200)(25 \times 10^{-3}) = 31.42 \ \Omega$$
$$\mathbf{Z_T} = R + jX_L = 50 + j31.42 = 59.05 \ \Omega \angle 32.15°$$

To calculate the circuit impedance using MultiSIM you must measure the current in the circuit and use Ohm's Law and the source voltage.

Select **AC Analysis** and set the **Frequency Parameters** as shown in Figure MSM-4-2(b). The AC Analysis will sweep the source frequency between the **FSTART** and **FSTOP** values and measure the magnitude and the phase angle of the selected output variable over the frequency range. Normally, the AC Analysis is run over a wider frequency

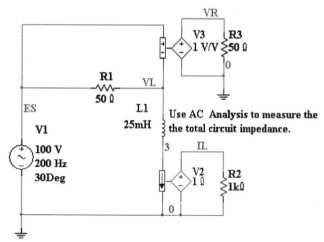

FIGURE MSM-4–2(a)
MultiSIM circuit for Example MSM-4-2.

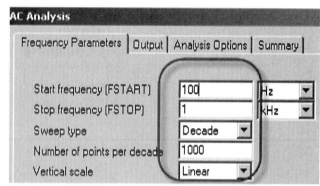

FIGURE MSM-4–2(b)
AC Frequency analysis 'frequency parameters' setup window for Example MSM-4-2.

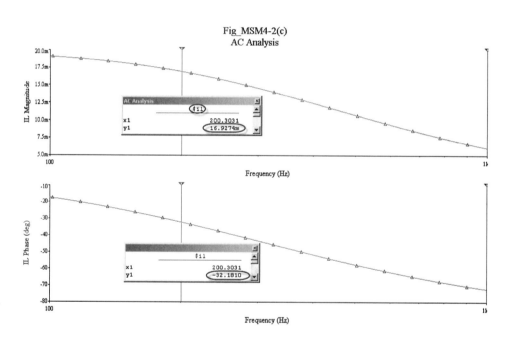

FIGURE MSM-4–2(c)
AC Frequency analysis plot for Example MSM4-2.

band, but for this example you are interested in only a single frequency, so a very narrow-frequency band was used. Choose **Linear** for the **Vertical Scale** and increase the **Number of points per decade** to 1000 to smooth out the waveform. This will also help you get the measurement cursor closer to 200 Hz.

To use MultiSIM's AC analysis, the circuit *must* be driven by an ac source. The magnitude and phase settings of the source do not affect the ac analysis. The source is assumed by the AC Analysis routine to be 1 V $\angle 0°$.

Select **IL** in the **Output variables** window as in previous examples and perform the simulation. Figure MSM4-2(c) shows the resulting plot after some minor format changes. The measured value of the current is $\mathbf{I_L} = 16.93$ mA $\angle -32.18°$.

$$\mathbf{Z_T} = \frac{\mathbf{V_S}}{\mathbf{I_L}} = \frac{1 \text{ V} \angle 0°}{16.93 \text{ mA} \angle -32.18°}$$

Using Ohm's Law you get:

$$= 59.07 \text{ }\Omega \angle 32.18°$$
$$= 50.00 \text{ }\Omega + j31.46 \text{ }\Omega$$

A comparison between calculated and measured values follows:

	Calculated	Measured/Calculated
Resistance	50 Ω	50.00 Ω
Inductive Reactance	31.42 Ω	31.46 Ω

MultiSIM Examples for Chapter 5

EXAMPLE MSM-5-1 Build a MultiSIM circuit which will simulate the circuit in Example 5-3 (Figure 5-5). Measure the real power and the power factor.

SOLUTION Build the circuit in Figure MSM-5-1 in MultiSIM. Connect the Wattmeter as shown and measure the real power and the power factor.

The Wattmeter has a Power Factor display along with the power in watts.

The results are as shown in the table below.

	Calculated	Simulated
Power (W)	230.4	230.375
Power Factor	.8	.8

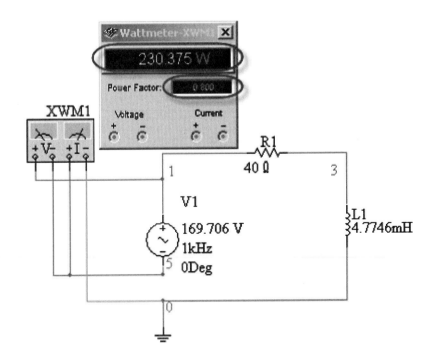

FIGURE MSM-5–1
MultiSIM circuit for Example MSM-5-1.

EXAMPLE MSM-5-2

Build the series resonant circuit of Example 5-9 in MultiSIM and determine the circuit's resonant frequency.

SOLUTION A **Bode Plot** is a plot of Magnitude and Phase versus frequency. In MultiSIM you have the choice between using a **virtual Bode Plotter** and performing a **Frequency Analysis**. The resonant frequency of the series resonant circuit will be the point at which the peak current amplitude is achieved.

Build the MultiSIM circuit shown in Figure MSM-5-2(a). Open the Bode Plotter and simulate the circuit. As you can see from the Bode Plot shown in Figure MSM-5-2(a), making measurements of any accuracy is difficult because the plot is so small. However, the Grapher will also have a copy of the Bode Plotter output.

Open the Grapher and note that there is a phase graph in addition to the amplitude graph. We will not use the phase graph here, so click inside the phase graph and then select **CUT** on the toolbar. This will make room for a larger-amplitude graph. One advantage of the Grapher screen is a **ZOOM IN** feature which allows you to make more accurate measurements. Use the mouse to draw a box around the peak of the waveform. This will zoom in on that section of the Bode Plot. Now use the cursors to measure center frequency. The resulting value is listed in the table on page 433.

The vertical axis of the plot is in decibels (dB). The point at which the amplitude is 3 dB down from the peak amplitude is known as the **Half Power Point**, or the **3 dB Down Point**. Because the peak value is basically 0 dB, the 3-dB down points will be at a level of −3 dB on each side of the peak. Place a cursor at each of the −3-dB down points and measure the difference in the frequencies between the two points. This value will be shown as dx in the measurement window. Figure MSM-5-2(c) shows the measured bandwidth

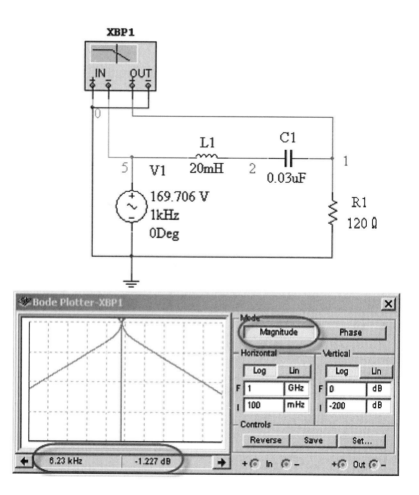

FIGURE MSM-5-2(a)
MultiSIM circuit and Bode Plotter output for Example MSM-5-2.

FIGURE MSM-5–2(b)
MultiSIM Grapher plot of the Bode Plot for Example MSM5-2 showing the resonant frequency.

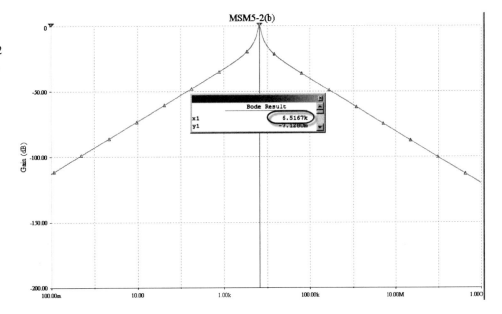

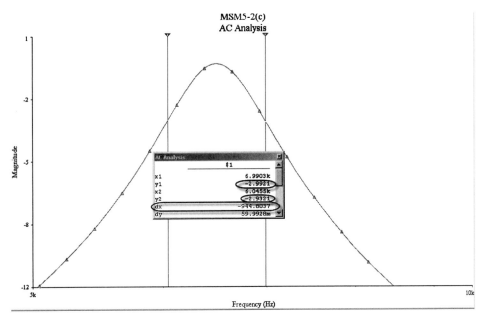

FIGURE MSM-5–2(c)
Bode Plot showing bandwidth measurement.

(the difference between the two 3-dB down points). The difference between the measured value and the calculated value is partly because it was not possible to select the exact 3-dB down points. Note that the higher the **number of points per decade** setting is, the closer to exact values you can get.

	Predicted	Simulated
Resonant Frequency	6.497 kHz	6.517 kHz
Bandwidth	950 Hz	944.8 Hz

EXAMPLE MSM-5-3   Return to the circuit of Example 5-3 (Figure 5-5) that you built in Example MSM-5-1. This time perform an AC Analysis on the voltages across the resistor and the inductor and analyze the phase angle at the 1-kHz source frequency. Compare the results with the calculated results from Example 5-3.

FIGURE MSM-5-3(a)
MultiSIM circuit for Example MSM-5-3.

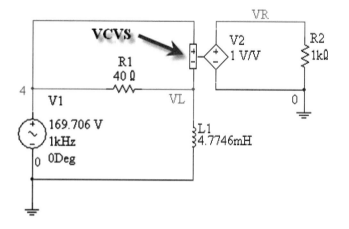

FIGURE MSM-5-3(b)
AC Analysis settings for Example MSM5-3.

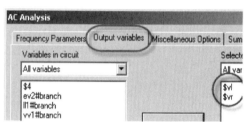

FIGURE MSM-5-3(c)
AC Analysis phase measurements for Example MSM5-3.

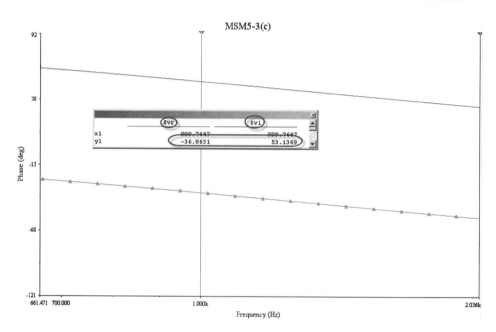

SOLUTION Build the circuit in Figure MSM-5-3(a). The voltage across the resistor is required to be analyzed, so a Voltage Controlled Voltage Source (VCVS) will be required to analyze the resistor's voltage waveform in its particular circuit location.

Once the circuit has been built, select **AC Analysis** from the analysis menu. Make the changes in the **Frequency Parameters** window and the **Output variables window** as shown in Figure MSM-5-3(b). Note that the **Number of points per decade** option has been changed from the default of 10 to 1000. This will allow more accuracy but it will slow the simulation and use more memory. We are only interested in the area around 1 kHz; therefore the frequency range has been reduced to a single decade.

Once the analysis has been executed, the Grapher will appear with the Amplitude and Phase graphs (Figure MSM-5-3(c)). Select the **Phase graph** and place a cursor at 1 kHz. Use the ZOOM IN feature as discussed in Example MSM-5-3 to increase the accuracy of the measurement.

The voltage across the inductor was not calculated in Example 5-3. The inductor voltage can be found by the voltage divider rule:

$$\overline{V}_L = \frac{169.706 \text{ V} \angle 0° \ (30 \ \Omega \angle 90°)}{40 \ \Omega + j30 \ \Omega}$$

$$= \frac{169.706 \text{ V} \angle 0° \ (30 \ \Omega \angle 90°)}{50 \ \Omega \angle 36.87°} = 101.82 \text{ V} \angle 53.13°$$

The phase relationships of the waveforms are as follows:

	Calculated Phase	Simulated Phase
VR	$-36.87°$	$-36.86°$
VL	$53.13°$	$53.14°$

MultiSIM Examples for Chapter 6

EXAMPLE MSM-6-1

Build the circuit in Example 6-3 (Figure 6-9(a)) using the parameters given in the example. Simulate the circuit to determine the peak value of the primary and secondary voltages, the peak load voltage, the dc load voltage, and the dc load current.

SOLUTION Build the circuit in Figure MSM-6-1(a) in MultiSIM. The **TS_Virtual transformer model** is found in the **Basic** parts bin. Double-click on the Transformer and set the **Primary to Secondary Turns Ratio** to a value of 5.

Open the o-scope and set the Timebase and the scales of Channel A and B as shown in Figure MSM-6-1(b). Simulate the circuit and observe the output of the circuit versus the input signal. To see both waveforms, use the Channel B Y position control to shift the output waveform up.

Use the o-scope measurement cursor to measure the positive peak values for the two signals. The DC meters will provide the measured values for the load voltage and current.

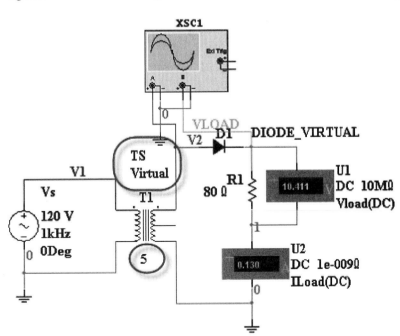

FIGURE MSM-6–1(a)
MultiSIM circuit used for Example MSM-6-1.

FIGURE MSM-6–1(b)
Transient analysis output for Example MSM-6-1.

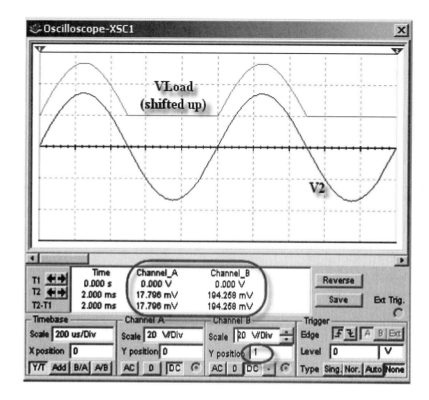

Note the approximate 0.7-V difference between **V2** and **VLoad** caused by the forward voltage drop across the diode.

	Predicted	Measured
V_2 (Peak)	33.94 V	33.9361 V
V_{Load} (Peak)	33.24 V	33.1252 V
V_{Load} (DC)	10.58 V	10.408 V
I_{Load} (DC)	132 mA	130 mA

■ EXAMPLE MSM-6-2

Build the circuit in Example 6-4 (Figure 6-10(a)) using the parameters given in the example. Simulate the circuit to determine the peak value of the primary and secondary voltages, the peak load voltage, the dc load voltage, and the dc load current.

SOLUTION Build the circuit in Figure MSM-6-2(a) in MultiSIM. The **TS_Virtual** transformer model is found in the **Basic** parts bin. Because the model does not have a setting for the secondary turns, the Primary to Secondary turns ratio will have to be set to 2.5:1 instead of 5:2. Note that the polarities of the two dc meters were chosen based on the predicted direction of the load current.

Open the o-scope and simulate the circuit. Use the o-scope cursor to measure the positive peak values for the two voltages (Figure MSM-6-2(b)). The DC meters will provide the measured values for the load dc voltage and dc current. Note the approximate 0.7-V difference between **V2A** and **VLoad** caused by the forward voltage drop across the diode.

	Predicted	Measured
V_{2A} (Peak)	33.94 V	33.9144 V
V_{Load} (Peak)	33.24 V	33.1035 V
V_{Load} (DC)	21.16 V	20.816 V
I_{Load} (DC)	265 mA	260 mA

FIGURE MSM-6–2(a)
MultiSIM circuit used for Example MSM-6-2.

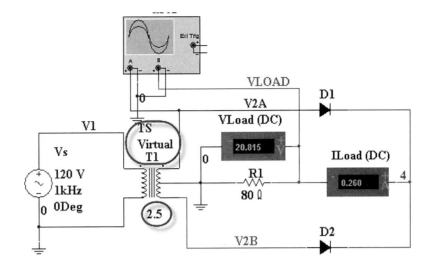

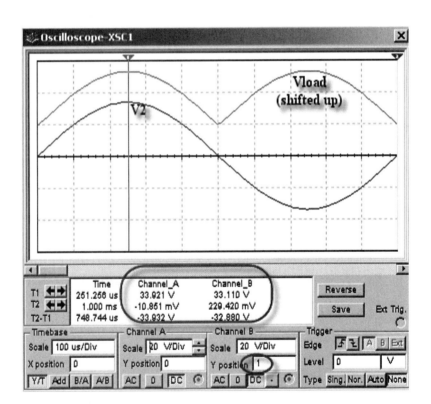

FIGURE MSM-6–2(b)
Transient analysis output for Example MSM-6-2.

■ EXAMPLE MSM-6-3   Build the circuit in Example 6-5 (Figure 6-12) using the parameters given in the example. Simulate the circuit to determine the peak value of the primary and secondary voltages, the peak load voltage, the dc load voltage, and the dc load current.

SOLUTION   Build the circuit in Figure MSM-6-3(a) in MultiSIM. The **TS_Virtual** transformer model is found in the **Basic** parts bin. Set the Primary to Secondary turns ratio to 5.

FIGURE MSM-6-3(a)
MultiSIM circuit used for Example MSM-6-3.

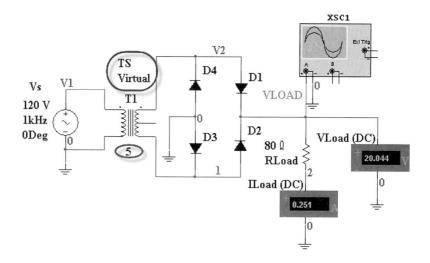

FIGURE MSM-6-3(b)
Transient analysis output for Example MSM-6-3.

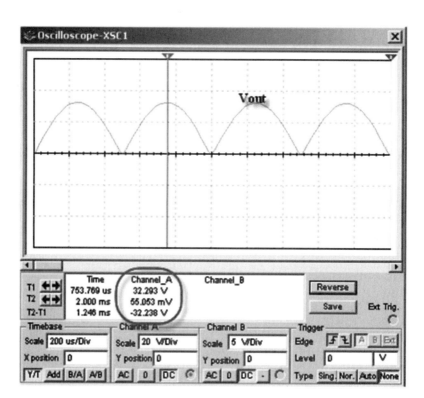

Simulate the circuit and use the o-scope to measure the positive peak values for the three voltages (Figure MSM-6-3(b)). The dc meters will provide the measured values for the load dc voltage and dc current. Note the difference of approximately 1.4 V between **V2** and **VLoad** due to the forward voltage drops of the two diodes.

	Predicted	Measured
V_{2A} (Peak)	33.94 V	33.9144 V
V_{Load} (Peak)	32.54 V	33.1035 V
V_{Load} (DC)	20.72 V	20.816 V
I_{Load} (DC)	259 mA	260 mA

EXAMPLE MSM-6-4

Build the circuit in Example 6-10 (Figure 6-24) using the parameters given in the example. Simulate the circuit to determine the circuit voltages and currents.

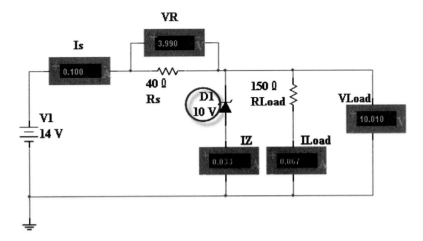

FIGURE MSM-6-4
MultiSIM circuit used for Example MSM-6-4.

SOLUTION Build the circuit in Figure MSM-6-4 in MultiSIM. Choose a **virtual Zener diode** from the **Diodes_Virtual** parts bin and change its **breakdown voltage** to 10 V. Simulate the circuit and observe the measured voltages and currents.

	Predicted	Measured
V_{Rs}	4 V	3.99 V
V_{Load}	10.0 V	10.01 V
I_{Load}	66.7 mA	67 mA
I_Z	33.3 mA	33 mA
I_s	100 mA	100 mA

MultiSIM Examples for Chapter 7

EXAMPLE MSM-7-1

Build the circuit of Example 7-5 (Figure 7-15). Observe the relationship between the calculated values of the currents and the simulated values.

SOLUTION Build the circuit in Figure MSM-7-1 in MultiSIM. The **BJT_NPN_VIRTUAL** is located in the **Transitors_Virtual** parts bin. The available Virtual NPN BJT has a beta value of 100 instead of 50 as stated in Example 7-5. In addition, the selected BJT is not ideal. For these reasons some of the values calculated in Example 7-5 will have to be recalculated.

$$I_L = I_C = \frac{V_{CC}}{R_L} = \frac{24 \text{ V}}{200 \text{ }\Omega} = 120 \text{ mA}$$

$$I_{B(\max)} = \frac{I_C}{\beta_{dc}} = \frac{120 \text{ mA}}{100} = 1.20 \text{ mA}$$

Observe the results of the simulations shown in Figure MSM-7-1. From the values shown, we note that there are some nonideal results. For instance, the ideal value of the current was based on the assumption that the BJT saturation voltage was zero. In this case it was 0.511 V; therefore the actual load current is:

$$I_L = I_C = \frac{V_{CC}}{R_L} = \frac{(24 - .511) \text{ V}}{200 \text{ }\Omega} = 117.4 \text{ mA}$$

FIGURE MSM-7–1
MultiSIM circuit used by Example MSM-7-1.

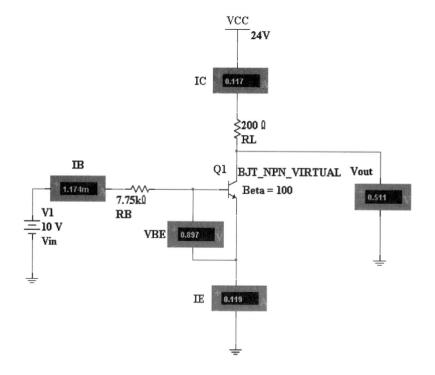

EXAMPLE MSM-7-2 Modify the circuit of Example MSM-7-1 by changing the input voltage to 1.5 V. This will bring the BJT out of saturation. Simulate the circuit and observe the collector and base currents as they relate to beta.

SOLUTION Build the circuit in Figure MSM-7-2 in MultiSIM. Simulate the circuit and note that the circuit is no longer in saturation because the collector-to-emitter voltage now

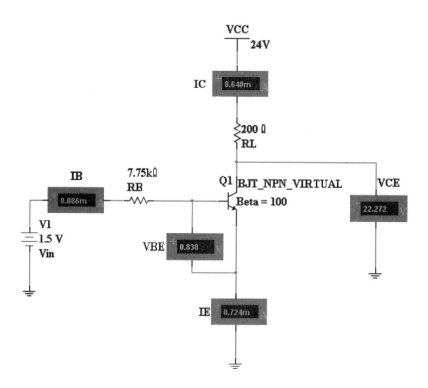

FIGURE MSM-7–2
MultiSIM circuit used for Example MSM-7-2.

represents a significant portion of the supply voltage. Now note that the collector current is equal to:

$$I_C = \beta I_B = 100\,(0.086\text{ mA}) = 8.6\text{ mA}$$

As seen from this equation, the collector current should be 100 times the base current. This is equal to the observed collector current within the attainable accuracy of the calculation compared with the measurement. In other words, the base current has been amplified by the beta value.

EXAMPLE MSM-7-3

Build a MultiSIM circuit which will drive a 100-V, 100-W lamp with a **2n2222 NPN BJT** and measure the on and off voltages and currents.

SOLUTION Build the circuit in Figure MSM-7-3(a) in MultiSIM. This circuit uses a **2n2222 BJT** as an electronic switch, which can be found in the **Transistors/BJT_NPN** parts bin. In this example, the BJT circuit has been turned into a current source which will drive a high-voltage lamp.

If the lamp were rated at 12 V or lower, it would be placed in the collector circuit of the BJT. In this case, the lamp is rated at 100 V, 100 W, which is significantly beyond the capability of the 2n2222 to drive. For this reason, an electronic relay has been placed on the collector circuit instead. The **RELAY 1A_VIRTUAL** can be found in the **BASIC_VIRTUAL** parts bin. This relay requires 50 mA to activate the secondary; therefore, the BJT needs to be biased so that that there will be at least 50 mA of collector current available when the BJT is in its ON condition. When the coil (inductor) on the relay's primary is excited by at least 50 mA, the electromagnetic field (EMF) around it will activate the secondary and the circuit will be closed, thus allowing current to flow through the secondary, turning the lamp ON. There is no physical connection between the primary and the secondary. In effect, the secondary is ISOLATED from the primary.

The diode across the coil is reverse biased when the relay coil is excited; therefore it has no effect on the circuit. When the coil is excited and then the BJT turns off, the current flow through the coil will be abruptly terminated. This will cause a very large reverse EMF voltage to be generated, thus destroying the BJT. The diode prevents this, because it will be forwardbiased by the large negative EMF voltage and will provide a discharge path for the coil.

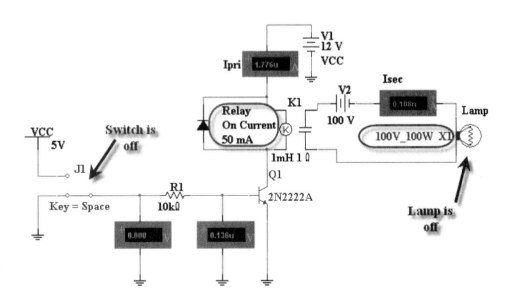

FIGURE MSM-7-3(a)
Circuit for Example MSM-7-3; transistor is OFF, lamp is OFF.

FIGURE MSM-7-3(b)
Circuit for Example MSM-7-3; transistor is ON, lamp is ON.

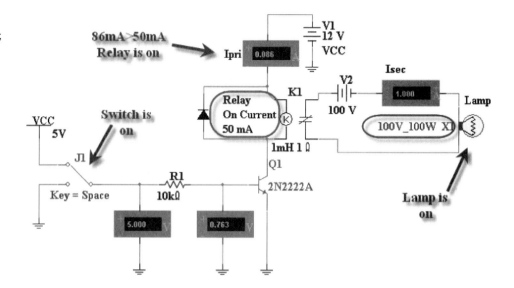

Observe Figure MSM-7-3(a). Note that when 0 V is applied to the base, the base-to-emitter voltage is too low to turn the BJT on; therefore, the collector current is limited to leakage current, which is not enough to excite the relay coil. As demonstrated in Figure MSM-7-3(b), when 5 V is applied to the base, the BJT turns on, and enough collector current flows to excite and activate the coil, thus turning the lamp on.

EXAMPLE MSM-7-4

Build a MultiSIM circuit which will invert 0- and 5-V signals applied to its input using MOSFET technology.

SOLUTION Build the circuit in Figure MSM-7-4(a) in MultiSIM. This device is simulating the actions of a digital **NOT gate** (an inverter), which will be discussed in a later chapter. The purpose of the device is to invert whatever logic signal is applied to it.

When 0 V (Logic 0) is applied to the gates of each MOSFET, the P-Channel MOSFET (**MOS_4TEP_VIRTUAL** found in the **Transistors / TRANSISTORS_VIRTUAL** parts bin) will be turned on while the N-Channel MOSFET (**MOS_4TEN_VIRTUAL**) will be turned off, thus causing an output of 5 V (Logic 1).

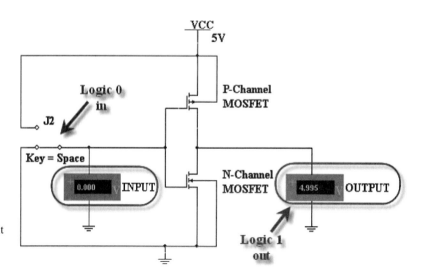

FIGURE MSM-7-4(a)
MultiSIM CMOS inverter circuit used for Example MSM-7-4: input LOW, output HIGH.

FIGURE MSM-7–4(b)
MultiSIM CMOS inverter circuit used for Example MSM-7-4: input HIGH, output LOW.

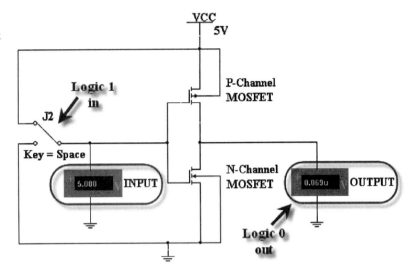

MSM-7-4(b) demonstrates that when 5 V (Logic 1) is applied to the MOSFET gates, the P-Channel MOSFET is OFF and the N-Channel MOSFET will be ON, thus causing an output of 0 V (Logic 0).

MultiSIM Examples for Chapter 8

EXAMPLE MSM-8-1

Use MultiSIM to build the inverting amplifier used in Example 8-7 (Figure 8-13) and measure the input and feedback currents and the output voltage. Calculate the voltage gain and the input resistance.

SOLUTION Build the circuit in Figure MSM-8-1. The 741 operational amplifier is found in the **Analog/OPAMP** parts bin. Once the circuit is activated, note that the measured value for the output voltage is $V_o = -9.963$ V.

The voltage gain A is:

$$A = \frac{V_o}{V_i} = \frac{-9.963 \text{ V}}{0.5 \text{ V}} = -19.926$$

Using Ohm's Law, the input current, and the input voltage, the input resistance Rin can be found to be:

$$R_{in} = \frac{V_{in}}{I_i} = \frac{0.5 \text{ V}}{.05 \text{ mA}} = 10 \text{ k}\Omega$$

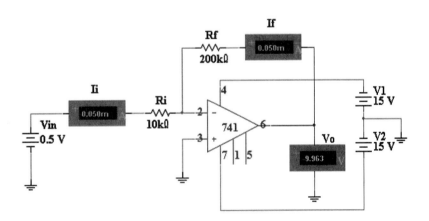

FIGURE MSM-8–1
MultiSIM inverting amplifier circuit used for Example MSM-8-1.

Finally, note that the values for the input current and the feedback current are essentially equal, which supports the ideal op-amp assumption that there is no current flowing into the op-amp itself.

	Calculated	Measured
A	−20	−19.926
Rin	10 kΩ	10 kΩ

EXAMPLE MSM-8-2

Use MultiSIM to build the circuit designed in Example 8-12. Select an input voltage in the linear range and measure the voltage gain.

SOLUTION Build the circuit in Figure MSM-8-2. The input voltage was arbitrarily selected as 0.3 V. The output voltage is measured as 7.526 V. The voltage gain A is:

$$A = \frac{Vo}{Vin} = \frac{7.526 \text{ V}}{0.3 \text{ V}} = 25.09$$

The current flowing in the noninverting input terminal is a small bias current and is not to be interpreted as the value that would result from the input voltage of 0.3 V. Indeed, the input impedance of a noninverting amplifier to a signal voltage is so large that it is difficult to measure.

A comparison of the ideal and actual gain values follows.

	Calculated	Measured
A	+25	+25.09

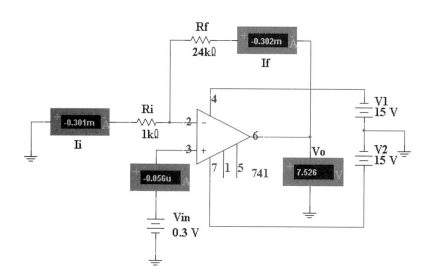

FIGURE MSM-8-2
MultiSIM noninverting circuit used for Example MSM-8-2.

EXAMPLE MSM-8-3

Demonstrate the ability of the voltage follower circuit to prevent a driving circuit from being loaded down by a small load resistance. Use the circuit of Figure MSM-8-3(a) as the driving circuit.

SOLUTION Build the circuit in Figure MSM-8-3(a) and note the value of the voltage across Rout_1 before the load resistance is attached.

$$VRout\_1 = 6 \text{ V}$$

Attach a 1-kΩ load resistor across Rout_1 (Figure MSM-8-3(b)). This load resistance is representative of the input resistance of a follow-on circuit which would be driven by the

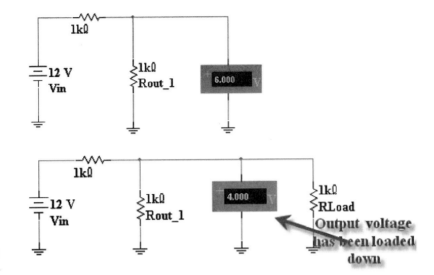

FIGURE MSM-8-3(a)
MultiSIM circuit used for Example MSM-8-3.

FIGURE MSM-8-3(b)
MultiSIM circuit for Example MSM-8-3 which has been loaded down by the load.

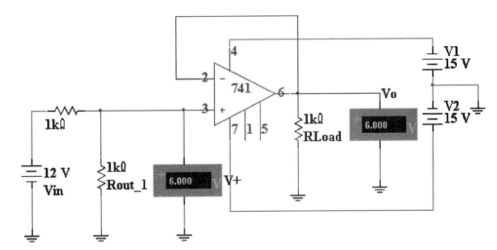

FIGURE MSM-8-3(c)
Voltage follower MultiSIM circuit which is isolating (buffering) the original circuit from the load.

circuit in Figure MSM-8-3(a). Simulate the circuit and note the voltage across the load resistance. Ideally, the output of a driving circuit should not change when the load is attached to it. If the load had been much larger than the 1-kΩ output resistance, the circuit would not have been loaded down and the load would have the original 6 V applied to it. However, in this case, the load resistance was low enough that it loaded the circuit down and the voltage across the load is 4 V instead of 6 volts.

This problem can be solved by placing a voltage follower op-amp circuit between the driving circuit and the load to isolate (buffer) the load from the driving circuit. This will prevent the load from affecting the driving circuit.

Place a voltage follower op-amp circuit between the circuit of Figure MSM-8-3(a) and the load resistor as shown in Figure MSM-8-3(c) and measure the new voltage across the load. Note that now the voltage across the load is the original 6 V that was available from the driving circuit. The load has been isolated (buffered) from the driving circuit.

■ **EXAMPLE MSM-8-4**

Build the circuit used in Example 8-14 (Figure 8-19) and prove that the load current is 5 mA when the circuit is driven by a 5-V dc supply.

SOLUTION Build the circuit in Figure MSM-8-4 in MultiSIM and simulate the circuit. This circuit differs from the previous examples in that the output voltage is not being monitored. This time, the current in the feedback loop is considered to be the output. Because the ideal op-amp assumptions indicate that $\mathbf{I_i = I_f}$, the load current, I_f, is being controlled by the input current I_i, which is being controlled by the voltage across R_i. Again, because

FIGURE MSM-8–4
MultiSIM VCIS circuit used for Example MSM-8-4.

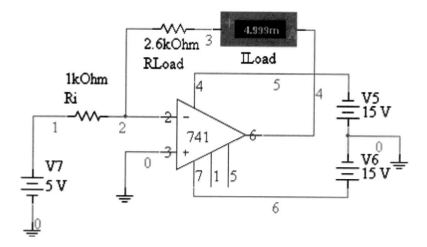

the ideal op-amp assumptions assume that the voltage at node 2 in Figure MSM-8-4 is zero volts due to the virtual short to ground, the input current is:

$$Ii = \frac{Vi - Vnode2}{Ri}$$
$$= \frac{5 - 0}{1 \text{ k}\Omega}$$
$$= 5 \text{ mA}$$
$$= If = ILoad$$

Note that when the circuit of Figure MSM-8-4 is simulated, the measured load current is 4.999 mA. This circuit is a member of a class of devices known as **Voltage Controlled Current Sources** or VCIS.

EXAMPLE MSM-8-5

Using the circuit from Example MSM-8-4, vary the load resistance from 0 Ω to 5 kΩ and show at what value of load resistance the circuit becomes nonlinear.

SOLUTION Build the circuit of Figure MSM-8-5(a). This circuit uses a **Voltage Controlled Resistor** (found in the **Basic/Basic_Virtual** parts bin), a Current Controlled

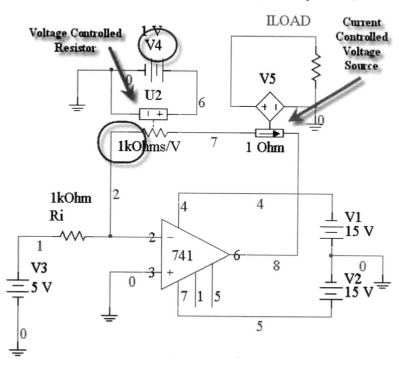

FIGURE MSM-8–5(a)
MultiSIM circuit for Example MSM-8-5.

MultiSIM Examples for Chapter 8   447

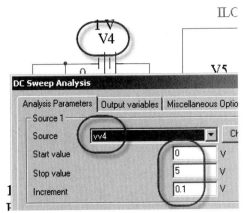

FIGURE MSM-8–5(b)
MultiSIM DC Sweep Analysis parameters window used for Example MSM-8-5.

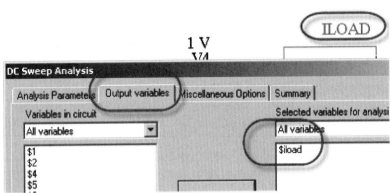

FIGURE MSM-8–5(c)
MultiSIM DC Sweep output variables window for Example MSM8-5.

FIGURE MSM-8–5(d)
MultiSIM DC Sweep–load current versus load resistance plot.

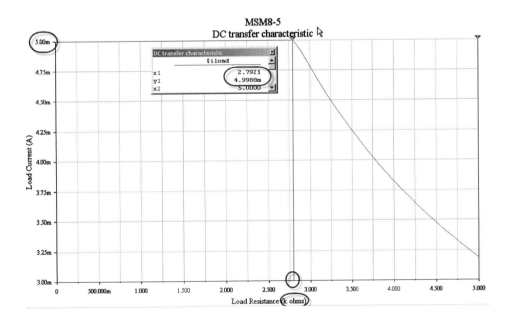

Voltage Source (used previously), and DC Sweep Analysis to vary the load resistance and plot the load current versus load resistance.

The Voltage Controlled Voltage Source, when combined with a battery (as shown in Figure MSM-8-5(a)), will be used to vary the load resistance. The resistor portion of the device is set up to vary 1000 ohms for every volt that the battery changes.

The Current Controlled Voltage Source is used to plot the load current as if it were voltage. The value of the resistor attached to the device is of no importance. It is important that the node number be known. This node was renamed ILOAD.

The DC Sweep Analysis is set up to vary vv4 (stands for voltage V4, where V4 is the battery attached to the Voltage Controlled Resistor (VCR) from 0 V to 5 V). The VCR is set to vary 1 kΩ/V, so this will in turn vary the load resistance from 0 ohms to 5 kΩ (Figure MSM-8-5(a)). The load current node name ($iload) is set into the output variable window (Figure MSM-8-5(b)). Set up the sweep as shown in Figure MSM-8-5(c).

As can be seen in the **DC Sweep** output plot (Figure MSM-8-5(d)), the load current is 5 mA until the load resistance increases above 2.79 kΩ, at which point the output becomes nonlinear. This value is higher than the calculated value of 2.6 kΩ because that value is based on saturation voltages of ±13 volts while *MultiSIM uses saturation voltages of ±14 volts*. Therefore,

$$R_L(\text{max}) = \frac{|V_{\text{sat}}|}{I_L} = \frac{|\pm 14 \text{ V}|}{5 \text{ mA}} = 2.8 \text{ k}\Omega$$

▌ EXAMPLE MSM8-6

Build the circuit from Example 8-15 (Figure 8-23) and then add the necessary components to modify the output equation so that the new equation looks like:

$$V_0 = 10V_1 + 5V_2 - 2V_3$$

SOLUTION The original equation, $V_0 = -10V_1 - 5V_2$ ended up with a circuit which looks like Figure MSM8-6(a). If V1 = 0.1 V and V2 = 0.2 V, the output should be:

$$\begin{aligned} V_0 &= -10V_1 - 5V_2 \\ &= -10(.1) - 5(.2) \\ &= -1 - 1 \\ &= -2 \text{ V} \end{aligned}$$

Now note that in the new equation, not only has the $-2V_3$ term been added, but the sign of the 10 and the 5 also has been changed. This means that the original equation's output must have been passed through an input of a second summing circuit with a gain of -1. The feedback resistance R_{f2} of the second summer is set at 20 kΩ, so R_{i3} can be determined as follows:

$$A = \frac{-R_{f2}}{R_{i3}}$$

$$-1 = \frac{-20 \text{ k}\Omega}{R_{i3}}$$

$$R_{i3} = 20 \text{ k}\Omega$$

This gain of -1 will invert the terms of the original equation. The gain of the third term is -2, so the value of R_{i4} is determined as follows:

$$A = \frac{-R_{f2}}{R_{i4}}$$

$$-2 = \frac{-20 \text{ k}\Omega}{R_{i4}}$$

$$R_{i4} = 10 \text{ k}\Omega$$

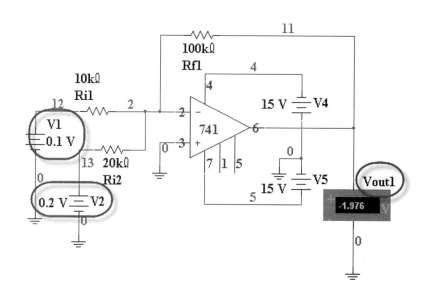

FIGURE MSM-8-6(a)
Beginning MultiSIM circuit building block for Example MSM-8-6.

FIGURE MSM-8–6(b)
Final two-stage MultiSIM summing circuit.

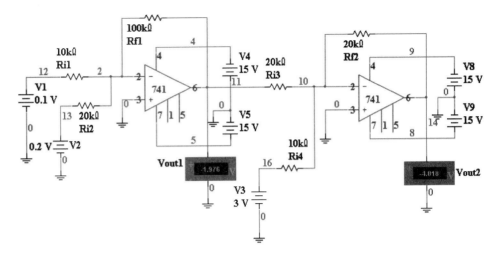

The final circuit is shown in Figure MSM8-6(b). If $V_3 = 3$ V, the final output voltage will be:

$$V_0 = 10V_1 + 5V_2 - 2V_3$$
$$= 10(0.1) + 5(0.2) - 2(3)$$
$$= 1 + 1 - 6$$
$$= -4 \text{ V}$$

MultiSIM Examples for Chapter 9

EXAMPLE MSM-9-1

Use MultiSIM to verify the result of Example 9-10.

SOLUTION Open MultiSIM and select the **Virtual Logic Converter**. The original switching function in Example 9-10 is:

$$f(a,b,c) = a\bar{b}\,\bar{c} + a\bar{b}c + ab\bar{c} + abc$$

For this equation to be entered into the Logic Converter, the switching function must be modified by replacing the bars representing inversion by a (′). The expression will now look like:

$$f(A, B, C) = AB'C' + AB'C + ABC' + ABC$$

Figure MSM-9-1(a) shows the Logic Converter with the expression entered into the expression entry area at the bottom of the converter. Once the expression has been entered, click on the **Algebra to Table** button (Figure MSM-9-1(a)). The result of this action will be a logic table with the min-terms of the expression.

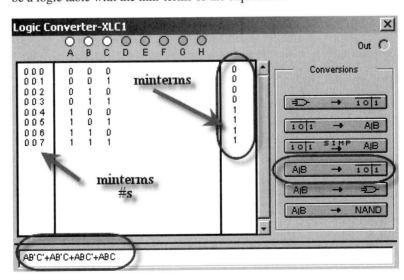

FIGURE MSM-9–1(a)
MultiSIM Logic Converter for Example MSM-9-1.

FIGURE MSM-9–1(b)
MultiSIM Logic Converter with the simplified Boolean expression for Example MSM-9-1.

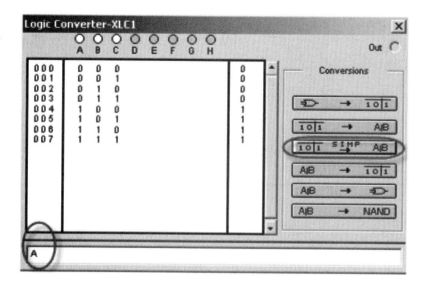

Once the table has been created, click on the **Table to Simplified Algebra** button (Figure MSM-9-1(b)). The simplified expression will appear in the expression entry area. As you can see from the figure, the resulting expression,

$$f(A, B, C) = A$$

matches the result of Example 9-10.

EXAMPLE MSM-9-2

Equation 9-50 lists the following switching function:

$$f(A, B, C, D) = \sum m(3, 4, 7, 10, 11) + \sum d(5, 9, 15)$$

Use MultiSIM to verify the simplified result shown in Equation 9-51.

SOLUTION Select and open the virtual Logic Converter in MultiSIM. When the device opens, select the A, B, C, and D variables at the top of the table area. Figure MSM-9-2(a) shows what the converter looks like at this point.

Note in Figure MSM-9-2(a) that there is a question mark in the result column of the logic table. If you click on a question mark, you can choose among a 0, 1, or an X.

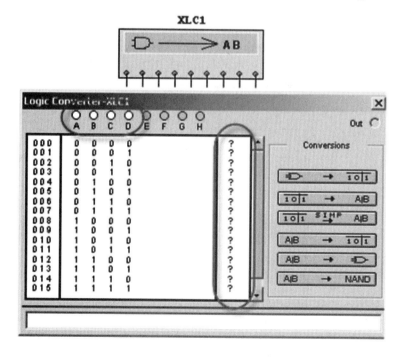

FIGURE MSM-9–2(a)
MultiSIM Logic Converter after selecting the switching variables for Example MSM-9-2.

FIGURE MSM-9–2(b)
MultiSIM Logic Converter with the resulting simplified boolean expression for Example MSM-9-2.

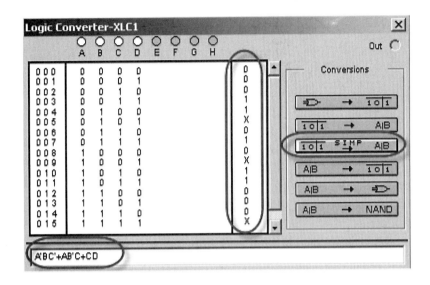

Figure MSM-9-2(b) shows the converter after Equation 9-50 has been entered. Once the table has been completed, select the **Table to Simplified Algebra** button and the simplified expression will appear in the expression entry area. This result matches the result shown in Equation 9-51.

EXAMPLE MSM-9-3

Use MultiSIM to create a logic diagram for Example 9-14 and to simplify the expression.

SOLUTION Select a Logic Converter in MultiSIM and enter the switching function:

$$f(x, y, z) = \sum m(0, 1, 3, 7)$$

(Allowance must be made for the fact that the converter requires substitution of the variables A, B, and C.)

Figure MSM-9-3(a) shows the converter after the switching function has been entered and simplified, as discussed in Example MSM-9-1. You must convert the expression back to one with the x, y, and z variables to compare it with the results of Example 9-14.

To create the logic diagram, click on the **Table to Circuit** button. Figure MSM-9-3(b) shows the logic diagram result.

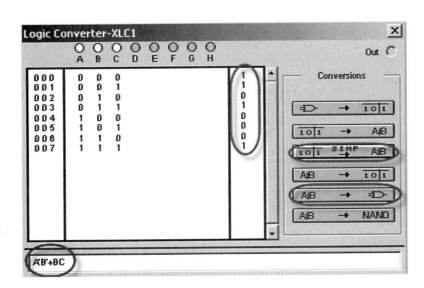

FIGURE MSM-9-3(a)
MultiSIM Logic Converter with the switching function for Example MSM-9-3 entered and simplified.

FIGURE MSM-9–3(b)
Logic diagram built by the MultiSIM Logic Converter for Example MSM-9-3.

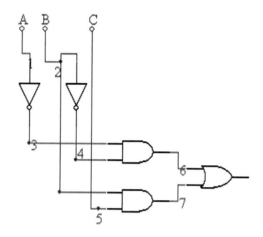

■ **EXAMPLE MSM-9-4**

Use MultiSIM to determine the min-term list and simplified Boolean expression for the circuit in Figure MSM-9-4(a).

SOLUTION Build the circuit in Figure MSM-9-4(a) in MultiSIM. All the logic gates in the circuit can be found in the **TTL / 74LS** parts bin. Each logic gate is actually a member of a set of like gates built on an integrated circuit (IC) chip. For instance, the 74LS00 is a NAND chip containing four 2-input NAND gates. The gates are numbered counter-clockwise starting with Gate A at pins 1, 2, and 3 and ending up with Gate D at pins 11, 12, and 13. If the chip is the first one placed on the schematic it will have the designation U1A, meaning chip #1, gate A. Each time you select a gate to be placed on the schematic, you will have the opportunity to select which of the gates on that chip you want to use or if you want to start with a new chip. Note the gate designations in Figure MSM-9-4(a).

Ensure that the MSB (wire A in Figure MSM-9-4(a)) of the circuit inputs is connected to the far left input of the Logic Converter while keeping the rest of the inputs in order. The input which is on the far right end of the input area of the Logic Converter is reserved for the circuit output. If these inputs are connected in any other order, the answer will be significantly different and the device will not work at all if the output wire is connected to any other input other than the one shown.

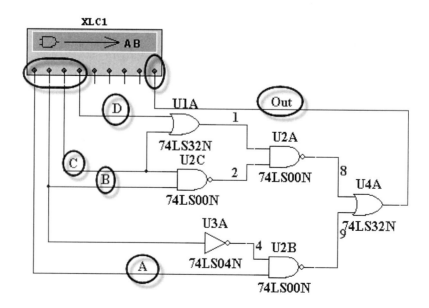

FIGURE MSM-9–4(a)
MultiSIM Logic Converter for Example MSM-9-4.

FIGURE MSM-9–4(b)
Logic Converter with the table and Boolean solutions to the circuit.

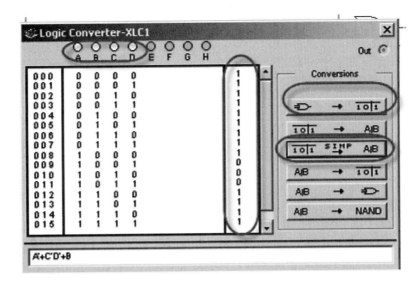

Open the Logic Converter and select the **Gate to Table** button (the uppermost button). The 1s and 0s will appear in the table area. As seen in Figure MSM-9-4(b), the resulting min-term and max-term lists are:

$$f(A, B, C, D) = \sum m(0 - 8, 12 - 15) \Leftarrow \textit{Min-term list}$$
$$= \prod M(9, 10, 11) \quad \Leftarrow \textit{Max-term list}$$

Once the table has been created, select the **Table to Simplified algebra** button (the third one down) and note that the resulting simplified expression is:

$$f(A, B, C, D) = A' + C'D' + B \Leftarrow \textit{MultiSIM's output form}$$
$$= \overline{A} + \overline{C}\,\overline{D} + B \Leftarrow \textit{Normal Boolean result}$$

MultiSIM Examples for Chapter 10

EXAMPLE MSM-10-1

Use MultiSIM to build the full adder introduced in Figure 10-4(a). Use the circuit to validate Table MSM-10-1.

Table MSM-10–1 Full Adder Logic Table for Example MSM-10-1.

Input			Output	
A_i	B_i	C_{i-1}	Sum	Carry
0	0	0	0	0
0	0	1	1	0
0	1	0	1	0
0	1	1	0	1
1	0	0	1	0
1	0	1	0	1
1	1	0	0	1
1	1	1	1	1

SOLUTION Open MultiSIM and build the circuit in Figure MSM-10-1. As in Example MSM-9-4, the gates can be found in the **TTL / 74LS** parts bin. The **Single Pole Double Throw (SPDT)** switches can be found in the **Basic/SWITCH** parts bin. When each one is placed on the schematic, double left-click on it, select the **VALUE** tab, and then select the

FIGURE MSM-10-1
MultiSIM Full Adder circuit for Example MSM-10-1.

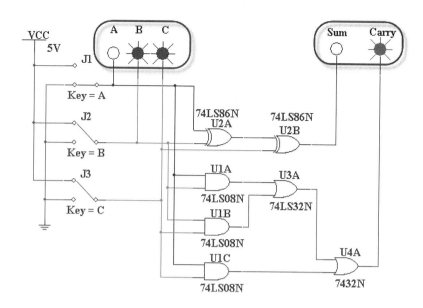

key which will be used to actuate that switch. The three switches in Figure MSM-10-1 are set to be actuated by keys A, B, and C.

The probes can be found in the **Indicators / PROBE** parts bin. There are several colors of probes available. Choose whichever color pleases you. Double-click on each probe and place a descriptive label name on each one, and then ensure that each probe is set up to **Display labels** under the **DISPLAY** tab.

Once the circuit has been constructed, start the simulation and then select the inputs for each row on the input side of Table MSM-10-1 and verify the output columns of the table with the condition of the two output probes for each row.

EXAMPLE MSM-10-2

Use MultiSIM to analyze the Nibble-Adder circuit in Figure 10-6.

SOLUTION The basic unit in binary is the **BI**nary digi**T** or **BIT**. A binary **byte** is 8 bits. A **nibble** is half of a byte, or 4 bits. Because the highest HEX (base 16) digit is F_{16}, which can be represented in binary as 1111_2, a nibble can be represented by a single hex digit per nibble.

Figure MSM-10-2(a) is a circuit which will add two nibbles or hex digits and will output a hex **Sum** digit and a **Carry** digit, each of which will either be a 0 or a 1. The maximum result from this machine is $1F_{16} (31_{10})$.

The **Half-Adder** and **Full-Adder** in Figure MSM-10-2(a) can be found in the **Misc. Digital / TIL** parts bin. This parts bin contains virtual digital parts instead of specific parts from specific manufacturers. Some of the parts in this bin are not available as standard components, but they are useful for analyzing digital processes.

The **Word Generator (WG)** in Figure MSM-10-2(a) can be found with the Virtual instruments. It is arguably one of the most useful digital instruments in MultiSIM. It can be programmed to provide multiple 32-bit binary words (4 bytes or 8 nibbles) to a circuit to be used to drive a digital circuit. The WG will also provide a clock signal to which each of the output words is synchronized. To fully understand all the capabilities of the WG, it is important to study the MultiSIM user manual. This example will use only a small part of the device.

Figure MSM-10-2(b) is a snapshot of a small portion of the WG programming window. The right portion of the window is the actual programming window and is currently programmed to provide four hex pairs to the Nibble-Adder circuit. Once the program is entered, identify the final program line so that the program will not continue into the unprogrammed area. Do this by right-clicking in the left column of the last programmed row. When this is done, the window in Figure MSM-10-2(c) will appear. Select **Set Final Position** from this menu. This window can also be used to set the current location of the program pointer (or cursor).

MultiSIM Examples for Chapter 10   455

FIGURE MSM-10–2(a)
MultiSIM Nibble-Adder circuit used for Example MSM-10-2.

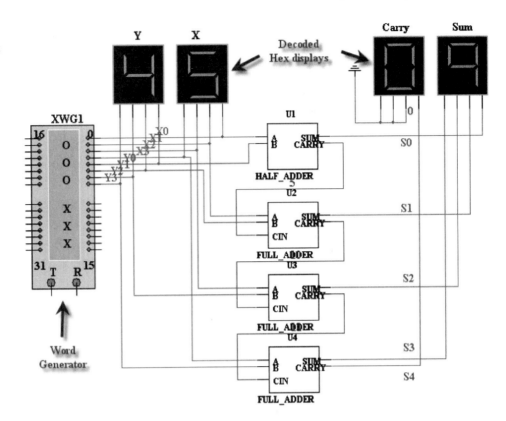

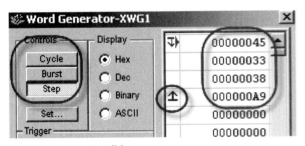

FIGURE MSM-10–2(b)
Continuation of Example MSM-10-2.

FIGURE MSM-10–2(c)
Continuation of Example MSM-10-2.

FIGURE MSM-10–2(d)
Continuation of Example MSM-10-2.

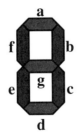

FIGURE MSM-10–2(e)
Continuation of Example MSM-10-2.

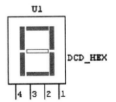

FIGURE MSM-10–2(f)
Continuation of Example MSM-10-2.

The leftmost region in Figure MSM-10-2(b) is the **Controls** section. This area defines how the program will run. The current setting is **Step** mode. In this mode the user controls the speed of the program by being able to **single-step** through the program by using the **Pause/Resume** button next to the simulation switch (Figure MSM-10-2(d)).

Many of the uses of digital circuits require the display of digital or hex digits. One way is through the use of the seven segment display (Figure MSM-10-3(e)). This device consists of seven individual LED (light emitting diode) segments which can be used to

FIGURE MSM-10–2(g)
Result from the second program line.

FIGURE MSM-10–2(h)
Result from the third program line.

FIGURE MSM-10–2(i)
Result from the fourth program line.

represent alphanumeric characters by selecting which segments are on or off. The basic seven segment display has seven inputs (one for each segment) and a common connection to which either a ground or V_{CC} is connected (depending on the LED configuration used). This device normally requires a seven-segment decoder/driver to drive each digit and supply the necessary logic signals to control each segment. Because the largest digit that can be displayed is a hex F (15_{10}) and it takes 4 bits to represent a 15_{10}, each driver would have four inputs.

Although this type of device is available in MultiSIM, it is easier to use a seven-segment display which has already been decoded. The Hex displays shown in Figure MSM-10-2(a) are examples of the decoded variety. They will be the primary display used in the digital examples in this book. Note that the device has four inputs instead of the seven inputs of the undecoded display.

Build the circuit in Figure MSM-10-2(a) and program the word generator as shown in Figure MSM-10-2(b). Single-step through each of the four program lines and observe the relationship between the input nibbles and the output as shown in Figures MSM-10-2(a), MSM-10-2(g), MSM-10-2(h), and MSM-10-2(i).

Note the way in which the display represents hex digits, which can be confused with other digits. For instance, compare the 6_{16} represented in Figure MSM-10-2(g) with the b_{16} represented in Figure MSM-10-2(h). Note that segment a is used for the 6 but it isn't for the b. If the upper-case B were used instead of the lower-case b, there would be no way to tell the difference between the B and an 8. This problem exists for other hex digits so the display represents the hex numbers 0 – F as (0, 1, 2, 3, 4, 5, 6, 7, 8, 9, A, b, C, d, e, F).

EXAMPLE MSM-10-3

Implement a 3:8 decoder with active low outputs and an active low chip enable by using two 2:4 decoders and an inverter.

SOLUTION Construct the circuit in Figure MSM-10-3(a) in MultiSIM. The **74LS139N** is a dual 2:4 decoder with active low chip enable and outputs. The **74LS139N** and the **74LS04N** can be found in the **TTL/74LS** parts bin. Note that the circuit is set up so that both decoders cannot be enabled at the same time because of the NOT gate.

Program the Word Generator (WG) as shown in Figure MSM-10-3(b) by selecting the **Set** button, at which point the window shown in Figure MSM-10-3(c) will open. Because it is desired to program in a count sequence from 0 to 7, it will be necessary to have 8 programmed rows. Set 08 in the **Buffer Size** block and select **Up Counter** from the **Pre-set Patterns**. Once these two settings have been made, select **Accept** and the window will close and the count sequence will be automatically programmed into the **WG** between rows 0 and 7. Finally, select **Step** mode and close the window.

Simulate by using the **Pause/Resume** button to step through the preprogrammed sequence. Note that only one output is active (LOW) at a time.

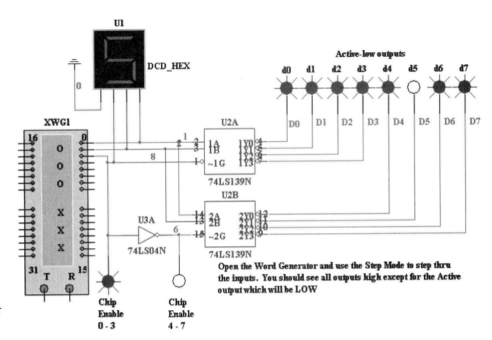

FIGURE MSM-10-3(a)
MultiSIM 3:8 decoder circuit for Example MSM-10-3.

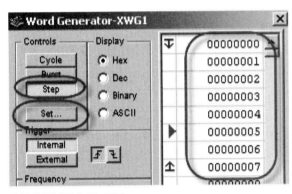

FIGURE MSM-10-3(b)
Continuation of Example MSM-10-3.

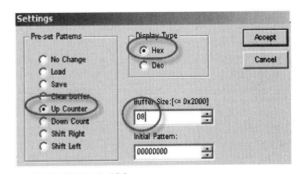

FIGURE MSM-10-3(c)
Continuation of Example MSM-10-3.

EXAMPLE MSM-10-4

Analyze how a **74SL147 Priority Encoder** responds to various inputs. Use a hex display to display the status of the output. Observe the relationship between the output and the inputs.

SOLUTION Construct the circuit in Figure MSM-10-4(a) in MultiSIM. The **74LS147N Priority Encoder** can be found in the **TTL/74LS** parts bin. Note that because the 74LS147 has active low outputs, it is not possible to directly interface a decoded hex display with the output. To create the interface, place a **NOT gate** on each output, which in effect will convert each output to active high from the viewpoint of the display.

This device has nine inputs to drive, so it will be necessary to program the two least significant nibbles in the Word Generator (WG), plus the least significant bit on the next nibble. This third nibble will be limited to either a 1 or a zero. The two lower nibbles can be programmed with anything between a 0_{16} and an F_{16}. Program the **WG** as shown in Figure MSM-10-4(b).

Finally, it would be nice to have the program cycle automatically through the program but stop at points of specific interest. This can be done by placing a **breakpoint** at any program line which you want to observe in detail. This is done by right-clicking in the left column beside the row of interest and selecting **Set Breakpoint** from the list that appears. Figure MSM-10-4(b) has three breakpoints set.

Finally select **Cycle** mode and the program will automatically start to cycle through the program steps. The speed at which the program cycles can be controlled by changing the frequency setting in the Word Generator. When a line with a breakpoint is reached, observe the input/output relationship, and then select the **Pause/Resume** button to allow the program to continue.

Note in Figure MSM-10-4(a) that while inputs 5, 6, and 7 are all Active (low), the highest-priority active input is input 8, which is displayed on the output display.

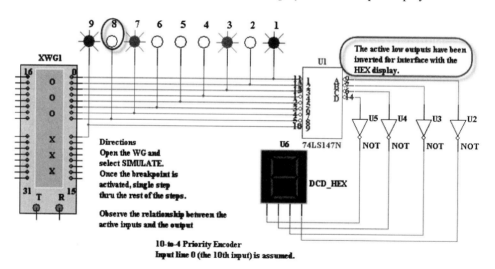

FIGURE MSM-10-4(a)
MultiSIM priority encoder circuit for Example MSM-10-4.

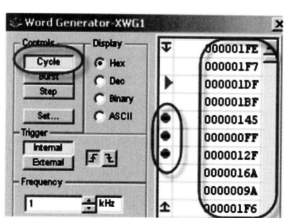

FIGURE MSM-10-4(b)
Continuation of Example MSM-10-4.

EXAMPLE MSM-10-5

Use a **74LS153 Multiplexer** to create a device which allows the user to select between four available 1-kHz, TTL clock sources, each with its own duty cycle, and route it to the output for use by other circuits.

SOLUTION The **duty cycle** of a rectangular TTL-compatible waveform is defined as the time that the waveform is at its high level (T_H) divided by the period (T) of the waveform. The duty cycle is usually provided as a percentage.

$$\text{Duty Cycle}(\%) = \frac{T_H}{T} \cdot 100\%$$

Build the circuit in Figure MSM-10-5(a). The four sources provide a special type of rectangular waveform known as a TTL compatible clock waveform. It can be found in the **Sources/SIGNAL_VOLTAGE_SOURCES** parts bin and is called a **CLOCK_VOLTAGE**. Once each source is placed in the schematic area, double-click on it. The window shown in Figure MSM-10-5(b) opens up. This device is designed to provide a clock signal to TTL devices; thus, it defaults to a rectangular waveform which cycles between 0 V (logic 0) and 5 V (logic 1) with a duty cycle of 50 percent. If the duty cycle remains at 50 percent, this waveform could be described further as a **Square Wave**. Set the duty cycle of the four sources to 50 percent, 25 percent, 75 percent, and 10 percent as shown in Figure MSM10-5(a).

Recall from earlier work that an Oscilloscope (o-scope) is a device which will measure a voltage waveform with respect to time and display the waveform on a CRT display. Multi-SIM has two o-scopes: a two channel and a four channel. The circuit in Figure MSM10-5(a) uses the two-channel o-scope to analyze the duty cycle of the signal on the output of the multiplexer (MUX). When you double-click on the o-scope, the signal display window will open. The bottom portion of the window is shown in Figure MSM-10-5(c). The time scale is adjusted in the time base section and the amplitude of each channel is adjusted in the Channel A and B sections. The Channel sections each have an AC, 0, and DC button.

- AC: Blocks all dc from the signal being displayed. The resulting signal will rotate around the signal base line.
- 0: Grounds out the channel input so that the signal base line can be adjusted.
- DC: Allows the signal to remain unchanged. This is the option that is used in this example.

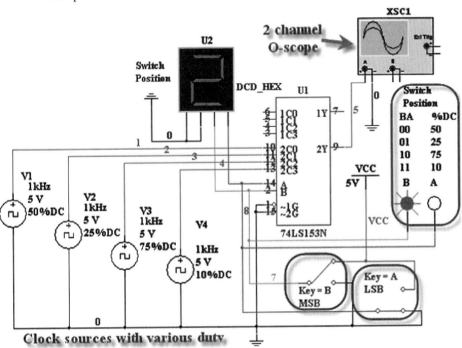

FIGURE MSM-10-5(a)
TTL Duty cycle selector circuit for Example MSM-10-5.

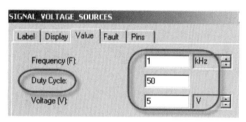

FIGURE MSM-10–5(b)
Continuation of Example MSM-10-5.

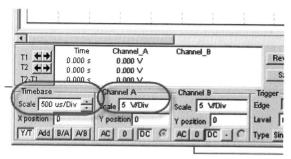

FIGURE MSM-10–5(c)
Continuation of Example MSM-10-5.

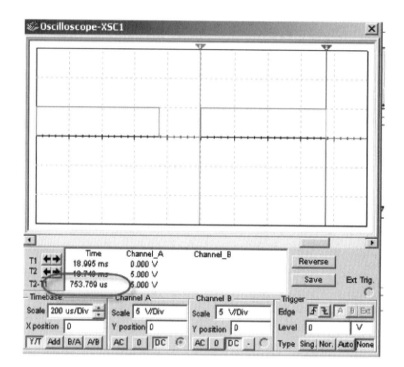

FIGURE MSM-10–5(d)
O-scope measurement of time low portion of duty cycle.

Further information about the o-scope can be found from the MultiSIM **Help** file or from the program documentation.

Simulate the circuit and select the output to be routed to the MUX output with the two switches. Figure MSM-10-5(a) indicates that signal number 2, the one with the 75 percent duty cycle waveform, has been selected.

Figures MSM-10-5(d) and (e) show the selected waveform with cursors set to measure T_H and T. The measured values are:

	Measured
Time High, T_H	753.769 μs
Period, T	1.005 ms

$$\text{Duty Cycle}(\%) = \frac{753.769 \ \mu s}{1.005 \ ms} \times 100\% = 75.00\%$$

The simulation result supports the expected duty cycle of 75 percent.

FIGURE MSM-10–5(e)
O-scope measurement of time high portion of duty cycle.

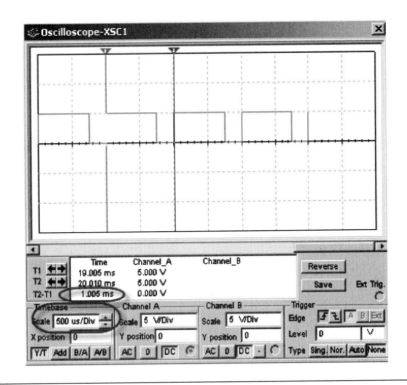

MultiSIM Examples for Chapter 11

EXAMPLE MSM-11-1

Use MultiSIM to create three separate circuits:

- A **D-FF** circuit
- A **T-FF** circuit
- A **JK-FF** circuit

Use a **74LS76 JK-FF** and any other necessary gates and devices to implement each circuit. Analyze each circuit's response to different sequences of inputs. Pay special attention to the relationship between the input signal and the clock.

SOLUTION

(1) Implement the **T-FF** circuit in Figure MSM-11-1(a). The two inputs of the **JK-FF** have been shorted together to create a single leading edge triggered **T-FF**. The frequency of the

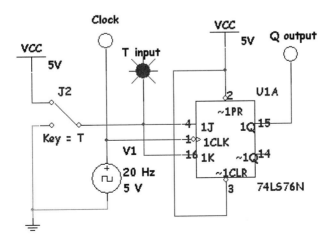

FIGURE MSM-11–1(a)
MultiSIM T-FF tutor for Example MSM-11-1.

FIGURE MSM-11-1(b)
MultiSIM D-FF tutor for Example MSM-11-1.

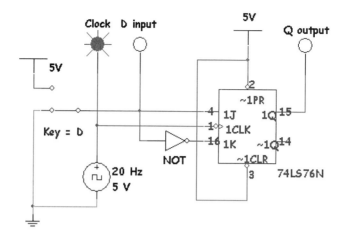

clock has been set low enough so that there is time to analyze the response of the circuit to changes in the **T** input. The **PRE** and **CLR** inputs are kept inactive for this example.

Simulate the circuit and observe the output as you change the condition of the **T** input. Note that the output will change with the **trailing (High-to-Low) edge** of the clock as long as the **T** input is set to a logic 1. This condition is called the device's **TOGGLE** mode. Once the **T** input is set to a logic 0, the output remains in a **HOLD** mode and the output will remain unchanged with each clock pulse. Because the effect of the **T** input on the circuit is controlled by the clock edge, the **T** input is known as a **synchronous** input.

(2) Implement the **D-FF** circuit in Figure MSM-11-1(b). A **NOT** gate is combined with the **JK-FF** to implement the **D-FF**. Again, the **PRE** and **CLR** inputs are kept inactive in this circuit.

Simulate the circuit and observe the output as you change the condition of the **D** input. Note that the output will **follow the input** at each **trailing edge** of the clock. When $D = 1$ the output will **SET** on the trailing edge of the clock, while when $D = 0$, the output will **RESET**. Because the effect of the **D** input on the circuit is controlled by the clock edge, the **D** input is a synchronous input.

(3) Implement the **JK-FF** circuit in Figure MSM-11-1(c). In addition to observing the effect of the **J** and **K** inputs on the output, the effect of the **PRE** and the **CLR** inputs on the output also will be analyzed.

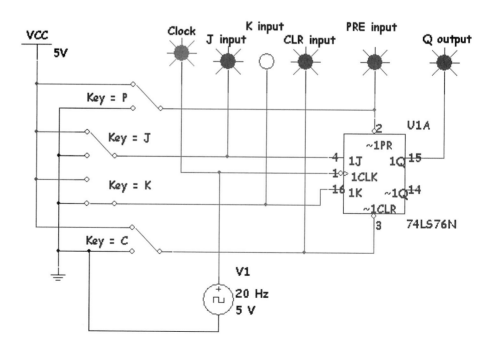

FIGURE MSM-11-1(c)
MultiSIM JK-FF tutor for Example MSM-11-1.

Simulate the circuit. The **PRE** and **CLR** are active low inputs; therefore, start out with both of them in their inactive states by placing a logic 1 on each of them. Observe the condition of the output as each of the four combinations of the **J** and the **K** inputs is applied to the circuit. The results of these inputs can be seen in the table below. Also note that the effect of these two inputs is again delayed until the trailing edge of the clock. Because the **J** and the **K** inputs depend on the clock to affect the output, they are synchronous inputs.

J	K	Present State Q_P	Next State Q_N	Mode
0	0	0	0	HOLD
0	0	1	1	
0	1	0	0	RESET
0	1	1	0	
1	0	0	1	SET
1	0	1	1	
1	1	0	1	TOGGLE
1	1	1	0	

Note that besides being controlled by the clock, the effect of the two inputs on the circuit's output also depends on the current or **PRESENT STATE**, Q_P of the output. The table shows that an accurate prediction of the state that the output will be in after the trigger event (the trailing clock edge) occurs based on the condition of the **J** and **K** inputs and the present state of the output. Also note that there are conditions under which either the **J** or the **K** input has no effect on the output. These conditions also depend on Q_p. When a FF input has no effect on the output, it is considered to be in a **Don't Care** state. The previous table can be modified into what is known as a State Transition Table, as shown in the table below. The Don't Care condition is represented by an **X**.

JK Transition Table

Q_P	→	Q_N	J	K
0	→	0	0	X
0	→	1	1	X
1	→	0	X	1
1	→	1	X	0

Simulate the circuit again with the **JK Transition** table in mind and verify the concept.

Now, place a logic 1 on both **J** and **K** and then toggle either the **PRE** or the **CLR** inputs. Observe the effect of this action on the output. Pay special attention to the relationship of any state change with the clock. Note that the effect of these inputs on the output is *independent of the clock*; therefore, they are asynchronous inputs. The results are summarized in the the table that follows.

PRE*	CLR*	Present State Q_P	Next State Q_N
0	1	0	1
0	1	1	1
1	0	0	0
1	0	1	0
1	1	0	0
1	1	1	1

*Based on active low PRE and CLR inputs

■ EXAMPLE MSM-11-2   Modify the 4-bit serial shift register in Figure 11-20 by adding an inverted feedback loop from the last stage of the shift register to the **D** input of the first stage. Use MultiSIM's Logic Analyzer to observe how the circuit responds.

SOLUTION   Implement the circuit in Figure MSM-11-2(a). The original circuit in Figure 11-20 required some kind of input device for the first **D** input. Once the value placed on **D** had rippled down to the end of the register, the sequence was complete and all Q outputs were high. By adding the inverted feedback loop into the circuit, the need for a separate input device (like a switch) is no longer needed.

This example is the first occasion where the Logic Analyzer is used. This device has 16 input channels and is designed only to measure digital signals. When it is connected to the circuit, double-click on it to open the display/setup window. At the bottom of the window in the Clock section there is a button labeled **Set**. Once this button is selected, the **Clock Setup** window shown in Figure MSM-11-2(b) opens. The main item of interest in this window is the **Clock rate** selection. *This setting MUST be set to 10 times the frequency of the circuit clock.* For example, because the frequency of the circuit clock in Figure MSM-11-2(a) is 100 Hz, this setting must be changed from its default value of 10 kHz to 1 kHz. Not performing this change is a common mistake, and it may introduce errors if not performed.

The traces on the Logic Analyzer will take on the node name and wire color of the wire that is connected to each input, as seen on the right side of Figure MSM11-2(c). Also note the **Clocks/Div** setting of 12 in the **Clock** section of the setup area at the bottom of the display screen. This value defaults to 1 but can be increased to compact the displayed window to encompass more of the signal. Once you have simulated this circuit, observe the screen as this value is increased and decreased.

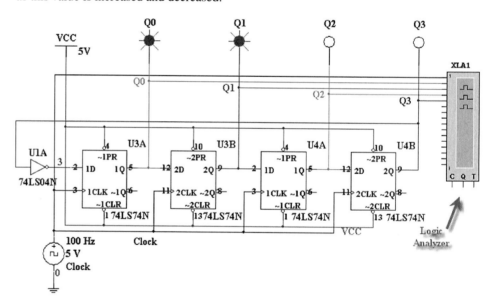

FIGURE MSM-11-2(a)
MultiSIM 4-bit serial shift register with feedback.

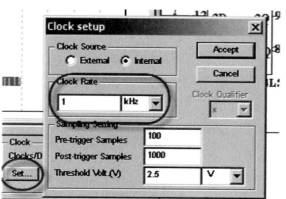

FIGURE MSM-11-2(b)
Logic Analyzer clock setup window.

FIGURE MSM-11–2(c)
Logic Analyzer output for Example MSM-11-2.

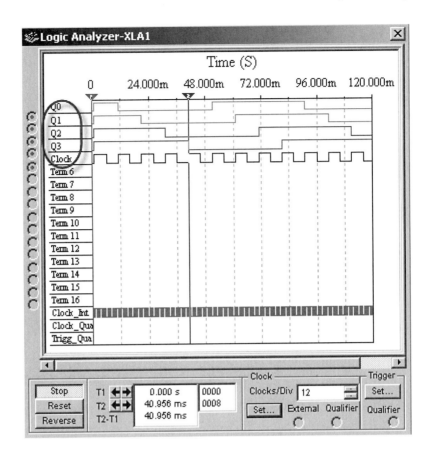

Simulate the circuit and observe the sequence of lights on the output probes. The circuit starts out with all outputs at logic 0. This means that the value on the first **D** input is a logic 1 (the logic 0 on **Q3** has been inverted and fed back to the first **D** input). This logic 1 value will shift to the right, turning each Q output high until all outputs are high. At this point the logic 1, which is now on **Q3**, is fed back and inverted to place a logic 0 on the first **D** input and then a logic 0 begins to shift to the right, filling each output with a logic 0 until all outputs are low. At this point, the process starts all over again.

Note that this circuit consists of leading edge **D-FFs** so, as Figure MSM-11-2(c) indicates, the logic transitions occur on the leading edge of each clock pulse instead of the trailing edge, as observed in the previous example.

To observe or modify the format of the Analyzer graph, use the Grapher, which was used earlier in the MultiSIM example section, to observe all graphs produced by any device which produces a plot or graph.

EXAMPLE MSM-11-3

Use MultiSIM to analyze the 4-bit binary ripple counter shown in Figure 11-22 but use **JK-FFs** configured as **T-FFs**.

SOLUTION The **JK FFs** in Figure MSM-11-3(a) are configured as **T-FFs** with their T inputs permanently set to logic 1s. As can be seen, each succeeding **T-FF** stage is clocked by the $\overline{Q}$ output of the preceding stage. The only stage which is directly driven by the circuit clock is the first stage. Circuits which share this characteristic are asynchronous circuits.

When connecting the circuit, ensure that the Clock Rate of the Logic Analyzer is set to 10 times the circuit clock rate, or 1 kHz as shown in Figure MSM-11-3(b). Simulate the circuit and observe the count sequence on the hex display. As seen in Figure MSM-11-3(b), this counter counts down from F to 0 and then starts over. If it were desired to have the counter count up instead of down, all that would be necessary would be to have the last three stages clocked by the Q output instead of the $\overline{Q}$ output.

FIGURE MSM-11–3(a)
MultiSIM 4-bit binary ripple counter.

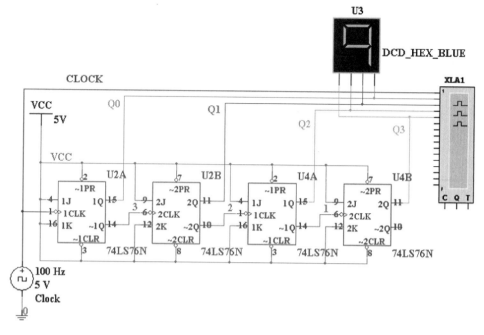

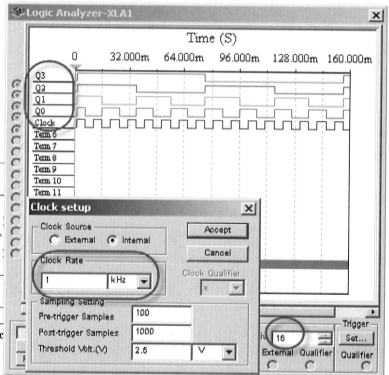

FIGURE MSM-11–3(b)
Logic Analyzer output for Example MSM-11-3.

Open the Logic Analyzer and set the **Clocks/Div** setting to 16. Note that the transitions in this circuit occur on the trailing edge of the clock since these **JK-FF**s are trailing edge devices.

To create a formatted graph of this output, open the Grapher and note that all simulation runs which have been performed on this example are also available there.

EXAMPLE MSM-11-4 Use MultiSIM to analyze the 4-bit synchronous binary counter shown in Figure 1-24.

SOLUTION Implement the circuit in Figure MSM-11-4(a). The **T-FF**s and the **AND** gates can be found in the **Misc. Digital / TIL** parts bin. This **T-FF** has active high **SET** and

FIGURE MSM-11–4(a)
MultiSIM 4-bit synchronous binary counter.

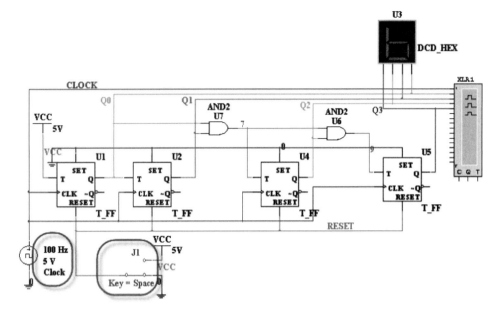

RESET (also known as **PRE** and **CLR** respectively) inputs. This example will use the **RESET** inputs to reset the counter. As seen in the circuit diagram, all the clock inputs are driven by the single clock source. Circuits which exhibit this characteristic are synchronous circuits.

Simulate the circuit and set the switch to place a logic 0 on the **RESET** line. Simulate the circuit and note that the circuit counts from **0** to **F** and then starts over. The circuit will reset to a 0 when a logic 1 is placed on the reset line.

Open the Logic Analyzer and note that the state transitions occur on the leading edge of the clock (Figure MSM-11-4(b)).

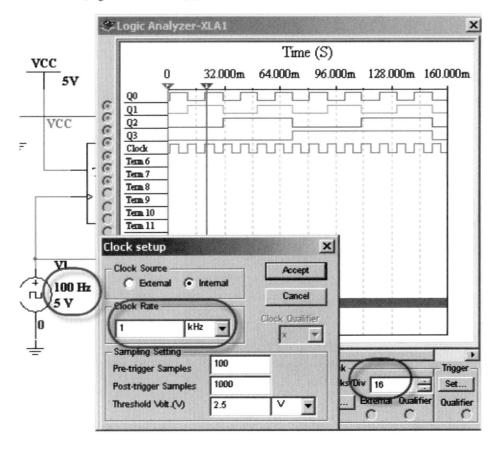

FIGURE MSM-11–4(b)
Logic Analyzer output for Example MSM-11-4.

EXAMPLE MSM-11-5

Modify the circuit in Example MSM-11-4 to count from 0 to 5 and then start over again.

SOLUTION Implement the circuit in Figure MSM11-5(a). This circuit is essentially the same as in Example MSM-11-4 with the exception of the **AND** gate, which has replaced the switch. The purpose of the **AND** gate is to **DECODE** the output and produce a logic 1 when Q2 AND Q1 are logic 1s. These values were chosen because the counter needs to reset on a 6 (0110_2). Of course, you could use a 4-input unit with its inputs tied to $\overline{Q_3}Q_2Q_1\overline{Q_0}$, but it is unnecessary. In the count from 0000_2 to 0110_2, the very first time that both Q_2 and Q_1 are high is on 0110_2.

Simulate the circuit and note that the count sequence is 0–5 and then it starts over again. This is a total of six states. This type of circuit is known as a **MOD 6** counter. Open the Logic Analyzer and observe the timing diagram (Figure MSM-11-5(b)). The AND gate produces a pulse which is too short for the Logic Analyzer to pick up, so the **RESET** line was not monitored.

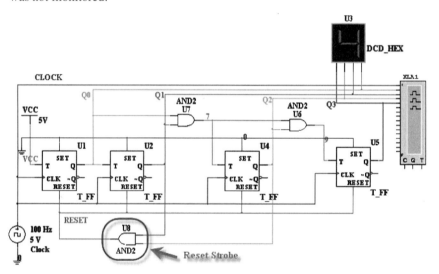

FIGURE MSM-11-5(a)
MultiSIM mod 6 synchronous counter.

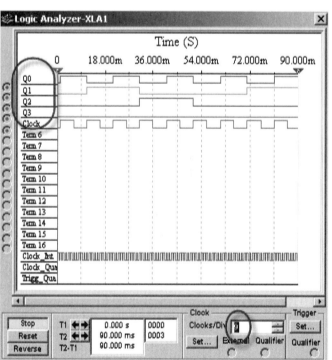

FIGURE MSM-11-5(b)
Logic Analyzer output for Example MSM-11-5.

MultiSIM Examples for Chapter 13

In Chapter 13, we learned that three-phase sources can be modeled by connecting three single-phase sinusoidal sources in either a delta or a wye configuration as desired and phasing them at 120° with respect to each other. When simulating three-phase sources in MultiSIM, it is often convenient to model them in the same way, that is, as a connection of three single-phase sources, phased at 0°, 120°, and 240°.

MULTISIM EXAMPLE 13-1 A three-phase 4-wire wye source is constructed using three 120-V ac sources, each phased at 120° with respect to the other two. Construct a MultiSIM circuit that will demonstrate the phase voltages and line voltages.

SOLUTION Figure MSM-13-1 shows three 120-V sources Va, Vb, and Vc, connected in a wye configuration. The sources have phase angles of 0°, 120°, and 240°, respectively. Voltmeters have been connected to measure the three phase voltages VPHASEa, VPHASEb, and VPHASEc, and the three line voltages VLINEab, VLINEbc, and VLINEca.

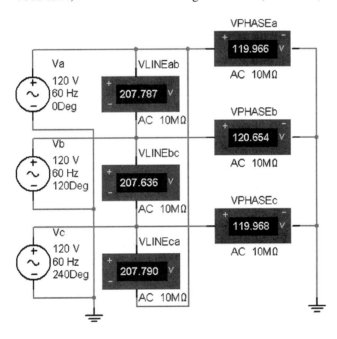

FIGURE MSM-13-1
MultiSIM Example 13-1.

MULTISIM EXAMPLE 13-2 A delta-connected 416-V three-phase source is powering a wye-connected balanced load consisting of three 10-Ω resistors. Devise a MultiSIM simulation of this circuit to verify the three-phase power equations

$$P = 3V_{\text{phase}}I_{\text{phase}}\cos\theta$$

and

$$P = \sqrt{3}\, V_{\text{line}}I_{\text{line}}\cos\theta$$

SOLUTION Figure MSM-13-2 shows the MultiSIM circuit. In this case (as with other three-phase delta sources in later MultiSIM examples), notice that a small (1 milliohm) resistance has been added in series with each of the three sources. In MultiSIM, ac sources are considered to be perfect; that is, they have zero internal impedance. When sources are connected in a delta configuration, the phasor-sum of any two of the sources must be equal to the third. When a program such as MultiSIM performs these calculations, small roundoff errors can cause these sums to be slightly unequal, thereby creating a mathematical error and causing the simulation to display an error message. To avoid these types of problems, a small value resistor is inserted in series with each of the

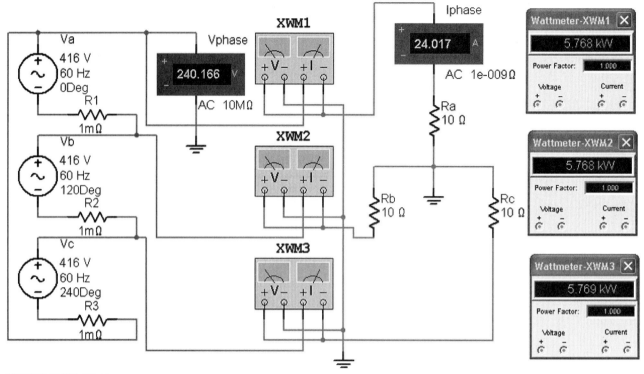

FIGURE MSM-13-2
MultiSIM Example 13-2.

sources, thereby providing a place for this error voltage to be dropped. To assure that the resistors do not adversely affect the theoretical results, the resistor values have been made very small.

In the MultiSIM simulation shown in Figure MSM-13-2, wattmeters have been connected to each of the three load resistors. Therefore, the total power is the sum of the wattmeter indications, or 17.29 kW. Notice that we have grounded the neutral of the wye load. Because the source does not have a neutral, this is normally not necessary. However, because MultiSIM requires at least one ground point in every circuit, this is a convenient place to locate the ground.

The loads are resistive, so the power factor is $\cos \theta = 1$. Using the phase current and phase voltage indications on the ammeter and voltmeter, the total power is

$$P = 3V_{\text{phase}} I_{\text{phase}} \cos \theta = 3 \times 240.17 \times 24 \times 1 = 17.29 \text{ kW}$$

The line voltage is given (416 V), and because the load is wye connected, the phase currents and line currents are equal. Therefore, the total power is also

$$P = \sqrt{3} \, V_{\text{line}} I_{\text{line}} \cos \theta = \sqrt{3} \times 416 \times 24 \times 1 = 17.29 \text{ kW}$$

MultiSIM Examples for Chapter 14

When performing MultiSIM analyses of transformer circuits, we generally use MultiSIM's virtual transformer. Once the transformer has been placed on the MultiSIM screen, we can double-click the mouse on the transformer, which will open the transformer's parameter window. In this window, the turns ratio can be set to any desired value. Although other transformer parameters can also be adjusted using this window, generally they are used for applications requiring specific winding inductance and resistance values. For the applications in this text, these values are left at their default values.

MULTISIM EXAMPLE 14-1

A transformer has a turns ratio of 8.49. It has a 240-V source connected to the high side winding and a 2-Ω load connected to the low side winding. Find the high side current and the low side current and voltage.

SOLUTION In the MultiSIM simulation in Figure MSM-14-1, notice that the high side and low side currents are 1.67 A and 14.1 A, respectively, and the low side voltage is 28.2 V ac.

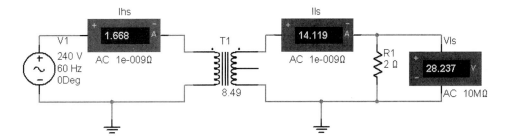

FIGURE MSM-14-1
MultiSIM Example 14-1.

MULTISIM EXAMPLE 14-2

A three-phase transformer with a turns ratio of 2 is excited by a 13.4-kV delta-connected source. The transformer converts a delta system to a wye system. Devise a MultiSIM simulation to find the low side line voltage and phase voltage.

SOLUTION The MultiSIM implementation of this circuit is shown in Figure MSM-14-2. In this case, notice that, as in earlier examples, a small (1 milliohm) resistance has been added in series with each of the three sources. In MultiSIM, ac sources are considered to be perfect; that is, they have zero internal impedance. As explained in an earlier example, when sources are connected in a delta configuration, the phasor-sum of any two of the sources

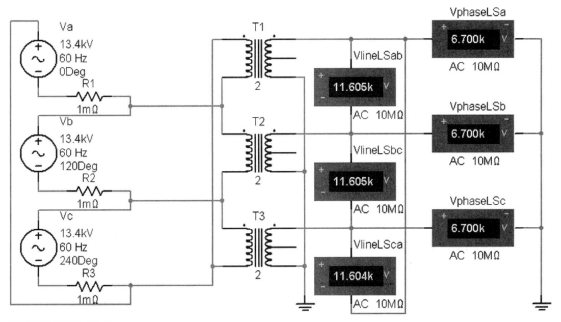

FIGURE MSM-14-2
MultiSIM Example 14-2.

must be equal to the third. When a program such as MultiSIM performs these calculations, small roundoff errors can cause these sums to be slightly unequal, thereby creating a mathematical error and causing the simulation to error out. To avoid these types of problems, a small value resistor is inserted in series with each of the sources, thereby providing a place for this error voltage to be dropped. To assure that the resistors do not adversely affect the theoretical results, the resistor values have been made very small.

The circuit uses three individual 13.4 kV sources with phase angles of 0°, 120°, and 240°, connected in a delta configuration. The three-phase transformer is represented by three individual single-phase transformers, with the high sides connected in delta and the low sides connected in wye with a neutral (which is grounded). Six ac voltmeters are connected to the low side to indicate the three line voltages and the three phase voltages.

■ MULTISIM EXAMPLE 14-3

The low side winding of a standard 120 V − 24 V (a = 5) transformer is connected in series (aiding) with the high side winding to form a boost-type autotransformer. The autotransformer is powering a 36-Ω load. Construct a MultiSIM simulation to find total input current, primary current, secondary current, apparent power delivered to the load, and the power rating of the transformer.

SOLUTION In the MultiSIM simulation shown in Figure MSM-14-3, the transformer T1 is connected in an autotransformer boost configuration. Three ammeters are connected on the input side to indicate the total input current Itotal, and the currents flowing into the primary (Ipri) and secondary (Isec) windings. On the load side, meters have been connected to measure the load current Iload) and load voltage (Vload). Two wattmeters are used in this circuit to measure the power into the primary of the transformer (XWM2) and the power dissipated by the load (XWM1). Because the load is resistive, the power factor will be unity (1), thereby making the real power and apparent power equal in this circuit. The primary circuit wattmeter XWM2 measures the power passing through the transformer core, which is also the minimum power rating of the transformer.

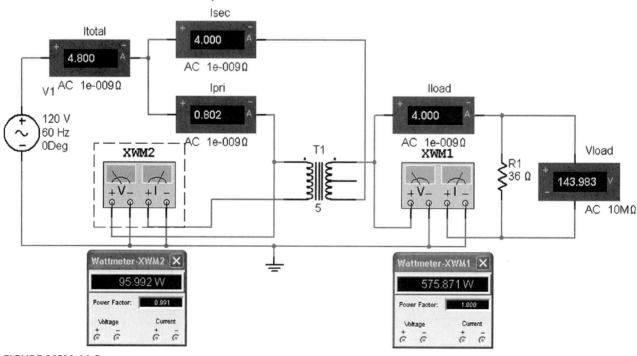

FIGURE MSM-14-3
MultiSIM Example 14-3.

■ MULTISIM EXAMPLE 14-4 This example illustrates the current conversion that is possible using a current transformer (CT). A CT can be simulated in MultiSIM using a transformer with an ammeter connected to the high side winding (note that the low-current winding in a CT is the high side voltage winding). In this example, a 500:5 CT is connected in series with a 1-Ω load that is powered by a 175-V ac source. We will use a MultiSIM simulation to show the current ratios in the CT.

SOLUTION Figure MSM-14-4 shows a transformer connected as a CT. The CT ratio of 500:5 equates to a turns ratio of 100. With an ammeter connected directly across the high side of the CT, the current is 1/100 of the low side current. In this simulation, a low side current of 175 A results in an ammeter reading of 175/100, or 1.75 A to the accuracy that can be expected.

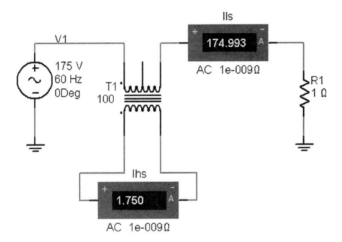

FIGURE MSM-14–4
MultiSIM Example 14-4.

MultiSIM Examples for Chapter 15

The dc generators and dc motors that are discussed in Chapter 15 can be modeled in MultiSIM using the components that were used in the models of the machines in the chapter discussions. That is, generated voltages, counter-emfs, and brush drops are modeled as dc voltage sources, and field windings are modeled as resistances.

■ MULTISIM EXAMPLE 15-1 A self-excited dc generator has the following characteristics: internally generated armature voltage Ea = 123.55 V, armature resistance Ra = 0.1 Ω, brush drop Vb = 2.0 V, shunt field resistance Rf = 200 Ω. It is powering load resistance of RL = 10 Ω. Develop a MultiSIM model of the circuit that will show the values of (a) armature current Ia, (b) armature power delivered Pa, (c) field current If, (d) field power Pf, (e) load current IL, (f) load voltage Vt, and (g) load power PL.

SOLUTION Figure MSM-15-1 shows the MultiSIM model for this example. Note that ammeters have been inserted to measure armature current, field current, and load current; wattmeters are used to measure armature power, field power, and load power; and a voltmeter is monitoring the load voltage. Also, notice that the armature current is equal to the sum of the field and load currents, and the armature power delivered is equal to the sum of the field and load powers. The reader is invited to solve the circuit manually to verify these values.

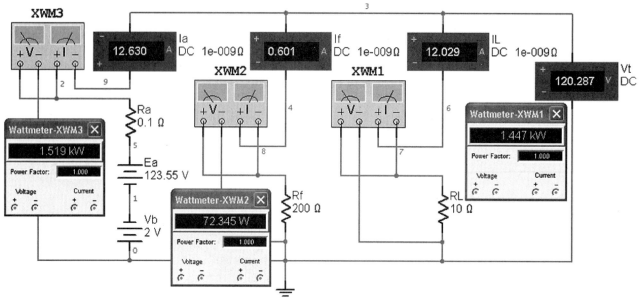

FIGURE MSM-15-1
MultiSIM Example 15-1.

MULTISIM EXAMPLE 15-2 A shunt dc motor has the following characteristics: field resistance Rf = 184 Ω, armature resistance Ra = 0.3 Ω, generated counter emf Vcemf = 132.28 V, brush drop Vb = 2.0 V. The motor is powered from a voltage source Vt = 140 V. Create a MultiSIM model that will find the armature current Ia, field current If, and total motor current It.

SOLUTION Figure MSM-15-2 illustrates a MultiSIM model of the motor with ammeters placed in the source, field, and armature circuits. Note that the total current is the sum of the field and armature currents.

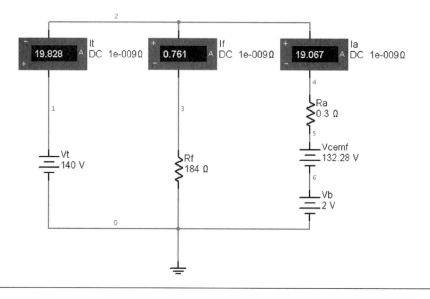

FIGURE MSM-15-2
MultiSIM Example 15-2.

MultiSIM Examples for Chapter 16

Generally speaking, ac motors are modeled in MultiSIM as impedances. Additionally, these impedances can be broken into their resistive and reactive (inductive) parts. Although MultiSIM does not have the capability to directly measure apparent power, as we will see in these examples, these quantities are easily calculated using the current and voltage.

■ **MULTISIM EXAMPLE 16-1** Each phase of a 60-Hz wye-connected three-phase motor is a 10-Ω impedance at 0.8 lagging power factor. The motor is connected to a 208-V three-phase wye source. Construct a MultiSIM circuit that will measure the total apparent power and real power delivered to the motor.

SOLUTION Before constructing the MultiSIM circuit, we must first determine how to model the motor impedance. We know from the given data that the motor impedance is

$$\mathbf{Z}_{\text{motor(phase)}} = 10 \angle (\cos^{-1} 0.8) \text{ }\Omega$$
$$= 10 \angle 36.87° \text{ }\Omega$$
$$= 8 + j6 \text{ }\Omega$$

MultiSIM has available an impedance block, $A + jB$, that will allow us to input each leg of the motor impedance in rectangular form, with a resistance of 8 Ω and a reactance of 6 Ω. We also know that because the motor presents a balanced load, the voltage, current, and power will be identical in each phase. Therefore, we need to measure the current and power in only one phase.

Figure MSM-16-1 shows the circuit with the three-phase source as three single-phase 120-V 60-Hz sources connected in a wye configuration. The motor is modeled as three $A + jB$ impedance blocks. After placing the blocks, double-clicking on them allows us to enter the resistance, reactance, and frequency values. An ammeter monitors the phase-a current, and a wattmeter monitors the phase-c real power. The total apparent power is 3 times the phase voltage times the phase current, or

$$S_{\text{total}} = 3S_{\text{phase}} = 3V_{\text{phase}} I_{\text{phase}} = 3 \times 120 \times 12 = 4.32 \text{ kVA}$$

The total real power is 3 times the wattmeter indication, or

$$P_{\text{total}} = 3P_{\text{phase}} = 3 \times 1152 = 3.456 \text{ kW}$$

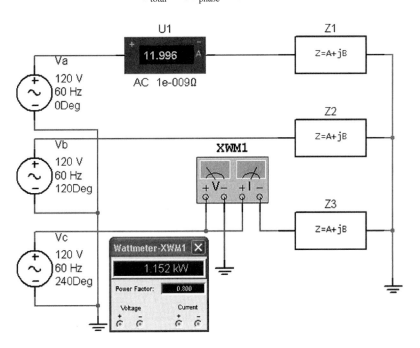

FIGURE MSM-16-1
MultiSIM Example 16-1.

■ **MULTISIM EXAMPLE 16-2** When operating at rated load, a 120-V 60-Hz single-phase induction motor can be modeled as a 10.8-Ω resistor in series with a 25.3-mH inductor. Construct a MultiSIM simulation that will determine the motor current, the real power, and the power factor.

FIGURE MSM-16-2
MultiSIM Example 16-2.

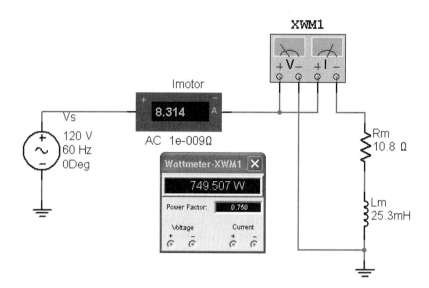

SOLUTION Before constructing the circuit, we should first devise a measurement strategy. We can easily monitor the motor current using an ac ammeter in series with the motor model. This will also allow us to calculate the apparent power S, which is simply V times I. We can also monitor the real power P and the power factor by connecting a wattmeter in the circuit.

Figure MSM-16-2 shows the MultiSIM circuit. Notice that with 120 V ac applied, the motor current is 8.314 A, and the real power is 749.5 W. Because the load is inductive, the power factor of 0.75 indicated by the wattmeter is lagging.

MultiSIM Examples for Chapter 17

Performing simulations of PLC programs in MultiSIM requires that the user construct both the PLC hardware and software (the ladder logic program). MultiSIM provides hardware I/O modules that are brought onto the screen and wired to external inputs and outputs. Then, in another area of the screen, the user builds the ladder logic program that connects the inputs and outputs. When the power switch is switched on, the simulated PLC begins executing the ladder logic program.

MultiSIM provides an assortment of I/O modules that will operate from various voltage levels. The user should carefully choose a module that best integrates into the external circuitry logic voltage levels. The input and output modules are chosen separately; therefore, the user has the flexibility to have differing input and output voltage levels. For the simulations that follow, 5-V input and output modules have been chosen.

I/O modules in MultiSIM are numbered in steps of 100 beginning with module number 100. Within a module, the I/O channels are numbered 1 through 8. The actual I/O number is constructed beginning with the module number, followed by a space, and ending with the I/O number. For example, the first input on module 200 would be input 2001.

MULTISIM EXAMPLE 17-1 A large manufacturing machine uses a PLC to control its operation. A portion of the PLC program controls the starting and stopping of the machine. The PLC has an input module 100 and an output module 200. Two normally open momentary pushbutton switches START and STOP are connected to inputs IN1 (100 1) and IN2 (100 2), respectively. An LED indicator named RUN is connected to output OUT1 (200 1), which simulates a large contactor that controls power to the machine. Construct and simulate a PLC program that will have START turn on the RUN LED. RUN will remain on even after START is released (that is, RUN is latched within the program), until STOP is pressed.

FIGURE MSM-17-1
MultiSIM Example 17-1.

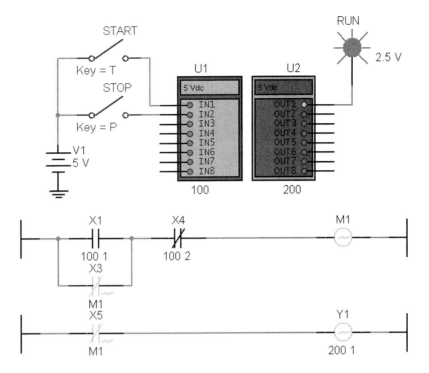

SOLUTION Figure MSM-17-1 illustrates this MultiSIM simulation. Prior to creating a PLC simulation with the switches on the circuit diagram, don't forget to turn the switch above the Circuit Window to the ON or 1 position. The top of the figure shows the hardware required for the simulation, and at the bottom of the figure are the two rungs required to implement the ladder logic program. The figure shows the program operation after the START switch has been switched on, then off. Notice in the first rung that internal relay M1 is on (indicated by an electrical spark symbol), and its NO contacts in parallel with input 100 1 keep the rung latched on. In the second rung, a second NO contact of M1 energizes output 200 1, which in turn energizes the RUN indicator connected to OUT1. When the STOP button is switched on, the NC contact 100 2 in the first rung will open, thereby de-energizing the first rung. In the second rung, relay contact M1 opens, which switches off output Y1.

■ **MULTISIM EXAMPLE 17-2** A manufacturing machine is used to punch and eject a metal bracket used to construct a computer power supply. The operations are to occur in the order PUNCH followed by EJECT. Each is timed, with 1 second for punch and 0.5 second for eject. The sequence of operation is started by a momentary START switch connected to input IN1. The outputs PUNCH and EJECT are connected to outputs OUT1, and OUT2, respectively. We will use indicator lamps on the outputs to simulate contactors that control the machine.

In operation, the machine is started in an idle condition. When START is switched on, the program begins sequencing each output for the required times. For safety reasons, the START switch must be held for the entire sequence. If the START switch is released, the sequence resets to the idle mode. However, even if the START switch is held on for a long period, the machine will perform only one sequence. To start another sequence, the START switch must be released and pressed again (this feature is called anti-repeat).

SOLUTION The MultiSIM simulation is illustrated in Figure MSM-17-2. Because MultiSIM is capable of timing very small increments of time, it runs very slowly with timers that are several tenths of a second or longer in duration. For this reason, we have chosen to speed up the simulation by having the timers run in milliseconds instead of seconds. Therefore, we will set the PUNCH time, T1, for 2 ms, and the EJECT time, T2, for 0.5 ms.

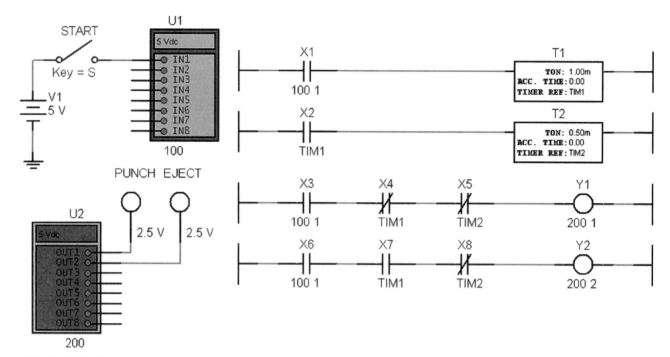

FIGURE MSM-17-2
MultiSIM Example 17-2.

On the left, the START switch is wired to input IN1, and two indicators, PUNCH and EJECT, are connected to outputs OUT1 and OUT2, respectively. The program on the right side uses cascaded timers T1 (named TIM1) and T2 (named TIM2) to perform the required timing functions. When START is switched on, timer T1 in the first rung is energized. Also, in rung 3, output Y1 is energized, which switches on OUT1 and the PUNCH lamp. When timer T1 times out, all NO contacts of TIM1 will switch on, which will switch off Y1 in rung 3, start timer T2, and energize output Y2 (EJECT) in rung 4. When timer T2 times out, output Y2 will switch off. If the START switch is switched off at any time during the sequence, the timers will reset to zero, and both outputs will switch off. If the START switch is held on for a long duration, then after the sequence is completed, it must be released (which resets the timers) and pressed again to start a new sequence.

MultiSIM Examples for Chapter 18

The first two examples presented here use the modules for analog-to-digital and digital-to-analog conversion. The last two examples will illustrate the sampling theorem.

MULTISIM EXAMPLE 18-1 Investigate the properties of the MultiSIM generic analog-to-digital and digital-to-analog converter modules by obtaining a quantization characteristic based on the four most significant bits, a unipolar characteristic, and a full-scale voltage of 10 V.

SOLUTION The generic analog-to-digital (A/D) and digital-to-analog (D/A) converters are obtained from the **Mixed** parts bin. They use 8 bits. This permits 256 possible levels, so a quantization curve based on this precision would be difficult to observe. However, by choosing the four most significant bits, a 4-bit process may be observed.

The circuit used in this experiment is shown in Figure MSM-18-1(a). A piecewise linear voltage source is programmed (Figure MSM-18-1(b)) to provide a ramp (linearly increasing) waveform that is used as Vin. It is set to increase from 0 to 10 V in 1 second as

FIGURE MSM-18–1(a)
MultiSIM A-to-D converter circuit for Example MSM-18-1.

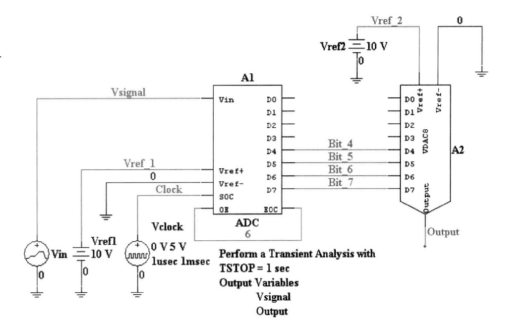

FIGURE MSM-18–1(b)
Piecewise linear voltage source (PWL) setup window.

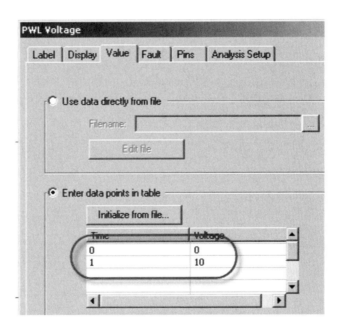

shown. A pulse generator is set up to control the sampling process and is programmed as shown in Figure MSM-18-1(c). The positive reference voltage (Vref) is set to 10 V, and the negative reference is set to 0; that is, it is grounded. The conversion process is initiated by setting the **start of conversion (SOC)** terminal to 5 V. The conversion is accomplished in 1 μs, which should be negligible on the time scale used. The **end of conversion (EOC)** terminal assumes a high state at the end of the conversion and this terminal is connected directly to the **OE** terminal, which outputs the values on the digital terminals (D0 through D7). The result, of course, is a digital word. By connecting the output directly to the D/A converter, the cycle of conversion and reconstruction can be viewed as an overall process.

Using the **Transient** mode set for a time span of 0 to 1 second, the output of the D/A converter is shown in Figure MSM-18-1(d). It is immediately seen that the strategy of the two converters is based on a lower **truncation** strategy rather than a **rounding** strategy. A rounding strategy could be achieved by decreasing the positive reference voltage by 1/2 of 1 LSB and providing a negative voltage of 1/2 of 1 LSB.

FIGURE MSM-18-1(c)
Pulse voltage source setup window.

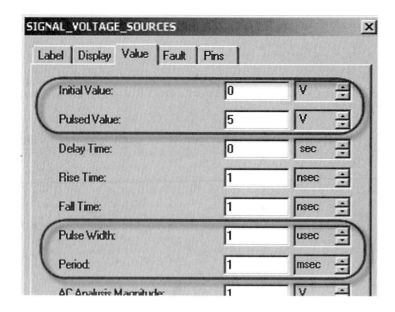

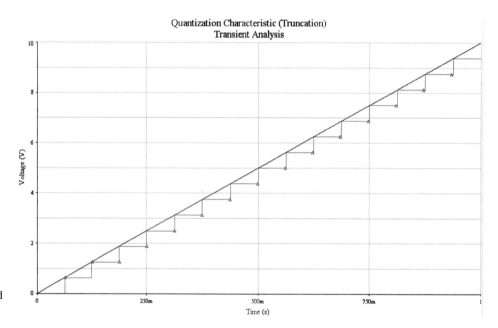

FIGURE MSM-18-1(d)
Quantization characteristic based on 4 bits and with truncation.

MULTISIM EXAMPLE 18-2 Modify the circuit of MultiSIM Example 18-2 three ways: (a) Use 8 bits in the conversion and reconstruction process. (b) Adapt the converters to bipolar operation over the range from -10 V to 10 V. (b) Use a sine wave with frequency 1 Hz and a peak value of 10 V (peak-to-peak value of 20 V).

SOLUTION The modified circuit is shown in Figure MSM-18-2(a). Note that the negative reference voltages for the A/D and D/A converters have been set to -10 V. Moreover, the input source is a sine wave with peak value 10 V and a frequency of 1 Hz.

The input and output waveforms are shown in Figure MSM-18-2(b). It is virtually impossible to discern the two separate waveforms because they appear as one waveform. Thus, the D/A converter is converting the digital waveform back to an analog waveform and the difference never exceeds 1 LSB.

FIGURE MSM-18–2(a)
MultiSIM circuit used for Example MSM-18-2.

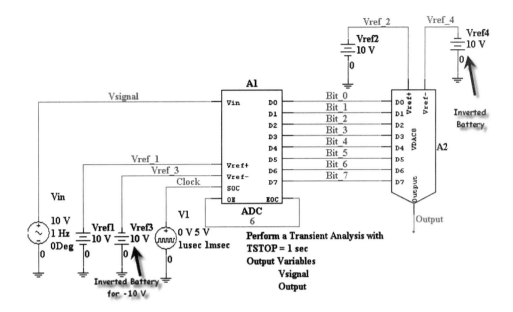

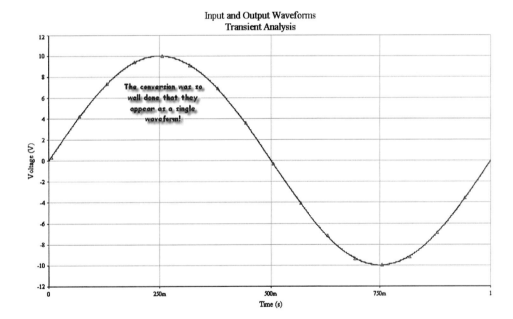

FIGURE MSM-18–2(b)
Input and output waveforms for Example MSM-18-2 (they appear as one waveform).

MULTISIM EXAMPLE 18-3 Develop a MultiSIM circuit for generating a sampled signal. The input signal is to be a 1-kHz sinusoid with an amplitude of 1 V. The sampling rate is to be 10 kHz and the width of the samples should be 10 percent of the period of the sampling function.

SOLUTION Refer to Figure MSM-18-3(a) for the discussion that follows. The sampling process can be readily implemented with a Multiplier, which is obtained from the **CONTROL_FUNCTIONS** option in the **Sources** toolbar. The sampling function is the **Pulse Voltage Source**. The **Initial Value** is set to **0 V**, the **Pulsed Value** is set to **1 V**, and the **Delay Time** is set to **0**. The values of **Rise Time** and **Fall Time** should be small compared with the pulse width. They have been left at their default values of **1 ns**, which should be acceptable. The **Period** is set to 1/1e4 = **0.1 ms**. Because the sample width is to occupy 10 percent of the period, the **Pulse Width** is set to **0.01 ms**. Of course, the signal to be sampled is an ac sinusoidal voltage set to a frequency of 1 kHz and an amplitude of 1 V. A **Junction** is used at the output of the multiplier.

FIGURE MSM-18–3(a)
MultiSIM circuit used for Example MSM-18-3.

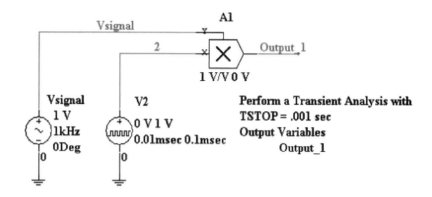

FIGURE MSM-18–3(b)
Sampled signal for Example MSM-18-3.

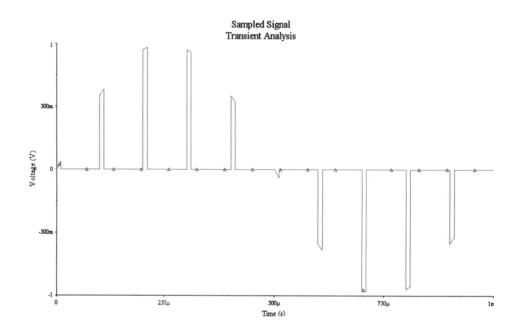

A **Transient Analysis** is performed over the time interval of 1 cycle of the input sinusoid, and the result is shown in Figure MSM-18-3(b). In a digital system, the samples could be converted to digital words at this point.

MULTISIM EXAMPLE 18-4

The circuit of Figure MSM-18-4(a) represents a combination of the sampled signal of MultiSIM Example 18-3 followed by an active three-pole Butterworth low-pass filter with a 3-dB bandwidth of 1.5 kHz. Perform a transient analysis of the circuit and show that the form of the original signal is reconstructed.

SOLUTION From the sampling theorem, it can be deduced that a bandwidth somewhat greater than 1 kHz is required to reconstruct the signal. Thus, the filter used should easily pass the original signal (the 1-kHz component). But it should essentially reject all the other superfluous components obtained in the sampling process. A Transient Analysis is performed over an interval from $t = 0$ to $t = 4$ ms and the result is shown in Figure MSM-18-4(b).

As is the case with any filter when it is initially excited, there is a transient interval or settling time. However, the steady-state output does appear to be sinusoidal and the period can be readily measured as 1 ms, which means that the frequency is 1 kHz as expected. There is a significant loss in amplitude because no holding circuit was employed, but the form of the input signal has been definitely recovered.

FIGURE MSM-18–4(a)
MultiSIM circuit used for Example MSM-18-4.

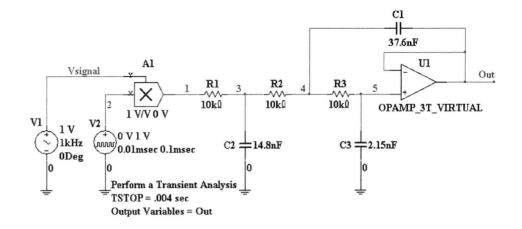

FIGURE MSM-18–4(b)
Reconstruction of the sampled signal using a low-pass filter.

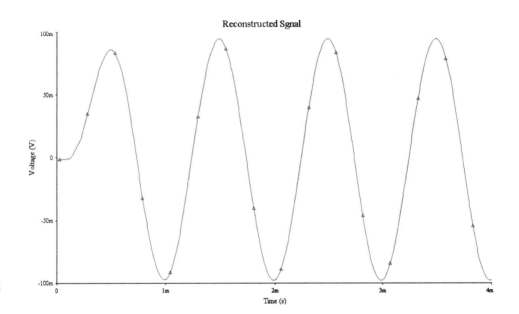

Appendix A

Solving Simultaneous Equations

The study of electrical engineering technology at the level of this text seldom involves the solution of more than two or three simultaneous equations. Moreover, many readers will have access to scientific calculators or computer software that can be used for the purpose. However, some readers who may have occasion to solve simultaneous equations by hand can benefit from a standard procedure that can be used, and this appendix will address that need. The procedure to be given here is known as **Cramer's rule**. It involves the use of **determinants**.

Determinants

A determinant A is a square array of numbers that can be evaluated to yield a specific value. It will be initially expressed as

$$A = \begin{vmatrix} a_{11} & a_{12} & \ldots & a_{1m} \\ a_{21} & a_{22} & \ldots & a_{2m} \\ \vdots & & & \vdots \\ a_{m1} & a_{m2} & \ldots & a_{mm} \end{vmatrix} \quad (A\text{-}1)$$

The determinant is characterized by a size consisting of the number of rows and columns. The form given by Equation A-1 is said to be of size $m \times m$. The process of evaluating a determinant can be a messy process for anything more than about a 2×2 or a 3×3. There are many "tricks" that appear in mathematics texts to simplify the process. The problem with some of these tricks is that they work fine when the values in the matrix are nice "clean" numbers. However, the processes become more burdensome when the numbers represent "real" data.

With the ready access of extensive mathematical software packages, there is very little reason or justification to try to accomplish the chore by hand for anything but the simplest cases. For that reason, much of the development here will focus on the 2×2 and the 3×3 determinants. In showing the development of the latter, we will introduce the concept of **expansion of minors**, which in theory could be used for any determinant.

The simplest matrix is a trivial 1×1 matrix, which is called a scalar. Its determinant is simply its value. Therefore, we need not consider it any further.

2 × 2 Case

A 2 × 2 determinant can be represented as

$$A = \begin{vmatrix} a_{11} & a_{12} \\ a_{21} & a_{22} \end{vmatrix} = a_{11}a_{22} - a_{12}a_{21} \tag{A-2}$$

In words, a 2 × 2 determinant is evaluated as the product of the first element of the first row and the second element of the second row minus the product of the other two terms. A common error in evaluating determinants is that of keeping the signs straight, because some of the elements may have negative signs.

Minors

Before we consider 3 × 3 determinants, it is helpful to define the term **minor**. Let M_{ij} represent the minor corresponding to the ith row and jth column of the given determinant. If the determinant A has a size $m \times m$, M_{ij} is a determinant of size $m - 1$ by $m - 1$ formed from the original determinant by crossing out the ith row and jth column and evaluating the resulting determinant.

Cofactors

A cofactor will be denoted as A_{ij}. It is directly proportional to the corresponding minor M_{ij} by the relationship

$$A_{ij} = (-1)^{i+j} M_{ij} \tag{A-3}$$

Thus, cofactors are alternately equal to the minors or the negatives of the minors.

The significance of minors and cofactors is that a higher-order determinant can be expanded into the sum of lower-order determinants. For example, a 3 × 3 determinant can be expanded into the sum of three 2 × 2 determinants, and a 4 × 4 determinant can be expanded into the sum of four 3 × 3 determinants. One can readily see how the number of smaller determinants can quickly grow as successive expansion is performed.

3 × 3 Case

Consider now a 3 × 3 determinant. As indicated in the previous paragraph, it can be expanded into three 2 × 2 determinants. One can expand along a row or a column. We will choose the former and express the determinant as

$$A = \begin{vmatrix} a_{11} & a_{12} & a_{13} \\ a_{21} & a_{22} & a_{23} \\ a_{31} & a_{32} & a_{33} \end{vmatrix} = a_{11}(A_{11}) + a_{12}(A_{12}) + a_{13}(A_{13}) \tag{A-4}$$

The individual cofactors can be expressed as

$$A_{11} = \begin{vmatrix} a_{22} & a_{23} \\ a_{32} & a_{33} \end{vmatrix} = a_{22}a_{33} - a_{23}a_{32} \tag{A-5}$$

$$A_{12} = -\begin{vmatrix} a_{21} & a_{23} \\ a_{31} & a_{33} \end{vmatrix} = -a_{21}a_{33} + a_{23}a_{31} \tag{A-6}$$

$$A_{13} = \begin{vmatrix} a_{21} & a_{22} \\ a_{31} & a_{32} \end{vmatrix} = a_{21}a_{32} - a_{22}a_{31} \tag{A-7}$$

The determinant can then be expressed as

$$\det(\mathbf{A}) = a_{11}(a_{22}a_{33} - a_{23}a_{32}) + a_{12}(-a_{21}a_{33} + a_{23}a_{31}) + a_{13}(a_{21}a_{32} - a_{22}a_{31}) \tag{A-8}$$

Solution of Simultaneous Equations

Assume the following form for a set of simultaneous linear equations:

$$\begin{aligned} a_{11}x_1 + a_{12}x_2 + \cdots a_{1m}x_m &= b_1 \\ a_{21}x_1 + a_{22}x_2 + \cdots a_{2m}x_m &= b_2 \\ &\vdots \\ a_{m1}x_1 + a_{m2}x_2 + \cdots a_{mm}x_m &= b_m \end{aligned} \quad (A\text{-}9)$$

The unknown variables are assumed to be $x_1, x_2, \ldots, x_n$ and the a and b constants are assumed to be known. In most applications in this book, the variables will likely be voltages or currents.

To solve the set of equations, first form a determinant consisting of the coefficients of the variables. This determinant will be denoted as Δ.

$$\Delta = \begin{vmatrix} a_{11} & a_{12} & \ldots & a_{1m} \\ a_{21} & a_{22} & \ldots & a_{2m} \\ \vdots & & & \vdots \\ a_{m1} & a_{m2} & \ldots & a_{mm} \end{vmatrix} \quad (A\text{-}10)$$

Next, beginning with the leftmost column, replace each column in turn with the column of terms on the righthand side of the equation array. These determinants will be denoted as Δ_1, Δ_2, and so on.

$$\Delta_1 = \begin{vmatrix} b_1 & a_{12} & \ldots & a_{1m} \\ b_2 & a_{22} & \ldots & a_{2m} \\ \vdots & & & \vdots \\ b_m & a_{m2} & \ldots & a_{mm} \end{vmatrix} \quad (A\text{-}11)$$

$$\Delta_2 = \begin{vmatrix} a_{11} & b_1 & \ldots & a_{1m} \\ a_{21} & b_2 & \ldots & a_{2m} \\ \vdots & & & \vdots \\ a_{m1} & b_m & \ldots & a_{mm} \end{vmatrix} \quad (A\text{-}12)$$

$$\vdots$$

$$\Delta_m = \begin{vmatrix} a_{11} & a_{12} & \ldots & b_1 \\ a_{21} & a_{22} & \ldots & b_2 \\ \vdots & & & \vdots \\ a_{m1} & a_{m2} & \ldots & b_m \end{vmatrix} \quad (A\text{-}13)$$

The unknown values can then be determined as follows:

$$x_1 = \frac{\Delta_1}{\Delta} \quad (A\text{-}14)$$

$$x_2 = \frac{\Delta_2}{\Delta} \quad (A\text{-}15)$$

$$\vdots$$

$$x_m = \frac{\Delta_m}{\Delta} \quad (A\text{-}16)$$

We will provide final results for the two most common cases of two and three simultaneous equations.

Two Simultaneous Equations

Assume that there are two variables indicated here as x_1 and x_2. The forms of the two equations are

$$a_{11}x_1 + a_{12}x_2 = b_1$$
$$a_{21}x_1 + a_{22}x_2 = b_2 \tag{A-17}$$

$$\Delta = a_{11}a_{22} - a_{12}a_{21} \tag{A-18}$$

The values of the two variables are

$$x_1 = \frac{a_{22}b_1 - a_{12}b_2}{\Delta} \tag{A-19}$$

$$x_2 = \frac{-a_{21}b_1 + a_{11}b_2}{\Delta} \tag{A-20}$$

Three Simultaneous Equations

The forms of the three equations are

$$a_{11}x_1 + a_{12}x_2 + a_{13}x_3 = b_1$$
$$a_{21}x_1 + a_{22}x_2 + a_{23}x_3 = b_2$$
$$a_{31}x_1 + a_{32}x_2 + a_{33}x_3 = b_3 \tag{A-21}$$

$$\Delta = a_{11}(a_{22}a_{33} - a_{23}a_{32}) + a_{12}(a_{23}a_{31} - a_{21}a_{33}) + a_{13}(a_{21}a_{32} - a_{22}a_{31}) \tag{A-22}$$

The three variables may be determined from the following equations:

$$x_1 = \frac{(a_{22}a_{33} - a_{23}a_{32})b_1 + (a_{13}a_{32} - a_{12}a_{33})b_2 + (a_{12}a_{23} - a_{13}a_{22})b_3}{\Delta} \tag{A-23}$$

$$x_2 = \frac{(a_{23}a_{31} - a_{21}a_{33})b_1 + (a_{11}a_{33} - a_{13}a_{31})b_2 + (a_{13}a_{21} - a_{11}a_{23})b_3}{\Delta} \tag{A-24}$$

$$x_3 = \frac{(a_{21}a_{32} - a_{31}a_{22})b_1 + (a_{12}a_{31} - a_{11}a_{32})b_2 + (a_{11}a_{22} - a_{12}a_{21})b_3}{\Delta} \tag{A-25}$$

Appendix B

Complex Numbers

In a sense, complex numbers are two-dimensional vectors. In fact, some of the basic arithmetic operations such as addition and subtraction are the same as those that could be performed with space vectors such as displacement, velocity, and force. However, once the process of multiplication is reached, the theory of complex number operations diverges significantly from that of space vectors. Therefore, accept the fact that there are similarities, but be aware that there are major differences in the meanings and operations associated with complex numbers.

Rectangular Coordinates

We begin the development with a two-dimensional rectangular coordinate system shown in Figure B-1. The two axes have the traditional labels of x and y, respectively. However, in complex variable theory, the horizontal or x-axis is called the **real axis** and the vertical or y-axis is called the **imaginary axis**.

Imaginary Number

The concept of an imaginary number arises from forming the square root of a negative number. We denote the square root of -1 as j as defined by

$$j = \sqrt{-1} \tag{B-1}$$

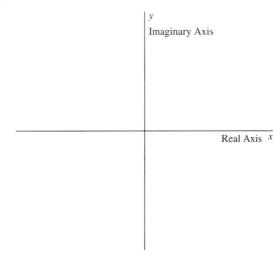

FIGURE B–1
Complex plane showing real and imaginary axes.

FIGURE B-2
Form of a complex number in the complex plane.

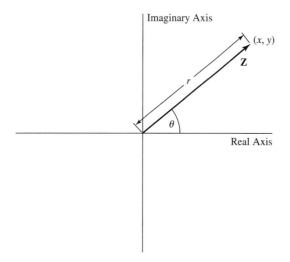

The y-coordinate of any point will be accompanied by the imaginary number j when expressed in complex form. For example, a point 4 units above the origin on the vertical axis could be denoted as $4j$ or $j4$.

Rectangular Form of a Complex Number

We will use boldface notation for complex numbers. Let **z** represent an arbitrary complex number. It can be interpreted as a vector-like quantity extending from the origin to a point with coordinates (x, y) as shown in Figure B-2. The **rectangular form** of **z** is

$$\mathbf{z} = x + jy \tag{B-2}$$

The value x is called the **real part** of **z** and the value y is called the **imaginary part** of **z**. We don't include the j with y when we refer to the imaginary part; that is, the imaginary part is actually the real number y. Moreover, we won't use boldface with x and y because they are real numbers.

When it is necessary to identify the real part of a complex value **z**, the notation Re(**z**) is often used. When it is necessary to identify the imaginary part, the notation Im(**z**) is used.

Let r represent the length of the complex number from the origin to the terminal point. From the trigonometry associated with the angle θ, we can say that

$$x = r \cos \theta \tag{B-3}$$

$$y = r \sin \theta \tag{B-4}$$

The quantity r is the **magnitude** of the complex number and θ is the **angle**. The basic units for angle are radians (rad). We will employ them for most of the work that follows.

Polar Form of a Complex Number

The inverse relationships to the previous two are as follows:

$$r = \sqrt{x^2 + y^2} \tag{B-5}$$

$$\theta = \text{ang}(\mathbf{z}) = \tan^{-1} \frac{y}{x} \tag{B-6}$$

where ang( ) represents the angle of the complex number and is calculated by forming the inverse tangent function as shown.

Euler's Formula

Take the formulas for x and y from Equations B-3 and B-4 and substitute in Equation B-2. The result is

$$\mathbf{z} = r\cos\theta + jr\sin\theta = r(\cos\theta + j\sin\theta) \tag{B-7}$$

A very important mathematical identity is that of Euler's formula, which reads

$$e^{j\theta} = \cos\theta + j\sin\theta \tag{B-8}$$

This formula, which is proven in more advanced mathematics texts, indicates that when a purely imaginary argument is used for the constant e, it is equivalent to a complex number having both real and imaginary parts, each of which is a sinusoidal function.

When θ is replaced by $-\theta$ in the equation, we obtain

$$e^{-j\theta} = \cos(-\theta) + j\sin(-\theta) = \cos\theta - j\sin\theta \tag{B-9}$$

because $\cos(-x) = \cos(x)$ and $\sin(-x) = -\sin(x)$.

By using Euler's formula, the polar form of the complex number can be expressed as

$$\mathbf{z} = re^{j\theta} \tag{B-10}$$

This concept serves as an important step in the development of complex variable theory. It permits a rigorous approach to the operations of multiplication, division, and exponentiation with complex numbers.

Common Engineering Form

It is common in many engineering applications to express the polar form as follows:

$$re^{j\theta} \triangleq r\angle\theta \tag{B-11}$$

This notation is especially popular in ac circuit theory, but one should remember what it really means so that some of the operations are better justified. We will use the exponential form more frequently and use radians for the angular measurement in the steps that follow. For the few cases where the other form is used, degrees will be used.

EXAMPLE B-1

A complex number is given by

$$\mathbf{z} = 4 + j3 \tag{B-12}$$

Determine the polar form.

SOLUTION The magnitude is given by

$$r = \sqrt{x^2 + y^2} = \sqrt{(4)^2 + (3)^2} = 5 \tag{B-13}$$

The angle is given by

$$\theta = \tan^{-1}\frac{3}{4} = 36.87° = 0.6435 \text{ rad} \tag{B-14}$$

The common engineering form with the angle expressed in degrees is

$$\mathbf{z} = 5\angle 36.87° \tag{B-15}$$

However, the more "proper" mathematical form is

$$\mathbf{z} = 5e^{j0.6435} \tag{B-16}$$

EXAMPLE B-2

A complex number is given by

$$\mathbf{z} = -4 + j3 \tag{B-17}$$

Determine the polar form.

SOLUTION The magnitude is the same as before, namely 5. Be careful when one or more of the real part and imaginary part signs are negative. If you are using a calculator with rectangular-to-polar conversion, it might take care of the sign differences, but for manual computations, the signs of both terms must be noted carefully. In this case, the real part is negative and the imaginary part is positive, so the angle must be in the second quadrant. These deductions lead to

$$\theta = \tan^{-1}\left(\frac{3}{-4}\right) = 180° - \tan^{-1}\frac{3}{4} = 180° - 36.87° = 143.13° = 2.498 \text{ rad} \quad \text{(B-18)}$$

The form of the complex number is then

$$\mathbf{z} = 5e^{j2.498} \quad \text{(B-19)}$$

EXAMPLE B-3

The polar form of a complex number is given by

$$\mathbf{z} = 4e^{j2} \quad \text{(B-20)}$$

Determine the rectangular form.

SOLUTION The angle is 2 rad and the conversion process follows.

$$x = 4\cos 2 = -1.6646 \quad \text{(B-21)}$$
$$y = 4\sin 2 = 3.6372 \quad \text{(B-22)}$$

The rectangular form is then

$$\mathbf{z} = -1.6646 + j3.6372 \quad \text{(B-23)}$$

EXAMPLE B-4

The polar form of a complex number is given by

$$\mathbf{z} = 10e^{-j} \quad \text{(B-24)}$$

Determine the rectangular form.

SOLUTION The angle is −1 rad and the conversion process follows.

$$x = 10\cos(-1) = 5.4030 \quad \text{(B-25)}$$
$$y = 10\sin(-1) = -8.4147 \quad \text{(B-26)}$$

The rectangular form is then

$$\mathbf{z} = 5.4030 - j8.4147 \quad \text{(B-27)}$$

B-1 Addition and Subtraction of Complex Numbers

The addition and subtraction of complex numbers is very much like that of vectors; that is, the real parts are added or subtracted and the imaginary parts are added or subtracted. If the process is being performed manually, the complex numbers should be expressed in or converted to rectangular forms.

For the purpose of illustration, assume the following two complex numbers in rectangular form:

$$\mathbf{z_1} = x_1 + jy_1 \quad \text{(B-28)}$$
$$\mathbf{z_2} = x_2 + jy_2 \quad \text{(B-29)}$$

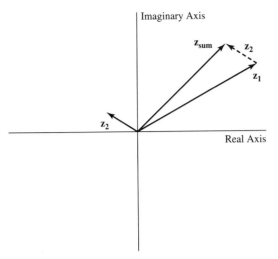

FIGURE B–3
Addition of two complex numbers.

FIGURE B–4
Subtraction of two complex numbers.

Addition

Let z_{sum} represent the sum of the two complex numbers. It is given by

$$z_{sum} = z_1 + z_2 = x_1 + jy_1 + x_2 + jy_2 = x_1 + x_2 + j(y_1 + y_2) \tag{B-30}$$

Geometric Interpretation of Addition

An interesting geometric interpretation of addition is shown in Figure B-3 for two arbitrary complex numbers z_1 and z_2. Imagine that z_2 is translated to the end of z_1 while retaining its direction and magnitude. A vector drawn from the origin to the tip of z_2 then represents the sum z_{sum}.

Subtraction

Let z_{diff} represent the complex number of Equation B-29 subtracted from that of Equation B-28. The result is

$$z_{diff} = z_1 - z_2 = x_1 + jy_1 - (x_2 + jy_2) = x_1 - x_2 + j(y_1 - y_2) \tag{B-31}$$

Geometric Interpretation of Subtraction

A geometric interpretation of subtraction can also be developed. It is shown in Figure B-4 for $z_1 - z_2$. First, assume that $-z_2$ is formed and this amounts to rotating z_2 by $180°$ or π rad. The operation can then be visualized as forming the sum of $-z_2$ to z_1 and the principle employed in the geometric addition process is then followed.

Additional Comments

The normal intuition associated with the addition and subtraction of real numbers does not always apply with complex numbers. For example, the sum of two complex numbers may yield a new complex number that has a smaller magnitude than either of the two being added. Conversely, the difference between two complex numbers may yield a new complex number that might have a larger magnitude than either of the two whose difference is being formed.

■ EXAMPLE B-5

Consider the following two complex numbers in rectangular form:

$$z_1 = 5 + j3 \tag{B-32}$$

$$z_2 = 2 - j7 \tag{B-33}$$

Determine the sum of the two numbers z_{sum}.

SOLUTION The sum is

$$z_{sum} = z_1 + z_2 = 5 + j3 + 2 - j7 = 7 - j4 \tag{B-34}$$

■ EXAMPLE B-6

For the two complex numbers of Example B-5, determine $z_{diff} = z_1 - z_2$.

SOLUTION The difference is

$$z_{diff} = z_1 - z_2 = 5 + j3 - (2 - j7) = 3 + j10 \tag{B-35}$$

■

B-2 Multiplication and Division

Multiplication and division are performed more easily when the complex numbers are expressed in polar form. Define two complex numbers as follows:

$$z_1 = r_1 e^{j\theta_1} \tag{B-36}$$

$$z_2 = r_2 e^{j\theta_2} \tag{B-37}$$

Multiplication

Let z_{prod} represent the product of the preceding complex numbers. By the law of exponents, multiplication is easily performed with the polar forms.

$$z_{prod} = z_1 z_2 = (r_1 e^{j\theta_1})(r_2 e^{j\theta_2}) = r_1 r_2 e^{j(\theta_1 + \theta_2)} \tag{B-38}$$

There is a simple interpretation to this result: *To multiply two complex numbers, multiply the magnitudes and add the angles.*

Division

Let z_{div} represent z_1 divided by z_2. Again using exponent operations, we have

$$z_{div} = \frac{z_1}{z_2} = \frac{(r_1 e^{j\theta_1})}{(r_2 e^{j\theta_2})} = \frac{r_1}{r_2} e^{j(\theta_1 - \theta_2)} \tag{B-39}$$

The interpretation in this case is as follows: *To divide two complex numbers, divide the numerator magnitude by the denominator magnitude and subtract the denominator angle from the numerator angle.*

Multiplication in Rectangular Form

It is also possible to perform the product of two complex numbers in rectangular form. Consider two complex numbers defined as

$$z_1 = x_1 + jy_1 \tag{B-40}$$

$$z_2 = x_2 + jy_2 \tag{B-41}$$

The product can be expressed as

$$\mathbf{z}_{prod} = (x_1 + jy_1)(x_2 + jy_2) = x_1 x_2 + jx_1 y_2 + jx_2 y_1 + j^2 y_1 y_2 \tag{B-42}$$

Regrouping and recognizing that $j^2 = -1$, we have

$$\mathbf{z}_{prod} = x_1 x_2 - y_1 y_2 + j(x_1 y_2 + x_2 y_1) \tag{B-43}$$

Don't try to memorize the result. Work it out as you need it.

Complex Conjugate

Before showing an alternate way for dividing, it is necessary to introduce the **complex conjugate**. Let **z** represent an arbitrary complex number and denote the corresponding complex conjugate as $\bar{\mathbf{z}}$. Assume that the form of **z** can be represented in both rectangular and polar forms as

$$\mathbf{z} = x + jy = re^{j\theta} \tag{B-44}$$

The complex conjugate can be expressed as

$$\bar{\mathbf{z}} = x - jy = re^{-j\theta} \tag{B-45}$$

Thus, *the complex conjugate of a complex number is formed by reversing the sign of the imaginary part in the rectangular form.* Alternately, *it is formed by reversing the sign of the angle in the polar form.*

An important result is obtained by forming the product of a complex number and its conjugate. It can be shown using either form that

$$(\mathbf{z})(\bar{\mathbf{z}}) = x^2 + y^2 = r^2 \tag{B-46}$$

Stated in words, *the product of a complex number and its conjugate is the magnitude squared*. The process of multiplying a complex number by its conjugate is referred to as **rationalization**.

Division in Rectangular Form

Based on the rectangular forms of the two complex numbers, consider the division

$$\mathbf{z}_{div} = \frac{\mathbf{z}_1}{\mathbf{z}_2} = \frac{x_1 + jy_1}{x_2 + jy_2} \tag{B-47}$$

The trick in this case is to multiply both numerator and denominator by the complex conjugate of the denominator. The process follows.

$$\mathbf{z}_{div} = \frac{(x_1 + jy_1)(x_2 - jy_2)}{(x_2 + jy_2)(x_2 - jy_2)} = \frac{x_1 x_2 + y_1 y_2 + j(x_2 y_1 - x_1 y_2)}{x_2^2 + y_2^2}$$
$$= \frac{x_1 x_2 + y_1 y_2 + j(x_2 y_1 - x_1 y_2)}{r^2} \tag{B-48}$$

The denominator is now a positive real number that can be divided into each of the numerator terms to simplify the result. Again, don't bother memorizing the result, but work it out when needed.

EXAMPLE B-7

Two complex numbers are given by

$$\mathbf{z}_1 = 8e^{j2} \tag{B-49}$$

$$\mathbf{z}_2 = 5e^{-j0.7} \tag{B-50}$$

Determine the product of the two complex numbers and denote it as $\mathbf{z}_3$.

SOLUTION Because both numbers are given in polar form, the product is easily formed as

$$z_3 = z_1 z_2 = (8e^{j2})(5e^{-j0.7}) = 40e^{j1.3} \tag{B-51}$$

For comparison in the next example, the corresponding rectangular form is

$$z_3 = 40(\cos 1.3 + j \sin 1.3) = 40(0.2675 + j0.9636) = 10.70 + j38.54 \tag{B-52}$$

EXAMPLE B-8

Repeat the multiplication of Example B-7 by first converting the two complex numbers to rectangular forms and using the rectangular multiplication process.

SOLUTION The first step is to convert the two values to rectangular forms.

$$z_1 = 8e^{j2} = 8(\cos 2 + j \sin 2) = 8(-0.4162 + j0.9093) = -3.329 + j7.274 \tag{B-53}$$

$$z_2 = 5e^{-j0.7} = 5(\cos 0.7 - j \sin 0.7) = 5(0.7648 - j0.6442) = 3.824 - j3.221 \tag{B-54}$$

The product can now be expressed as

$$\begin{aligned} z_3 = z_1 z_2 &= (-3.329 + j7.274)(3.824 - j3.221) \\ &= -12.73 + 23.43 + j(27.82 + 10.72) = 10.70 + j38.54 \end{aligned} \tag{B-55}$$

Even with possible roundoff, the answer comes out exactly the same as with the polar forms. Clearly, the simpler approach in this case was to use the polar forms. However, suppose that the two complex numbers had been given in rectangular forms. In that case, they would have had to be individually converted to polar forms before proceeding with the polar multiplication. This means that it might have been just as easy to carry out the multiplication with the rectangular forms. The choice could depend on the form in which the final answer is desired.

EXAMPLE B-9

For the two complex numbers of Examples B-7 and B-8, determine the quotient z_1/z_2 and denote it as z_4.

SOLUTION The quotient is readily determined in polar form as

$$z_4 = \frac{z_1}{z_2} = \frac{8e^{j2}}{5e^{-j0.7}} = 1.6e^{j2.7} \tag{B-56}$$

The result will now be converted to rectangular form as a matter of interest.

$$z_4 = 1.6(\cos 2.7 + j \sin 2.7) = 1.6(-0.9041 + j0.4274) = -1.447 + j0.6838 \tag{B-57}$$

Answers to Selected Odd-Numbered Problems

Note 1: Many values are provided as lists without identifiers to ensure that a student carefully correlates these values with those obtained from a complete solution.

Note 2: Many values are listed to a much higher accuracy than is normally attainable in practical circuits to assist in checking the results.

Chapter 1

1-1	1.8 mA
1-3	−3 mA
1-5	64.8 mW
1-7	18 mW
1-9	6 V
1-11	7.2 Ω
1-13	4.472 A, 223.6 V
1-15	18 V
1-17	30 kΩ
1-19	$7.88
1-21	10 V, −42 V
1-23	8 mA
1-25	8 Ω
1-27	(a) 4 A (b) 8 V, 12 V, 20 V (c) Sources: 64 W and 96 W, both delivered. Resistors: 32 W, 48 W, 80 W.
1-29	(a) 2 mA (b) 9.4 V, 4 V (c) 25.4 V (d) Sources: 50.8 mW delivered, 24 mW absorbed, Resistors: 18.8 mW, 8 mW.
1-31	(a) 60 V (b) 30 mA, 60 mA, 6 mA (c) 96 mA (d) Source: 5.76 W, Resistors: 1.8 W, 3.6 W, 0.36 W.
1-33	self-checking
1-35	self-checking

Chapter 2

2-1	4.3 A, −0.95 A
2-3	5.578 mA, 2.367 mA, 2.977 mA
2-5	13.65 mA, 7.079 mA, 3.371 mA
2-7	−9.5 V
2-9	6.422 V, −3.047 V
2-11	10.28 V, 15.84 V
2-13	2198 Ω
2-15	24 V in series with 12 Ω. 2 A in parallel with 12 Ω.
2-17	48 V in series with 9 Ω. 5.333 A in parallel with 9 Ω.
2-19	self-checking
2-21	$R_L v_{s1} = (R_1 + R_L)v_{s2}$

Chapter 3

3-1

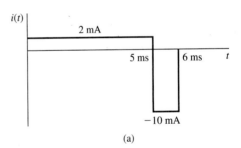

3-3

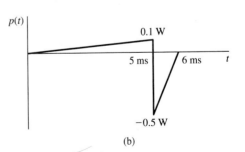

497

- 3-5   250 μJ
- 3-7   900 μJ
- 3-9 $\omega C V_p \cos \omega t$
- 3-11 $i_L = 0, v_C = 0, v_L = 16\text{ V}, i_C = 2\text{ A}$
- 3-13 $v_L = 0, i_C = 0, i_L = 6\text{ A}, v_C = 0$
- 3-15 $i_L = 0, v_C = 6\text{ V}, i_C = 1.5\text{ A}, v_L = 18\text{ V}$
- 3-17 $v_L = 24e^{-200t}$ $i = 3(1 - e^{-200t})$
- 3-19 $v_C = 100\left(1 - e^{-10^5 t}\right)$ $i = 2e^{-10^5 t}$
- 3-21   2.105 μF
- 3-23   0.004751 μF

Chapter 4

- 4-1   (a) 170 V   (b) $120\pi = 377.0$ rad/s   (c) 60 Hz   (d) 16.67 ms   (e) 162.7 V
- 4-3   20 V ∠45°
- 4-5 $2\sin(2\pi \times 1000t - 30°)$
      $2\cos(2\pi \times 1000t - 120°)$
- 4-7 $13\sin(1000t + 52.63°)$
- 4-9 $17.44\sin(200t + 6.587°)$
- 4-11   (b) 2.683 A ∠−63.43°, 26.83 V ∠−63.43°, 53.66 V ∠26.57°
      (c) $2.683\sin(1000t - 63.43°)$,
      $26.83\sin(1000t - 63.43°)$,
      $53.66\sin(1000t + 26.57°)$
- 4-13   (b) 21.28 mA ∠57.87°, 21.28 V ∠57.87°, 33.88 V ∠−32.13°
      (c) $21.28 \times 10^{-3}\sin(2000\pi t + 57.87°)$,
      $21.28\sin(2000\pi t + 57.87°)$,
      $33.88\sin(2000\pi t - 32.13°)$
- 4-15   (b) 37.14 mA ∠−68.20°, 7.428 V ∠−68.20°, 37.14 V ∠21.80°, 18.57 V ∠−158.2°
- 4-17   (a) 1 A ∠36.87°, 80 V ∠36.87°, 120 V ∠126.87°, 180 V ∠−53.13°

Chapter 5

- 5-1   (a) 120.2 V rms   (b) 96.33 W
- 5-3   (a) 3.536 A rms   (b) 100 W
- 5-5   (a) 4 A rms ∠−53.13°   (b) 160 VA   (c) 0.6 lagging   (d) 96 W   (e) 128 VAR
- 5-7   (a) 1200 VA   (b) 1039 W   (c) 600 VAR
- 5-9   (a) 53.13°, voltage leading current   (b) 1200 VA   (c) 720 W   (d) 960 VAR
- 5-11   (a) 960 VA   (b) 0.5208   (c) 819.5 VAR
- 5-13   123.4 μF
- 5-15   (a) 29.76 A   (b) 23.15 A
- 5-17   1.257 A rms ∠90°
- 5-19   16.67 Ω in parallel with $j12.5$ Ω
- 5-21   1.6 Ω in series with $j1.2$ Ω
- 5-23   (a) 15.92 kHz   (b) 10   (c) 1.592 kHz
- 5-25   (a) 15.92 kHz   (b) 20   (c) 796.0 Hz
- 5-27   self-checking
- 5-29   self-checking
- 5-31   self-checking

Chapter 6

- 6-1   7.767 mA, 0.7 V, 23.3 V
- 6-3   0, −12 V, 0
- 6-5   (a) 162.6 V   (b) 40.66 V   (c) 39.96 V   (d) 12.72 V   (e) 254.4 mA
- 6-7   The first three answers are the same.   (d) 25.44 V   (e) 508.8 mA
- 6-9   The first two answers are the same.   (c) 39.26 V   (d) 24.99 V   (e) 499.9 mA
- 6-11   (a) 1.389 V   (b) 41.04 V
- 6-13   3 V, 45 mA, 60 mA, 15 mA
- 6-15   3 V, 36 mA, 60 mA, 24 mA
- 6-17   4.340:1
- 6-19   (a) 20.83 mF = 20,830 μF   (b) 12.1 V   (c) 13.5 V   (d) 12.57:1
- 6-21   35.71 Ω
- 6-23 $v_L^+ = 0.5\dfrac{N_2}{N_1}v_1 - 0.7$ $v_L^- = -0.5\dfrac{N_2}{N_1}v_1 + 0.7$

Chapter 7

- 7-1   (a) 119 mA   (b) 40 μA   (c) 60.25 mA
- 7-3   (a) cutoff   (b) saturation   (c) active   (d) saturation
- 7-5   125
- 7-7   (a) 12.5 mA   (b) 12.6 mA
- 7-9   (a) 100   (b) 95   (c) 105
- 7-11   2150 Ω
- 7-13   62.5 Ω, 83.33 Ω, 125 Ω, 250 Ω, ∞
- 7-15 $16 \times 10^{-3}\left(1 + \dfrac{V_{GS}}{2}\right)^2$
- 7-17   5.76 mA
- 7-19   −0.4189 V
- 7-21 $\dfrac{62.5}{1 - |V_{GS}|/2}$
- 7-23   1 to 0.0204

Chapter 8

- 8-1   The circuit has the form of Figure 8-4 with $R_{\text{in}} = 4\text{ k}\Omega$, $R_{\text{out}} = 1\text{ k}\Omega$, and $A = 80$.

8-3   48
8-5   325,000
8-7   −12.31 μV
8-9   (a) −22   (b) 10 kΩ   (c) 0
8-11 (a) 0.6364 V   (b) 4.4 V, −13.2 V, 14 V
8-13 $R_i = 30$ kΩ, $R_f = 150$ kΩ
8-15 (a) 17.5   (b) ∞   (c) 0
8-17 (a) 0.8 V   (b) −3.5 V, 10.5 V, −14 V
8-19 There are many solutions, but the following is typical: $R_i = 1$ kΩ and $R_f = 39$ kΩ
8-21 $R_f = 10$ kΩ, $|i_{i,\max}| = 1.3$ mA
8-23 $R_i = 10$ kΩ, $R_{L,\max} = 10.83$ kΩ
8-25 $R_2 = 3$ kΩ, $R_{L,\max} = 2750$ Ω
8-27 $R_1 = 3$ kΩ, $R_2 = 7.5$ kΩ
8-29 The circuit has the form shown in Figure 8-22 with all resistance values equal. A typical value for each resistance is 10 kΩ.
8-31 A possible design has $R = 100$ kΩ and $C = 1$ nF.
8-33 $C_1 = 2.251$ nF, $C_2 = 1.125$ nF
8-35 $C_1 = 1.722$ nF, $C_2 = 1.471$ nF, $C_3 = 4.159$ nF, $C_4 = 608.8$ pF
8-37 self-checking
8-39 self-checking
8-41 3
8-43 The circuit has the form shown in Figure 8-28. $R = 28.22$ kΩ, $C_1 = 0.01$ μF, $C_2 = 3.925$ nF, $C_3 = 570.7$ pF
8-45 The circuit has the form shown in Figure 8-31. $C_1 = C_2 = 15.92$ nF, $R_1 = 100$ kΩ, $R_2 = 502.5$ Ω, $R_3 = 200$ kΩ

Chapter 9

9-1   (a) 11_{10} (b) 136_{10} (c) 1111110_2
      (d) 10000000001_2
9-3   (a) 1100_2 (b) 101001111_2 (c) 1111_2
      (d) -1000110_2
9-5   (a) 00000101_2 (b) 11101001_2
      (c) out of range   (d) 01110111_2
      (e) 11111111_2 (f) 00000001_2
      (g) 11100001_2 (h) 10000101_2
9-7   (a) $000111 = 7_{10}$ (b) $111001 = -7_{10}$
      (c) $111111 = -1_{10}$ (d) $010110 = 22_{10}$
      (e) $000001 = 1_{10}$ (f) $101110 = 18_{10}$
      (g) $111011 = -5_{10}$ (h) $111111 = -1_{10}$
      (i) $111011 = -5_{10}$ (j) out of range
9-9   (a) 224_8 (b) 11101100_2
      (c) 111110011_2 (d) 59_{16}
      (e) 730_{16} (f) $35,271_8$

9-11

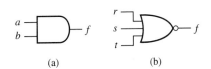

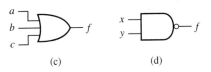

9-13 (a) x
      (b) $ab + ac$ or $a(b + c)$
      (c) $\overline{w}xz + xy$
      (d) $\overline{a}\,\overline{b}\,\overline{d} + \overline{b}\,\overline{c}\,d$
9-15 (a) $\overline{a}\,\overline{c} + b\overline{c} + ac$
      (b) $\overline{a}\,\overline{b}\,c + \overline{b}\,d + ab\overline{d} + acd$ or $\overline{a}\,\overline{b}\,c + \overline{b}\,d + ab\overline{d} + abc$
      (c) $\overline{x}\,y + wz$
      (d) $r + tu + \overline{r}\,\overline{s}\,t$ or $r + tu + \overline{s}\,t\,\overline{u}$
9-17 (a) $\Sigma m(2, 3)$
      (b) $\Sigma m(1, 5)$
      (c) $\Sigma m(0, 1, 2, 3, 4, 5, 6)$
      (d) $\Sigma m(0, 1, 4, 5, 7, 8, 9, 12, 13)$

9-19

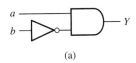

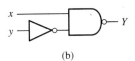

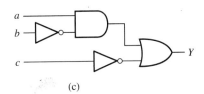

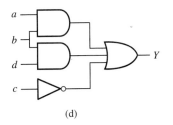

9-21 $Y = \overline{A}\,B + CD$

Chapter 10

10-1

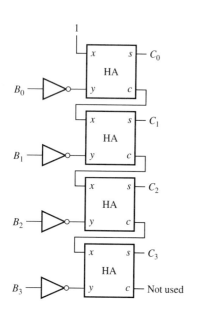

10-3

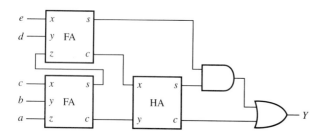

10-5

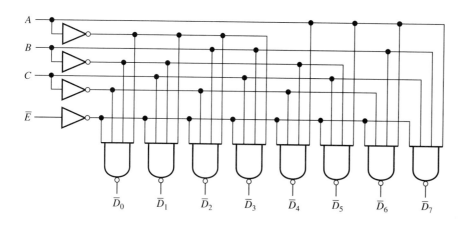

10-7

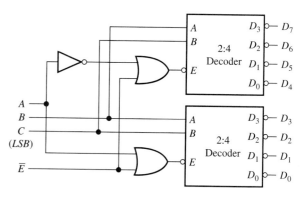

10-9

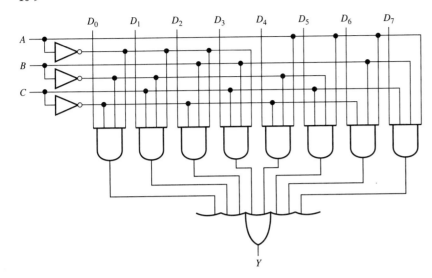

10-11 It becomes an exclusive-OR gate with S_0 and S_1 as the inputs and Y as the output.

10-13 Lower power consumption, runs cooler, higher reliability, longer life.

Chapter 11

11-1 $Q = 1$

11-3 $Q = 1$

11-5 A level-triggered flip flop samples and stores the value of D at all times when the clock = 1. An edge-triggered flip flop samples and stores the value of D only when the clock changes.

11-7 $Q = 1$

11-9 $Q = 1$

11-11 $Q = 0$

11-13 00100100_2

11-15 01010110_2

11-17 0001_{BCD}

11-19 8 address pins

11-21 speed, small size, low power consumption

Chapter 12

12-1 1.67 T

12-3 400 A-t

12-5 1,125 Wb

12-7 2,000 A-t/m

12-9 0.48×10^6 A-t/Wb

12-11 (a) 909.1×10^{-6} Wb/A-t-m (b) 723.4

Chapter 13

13-1 $v_c = 1,177 \sin(\omega t - 240°)$ or
$v_c = 1,177 \sin(\omega t + 120°)$

13-3 (a) 120.1 V (b) 5 A

13-5 (a) 7,736 V (b) 25 A (c) 580.2 kW

13-7 (a) 13,400 V (b) 14.43 A (c) 580.2 kW

13-9 8 kW

13-11 (a) 4.326 kW (b) 12.01 A

13-13 delta

Chapter 14

14-1 (a) 2.56 (b) 613.3 V

14-3 200 V

14-5 5 W

14-7 0.97, or 97%

14-9 4.8 Ω

14-11 (a) 606.2 V (b) 15 A (c) 183.7 V
(d) 183.7 V (e) 49.5 A (f) 28.6 A

14-13 (a) 168 V (b) 1.12 A
(c) 1.57 A (d) 53.8 VA

14-15 28,800 V, 240:1, or 28,800:120

14-17 300:5, or $a = 60$

Chapter 15

15-1 1.5 mV

15-3 1.04 m/s

15-5 (a) 0.8 A (b) 15.8 A

15-7 121.6 V

15-9 0.68, or 68%

15-11 0.67, or 67%

15-13 415.8 V

15-15 0.028 N

15-17 0.0743 N

15-19 (a) 47.2 A (b) 1.2 A

15-21 0.656, or 65.6%

15-23 (a) compound (b) series (c) shunt
(d) series (e) shunt (f) series
(g) permanent magnet

Chapter 16

16-1 42.5 Hz
16-3 0.62, or 62%
16-5 1500 rpm
16-7 0.0417, or 4.17%
16-9 8 poles
16-11 0.628, or 62.8%
16-13 PSC motor
16-15 1200 rpm

Chapter 17

17-1

17-3

17-5

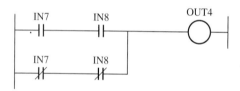

17-7 The rung below can be reduced to minimize the number of contacts. It has been left nonreduced for clarity.

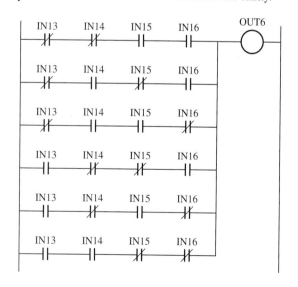

17-9

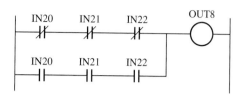

17-11

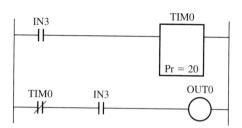

17-13

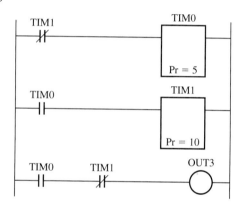

Chapter 18

18-1 (a) 64 (b) 1024
18-3 9
18-5 640 s = 10.67 minutes
18-7 0000, 0010, 0011, 0110, 1001, 1100, 1111
18-9 1110, 1100, 1100, 1000, 0101, 0010, 0000
18-11 (a) 15.26×10^{-6} (b) 305.2 µV (c) 0.999985
(d) 19.9997 V (e) 7.63×10^{-6} (f) 152.6 µV
18-13 (a) 30.52×10^{-6} (b) 305.2 µV (c) 0.999969
(d) 9.99969 V (e) 15.26×10^{-6} (f) 152.6 µV
18-15 1.544 Mb/s
18-17 (a) 16 kHz (b) 62.5 µs
18-19 (a) 20 kHz (b) 50 µs
18-21 5

Index

A

A/D conversions. *See* analog to digital (A/D) conversions
AC. *See* alternating currents (AC)
AC generators. *See* alternators
acceptors, 120
accumulator value, 369
active filters, 199–204
    Butterworth design forms, 201–202
    Butterworth low-pass characteristics, 200
    characteristic forms, 199
    decibel response forms, 200
    examples, 203–204
    two-pole band-pass responses, 202
active region, BJT, 147, 151
adders, 235–237
    full-adder, 236
    half-adder, 236
    symmetrical circuit, 237
    symmetrical function, 237
admittance, 106–110
    AC impedance, 88
    circuit parameters, 106–107
    conversion from parallel to series circuit, 108–109
    conversion from series to parallel circuit, 107–108
    examples, 109—110
    and Ohm's law, 106
    quantities, 106
aliasing, 385
alternating currents (AC)
    AC machines, 329–348
    admittance, 106–110
        AC impedance, 88
    alternators, 330–337
        characteristics, 331
        connecting and disconnecting, 334
        efficiency, 332, 335–336
        examples, 333–334, 336
        Faraday's Law, 330
        losses within alternators, 336
        synchronous motors, 336–337
    amplifiers, 181
    capacitance, AC voltage-current relationships, 86–87
    capacitive impedance, 88–89
    circuits, 77–94, 95–116
        AC impedance, 87–92
        AC voltage-current relationships, 85–87
        admittance, 106–110
        complex power in AC circuits, 99–103
        DC and RMS values, 95–99
        multiSIM examples, 427–431, 431–435
        phasors, 80–84
        power factor correction, 103–106
        problems, 92–94, 115–116
        resonance, 110–114
        sinusoidal functions, 78–80

complex power in AC circuits, 99–103
    apparent, real and reactive power, 99–101
    examples, 102–103
    power development, 99–100
    power triangle, 101
connecting and disconnecting AC generators, 334
DC and RMS values, 95–99
differentiation circuits, 197
diode rectifier circuits, conversion from AC to DC, 126
direct current (DC) compared, 4
effective values of AC voltages and currents, 10
efficiency of AC generators, 332, 335–336
Faraday's Law, 330
impedance, 87–92
    admittance, 88
    capacitive impedance, 88–89
    examples, 89–92
    inductive impedance, 88
    resistance and reactance, 87–88
    steady-state AC model, 89
inductance, AC voltage-current relationships, 86
inductive impedance, 88
integrators and integration, 197
losses within alternators, 336
Ohm's law, 85
phasors, 80–84
power
    complex, in AC circuits, 99–103
    power factor correction, 103–106
reluctance motors, 346–348
resistance and reactance, 87–88, 88
    AC voltage-current relationships, 85
resonance, 110–114
single-phase induction motors, 341–346
sinusoidal functions, 78–80
steady-state
    AC models, 89
    applications, AC components, 61
    response in AC circuits, 80
synchronous motors, 336–337
three-phase induction motor, 337–341
universal motors, 348
voltage-current relationships, 85–87
    capacitance, 86–87
    differentiation and integration of complex exponential function, 85–86
    inductance, 86
    and Ohm's law, 85
    resistance, 85
alternators (AC generators), 330–337
    characteristics, 331
    connecting and disconnecting, 334
    efficiency, 332, 335–336
    examples, 333–334, 336
    Faraday's Law, 330
    losses within alternators, 336
    synchronous motors, 336–337

American Standard Code for Information Interchange (ASCII), 381–382
ammeter, 12
amplifiers
    amplifier circuits
        BJT, 153–154
        inverting, 184–188
        JFET, 163
        noninverting, 188–191
    BJT amplifier circuits, 153–154
    connecting to source and load, 176–177
    DC coupled and AC coupled, 181
    difference amplifiers, 195
    equivalent circuit models, 175–176
    examples, 178–179
    general properties, 174–175
    ideal voltage amplifiers, 177
    inverting. *See* inverting amplifiers and inverting amplifier circuits
    JFET amplifier circuits, 163
    noninverting. *See* noninverting amplifiers and noninverting amplifier circuits
    open-circuit voltage gain, 176
    operational. *See* operational amplifiers (op-amp)
    properties, 173–179
        connecting to source and load, 176–177
        equivalent circuit model, 175–176
        examples, 178–179
        ideal voltage amplifier, 177
        open-circuit voltage gain, 176
        resistance, input and output, 176
        voltage amplifier, 174
    resistance, input and output, 176
amplitude-shift keying (ASK), 395
analog signals, 167–168
    aliasing, 385
    analog to digital (A/D) conversions, 387, 388–394
    anti-aliasing analog filter, 386
    bandwidth, 384–385
    bipolar offset encoding, 390
    description, 384
    digital-to-analog (D/A) conversions, 387, 388–394
    examples, 392–394
    normalization, 389
    quantization curves, 390–391
    quantization error, 392
    sampling, 384–386
    sampling theorem, 385
    unipolar encoding, 389–390
AND circuit (AND gate), 218
apparent power, 99–100
armature, 321–322
    definition, 313, 330
    impedance-protected armature coils, 345
    shunt and compound DC generators, 313, 314, 315
ASCII code (American Standard Code for Information Interchange), 381–382
ASK (amplitude-shift keying), 395

503

504   Index

asynchronous serial data transmission, 381–383
asynchronous communications, 378–379
asynchronous counters, 252–253
asynchronous devices, 246
autotransformers, 297–300
    buck configuration, 299–300
    conducted power, 299
    examples, 299, 300
    rules, 298
    transformed power, 299
    variable autotransformer (Variac ©), 300
auxiliary winding (start winding), 342, 344
average power, 96–97
axis, real and imaginary, 81

B

B-H curve and permeability, 267
back emf, 321
balanced load, wye and delta load connections, 278–279
band-pass filters, 113, 199
band-pass responses, 202
band-rejection filters, 199
bandwidth, 114, 384–385
    baseband signals, 385
    Fourier analysis, 385
    frequency spectrum, 384–385
    spectral analysis, 385
base, in BJT, 144–146
base characteristics, BJT, 149
base-collector diodes, 145
base-collector junction, 145, 147
base current and saturation, BJT switches, 155–156
base-emitter diodes, 145
base-emitter junction, 145, 147
baseband signals, 385
basic circuit parameters, 52–58
    capacitance, 53–54
    energy and energy storage, 54
    examples, 54–58
    inductance, 52–53
    and Ohm's law, 52
    passive circuit parameters, 52
    resistance, 52
    time-varying variables, 52
basic DC circuits, 3–32
    DC circuit variables, 3–8
    equivalent resistance, 17–19
    Kirchhoff's laws, 15–17
    measuring DC circuit variables, 12–13
    multiSIM examples, 410–415
    power and energy, 14–15
    problems, 30–32
    simple DC circuit, 8–12
    single loop circuit (series circuit), 19–23
    single node-pair circuit (parallel circuit), 23–28
    voltage and current divider rules, 28–30
baud rates, 397
BCD (binary coded decimal), 217
BER (bit error rate), 398
beyond pinchoff region, 158, 161, 163
bias compensation resistors, 186
bias supply, 175
bidirectional counters, 253
binary code, 209–212
    binary coded decimal (BCD), 217
        conversion from decimal to BCD, 217
        examples, 217
    binary vs M-ary communication, 378
    circuits, 209–212, 217
    conversion between binary and decimal systems, 210
    examples, 210–212
    See also hexadecimal number system (hex)
binary digit (bit), 210
binary phase-shift keying (BPSK), 395–396
binary point, 211
binary ripple counter, 252
binary vs M-ary communication, 378
bipolar junction transistors. See BJTs (bipolar junction transistors)
bipolar offset encoding, 390
bipolar offset representation in A/D conversion, 389
bit (binary digit), 210
    bits and bytes, digital communications, 378
    bits and levels, digital communications, 378
bit error rate (BER), 398
BJTs (bipolar junction transistors), 143, 144–146
    amplifier circuits, 153–154
    base, emitter and collector, 144–146
    characteristic curves, 148–153
        base characteristics, 149
        collector characteristics, 150
        common-emitter form, 148
        DC current gain, 151
        examples, 152–153
        measurement circuit, 148–149
        operating regions, 150–151
        polarities and current directions, 151
        variations in current gain, 151–152
    currents, 146
    examples, 146
    NPN and PNP, 144
    operating regions, 146–148
        active region, 148
        base-collector junction, 145, 147
        base-emitter junction, 145, 147
        cutoff region, 147
        examples, 148
        saturation region, 147
    operation, 145
    switches, 154–157
        base current and saturation, 155–156
        examples, 156–157
        NPN switch, 154–155
        PNP switch, 155
        saturation and cutoff, 154
block diagrams of op-amp, 179–180
Boole, George, 220, 228
Boolean algebra and switching functions, 220–222
    examples, 222
    postulates, 220–221
    theorems, 221
Boolean expansion, 229–230
    direct expansion, 229–230
    K-map expansion, 229
Boolean expression, 218, 219, 220, 360
Boolean reduction and Karnaugh maps, 222–228
    canonical form and reduced form, 222–223
    examples, 223, 224, 226, 227
    Karnaugh maps (K-maps), 224–228
    OR, ANDs, and SOP, 224
boundary conditions for energy storage elements, 58–64
    examples, 61–64
    inductive current and capacitive voltage boundary behavior, 58–59
    initial equivalent circuits, no energy storage, 59
    initial equivalent circuits with energy storage, 59–60
    skin effect, 61
    steady-state applications, 61
    steady-state DC conditions, 60
    steady-state DC models, 60–61
BPSK (binary phase-shift keying), 395–396
branches, Boolean expression, 360
breakdown speed, 339
brush drop, 310
brush loss, 317
brushes, 310
buck configuration, 299–300
buses, 24
Butterworth filter design forms, 201–202
Butterworth low-pass characteristics, 200–201
bytes, 212

C

canonical form, Boolean, 222–223
    minterm, definition, 223
    OR, ANDs, and SOP, 224
capacitance
    AC voltage-current relationships, 86–87
    definition, 53
    and inductance combinations, 69–70
    and inductance examples, 70–71
    passive circuit parameter, 53–54
capacitive energy, 54
capacitive impedance, 88–89
capacitive susceptance, 107
capacitive voltage-current relationships, 53
capacitor-start, capacitor-run induction motor (CSCR) (two value capacitor), 344
capacitor-start split-phase induction motor, 344
capacitors, 53
    in parallel, 70
    in a series, 70
    uncharged, 59
cemf, 321–322
channel, JFETs, 158
charge
    definition, 5
    symbol and unit, 4
check valve, 120
chip, 242
choke, 53
circuits
    AC, 77–94, 95–116
    admittance, 106–110
    admittances, 53–54, 106–107
    capacitance, 53–54
    circuit analysis
        basic electrical variables, 4
        operational amplifier, 183–184
    combining signals, 194–196
        difference amplifier, 195
        examples, 195–196
        linear combination circuit, 194–195
    complex power in AC circuits, 99–103
    DC, 3–32
    digital. See digital circuits
    energy and energy storage, 54
    impedance, 87–92
    inductance, 52–53
    initial equivalent, 59–60
    integration and differentiation, 196–198
    inverting amplifier, 184–188
    italics, use in basic electrical variables, 4
    latch circuits, 361–362
        safety considerations, 361
        sealing contact (seal-in contact), 361
        storage operations, 361–362
    magnetic circuits
        See magnetic circuits
    multiSIM examples, 427–431, 431–435
    noninverting amplifier circuits. See noninverting amplifiers and noninverting amplifier circuits
Ohm's law, 52

Index

parallel circuits. *See* parallel circuit (single node-pair circuit)
parameters, 52–58
    admittances, 53–54, 106–107
    capacitance, 53–54
    energy and energy storage, 54
    examples, 54–58
    inductance, 52–53
    and Ohm's law, 52
    passive circuit parameters, 52
    resistance, 52
    time-varying variables, 52
passive circuit parameters, 52
phasors, 80–84
power factor correction, 103–106
problems, 92–94, 115–116
rectifier circuits with filtering. *See* rectifier circuits with filtering
resistance, 52
resonance, 110–114
RMS values, 95–99
series circuit (single loop circuit). *See* series circuit (single loop circuit)
single-loop. *See* single-loop circuit
single node-pair. *See* parallel circuit (single node-pair circuit)
sinusoidal functions, 78–80
symmetrical circuit, 237
table of basic electrical variables
    table, 4
three-phase. *See* three-phase circuits
time-varying variables, 52
transient, 51–73
    basic circuit parameters, 52–58
    boundary conditions for energy storage elements, 58–64
    first-order circuits with DC excitations, 64–69
    inductance and capacitance combinations, 69–71
    multiSIM examples, 422–427
    problems, 71–73
    voltage-current relationships, 85–87
zener regulator circuits. *See* zener regulator circuits
circular (closed) cores, 265
clock input, 247
closed (circular) cores, 265
closed-loop gain, 182
CMOS (complementary metal oxide semiconductors), 243
coding. *See* encoding and encoders
coil, 53
collector, in BJT, 144–146
collector characteristics, BJT, 150
combinational logic and logic diagrams
    examples, 231
    steps in design of digital circuit, 230–231
common-emitter form, BJT, 148
common-source configuration, JFETS, 159
commutator and brush assembly, 310
complementary metal oxide semiconductors (CMOS), 243
complex exponential function, differentiation and integration, 85–86
complex plane, 81
complex power in AC circuits, 99–103
    apparent, real and reactive power, 99–101
    examples, 102–103
    power development, 99–100
    power triangle, 101
compound and shunt DC generators, 313–319
    armature, 313, 314, 315
    compound generator model, 318–319
    differentially compounded generator, 319
    efficiency, 316–317

examples, 315, 316, 317
flashing the field, 314
long shunt compound generator, 319
losses within the generator, 317
rheostat, 317
shunt field and series field, 313–314, 318
shunt generator model, 314–318
terminal voltage, 314
voltage buildup, 314
compound generator model, 318–319
compound motor
    advantages and disadvantages, 325
    reversing direction of rotation, 325
condenser, 53
conductance
    description, 7–8
    symbol and unit, 4
conducted power, 299
conductors, 120
connecting and disconnecting AC generator (alternator), 334
constant voltage model, diode circuit, 123
constant voltage transformer, 289
constellation diagrams, 397–398
contactors, 356
continuous time, 384
control transformers, 354
control voltage, 354
conventional current flow *vs* electron flow, 6
conversion of numbers
    from binary to decimal, 210
    from binary to hex, 215
    from decimal to BCD, 217
    from decimal to binary, 215
    from decimal to hex, 215
    from hex to decimal, 215
    from octal to binary, 216
    from two-complement to binary, 212–213
copper loss, 289–290
core laminations, 290
core loss, 289, 290
core material, flux, 262
core saturation
    desirable instances, 289
    and permeability, 267
counter-emf, 321–322
counters, 252–254
    asynchronous counters, 252–253
    bidirectional counters, 253
    binary ripple counter, 252
    look-ahead carry, 253
    for PLCs, 371
    propagation delay, 252–253
    synchronous counters, 253–254
    up-counters, 253
    up/down counters, 253
CPU
    digital communications, 379
    for PLCs, 363
CSCR (capacitor-start, capacitor-run induction motor) (two value capacitor), 344
CT (current transformer)
    instrument transformers, 301–302
Curie point, 262
current
    definition, 6
    ideal DC source, 7
    source, single-loop circuit, 20–21
    in terms of voltage and admittance, 106
current and voltage, representation as phasors, 82–83
current and voltage divider rules
    descriptions, 28–29
    examples, 29–30

current and voltage measurement, comparison, 13
current-controlled current sources (ICIS), 192–193
current-controlled voltage sources (ICVS), 192
current gain, DC, 151
current gain, variations, 151
current transformer (CT)
    instrument transformers, 301–302
cutoff frequency, 199, 200
cutoff region, 159
cutoff region, BJT, 147, 150

D

D/A conversions. *See* digital to analog (D/A) conversions
data rate, digital communications, 380
data rate (information rate), 397
data selectors or mux (multiplexers), 241
data valid output, 239
DC. *See* direct currents (DC)
decibel response forms, 200
decoders and decoding, 237–239, 387
    Boolean equations, 237
    definition, 237
    digital-to-analog (D/A) conversion, 387
degaussing, 262
delay-off (TOF) relays, 358–359
delay-on (TON) relay, 358
delta connection, 278
DeMorgan, Augustus, 228
DeMorgan's theorem, 228–229
depletion layer, JFETs, 158
depletion mode, MOSFETs, 157
dibit, 397
dielectric material, 61
difference amplifier, 195
differential input voltage, 181
differentially compounded generator, 319
differentiation and integration of complex exponential function, 85–86
digital circuits
    adders, 235–237
    advanced, 235–244
    basic, 209–233
    binary code. *See* binary code
    Boolean algebra and switching functions, 220–222
    Boolean expansion, 229–230
    Boolean reduction and Karnaugh maps, 222–228
    counters, 252–254
    decoders, 237–239
    DeMorgan's theorem, 228–229
    encoders, 239–241
    exclusive OR and exclusive NOR, 230
    flip flops, 247–251
    hexadecimal number system, 214–216
    integrated-circuit digital logic families, 242–243
    Karnaugh maps with don't care conditions, 228
    latch and flip flop initialization, 246
    latches, 246–247
    logic diagrams, 230–231
    logic operations, 218–219
    memories, 254–257
    multiplexers, 241
    parallel registers, 251
    sequential forms, 245–258
        counters, 252–254
        flip flops, 247–251
        latch and flip flop initialization, 246
        latches, 246–247
        memories, 254–257
        multiSIM examples, 461–468
        parallel registers, 251
        problems, 257–258
        serial (shift) registers, 251–252

digital circuits (*Continued*)
  serial (shift) registers, 251–252
  subtraction, twos-complement numbers, 213–214
  See also integrated-circuit digital logic families
digital communications, 377–399
  analog-digital and digital-analog conversion, 388–394
  analog signal, 384–386
  digital to analog (D/A) conversions, 387
  M-ary encoding, 396–398
  modulation methods, 395–396
  multiSIM examples, 478–484
  problems, 398–399
  pulse code modulation (PCM), 386–388
  terminology, 378–380
  textual data. *See* encoding and transmission of textual data
  See also integrated-circuit digital logic families
diodes, 119–142
  diode circuit models, 122–126
    constant voltage model, 123
    examples, 124–126
    forward biased and reversed biased, 124
    ideal diode model, 122–123
    other models, 123–124
  diode rectifier circuits, 126–132
    convert AC to DC, 126
    examples, 130–132
    full-wave bridge rectifier, 129–130
    full-wave rectifier (unfiltered), 128–129
    half-wave rectifier (unfiltered), 126–128
  diode terminal characteristics, 121–122
  junction diodes, 120–122
  multiSIM examples, 435–439
  other diode types, 136–137
  problems, 140–142
  rectifier circuits with filtering, 132–136
  semiconductor concepts, 120
  zener regulator circuits, 137–140
direct coupling, 13
direct current (DC)
  AC compared, 4
  ammeter, ideal, 12
  armature, 313, 314, 315
  back emf, 321
  basic DC circuits, 3–32
    DC circuit variables, 3–8
    equivalent resistance, 17–19
    Kirchhoff's laws, 15–17
    measuring DC circuit variables, 12–13
    multiSIM examples, 410–415
    power and energy, 14–15
    problems, 30–32
    simple DC circuit, 8–12
    single loop circuit (series circuit), 19–23
    single node-pair circuit (parallel circuit), 23–28
    voltage and current divider rules, 28–30
  BJT and DC current gain, 151
  brush drop, 310
  brushes, 310
  circuit variables, 3–8
    measuring, 12–13
  commutator and brush assembly, 310
  compound and shunt DC generators, 313–319
    armature, 313, 314, 315
    compound generator model, 318–319
    differentially compounded generator, 319
    efficiency, 316–317
    examples, 315, 316, 317
    flashing the field, 314
    long shunt compound generator, 319
    losses within the generator, 317
    rheostat, 317
    shunt field and series field, 313–314, 318
    shunt generator model, 314–318
    terminal voltage, 314
    voltage buildup, 314
  compound generator model, 318–319
  compound motor, 325
  counter-emf (cemf) (back emf), 321
  current gain, 151
  current source, ideal, 7
  DC machines, 307–327
  DC values, 95–96, 97
  definition, 6
  differentially compounded generator, 319
  dynamic braking of DC motors, 326
    plugging, 326
  efficiency, 316–317
  equivalent resistance, 17–19
    examples, 18–19
    parallel resistance, 17–18
    properties, 18
    series resistance, 17
    two resistances in parallel, 18
  examples, 309, 313
  Faraday's Law, 308, 309, 312
  field flux, cemf, armature current, motor speed relationship, 321–322
  field voltage, 311
  first-order circuits with DC excitations, 64–65
    examples, 65–69
    first-order differential equation, 64–65
    RL and RC circuits, 64
    time constant, 64–65
  first order differential equation, 64–65
  flashing the field, 314
  flux compression (flux bunching), 320
  general DC circuit analysis, 33–49
  generators and magnetic induction, 308–313
    brush drop, 310
    brushes, 310
    commutator and brush assembly, 310
    examples, 309, 313
    Faraday's Law, 308, 309, 312
    field voltage, 311
    prime mover, 311
    reversing output voltage polarity of generators, 311–312
    right-hand rule, 309
    self-excited generators, 311
    slip rings, 310
    tachogenerators, 313
  ideal DC ammeter, 12
  ideal DC current source, 7
  ideal DC voltage source, 7
  ideal DC voltmeter, 12
  Kirchhoff's laws, 15–17
  Lenz's Law, 319, 321
  long shunt compound generator, 319
  losses within the generator, 317
  magnetic induction and DC generators, 308–313
    brush drop, 310
    brushes, 310
    commutator and brush assembly, 310
    examples, 309, 313
    Faraday's Law, 308, 309, 312
    field voltage, 311
    prime mover, 311
    reversing output voltage polarity of generators, 311–312
    right-hand rule, 309
    self-excited generators, 311
    slip rings, 310
    tachogenerators, 313
  measuring DC circuit variables, 12–13
    examples, 13
  mesh current analysis, 33–36
  motors, 319–326
    efficiency, 326
    motor action, 319–322
      counter-emf (cemf) (back emf), 321
      field flux, cemf, armature current, motor speed relationship, 321–322
      flux compression (flux bunching), 320
      Lenz's Law, 319, 321
      motor generator, 320
      permanent magnet DC motor, 321
      right-hand rule, 319, 320
    motor generator, 320
  node voltage analysis, 36–39
  Ohm's law, 9
  permanent magnet DC motor, 321
  phasor magnitudes, RMS values as, 98
  power and energy, 14–15
    alternate expressions of power relationships in DC circuits, 9–10
    power relationships in DC circuits, 9
  prime mover, 311
  reversing output voltage polarity of generators, 311–312
  rheostat, 317
  right-hand rule, 309, 319, 320, 321
  RMS and DC values, 95–99
    examples, 99
    power in resistance, 96–97
  self-excited generators, 311
  series motor, 324–325
  shunt, series, and compound DC motor, 322–326
    compound motor, 325
    examples, 323, 325
    motor efficiency, 326
    series motor, 324–325
    shunt motor, 322–324
  shunt and compound DC generators, 313–319
    armature, 313, 314, 315
    compound generator model, 318–319
    differentially compounded generator, 319
    efficiency, 316–317
    examples, 315, 316, 317
    flashing the field, 314
    long shunt compound generator, 319
    losses within generators, 317
    rheostat, 317, 319
    short shunt compound generators, 318
    shunt field and series field, 313–314, 318
    shunt generator model, 314–318
    terminal voltage, 314
    voltage buildup, 314
  shunt field and series field, 313–314, 318
  shunt generator model, 314–318
  shunt motor, 322–324
  simple DC circuits, 8–12
    alternate expressions, 9–10
    description, 8–9
    examples, 11–12
    Ohm's law, 9
    power relationships, 9
  single loop circuit (series circuit), 19–23
  single node-pair circuit (parallel circuit), 23–28
  sinusoid values, 97–98
  slip rings, 310
  source transformations, 39–41
  symbols, 4
  tachogenerators, 313
  terminal voltage, 314
  Thevenin's and Norton's theorems, 41–47
  torque, 320

voltage
    buildup, 314
    current divider rules, 28–30
    electromotive force (emf), 6
    ideal sources, 6–7
    potential difference, 6
    symbols, 4, 6
    voltmeter, ideal DC, 12
direct flux (main flux), 341
discrete time, 384
distortion, signal, in amplifiers, 175
diverter, 319
donors, 120
don't care conditions in K-maps, 228
doping, 120
double-subscript notation, 12
drain, JFETs, 158, 159–160
drain current equation, 166
DRAM (dynamic random access memory), 257
duplex connection, 379
dynamic braking of DC motors, 326
    plugging, 326
dynamic random access memory (DRAM), 257

E

E-core transformer construction, 295
Early effect, 149
eddy current loss, 290
edge-triggered D flip flop, 248–249
Edison, Thomas, 286
EEPROM (electrical erasable programmable ROM), 257, 364
effective values
    AC voltage and current, 10
    of sinusoid, 97–98
effective values of AC voltage and current, 10
efficiency of generators
    shunt and compound DC generators, 316–317
electrical erasable programmable ROM (EEPROM), 257, 364
electrically programmable ROM (EPROM), 256–257
electromagnetism, 261–262
    Curie point, 262
    degaussing, 262
    and ferrous materials, 262
    magnetic field direction of electromagnet, 265
    magnetization curve and permeability, 267
    magnetomotive force (mmf), 264
    retentivity, 262
    See also magnetic circuits
electromotive force. *See* emf (electromotive force)
electron flow *vs* conventional current flow, 6
electrons, 120
emergency stop pushbutton switches, 355
emf (electromotive force), 6
    back emf, 321
    cemf, 321–322
emitter, in BJT, 144–146
encoding and encoders, 239–241
    bipolar offset, 390
    data valid output, 239
    examples, 240–241
    M-ary. *See* M-ary encoding
    priority encoders, 240
    textual data, 381–383
        ASCII Code (American Standard Code for Information Interchange), 381, 382
        asynchronous serial data transmission, 381–383
        conclusion, 383
        examples, 383
    unipolar, 389
energy
    capacitive, 54

definition, 14
inductive, 54
kinetic, 54
potential, 54
and power, 14–15
and power, example, 15
symbol and unit, 4
See also power
energy storage elements, boundary conditions, 58–64
    examples, 61–64
    inductive current and capacitive voltage boundary behavior, 58–59
    initial equivalent circuits, no energy storage, 59
    initial equivalent circuits with energy storage, 59–60
    skin effect, 61
    steady-state applications, 61
    steady-state DC conditions, 60
    steady-state DC models, 60–61
energy storage parameters, 52, 54
enhancement mode, MOSFETs, 157
EPROM (electrically programmable ROM), 256–257
equivalent circuit models, amplifiers, 175–176
equivalent resistance, DC circuits
    examples, 18–19
    parallel resistance, 17–18
    properties, 18
    series resistance, 17
    two resistances in parallel, 18
Euler's formula, 81
even parity, 383
examples
    AC generator (alternator), 333–334, 336
    AC impedance, 89–92
    active filters, 203–204
    admittance, 109—110
    amplifier properties, 178–179
    analog-digital and digital-analog conversion, 392–394
    analog signal, sampling, 385–386
    autotransformers, 299, 300
    basic circuit parameters, 54–58
    binary coded decimal (BCD), 217
    binary number system, 210–212
    bipolar junction transistor (BJT), 146
    BJT characteristic curves, 152–153
    BJT operating regions, 148
    BJT switches, 156–157
    Boolean algebra and switching functions, 222
    Boolean reduction and Karnaugh maps, 223, 224, 226, 227
    boundary conditions for energy storage elements, 61–64
    circuits, combining signals, 195–196
    complex power in AC circuits, 102–103
    DC and RMS values, 99
    diode circuit models, 124–126
    diode rectifier circuits, 130–132
    encoders, 240–241
    encoding and transmission of textual data, 383
    equivalent resistance, 18–19
    flux density, 263
    hexadecimal number system (hex), 215–216
    impedance matching transformers, 293–294
    impedance reflection, 292–293
    inductance and capacitance combinations, 70–71
    instrument transformers, 301, 302
    integration and differentiation circuits, 198
    inverting amplifier circuit, 187–188
    JFETs (junction field effect transistors), 161–163
    Karnaugh maps (K-maps), 223, 224, 226, 227
    Kirchhoff's laws, 16–17
    logic diagrams and combinational logic, 231

magnetic induction and DC generators, 309, 313
magnetomotive force (mmf), 264
measuring DC circuit variables, 13
memories, 255
mesh current analysis, 34–36
MOSFETs, 166
node voltage analysis, 37–39
noninverting amplifier circuit, 190–191
octal number system, 216–217
operational amplifier controlled sources, 193–194
permeability, 269–270
phase voltages and line voltages, 276
phasors, 83–84
PLC programming, 367–368
PLC programs with timers, 369–370
power and energy, 15
rectifier circuits with filtering, 136
reluctance, 265
reluctance motors, 347
resonance, 114
shunt, series, and compound DC motor, 323, 325
shunt and compound DC generators, 315, 316, 317
simple DC circuit, 11–12
single-loop circuit, 21–23
single node-pair circuit, 26–28
source transformations, 40–41
subtraction using two-complement numbers, 214
terminology, digital communications, 380
Thevenin's and Norton's theorems, 42–47
three-phase induction motor, 339–340
three-phase power calculations, 280–281
three-phase theory, 273
three-phase transformers, 296–297
transformer performance, effect of power factor, 291
transformers, ideal, 287–288
voltage and current divider rules, 29–30
zener regulator circuits, 138–140
exclusive NOR
    agreement gate, 230
    XNOR, 230
exclusive OR
    disagreement gate, 230
    XOR, 230
exponential form of complex number, 81
extended ASCII code (American Standard Code for Information Interchange), 381

F

fanout, 242
farad, 53
Faraday's Law
    AC generator, 330
    magnetic induction, 308, 309, 312
    three-phase induction motor, 338
ferromagnetic core, 52
ferroresonant transformer, 289
ferrous materials and magnetism, 262
FET (field effect transistor) family, 157
FET schematic summary, 168–169
FET switches, 166–168
    analog multiplexer, 167–168
    JFET resistance in ohmic region, 167
    N-channel switches, 167
field copper loss, 317
field effect transistor (FET) family, 157
field flux, cemf, armature current, motor speed relationship, 321–322
field loss, 336
field voltage, 311
filter circuits, 135
filter design forms, Butterworth, 201–202
filters. *See* active filters

first-order circuits with DC excitations
examples, 65–69
first-order differential equation, 64–65
RL and RC circuits, 64
time constant, 64–65
flashing the field, 314
flip flops, 247–251
clock input, 247
edge-triggered D flip flop, 248–249
JK flip flops, 250–251
synchronous devices, 247
T flip flops, 250
timing diagrams, 247–248
flux
core material, 262
leakage, 294
MAGLEV (magnetically levitated), 263
main (direct), 341
quadrature, 342
right-hand rule for flux, 262
solenoid, 262
flux bunching (flux compression), 320
flux compression (flux bunching), 320
flux density
definition, 263
examples, 263
ideal transformers, 288–189
forward-bias region, 122, 124
four-wire system, 276
Fourier analysis, 385
frequency-shift keying (FSK), 395
frequency spectrum, 384–385
friction, 317
friction and windage loss, 336
FSK (frequency-shift keying), 395
full-adder, 236
full duplex connection, 379
full-wave rectifiers with capacitor filter, 133–134
fuses, 354

G

gate-to-source cutoff voltage, 159
gates, JFETs, 158
general DC circuit analysis, 33–49
mesh current analysis, 33–36
multiSIM examples, 415–421
node voltage analysis, 36–39
problems, 47–49
source transformations, 39–41
Thevenin's and Norton's theorems, 41–47
General Motors, 352
generators
alternators. See alternators (AC generators)
compound. See compound and shunt DC generators
efficiency
AC generators, 332, 335–336
shunt and compound DC generators, 316–317
long shunt compound generator, 319
losses within generators, 317
magnetic induction. See magnetic induction and DC generators
motor generators, 320
reversing output voltage polarity, 311–312
reversing output voltage polarity of generators, 311–312
self-excited, 311, 314
short shunt compound generators, 318
shunt DC generators. See compound and shunt DC generators
tachogenerators, 313
ground
definitions, 10
reference (neutral), 274

symbols, 10
virtual, 185
ground rail, 354, 359

H

half-adder, 236
half duplex connection, 379
half-wave rectifier circuit (unfiltered), 126–128
half-wave rectifier with capacitor filter, 132–133
henry, 53
hex. See hexadecimal number system
hexadecimal number system (hex), 214–216
conversion from binary to hex, 215
conversion from decimal to hex, 215
conversion from hex to binary, 216
conversion from hex to decimal, 215
examples, 215–216
See also binary code
high-pass filters, 199
high side winding, 287
holes, 120
hot rail, 354, 359
hysteresis loss, 290

I

IC (integrated circuit), 144
ICIS (current-controlled current sources), 192, 193
ICVS (current-controlled voltage sources), 192
ideal DC ammeter, 12
ideal DC current source, 7
ideal DC voltage source, 7
ideal DC voltmeter, 12
ideal diode model, 122–123
ideal op-amp differentiator circuits, 197–198
ideal op-amp integrator circuits, 196–197
ideal sources, 6–7
ideal DC current source, 7
ideal DC voltage source, 7
ideal transformers, 286–289
constant voltage transformer, 289
examples, 287–288
ferroresonant transformer, 289
high side winding, 287
low side winding, 287
primary winding, 286
secondary winding, 286
turns ratio, 287
ideal voltage amplifier, 177
IGFETs (insulated gate field effect transistors), 157
imaginary axis, 81
impedance, 87–92
AC impedance, 87–92
admittance, 88
capacitive, 88–89
examples, 89–92
impedance matching transformers, 293–294
examples, 293–294
impedance-protected armature coils, 345
impedance reflection, 292–293
examples, 292–293
inductive, 88
resistance and reactance, 87–88
series and parallel resonances, 113
steady-state AC model, 89
impurities, 120
inductance
AC voltage-current relationships, 86
and capacitance combinations, 69–70
and capacitance exercises, 70–71
passive circuit parameter, 52–53, 54
induction motors
single-phase, 341–346
three-phase, 337–341

inductive energy, 54
inductive impedance, 88
inductive susceptance, 107
inductive voltage-current relationships, 53
inductor, unfluxed, 59
inductors, 52
in parallel, 70
in series, 69–70
information rate (data rate), 397
initial equivalent circuits, 59–60
initialization, latch and flip flop, 246
input and output resistances, 175–176
inverting amplifier circuit, 186
noninverting amplifier circuit, 189
input for PLCs
discrete inputs, 363, 372
register inputs, 363
instantaneous power, 54, 96
instantaneous waveforms, 273
instrument transformers, 300–303
current transformer (CT), 301–302
examples, 301, 302
potential transformer (PT), 301
ratings, 301
safety issues, 302
insulated gate field effect transistors (IGFETs), 157
insulators, 120
integrated-circuit digital logic families, 242–243
chip, 242
complementary metal oxide semiconductors (CMOS), 243
fanout, 242
pinouts, 243
quad two-input NAND, 242
quad two-input NOR, 242
transistor-transistor logic (TTL), 242–243
triple three-input AND, 242
integrated circuit (IC), 144
integration and differentiation circuits, 196–198
AC integrator, 197
examples, 198
ideal differentiator, possible problems, 198
ideal integrator, possible problems, 197
ideal op-amp differentiator circuit, 197–198
ideal op-amp integrator circuit, 196
low-frequency differentiator, 198
internal relays in PLCs, 368
INVERT (or NOT), 219
inverting amplifiers and inverting amplifier circuits, 174–175, 184–188
closed-loop gain, 185–186
examples, 187–188
input and output resistances, 186
and Ohm's law, 185
resistance values, 186–187
inverting input, 180
italics, use for electrical variables, 5

J

JFETs (junction field effect transistors), 157
amplifier circuit, 163
beyond pinchoff region, 158, 161, 163
channel layout, 158
characteristics, 159
cutoff region, 158, 159
drain current formula, 161
examples, 161–163
gate-to-source cutoff voltage and cutoff region, 159
N-channel layout, 158
ohmic region, 160–161, 166–167
operating regions, 158
switch, N-channel, 167
JK flip flops, 250–251

Index 509

junction diode, 120–122
  applications, 121
  and Ohm's law, 121
  schematic symbols, 121
  terminal characteristics, 121–122
junction field effect transistors (JFETs), 157
junctions
  bipolar junction transistors. *See* BJTs (bipolar junction transistors)

K

K-maps. *See* Karnaugh maps
Karnaugh, Maurice, 224
Karnaugh maps (K-maps), 224–228
  with don't care conditions, 228
  examples, 223, 224, 226, 227
  five- and six-variable, 228
  four-variable, 227
  three-variable, 225–226
  topology, 226–227
  two-variable, 224–225
KCL (Kirchhoff's current law), 16
keying
  amplitude-shift keying (ASK), 395
  binary phase-shift keying (BPSK), 395–396
  frequency-shift keying (FSK), 395
  quadriphase shift keying (QPSK), 397
kinetic energy, 54
Kirchhoff's current law (KCL), 16
Kirchhoff's laws
  examples, 16–17
  and single-loop circuit, 20–21
  and single node-pair circuit, 24–26
Kirchhoff's voltage law (KVL), 15–16
KVL (Kirchhoff's voltage law), 15–16

L

ladder diagrams, 359–361
  branches, 360
  ground rails, 359
  hot rails, 359
  loads, 359–360
  rungs, 359
  safety considerations, 359
ladder logic, 352, 359
lagging power factor, 101
lamps on PLCs, 356
large-scale integration (LSI), 144
latch circuits, 361–362
  safety considerations, 361
  sealing contact (seal-in contact), 361
  storage operations, 361–362
latches
  asynchronous devices, 246
  initialization, 246
  RS latch, 246
  $\overline{RS}$ latch, 246–247
leading power factor, 101
leakage flux, 294
least significant bit (LSB), 210, 381, 388
LED (light emitting diode), 137
Lenz's Law
  DC motor, 319, 321
  shaded pole, 345
  three-phase induction motor, 338
level-triggered D flip flop, 247–248
levels and bits, digital communication, 378
light emitting diode (LED), 137
limit switch, 354, 355–356
line current, 278
line-to-line voltages, 276
linear amplifier, 174
linear combination circuit, 194–195

loads, Boolean expression, 359–360
logic
  Boolean expression, 218, 219, 220
  AND circuit (AND gate), 218
  diagrams and combinational logic
    examples, 231
    steps in design of digital circuit, 230–231
  INVERT (or NOT), 219
  NAND and NOR operations, 219–220
  operations, 218–219
  OR gate, 219
  programmable logical controllers. *See* PLCs (programmable logic controllers)
  See also integrated-circuit digital logic families
long shunt compound generator, 319
look-ahead carry, 253
losses within alternators, 336
losses within generators, 317
low-frequency differentiator, 198
low-pass filters, 199
low side winding, 287
LSB (least significant bit), 210, 381, 388
LSI (large-scale integration), 144
lumped resistor, 9

M

M-ary encoding, 396–398
  binary *vs* M-ary communication, 378
  constellation diagrams, 397–398
  M-ary relationship, 396–397
  QAM (quadrature amplitude modulation), 397–398
  QPSK (quadriphase shift keying), 397
  signal-to-noise ratio, 398
  symbol rate and data rate, x397
machine control components, 352–359
  control transformer, 354
  delay-off, 358–359
  delay-on, 358
  fuses, 354
  lamps, 356
  reference designators, 352–353
  relays, 356–357
  switches, 354–356
  time-delay relays, 357–358
  wire numbering, 353
MAGLEV (magnetically levitated), 263
magnetic circuits, 261–270
  closed (circular) cores, 265
  flux, 262–263
  flux density, 263
  magnetic field direction of electromagnet, 265
  magnetic field intensity, 266
  magnetism and electromagnetism, 261–262
  magnetomotive force, 264
  permeability, 266–270
  problems, 270
  reluctance, 264–265
  See also electromagnetism
magnetic field direction of electromagnet, 265
magnetic field intensity, 266
magnetic induction and DC generators, 308–313
  brush drop, 310
  brushes, 310
  commutator and brush assembly, 310
  examples, 309, 313
  Faraday's Law, 308, 309, 312
  field voltage, 311
  prime mover, 311
  reversing output voltage polarity of generators, 311–312
  right-hand rule, 309
  self-excited generators, 311

slip rings, 310
tachogenerators, 313
magnetically levitated (MAGLEV), 263
magnetization curve and permeability, 267
magnetomotive force (mmf), 264
main flux (direct flux), 341
marks and spaces, digital communications, 379
mask-programmed ROM, 255–256
mathematics for PLCs, 371
measurement circuit, BJT, 148–149
measurements
  comparison of voltage and current, 13
  devices, 13
  resistance, 13
  summary, 13
measuring DC circuit variables, 12–13
  examples, 13
memories, 254–257
  digital communications, 379
  dynamic random access memory (DRAM), 257
  examples, 255
  for PLCs, 364
  read only memory (ROM), 255–257
  static random access memory (SRAM), 257
memory, volatile, 245
mesh current analysis, 33–36
  current source, 34
  definition, 33
  examples, 34–36
  and KVL, 34
metal-oxide semiconductor field effect transistors. *See* MOSFETs
minterm
  definition, 223
  and K-maps, 225
minuend and subtrahend, 213
mmf (magnetomotive force), 264
modular PLCs, 362
modulation methods
  amplitude-shift keying (ASK), 395
  binary phase-shift keying (BPSK), 395–396
  frequency-shift keying (FSK), 395
MOSFETs
  depletion mode and enhancement mode, 157, 164–165
  drain current equation, 166
  examples, 166
  N-channel and P-channel, 157, 164–165
  switch, N-channel, 167
  threshold voltage, 165
most significant bit (MSB), 210, 381, 388
motor action and DC motors, 319–322
  counter-emf (cemf) (back emf), 321
  field flux, cemf, armature current, motor speed relationship, 321–322
  flux compression (flux bunching), 320
  Lenz's Law, 319, 321
  motor generator, 320
  permanent magnet DC motor, 321
  right-hand rule, 319, 320
motor efficiency, 326
motor generator, 320
motor speed, field flux, cemf, armature relationship, 321–322
MSB (most significant bit), 210, 381, 388
multimeters, 12
multiplexers
  analog, 167–168
  data selectors or mux, 241
  time-division (TDM), 167–168
MultiSIM laboratory, 401–483
  analysis and simulation, 408
  circuit schematics, 404

Index

MultiSIM laboratory (*Continued*)
  components, 405
  grids, 404, 405
  grounds, 406
  node numbering, 407–408
  notational conventions, 408
  notational differences, 401
  opening windows, 407
  parts and bins, 405
  reference designation, 408–409
  standard components *vs* virtual components, 408
  toolbars, 404
  virtual oscilloscope, 409–410
  wiring, 406
mux or data selectors (multiplexers), 241

N

N-channel
  JFETs, 157, 167
  MOSFETs, 157, 167
N-type material, 120
negative binary numbers, 212–213
  conversion from twos-complement to binary, 212–213
  sign and magnitude, 212
  twos complement, 212
negative feedback, 152, 182
negative signs, 14
neutral, 272, 274, 278
nibbles, 212
no-load (open-circuit) test, 290
node voltage analysis, 36–39
  common ground, 36
  examples, 37–39
  and KCL, 37
non-feedback circuits, 42
noninverting amplifiers and noninverting amplifier circuits, 174–175, 188–191
  alternate configuration, 189
  examples, 190–191
  input and output resistances, 189
  voltage follower, 190
noninverting input, 180
nonretentive on-timer (TON), 369, 370
normalization, 389
normalized frequency, 200
Norton's and Thevenin's theorems, 41–47
  examples, 42–47
  non-feedback circuits, 42
  open-circuit voltage and Thevenin's theorem, 41–42
  resistance, 42
  short-circuit current and Norton's theorem, 42
  source transformations, 42
NPN and PNP, 144
NPN switch, 154–155

O

O-core (shell) transformer construction, 294–295
octal number system, 216–217
  conversion between octal and binary, 216
  conversion between octal and hex, 216
  examples, 216–217
odd parity, 383
ohmic region
  description, 160–161
  FET switches, 166–167
ohmmeter, 12, 13
Ohm's law
  AC voltage-current relationships, 85
  and admittance, 106
  and basic circuit parameters, 52
  definition, 9
  and junction diodes, 121
  and single-loop circuit, 20–21
  and single node-pair circuit, 24–25
on-off keying (OOK), 395
OOK (on-off keying), 395
op-amp differentiator circuits, 197–198
op-amp integrator circuits, 196–197
op-amps. *See* operational amplifiers
open-circuit (no-load) test, 290
open-circuit voltage and Thevenin's theorem, 41–42
open-circuit voltage gain, 175–176
open circuits, 8
open frame PLCs, 362
open-loop gain, 182
open-loop voltage, 181
operating regions, BJT, 150–151
operational amplifiers (op-amp), 173–206
  active filters, 199–204
  amplifier properties, 173–179
  block diagrams, 179–180
  circuit analysis, 183–184
  circuits that combine signals, 194–196
  controlled sources, 191–194
    current sources, voltage-controlled and current controlled, 192–193
    examples, 193–194
    voltage sources, voltage-controlled and current controlled, 192
  DC coupling and offsets, 181
  differential input voltage, 181
  examples, 182–183
  input-output characteristics, 181
  integration and differentiation circuits, 196–198
  inverting amplifier circuit, 184–188
  inverting and noninverting input, 180
  multiSIM examples, 443–449
  negative feedback, 182
  noninverting amplifier circuit, 188–191
  open-loop voltage, 181
  operational amplifiers (op-amps), 179–183
  power supply connections, 179–180
  problems, 205–206
  saturation voltages, 182
  symbol direction, 180
  voltage regulation with op-amp, 204
optocoupler, 137
OR gate, 219
oscilloscope, 13, 52, 309
output and input resistances, 175–176, 186
  inverting amplifier circuit, 186
  noninverting amplifier circuit, 189
output for PLCs
  comparison of types, 374–375
  discrete relay outputs, 363, 372–375
  discrete solid-state outputs, 363, 372–375
  triac outputs, 374–375
overflow, twos-complement, 214

P

P-channel
  JFETs, 157
  MOSFETs, 157
P-type material, 120
PAM (pulse amplitude modulated) signal, 168
parallel circuit (single node-pair circuit), 23–28
  application to representative circuit, 25, 26
  buses, 24
  constraint, 24
  conversion to series circuit, 108–109
  current sources and resistances, 25–26
  examples, 26–28
  and Kirchhoff's laws, 24–26
  and Ohm's law, 24–25
parallel registers, 251
parallel resistance, 17–18
parallel resonance, 112–113
parallel *vs* serial, digital communications, 379
parity check bit, 381
passband, 200
passive circuit parameters, 52–54
passive filters, 199
PCM (pulse code modulation, 386–388
  decoding, 387
  quantization, 386–387
  signal restoration, 387
  time-division multiplexing, 387–388
PCM word, 387
periodic waveform, 95
permanent magnet DC motor, 321
permanent split capacitor (PSC) motor, 344
permeability, 266–270
  B-H curve, 268
  core saturation, 267
  examples, 269–270
  magnetization curve, 268
  saturation curve, 268
phase currents, 278
phase-shift keying (PSK), 395
phase voltages and line voltages, 274–276
  conventions, 276
  examples, 276
  four-wire system, 276
  line-to-line voltages, 274
  line voltages, 274–276
  phase voltages, 274
  three-wire system, 276
phasor forms, three-phase theory, 273–274
phasor magnitudes, RMS values as, 98
phasor transform, 83
phasors, 80–84
  definitions, 81–82
  Eulers formula, 81
  examples, 83–84
  representation of current and voltage, 82–83
  steady-state response in AC circuits, 80
photodiode, 137
pinouts, 243
PLCs (programmable logic controllers)
  advanced features, x371
  advanced PLC features
    counters, 371
    mathematics, 371
    sequencers, 371
  configuration, 363–364
    CPU, 363
    input, 363
    output, 363
    power supply, 363
    programming unit, 363–364
  ladder diagrams, 359–361
  latch circuits, 361–362
  machine controls, components, 352–359
  machine controls, history, 351–352
  modular, 362
  multiSIM examples, 476–478
  open frame, 362
  operation, 364–365
  problems, 375
  programming, 365–369
    examples, 367–368
    internal relays, 368
  programs with timers, 369–370
  shoe box (brick), 362
  timers, PLC programs with, 369–370
    accumulator value, 369
    examples, 369–370

nonretentive on-timer (TON), 369–370
other timers, 370
preset value, 369
types, 362
wiring, 371–375
discrete inputs, 372
discrete outputs, 372–375
power connections, 372
plugging, 326
PNP and NPN, 144
PNP switch, 155
polar form of complex number, 82
polarities and current directions, BJT, 151
positive resistance, 14
postulates, Boolean algebra, 220–221
potential difference, 6
potential energy, 54
potential transformer (PT)
instrument transformers, 301
power
complex, in AC circuits, 99–103
definition, 14
and energy, 14–15
and energy, example, 15
instantaneous, 54, 96
reactive, 101
symbol and unit, 4
See also energy
power absorbed, 14
power connections in PLCs, 372
power delivered, 14
power development, 100
power factor, 100–101
power factor, effect on transformer performance, 291–292
power factor correction, 103–106
examples, 104–106
power triangle, 103–104
power flow in single- and three-phase systems, 281–283
single-phase instantaneous power, 281–282
three-phase instantaneous power, 282–283
power relationships in DC circuits, 9
alternate expressions, 9–10
power supplies
PLCs, 363
voltage regulator, 135
power triangle, 101, 103–104
prefixes and units for electrical variables
table and symbols, 4–5
use of italics, 5
preset value, 369
primary winding, 286
prime mover, 311
priority encoders, 240
problems
AC circuits, 92–94, 115–116
AC machines, 348
DC circuits, basic, 30–32
DC machines, 326–327
digital circuits, advanced, 243–244
digital circuits, basic, 231–233
digital circuits, sequential forms, 257–258
digital communications, 398–399
diodes and their applications, 140–142
general DC circuit analysis, 47–49
magnetic circuits, 270
operational amplifiers, 205–206
programmable logic controllers (PLCs), 375
three-phase circuits, 283–284
transformers, 303–304
transient circuits, 71–73
transistors, 169–171

programmable logical controllers. *See* PLCs (programmable logic controllers)
programmable ROM (PROM), 256
programming unit for PLCs, 363
PROM (programmable ROM), 256
propagation delay, 252–253
PSC (permanent split capacitor) motor, 344
PSK (phase-shift keying), 395
PT (potential transformer)
instrument transformers, 301
pulse amplitude modulated (PAM) signal, 168
pulse code modulation (PCM), 386–388
decoding, 387
quantization, 386–387
signal restoration, 387
time-division multiplexing, 387–388
pushbutton switches, 354–356
emergency stop, 355
momentary or maintained, 354–355
normally closed (NC), 354–355
normally open (NO), 354–355

Q
Q-factor, 113–114
QAM (quadrature amplitude modulation), 397–398
QPSK (quadriphase shift keying), 397
quad two-input NAND, 242
quad two-input NOR, 242
quadrature amplitude modulation (QAM), 397–398
quadrature field, 342
quadrature flux, 342
quadriphase shift keying (QPSK), 397
quantization, 386–387
analog-to-digital (A/D) conversion, 387
curves, 390–391
error, 392
PCM word, 387

R
RAM (random access memory), 364
random access memory (RAM), 364
reactance and resistance, 87–88
reactive parameters, 52
reactive power, 99–100
read only memory (ROM), 255–257
electrical erasable programmable ROM (EEPROM), 257
electrically programmable ROM (EPROM), 256–257
mask-programmed ROM, 255–256
for PLCs, 364
programmable ROM (PROM), 256
real axis, 81
real power, 99–100
rectangular form of complex number, 81
rectifier circuits with filtering
examples, 136
filter circuits, 135
full-wave rectifiers with capacitor filter, 133–134
half-wave rectifier with capacitor filter, 132–133
power supplies with voltage regulator, 135
reverse diode voltage, 135
ripple analysis, 134–135
rectifiers, 335
reduced form, Boolean, 222–223
reference designators, 352–353
registers
parallel, 251
serial (shift) registers, 251–252
regulator circuit, 204
relays
definition, 356
energized and de-energized conditions, 357

reasons for use, 357
relay output for PLCs, 372–375
reluctance, 264–265
examples, 265
reluctance motors, 346–347
characteristics of induction and synchronous motors, 347
examples, 347
representation of current and voltage as phasors, 82–83
resistance, 87–88
AC impedance, 88
AC voltage-current relationships, 85
amplifier circuit, resistance values, 186–187
definition, 7
equivalent, 17–18
examples, 18–19
parallel, 17–18
properties, 18
series, 17
input, 175–176, 186, 189
measurements, 13
in ohmic region, JFET, 85
and Ohm's law, 85
output, 175–176, 186, 189
passive circuit parameter, 52
positive, 14
resistance-start split-phase induction motor, 343–344
resistors, 7
bias compensation, 186
lumped resistor, 9
symbol and unit, 4
Thevenin's and Norton's theorems, 42
values, amplifier circuit, 186–187
voltage-controlled, 160, 166
resistance-start split-phase induction motor, 343–344
resonance, 110–114
bandwidth, 114
examples, 114
parallel resonance, 112–113
properties, 113
Q-factor, 113–114
series resonance, 111–112
resonant frequency, 111
retentivity, 262
reverse-bias region, 122, 124
reverse breakdown, 122
reverse diode voltage, 135
reversing output voltage polarity of generators, 311–312
rheostat, 317, 319
right hand rule
flux, 262
magnetic induction, 309, 320
solenoids, 265
ripple, 132, 135
ripple analysis, 134–135
RMS value (root-mean-square), 10, 95–99
roadmap, reference designators, 353
ROM (read only memory), 255–257
electrical erasable programmable ROM (EEPROM), 257
electrically programmable ROM (EPROM), 256–257
mask-programmed ROM, 255–256
for PLCs, 364
programmable ROM (PROM), 256
root-mean-square (RMS), 10, 95–99
rotor copper loss, 317
rotor core loss, 317
RS or $\overline{RS}$ latch, 246
rung, 359

S

safety considerations, 353, 359, 361
sampled-data signal, 384
saturation
 current increase after, 155
 and cutoff in switches, 154
 region, BJT, 147–150
 voltages, 182
saturation curve and permeability, 267
schematic symbols, junction diodes, 121
Schottky diode, 137
seal-in contact (sealing contact), 361
sealing contact (seal-in contact), 361
secondary winding, 286
selectivity factor, 113
selector switches, 354, 355, 356
self-excited generators, 311, 314
semiconductor concepts, 120
semiconductors, 120
sequencers for PLCs, 371
sequencing, three-phase, 277
serial (shift) registers, 251–252
series circuit (single loop circuit), 19–23
 application to representative circuit, 20–21
 constraint, 20
 conversion to parallel circuit, 107–108
 examples, 21–23
 and Kirchhoff's laws, 20–21
 and Ohm's law, 20–21
 presence of current source, 20
series field and shunt field, 313–314, 318
series motor, 324–325
series resistance, 17
series resonance, 111–112
shaded-pole induction motors, 344–345
shading bars, 344
shell (O-core) transformer construction, 294–295
shift (serial) registers, 251–252
shoe box PLCs, 362
short-circuit current and Norton's theorem, 42
short circuits, 8
short shunt compound generators, 318
shunt, series, and compound DC motor, 322–326
 compound motor, 325
 examples, 323, 325
 motor efficiency, 326
 series motor, 324–325
 shunt motor, 322–324
shunt and compound DC generators, 313–319
 armature, 313, 314, 315
 compound generator model, 318–319
 differentially compounded generator, 319
 efficiency, 316–317
 examples, 315, 316, 317
 flashing the field, 314
 long shunt compound generator, 319
 losses within generators, 317
 rheostat, 317, 319
 short shunt compound generators, 318
 shunt field and series field, 313–314, 318
 shunt generator model, 314–318
 terminal voltage, 314
 voltage buildup, 314
shunt field and series field, 313–314, 318
shunt generator model, 314–318
shunt motor, 322–324
sign and magnitude, 212
signal distortion, in amplifiers, 175
signal restoration, 387
signal-to-noise ratio, 398
simplex connection, 379
simplex vs duplex, digital communications, 379

single- and three-phase systems, comparison of power flow, 281–283
single loop circuit (series circuit), 19–23
 application to representative circuit, 20–21
 constraint, 20
 examples, 21–23
 and Kirchhoff's laws, 20–21
 and Ohm's law, 20–21
 presence of current source, 20
single node-pair circuit (parallel circuit), 23–28
 application to representative circuit, 25, 26
 buses, 24
 constraint, 24
 current sources and resistances, 25–26
 examples, 26–28
 and Kirchhoff's laws, 24–26
 and Ohm's law, 24–25
single-phase induction motors, 341–346
 advantages and disadvantages, 345–346
 capacitor-start split-phase induction motor, 344
 permanent split capacitor (PSC) motor, 344
 resistance-start split-phase induction motor, 343–344
 shaded-pole induction motors, 344–345
 two-value capacitor (capacitor-start, capacitor-run induction motor) (CSCR), 344
sinusoid and sinusoidal functions, 78–80
 cosine function, 78–79
 effective value of sinusoid, 97–98
 examples, 80
 general sinusoidal function, 79
 relative phase sequence, 79–80
 sine function, 78
 units for angles, 79
 voltages and currents, 4, 10
sinusoidal voltages and currents
 See also AC
slip rings, 310
slip speed, 338–339
solenoid
 definition, 262
 right-hand rule for, 265
solid-state output for PLCs, 372–375
solid-state switches, 154
SOP (sum of products form), 224
source transformations, 39–41
 applications to mesh and node analysis, 40
 conversions, 40
 examples, 40–41
 Thevenins and Norton's theorems, 42
sources, ideal. See ideal sources
spectral analysis, 385
squirrel-cage rotor, reliability, 337–338
SRAM (static random access memory), 257
Stanley, William Jr., 286
start-stop data transmission systems, 379
start winding (auxiliary winding), 342, 344
static random access memory (SRAM), 257
stator copper loss, 336
stator core loss, 336
steady-state AC model, 89
step-up and step-down transformers, 127, 354
stopband, 200
stray loss, 336
subtraction using two-complement numbers, 213–214
 examples, 214
 twos-complement overflow, 214
subtrahend and minuend, 213
sum of products form (SOP), 224
switches
 BJT, 154–157
 FET, 166–168
 JFET, N-channel, 167

 MOFSET, N-channel, 167
 NPN, 154–155
 for PLCs, 354–356
 PNP, 155
 saturation and cutoff, 154
 solid-state, 154
switching function, canonical form and reduced form, 222–223
symbol rate, 397
symbols and units for electrical variables, 4–5
symmetrical circuit, 237
symmetrical function, 237
synchronous capacitors, 337
synchronous communications, 378–379
synchronous condensers, 337
synchronous counters, 253–254
synchronous devices, 247
synchronous motors, 336–337
synchronous speed, 339
synchronous vs asynchronous, digital communications, 378–379

T

T flip flops, 250
tachogenerators, 313
TDM (time-division multiplexing), 167–168, 387–388
terminal characteristics, junction diode, 121–122
terminal voltage, 314
terminology, digital communications, 378–380
 binary vs M-ary communication, 378
 bits and bytes, 378
 bits and levels, 378
 data rate, 380
 examples, 380
 marks and spaces, 379
 parallel vs serial, 379
 simplex vs duplex, 379
 synchronous vs asynchronous, 378–379
 textual vs binary data, 379
 unipolar vs bipolar, 379–380
Tesla, Nikola, 286
textual data
 encoding and transmission. See encoding and encoding
 vs binary data, 379
theorems, Boolean algebra, 221
Thevenin's and Norton's theorems, 41–47
 examples, 42–47
 non-feedback circuits, 42
 open-circuit voltage and Thevenin's theorem, 41–42
 resistance, 42
 short-circuit current and Norton's theorem, 42
 source transformations, 42
three-phase circuits, 271–284
 multiSIM examples, 469–470
 phase voltages and line voltages, 274–276
 power flow, single- and three-phase systems, 281–283
 problems, 283–284
 three-phase power calculations, 279–281
 three-phase sequencing, 277
 three-phase theory, 271–274
 three-phase wye and delta connections, 277–278
 three-phase wye and delta load connections, 278–279
three-phase induction motor, 337–341
 breakdown speed, 339
 examples, 339–340
 Faraday's Law, 338
 Lenz's Law, 338
 slip speed, 338–339

squirrel-cage rotor, reliability, 337–338
wound rotor three-phase induction motor, 340
three-phase power calculations, 279–281
   examples, 280–281
three-phase sequencing
   ABC and CBA, 277
   phase sequence relay, 277
three-phase theory, 271–274
   examples, 273
   instantaneous waveforms, 273
   phasor forms, 273–274
   wye configuration, 272–273
three-phase transformers, 295–297
   delta-to-delta, 297
   delta-to-wye, 297
   examples, 296–297
   wye-to-delta, 297
   wye-to-wye, 295–296
three-phase wye and delta connections, 277–278
   delta connection, 278
   relationships, 278
   wye connection, 278
three-phase wye and delta load connections
   balanced load, 278–279
   phase currents, 278
three-wire system, 276
threshold voltage, 165
time-delay relays (TDR), 357–359
   delay-off (TOF), 358–359
   delay-on (TON), 358
time-division multiplexing (TDM), 167–168, 387–388
time-varying variables, 52
timing diagrams, flip flop, 247–248
TOF (delay-off), 358–359
TOF timers, retentive and nonretentive, 370
toggles, 250
TON (delay-on) relay, 358
TON timers, retentive and nonretentive, 369, 370
topology, K-map, 226–227
tracers, 353
transconductance, 192
transformed power, 299
transformers, 285–304
   autotransformers. *See* autotransformers
   construction, 294–295
      E-core, 295
      shell (O-core), 294
   definition and history, 285–286
   efficiency, 290–291
   ideal, 286–289
      constant voltage transformer, 289
      examples, 287–288
      ferroresonant transformer, 289
      high side winding, 287
      low side winding, 287
      primary winding, 286
      secondary winding, 286
      turns ratio, 287
   impedance matching transformers, 293–294
   impedance reflection, 292–293
   instrument transformers, 300–303
   multiSIM examples, 470–473
   performance, effect of power factor, 291–292
      examples, 291
   power factor, effect on performance, 291–292
   power losses, 289–290
      copper loss, 289–290
      core laminations, 290
      core loss, 289, 290
      eddy current loss, 290
      hysteresis loss, 290
      open-circuit (no-load) test, 290

   problems, 303–304
   step-up and step-down, 127
   three-phase, 295–297
      delta-to-delta, 297
      delta-to-wye, 297
      examples, 296–297
      wye-to-delta, 297
      wye-to-wye, 295–296
   useful tips, 303
transient circuits, 51–73
   basic circuit parameters, 52–58
   boundary conditions for energy storage elements, 58–64
   first-order circuits with DC excitations, 64–69
   inductance and capacitance combinations, 69–71
   multiSIM examples, 422–427
   problems, 71–73
transistor-transistor logic (TTL), 154, 242–243
transistors, 143–171
   bipolar junction transistor (BJT), 144–146
   bipolar junction transistors. *See* BJTs (bipolar junction transistors)
   BJT amplifier circuit, 153–154
   BJT characteristic curves, 148–153
   BJT operating regions, 146–148
   BJT switches, 154–157
   FET schematic summary, 168–169
   FET switches, 166–168
   field effect transistor family (FET), 157
   JFETs, 163
   junction field effect transistors (JFET), 158–163
   metal-oxide semiconductor field effect transistors. *See* MOSFETs
   MOSFETs, 164–166
   multiSIM examples, 439–443
   problems, 169–171
transition band, 200
transmission of textual data
   encoding
      and transmission, textual data. *See* encoding and transmission of textual data
transresistance, 192
triac outputs for PLCs, 374–375
triple three-input AND, 242
TTL (transistor-transistor logic), 154, 242–243
turns ratio, 287
two-pole band-pass responses, 202
two-value capacitor (capacitor-start, capacitor-run induction motor) (CSCR), 344
twos-complement numbers
   minuend and subtrahend, 213
   and negative binary numbers, 212
   subtraction using, 213–214
   twos-complement overflow, 214

U

uncharged capacitor, 59
unfluxed inductor, 59
unipolar encoding, 389
unipolar representation in A/D conversion, 389
unipolar *vs* bipolar, digital communications, 379–380
units and prefixes for electrical variables
   tables and symbols, 4–5
   use of italics, 5
universal motors, 348
up-counters, 253
up/down counters, 253

V

vacuum tubes, 143
varactor diode, 137

variable autotransformer (Variac ©), 300
variables
   basic electrical, in circuit analysis, 4
   time-varying, 52
Variac © (variable autotransformers), 300
variations in current gain, BJT, 151
VCIS (voltage-controlled current sources), 192
VCVS (voltage-controlled voltage sources), 192
very large-scale integration (VLSI), 144
virtual ground, 185
VLSI (very large-scale integration), 144
volatile memory, 245
voltage, differential input, 181
voltage amplifier input-output, 174
voltage and current, effective values, 10
voltage and current, representation as phasors, 82–83
voltage and current divider rules
   descriptions, 28–29
   examples, 29–30
voltage and current forms
   definitions, 5
   symbols and units, 4
voltage and current measurement, comparison, 13
voltage buildup, 314
voltage-controlled current sources (VCIS), 192
voltage-controlled resistance, 160, 166
voltage-controlled voltage sources (VCVS), 192
voltage-current relationships, 85–87
   AC, 85–87
   capacitance, 86–87
   capacitive, 53
   differentiation and integration of complex exponential function, 85–86
   inductance, 86
   inductive, 53
   and Ohm's law, 85
   resistance, 85
voltage follower, 190
voltage gain, 174, 175–176
voltage regulation with op-amp, 204
voltage regulator, 135, 204
voltage source, ideal DC, 7
voltages, saturation, 182
voltmeter, ideal DC, 12
Voyager I and *Voyager II*, 243

W

Westinghouse, George, 286
Westinghouse Electric Company, 286
Wheatstone bridge, 13
windage, 317
wire numbering, 353
word, digital communications, 378
worst-case design, 152
wound rotor three-phase induction motor, 340
wye and delta connections, relationships, 278
wye configuration, 272–273
wye connection, 278

Z

zener diodes, 122, 136, 204
zener regulator circuits, 137–140
   examples, 138–140
   and KCL, 138
   qualitative explanation, 138
zero-bias drain current, 161

IMPORTANT-READ CAREFULLY: This End User License Agreement ("Agreement") sets forth the conditions by which Delmar Learning, a division of Thomson Learning Inc. ("Thomson") will make electronic access to the Delmar Learning-owned licensed content and associated media, software, documentation, printed materials and electronic documentation contained in this package and/or made available to you via this product (the "Licensed Content"), available to you (the "End User"). BY CLICKING THE "I ACCEPT" BUTTON AND/OR OPENING THIS PACKAGE, YOU ACKNOWLEDGE THAT YOU HAVE READ ALL OF THE TERMS AND CONDITIONS, AND THAT YOU AGREE TO BE BOUND BY ITS TERMS CONDITIONS AND ALL APPLICABLE LAWS AND REGULATIONS GOVERNING THE USE OF THE LICENSED CONTENT.

1.0 SCOPE OF LICENSE

1.1 Licensed Content. The Licensed Content may contain portions of modifiable content ("Modifiable Content") and content which may not be modified or otherwise altered by the End User ("Non-Modifiable Content"). For purposes of this Agreement, Modifiable Content and Non-Modifiable Content may be collectively referred to herein as the "Licensed Content." All Licensed Content shall be considered Non-Modifiable Content, unless such Licensed Content is presented to the End User in a modifiable format and it is clearly indicated that modification of the Licensed Content is permitted. Audio products may be reproduced at the user's discretion.

1.2 Subject to the End User's compliance with the terms and conditions of this Agreement, Thomson Delmar Learning hereby grants the End User, a nontransferable, non-exclusive, limited right to access and view the enclosed software on an unlimited number of computers at one site, or on one network at one site, if applicable, The End User shall not (i) reproduce, copy, modify (except in the case of Modifiable Content), distribute, display, transfer, sublicense, prepare derivative work(s) based on, sell, exchange, barter or transfer, rent, lease, loan, resell, or in any other manner exploit the Licensed Content; (ii) remove, obscure or alter any notice of Thomson Delmar Learning's intellectual property rights present on or in the License Content, including, but not limited to, copyright, trademark and/or patent notices; or (iii) disassemble, decompile, translate, reverse engineer or otherwise reduce the Licensed Content.

2.0 TERMINATION

2.1 Thomson Delmar Learning may at any time (without prejudice to its other rights or remedies) immediately terminate this Agreement and/or suspend access to some or all of the Licensed Content, in the event that the End User does not comply with any of the terms and conditions of this Agreement. In the event of such termination by Delmar Learning, the End User shall immediately return any and all copies of the Licensed Content to Delmar Learning.

3.0 PROPRIETARY RIGHTS

3.1 The End User acknowledges that Thomson Delmar Learning owns all right, title and interest, including, but not limited to all copyright rights therein, in and to the Licensed Content, and that the End User shall not take any action inconsistent with such ownership. The Licensed Content is protected by U.S., Canadian and other applicable copyright laws and by international treaties, including the Berne Convention and the Universal Copyright Convention. Nothing contained in this Agreement shall be construed as granting the End User any ownership rights in or to the Licensed Content.

3.2 Thomson Delmar Learning reserves the right at any time to withdraw from the Licensed Content any item or part of an item for which it no longer retains the right to publish, or which it has reasonable grounds to believe infringes copyright or is defamatory, unlawful or otherwise objectionable.

4.0 PROTECTION AND SECURITY

4.1 The End User shall use its best efforts and take all reasonable steps to safeguard its copy of the Licensed Content to ensure that no unauthorized reproduction, publication, disclosure, modification or distribution of the Licensed Content, in whole or in part, is made. To the extent that the End User becomes aware of any such unauthorized use of the Licensed Content, the End User shall immediately notify Delmar Learning. Notification of such violations may be made by sending an Email to delmarhelp@thomson.com.

5.0 MISUSE OF THE LICENSED PRODUCT

5.1 In the event that the End User uses the Licensed Content in violation of this Agreement, Thomson Delmar Learning shall have the option of electing liquidated damages, which shall include all profits generated by the End User's use of the Licensed Content plus interest computed at the maximum rate permitted by law and all legal fees and other expenses incurred by Thomson Delmar Learning in enforcing its rights, plus penalties.

6.0 FEDERAL GOVERNMENT CLIENTS

6.1 Except as expressly authorized by Delmar Learning, Federal Government clients obtain only the rights specified in this Agreement and no other rights. The Government acknowledges that (i) all software and related documentation incorporated in the Licensed Content is existing commercial computer software within the meaning of FAR 27.405(b)(2); and (2) all other data delivered in whatever form, is limited rights data within the meaning of FAR 27.401. The restrictions in this section are acceptable as consistent with the Government's need for software and other data under this Agreement.

7.0 DISCLAIMER OF WARRANTIES AND LIABILITIES

7.1 Although Thomson Delmar Learning believes the Licensed Content to be reliable, Thomson Delmar Learning does not guarantee or warrant (i) any information or materials contained in or produced by the Licensed Content, (ii) the accuracy, completeness or reliability of the Licensed Content, or (iii) that the Licensed Content is free from errors or other material defects. THE LICENSED PRODUCT IS PROVIDED "AS IS," WITHOUT ANY WARRANTY OF ANY KIND AND THOMSON DELMAR LEARNING DISCLAIMS ANY AND ALL WARRANTIES, EXPRESSED OR IMPLIED, INCLUDING, WITHOUT LIMITATION, WARRANTIES OF MERCHANTABILITY OR FITNESS OR A PARTICULAR PURPOSE. IN NO EVENT SHALL THOMSON DELMAR LEARNING BE LIABLE FOR: INDIRECT, SPECIAL, PUNITIVE OR CONSEQUENTIAL DAMAGES INCLUDING FOR LOST PROFITS, LOST DATA, OR OTHERWISE. IN NO EVENT SHALL DELMAR LEARNING'S AGGREGATE LIABILITY HEREUNDER, WHETHER ARISING IN CONTRACT, TORT, STRICT LIABILITY OR OTHERWISE, EXCEED THE AMOUNT OF FEES PAID BY THE END USER HEREUNDER FOR THE LICENSE OF THE LICENSED CONTENT.

8.0 GENERAL

8.1 Entire Agreement. This Agreement shall constitute the entire Agreement between the Parties and supercedes all prior Agreements and understandings oral or written relating to the subject matter hereof.

8.2 Enhancements/Modifications of Licensed Content. From time to time, and in Delmar Learning's sole discretion, Thomson Delmar Learning may advise the End User of updates, upgrades, enhancements and/or improvements to the Licensed Content, and may permit the End User to access and use, subject to the terms and conditions of this Agreement, such modifications, upon payment of prices as may be established by Delmar Learning.

8.3 No Export. The End User shall use the Licensed Content solely in the United States and shall not transfer or export, directly or indirectly, the Licensed Content outside the United States.

8.4 Severability. If any provision of this Agreement is invalid, illegal, or unenforceable under any applicable statute or rule of law, the provision shall be deemed omitted to the extent that it is invalid, illegal, or unenforceable. In such a case, the remainder of the Agreement shall be construed in a manner as to give greatest effect to the original intention of the parties hereto.

8.5 Waiver. The waiver of any right or failure of either party to exercise in any respect any right provided in this Agreement in any instance shall not be deemed to be a waiver of such right in the future or a waiver of any other right under this Agreement.

8.6 Choice of Law/Venue. This Agreement shall be interpreted, construed, and governed by and in accordance with the laws of the State of New York, applicable to contracts executed and to be wholly preformed therein, without regard to its principles governing conflicts of law. Each party agrees that any proceeding arising out of or relating to this Agreement or the breach or threatened breach of this Agreement may be commenced and prosecuted in a court in the State and County of New York. Each party consents and submits to the non-exclusive personal jurisdiction of any court in the State and County of New York in respect of any such proceeding.

8.7 Acknowledgment. By opening this package and/or by accessing the Licensed Content on this Website, THE END USER ACKNOWLEDGES THAT IT HAS READ THIS AGREEMENT, UNDERSTANDS IT, AND AGREES TO BE BOUND BY ITS TERMS AND CONDITIONS. IF YOU DO NOT ACCEPT THESE TERMS AND CONDITIONS, YOU MUST NOT ACCESS THE LICENSED CONTENT AND RETURN THE LICENSED PRODUCT TO THOMSON DELMAR LEARNING (WITHIN 30 CALENDAR DAYS OF THE END USER'S PURCHASE) WITH PROOF OF PAYMENT ACCEPTABLE TO DELMAR LEARNING, FOR A CREDIT OR A REFUND. Should the End User have any questions/comments regarding this Agreement, please contact Thomson Delmar Learning at delmarhelp@thomson.com.